Transforming Noise

Transforming Noise

A History of Its Science and Technology from Disturbing Sounds to Informational Errors, 1900–1955

CHEN-PANG YEANG

Institute for the History and Philosophy of Science and Technology, University of Toronto, Canada

OXFORD
UNIVERSITY PRESS

Great Clarendon Street, Oxford, OX2 6DP,
United Kingdom

Oxford University Press is a department of the University of Oxford. It furthers the University's objective of excellence in research, scholarship, and education by publishing worldwide. Oxford is a registered trade mark of Oxford University Press in the UK and in certain other countries

Published in the United States of America by Oxford University Press
198 Madison Avenue, New York, NY 10016, United States of America

British Library Cataloguing in Publication Data
Data available

Library of Congress Control Number: 2023944408

ISBN 978–0–19–888776–8

DOI: 10.1093/oso/9780198887768.001.0001

Printed and bound by
CPI Group (UK) Ltd, Croydon, CR0 4YY

Preface

Writing a book on noise is both exciting and intimidating. On this topic, we can easily imagine a monograph in media studies, musicology, environmental humanities, cultural history, science, technology, and society (STS), or philosophy; a textbook in psychoacoustics, probability theory, telecommunication systems, semiconductor devices, bioinformatics, or quantum computing; and more. I have met artists, historians, literary scholars, mathematicians, physicists, and engineers who expressed a keen interest when they came to know about my project. In many senses, a story of noise is a story of modern life in a nutshell. But the extremely broad potential of noise as a subject also posed a challenge for carving out a sufficiently coherent and comprehensive yet manageable scope for a book. I ended up spending more than a decade researching and writing *Transforming Noise.*

I began to consider writing a history of noise when I worked on a PhD dissertation on radio-channel characterization in the early twentieth century. While my dissertation focused on the science and technology of electromagnetic-wave propagation in the ionosphere, I noticed that my historical actors were preoccupied with taming sonic disturbances in radio broadcasting and wireless telephony: crossband interference, atmospheric discharge, and current fluctuations in receivers. I had been aware of the two connotations of noise—one referred to sensory stimulations like street din and mechanical cacophony while the other referred to random errors plaguing data and signals in general. But why and how these two different realms were historically related were a puzzle to me. Perhaps the radio disturbances could offer clues to its answer?

This thought marked a point of entry leading to a book project on the historical transformation of noise from annoying sounds to informational errors. The project turned out to be a long intellectual journey beyond radio engineering. I learned a lot about the epistemology, technology, and culture of sound reproduction scrutinized in sound studies; probed—albeit briefly—musical theory; and uncovered a science and history of Brownian motion and stochastic processes that historians of physics and mathematicians had barely examined. When I ventured into the areas of noise as random fluctuations, my "previous life" as a student of electrical engineering was reincarnated in an unexpected way. My coursework, doctoral and postdoctoral research in stochastic detection and estimation, information transmission, and statistical mechanics prepared me for understanding the intricate technicality of informational noise in my historical inquiry. The connection could be personal, too. When I wrote about the Wiener filter, Shannon's "water pouring" channels, and radar target recognition, I recalled the graduate courses I had taken at MIT developed and taught by Wiener's and Shannon's disciples or grand-disciples, and even my military service as an instructor of radar systems at a naval technical school in Taiwan.

Encouragements, suggestions, and aids from some colleagues were important fuel to this book project. I would like to express my gratitude to Jed Buchwald, David Mindell, Martin Niss, and Roland Wittje. Jed and David were my PhD supervisors. I learned much about the histories of physics, electrical engineering, and computing from them when I was a graduate student at MIT's STS Program. This knowledge paved a ground for *Transforming Noise*. Moreover, they both read my book manuscript and gave me extremely useful advice on its structure, narrative, thematic framework, and concrete historical or technical contents. Jed also offered significant help in the publishing phase of my book project.

Martin and Roland share the same interest as mine in the technical history of noise. Martin taught me a lot about the history of statistical mechanics and prompted me to think of noise from the perspective of random fluctuations. Roland introduced me to sound studies and urged me to probe into sonic noise, sound reproduction, and music. In 2012, we co-organized a panel—with Allan Franklin and my former students Aaron Sidney Wright and Shawn Bullock at the University of Toronto—titled "Understanding Noise in Twentieth-Century Physics and Engineering" at the History of Science Society (HSS) Annual Meeting in San Diego. Joined by the late Joan Lisa Bromberg, we turned the outcome from the HSS panel into a special issue in *Perspectives on Science* in 2016. This collaboration played a crucial part in shaping my book project. I am grateful to all of them and really miss Joan. In addition, both Martin and Roland carefully went through my book manuscript and provided very valuable and detailed comments, suggestions, and corrections based on their scholarly, technical, and linguistic expertise.

The conceiving, planning, researching, and writing of *Transforming Noise* took place at a number of institutions where I worked or visited. I received warm help, support, and inspirations from researchers, scholars, staff, and students at these institutions. At the Institute for the History and

Philosophy of Science and Technology (IHPST), University of Toronto, the unit of my employment, I thank Craig Fraser's guidance on the scholarship in the history of stochastic processes, Joseph Berkovitz's introduction to the philosophy of probability, Mark Solovey's input on some early ideas of my book, and the friendly company of the late Janis Langins, the late Trevor Levere, Pauline Mazumdar, Paul Thompson, Marga Vicedo, Poly Winsor, and Denis Walsh. I am also blessed to have the opportunity to work with some enthusiastic and devoted graduate students, undergraduate students, and associates as my research assistants—Jeni Barton, Zhixiang Cheng, Chris Conway, Victoria Fisher, Quinn Harrington, Sophie LeBlanc, Nicolas Sanchez Guerrero, Ava Spurr, Noah Stemeroff, and Erich Weidenhammer—and teaching assistants for "Making Sense of Uncertainty," a course developed along with my book project—Atoosa Kasirzadeh, Aaron Kenna, William Rawleigh, and Fan Zhang. At the University of Toronto outside the IHPST, I appreciate the friendship of Willy Wong and Yurou Zhong.

Thanks to Otto Sibum's generous arrangement, I visited his research group at the Max Planck Institute for the History of Science in Berlin in spring 2006. I am grateful of the opportunities at Max Planck to receive valuable feedback from Otto, David Bloor, and Charlotte Bigg about Brownian motion and the material culture of experimentation. In 2012–13, I was a member of the School of Historical Studies, Institute for Advanced Study (IAS) in Princeton. The core structure and themes of *Transforming Noise* grew out of a colloquium talk I gave at the School of Historical Studies. During this visit, my book project benefited considerably from friendly and inspiring interactions with Yves-Alain Bois, Christer Bruun, Nicola Di Cosmo, Ingrid Maren Fumiss, Yonglin Jiang, Christian Meyer, Hyun Ok Park, Weirong Shen, Heinrich von Staden, Michael van Walt van Praag, Heidi Voskuhl, Aihe Wang, and Frédérique Woerther at the IAS and Janet Chen, Danian Hu, Michael Gordin, Matt Stanley, and Emily Thompson outside the IAS.

In the summers of 2014–16, I visited the Institute of Modern History (IMH), Academia Sinica in Taipei. In July 2019–June 2021, I stayed in Taipei as a visiting scholar at the IMH and a visiting associate professor at the Department of Electrical Engineering at National Taiwan University (NTUEE). These extended stays in Taipei were pivotal to my book project. In fact, I completed most of the writing of *Transforming Noise* at the IMH and NTUEE. I am deeply grateful to Hsiang-lin Sean Lei, who kindly hosted my visits to the IMH over the past few years; and Shi-Chung Chang and Shih-Yuan Chen, who made a generous arrangement for my visit to NTUEE. At Academia Sinica, my brother Chen-Hsiang Yeang, Chao-Huang Lee, Ku-ming Kevin Chang, Pingyi Chu,

Shang-jen Li, and Yu-chih Lai offered warm support, company, and intellectual stimulation. At Taiwan University, I was fortunate to have a research collaboration with Shih-Yuan and his excellent students Yun-Ying Jan, Kai-Hung Cheng, and Hung-Yu Tsao. I also appreciate the fruitful dialogues with Shyh-Kang Jeng, Ruei-Beei Wu, Yu-yu Cheng, and Kuang-Chi Hung.

Throughout the development of *Transforming Noise*, I presented research results at academic conferences, including the annual meetings of the HSS in 2012, 2018, and 2022, Society for the History of Technology (SHOT) in 2007 and 2016, Society for Social Studies of Science (4S) in 2017, International Committee for the History of Technology (ICOHTEC) in 2011, and International Congress of History of Science and Technology (ICHST) in 2021. I would like to thank Hans-Joachim Braun (who invited me to participate in his organized session at ICOHTEC 2011), Scott Walter and Alexei Kojevnikov (who invited me to participate in their organized session at ICHST 2021), as well as my fellow panelists Kenji Ito, Shaul Katzir, and Takashi Nishiyama, who helped me organize the session at SHOT 2016.

I am indebted to the anonymous reviewers of my book manuscript. Their positive but frank and useful comments and suggestions prompted me to think of the subjects and contents from different angles and helped me improve the scholarly quality of this work. Moreover, my editor Sonke Adlung's contribution to this project must be acknowledged. Without his enthusiastic support, significant effort, and efficient handling, *Transforming Noise* would not be published by Oxford University Press. I really appreciate his help. Manhar Ghatora was the book's project editor at Oxford University Press. Karthiga Ramu was the project manager from Integra Software Services for book production. Victoria Fisher assisted me in image production, acquisition of copyright permissions, and indexing. Andrew Stanley copy-edited my manuscript; Debbie Protheroe selected the image for the book cover; and Libby Holcroft designed the book cover on behalf of Oxford University Press. I am thankful to them.

My research for *Transforming Noise* was made possible by access to the following archives and libraries and the kind assistance from their staff: AT&T Archives (Warren, New Jersey, US); Claude Shannon Papers, Library of Congress (Washington, DC, US); DSIR Papers, National Archives (London, UK); George Uhlenbeck Papers, Manuscript Collection, Library of the University of Michigan (Ann Arbor, Michigan, US); Historical Records, Rutherford Appleton Laboratory, Space Science Department (Didcot, UK); Marian Smoluchowski Papers (MS9397), Manuscript Collections, Library of Jagellonian University (Krákow, Poland); Norbert Wiener Collection (MC22), MIT Archives (Cambridge, Massachusetts, US); Paul Langevin

Papers, Library of ESPCI Paris Tech (Paris, France); Thomas A. Edison Papers, Rutgers University (New Jersey, US, accessing mailed microfilms); Walter Schottky Papers (NL 100), Deutsches Museum Archives (Munich, Germany); University of Toronto Libraries; IAS Libraries; Princeton University Libraries; Libraries of Academia Sinica; and Libraries of National Taiwan University. In addition, I had the privilege to interview with Mortimer Rogoff in Boston in June 2004 (a few years before he passed away) and utilize David Mindell's interview transcript with Rogoff in November 2003. In particular, I am grateful for the generous help from Sheldon Hochheiser at AT&T Archives, Paul Israel at Edison Papers, Chris Davis at the Rutherford Appleton Laboratory, Grzegorz Fulara at the Manuscript Department of the Library of Jagellonian University, and Catherine Kounelis at the Library of ESPCI Paris Tech.

To research and write *Transforming Noise*, I received the following financial supports: Dibner Postdoctoral Fellowship (2004–05), postdoctoral fellowship at the Max Planck Institute for the History of Science (2006), Connaught New Staff Matching Grant at the University of Toronto (2007–09), Canada Social Sciences and Humanities Research Council (SSHRC) Standard Research Grant (2010–14), and Fellowship from the School of Historical Studies, Institute for Advanced Study (2012–13).

Portions of this book were adapted from three journal articles I published before. Parts of the contents in chapter 6 were drawn from "Tubes, randomness, and Brownian motions: Or, how engineers learned to start worrying about electronic noise," *Archives for the History of Exact Sciences*, 65:4 (2011), 437–470. Chapter 4 was rewritten and extended with a different conceptual framework and more archival research from "The sound and shapes of noise: Measuring disturbances in early twentieth-century telephone and radio engineering," *ICON: Journal of the International Committee for the History of Technology*, 18 (2012), 63–85. The structure of chapters 5, 7, and 9 was inspired by "Two mathematical approaches to random fluctuations," *Perspectives on Science*, 24:1 (2016), 45–72. Yet, the contents, analytical angle, historical narrative, and empirical claims in these chapters are significantly different from and much longer than those in the article. I acknowledge these journals' permissions for me to use the three articles in my book.

I want to dedicate this book to my family: my wife Wen-Ching Sung, my father Jia-Chu Yeang, my mother Lan-Chun Hsu, and my children Elena and Felix Yeang. Pursuing a career in academia can be lonely, tough, and materially less rewarding. Spending years researching and writing a book is against today's trend. Their love, support, dedication, and understanding have given me courage to carry on.

Contents

1
Introduction

On July 27, 1844, Nathaniel Hawthorne sat alone under the morning sun in the woods of Sleepy Hollow near Concord, Massachusetts. The place where he settled himself was a flat and circular meadow within the forest. Seeking literary inspirations, the forty-year-old writer recorded as many of his impressions as possible on an eight-page note. What caught his attention first were the sights and sounds of the wild: the Indian corns cluttering the meadow, the twigs and branches on the pathway in the woods, sunshine glimmering through shadow, birds chirping, leaves hissing in winds. Then he heard a village clock striking, a cowbell tinkling, and mowers cutting grasses with scythes. To him, these signs of human activities blended seamlessly with the tranquil pastoral scene and his inner calmness. Yet, what happened next broke the balance between landscape, society, and mind. As Hawthorne wrote:[1]

> But, hark! There is the whistle of the locomotive—the long shriek, harsh, above all other harshness, for the space of a mile cannot mollify it into harmony. It tells a story of busy men, citizens from the hot street, who have come to spend a day in a country village, men of business; in short of all uniqueness; and no wonder it gives such a startling shriek, since it brings the noisy world into the midst of a slumbrous peace.

The literary scholar Leo Marx characterized Hawthorne's 1844 encounter and the similar experiences of other American authors around the same period as "machines in the garden," a moment when industrial technology intruded into the hitherto quiet, peaceful, and harmonious life in nature. In the mid-nineteenth century, perhaps nothing was more symptomatic of technology's invasiveness than noise from a screaming whistle or a roaring engine in a locomotive or a factory. To Hawthorne and his contemporaries, noise was cacophony, din, disturbing and annoying sounds.

Almost one and a half centuries later, another fiction writer probed the impacts of modern technology on the human condition. In a farce-thriller,

[1] Quoted from Leo Marx, *The Machine in the Garden: Technology and the Pastoral Idea in America* (Oxford: Oxford University Press, 2000), 13–14.

Don DeLillo portrayed a mediocre academic's tedious and aimless life with his family until leaking of toxic gas from a neighboring chemical plant overturned the college town where they lived. Like Hawthorne's note, DeLillo's story was an allegory of technology's intrusion into daily life, pivoting on the metaphoric power of noise. In contrast to the locomotive and steam engine in Hawthorne's Industrial Revolution, the "machines" of DeLillo's late Cold War America were television, radio, and loudspeakers. And the noise of the latter was not the type of sharp and sudden sonic bursts that distracted Hawthorne. Rather, it was the non-stopping background hums that continued to hit everyone's ear—a phenomenon that DeLillo highlighted in the title of his book: *White Noise*—such as the voice from a supermarket loudspeaker announcing "Kleenex Softique, your truck's blocking the entrance," or sounds from a TV that had been left turned on for over two months.[2]

DeLillo's white noise was not just sounds, however. It also encompassed TV commercials, tabloid gossips, commodities' brand names, news images, and more. As a character in the story said, "the incessant bombardment of information" made people suffer from a "brain fade." "The flow is constant: words, pictures, numbers, facts, graphics, statistics, specks, waves, particles, motes."[3] A literature critique contended that the very idea of "white noise" in DeLillo's novel "implies a neutral and reified mediaspeech, but also a surplus of data and an entropic blanket of information glut which flows from a media-saturated society."[4]

The notion of noise underlying DeLillo's novel is all too familiar to us in the early twenty-first century. Today, we refer to noise when we talk about the transmission rate of COVID-19, the 5G wireless networks, self-driving cars, genomic pathways of *E. coli*, a fluctuating stock market, data for the global climate change, traces of reverberating gravitational waves from a remote galaxy, or polls before an election. The technoscience we are facing is clearly more ubiquitous and invasive than that at Hawthorne's or even DeLillo's time. In these broader contexts, the meaning of noise has undergone a salient change: from unwelcoming and irritating sounds to unwanted, erroneous, or interfering signals, messages, and information. Between Hawthorne's note in 1844 and DeLillo's novel in 1985, a semantic transformation took place.

[2] Don DeLillo, *White Noise* (New York: Penguin, 1985), 36, 50–51.

[3] Ibid., 66.

[4] Leonard Wilcox, "Baudrillard, DeLillo's White Noise, and the end of heroic narrative," in Hugh Ruppensburg and Tim Engles (eds.), *Critical Essays on Don DeLillo* (New York: G.K. Hall & Co., 2000), 197; quoted in Karen Weekes, "Consuming and dying: Meaning and the marketplace in Don DeLillo's *White Noise*," *Literature Interpretation Theory*, 18:4 (2007), 286.

Only, the transformation was not merely semantic. In the 1970s, the cultural studies scholar Raymond Williams probed the broader implications of a changing vocabulary beyond the linguistic sense. He observed that the everyday usage of some words "common in descriptions of wider areas of thought and experience" underwent conspicuous changes within a short span of time. He found that such alterations often accompanied a rapid transition of culture and society. To him, the semantic transformations of certain words were a particularly good barometer of social or historical processes. These terms, which Williams dubbed as "keywords," were either "significant, binding words in certain activities and their interpretation," or "significant, indicative words in certain forms of thought." Moreover, the language did not only reflect the processes of society and history; some important social and historical processes occurred "*within* language."[5]

Noise is a keyword in the Williamsian sense. Its semantic change from the late nineteenth to the mid-twentieth century indexes a profound transformation at large. In contrast to those in popular culture, general society, and political spheres that words like "capitalism," "class," "liberal," and "subjective" captivate, the evolving meaning of noise epitomizes a set of fundamental alterations in the cognitive, conceptual, epistemic, ontological, material, and technical aspects of modern science and technology. In the technoscientific cultures and practices, noise started as a specific kind of sounds that needed to be domesticated through materialization, abstraction, quantification, measurement, and mathematical representation. But it eventually became a synonym for errors, fluctuations, and deviations that affected all kinds of signals, messages, data, and information. Why and how did this transformation occur? What was it about? What were its implications for the science of sensory experiences, the technology of sound reproduction, standards of objectivity and exactitude, understanding of and coping with uncertainty, and the computational form of life? How were these developments embedded in the characteristically twentieth-century contexts of mass media, corporate research, world wars, and military-industrial-academic complexes, as well as the intellectual traditions of indeterminism, atomism, probabilistic thinking, and changing notions of mathematical rigor? This book aims to answer such questions.

[5] Raymond Williams, *Keyword: A Vocabulary of Culture and Society* (New York: Oxford University Press, 1985), 14–15, 22.

From Disturbing Sounds to Informational Errors

The historical transformation of noise proceeded in four stages. In the 1900s–30s, the perception of, understanding of, and grappling with noise underwent a fundamental change. Whereas noise was largely conceived as disturbing sounds by the end of the nineteenth century, it began to refer to the physical factors that gave rise to such sounds in the 1900s–30s. This was the time the German philosopher Walter Benjamin dubbed as the age of "technical reproducibility" (*technischen Reproduzierbarkeit*). The entrance of sound reproduction technologies into industry, military conflicts, and the social life of the general public in the transatlantic world during this period played a crucial part in this conceptual change. On the backdrops of the phonograph, telephone, and radio, noise started to gain wider meanings and significances through two avenues: materialization and abstraction. On the one hand, noise was materialized into something associated with the defective conditions of the substances used in sound reproduction—e.g. the "surface noise" on phonographic records. On the other hand, noise was abstracted into quantifiable entities via the development and deployment of various instruments and techniques to measure the effects of street din, radio static, electronic noise, and other kinds of unwanted sonic products in mass media and telecommunications. In metrology, whether noise should be measured in terms of its effect on aural perception or its physical intensity underwent continual deliberations.

Meanwhile, the studies of Brownian motion in the 1900s began to lay a foundation for theoretical understanding of noise, and such studies were applied in industrial researchers' investigations into electronic noise in radio and telephony in the 1910s–20s. Brownian motion refers to the irregular movements of suspended particles in a fluid. In the history of physics, this phenomenon is best known for its association with an intellectual breakthrough in 1905–08 when Albert Einstein developed a theory to explain it in terms of molecular collisions and the French physicist Jean Perrin devised an ultramicroscopic experiment to verify Einstein's predictions. The episode is usually interpreted—as part of Einstein's achievements in his Annus Mirabilis of 1905—to be a strong vindication of atoms' reality and a triumph of atomism. Situated in the history of noise, however, the booming research on Brownian motion in the 1900s had a different significance, for it marked a paradigmatic change of statistical mechanics from a preoccupation with explaining thermodynamically stable effects to an escalating interest in the phenomena of fluctuations. The formal structure of Einstein's, Marian Smoluchowski's, Paul Langevin's, and Louis Bachelier's works on Brownian motion would turn into the backbone of the random-process theory, the mathematical representation of noise in the following decades. Moreover, physicists working in telecommunication industrial research establishments—such as Siemens, AT&T, and General Electric—in the 1910s–20s found a similarity between Brownian motion and electronic noise—the current fluctuations in electrical circuits that engendered a hissing and cracking background in the speakers of the radio and telephone. This finding forged a connection between random fluctuations in statistical physics and noise from electrical sound reproduction technologies. It also promised a relevance of the fluctuation theory to further understanding and handling of noise.

The interwar years witnessed a further development for the theories of noise. In this period, a common view equating noise not only with disturbing sounds but also with random fluctuations in general was emerging. Yet, there was not a unified theoretical framework to deal with random noise. Instead, three approaches to noise became popular in different realms of problem situations. These approaches were marshalled by physicists, electrical engineers, and mathematicians, respectively. Continuing the line of research on Brownian motion at the turn of the century, statistical physicists in the 1910s–30s endeavored to construct the dynamics of Brownian particles subject to thermal molecular collisions, fluidal friction, and various kinds of external forces, and to find the temporal evolution of the fluctuating movements by solving the dynamic equations. This "time-domain" approach was in contrast to telecommunication engineers' "frequency-domain" approach that modeled the atmospheric statics and other types of radio noise with a set of harmonic oscillations with a prespecified spectral profile. Around the same time, "pure" mathematicians were more concerned with the rigorous conceptual foundation of the random-process theory. They used Brownian motion as an exemplar to formulate random fluctuations in the framework of modern mathematical apparatuses including measure theory, functional space, ergodic theory, and generalized harmonic analysis.

Finally, the emerging information sciences in the 1940s–50s converted the interwar theoretical studies of noise into tools for statistical detection, estimation, prediction, and information transmission. In so doing, they turned noise into an informational concept. The beginning of the technosciences in command, control, communication, computing, and information during World War II and the early Cold War has been one of the best-known episodes in the history of science and technology. In this episode, the technosciences at stake were developed for building artificial systems with functionally "smart" features—self-regulation, conveying meaning in communications, optimization, pattern recognition, decision making, etc.—or for understanding why humans, organisms, organizations, or other systems in the social or biological worlds possessed a similar "intelligence." Thus, the researchers in the information sciences often operated under the premise of an ontology for an intelligent being as an object of inquiry, be it an enemy, a prototype human mind, or an interconnected network of calculations. Here, we are nonetheless reminded of the importance of noise in the early development of the information sciences. In contrast to the ontology of an intelligent object, noise epitomized an "ontology of the ambience" that stood for all kinds of uncertain and disturbing factors from the environment. The science of smart detection, prediction, and estimation as part of a feedback control system and the science of information transmission were conducted precisely against a background filled with noise.

The treatments of noise in information sciences of the mid-twentieth century resulted from a synergy of the three interwar approaches. The urgency of national defense after the outbreak of World War II prompted physicists, mathematicians, and engineers to work together in the framework of the rapidly expanding military-industrial-academic complex in the US. The stochastic theory of Brownian motion and the generalized harmonic analysis were applied to the theory of radar target detection amidst a noisy background, the statistical prediction and filtering of enemy targets under various observational errors in antiaircraft gunfire directing, and the theory of information transmission through a noisy channel. At the onset of the Cold War, moreover, engineers utilized insights from the information sciences and turned noise from a plight to be suppressed into an asset for robust information transmission, leading to spread-spectrum communications. By the second half of the twentieth century, noise entered the conceptual and technical repertoire of statistics, physical sciences, life sciences, social sciences, and digital and computing technologies at large, and became entrenched into the modern technosciences' form of life.

Sound, Chance, Information

Three bodies of literature have engaged the history of noise. First, a set of scholarly works have examined noise from the perspectives of sound studies in the histories of technology, science, environment, and music. A focus of these works is the disconcerting sensibilities on the excessive din emerging from the ever-louder industrial soundscape since the nineteenth century, and the development of noise as a public environmental problem. Along this line of inquiry, historians and technology studies scholars have examined the assemblage of various attempts to deal with the problem of noise, from the legal means of introducing zoning laws and the social movements to abate noise to the architectural approach of acoustic building designs and the engineering methods of quantifying and measuring environmental noise. Some writings have also tried to connect the socio-technical history of ambient noise with the accumulation of acoustic and physio-psychological knowledge on noise. Another focus is to look into the cultural significances of noise as a marker of changing modern aesthetic standards that distinguished music from nonmusical sounds. To explore this aspect, researchers have studied the introduction, materialization, and reception of new types of avant-garde and popular music—such as Dadaistic and Jazz compositions—that were often deemed as noise by many contemporaries. These historical works have understood noise as a sonic attribute and have placed it in the technology, culture, politics, and science related to sounds.[6]

The second body of literature concerns the historical studies of the means and approaches to understand, represent, and tackle chance, uncertainty, and randomness in the natural or human world, as well as their social, political, intellectual, and philosophical implications. Here the histories of probability theory and statistics come to the fore. Historians and philosophers of science have produced works on the origin of "classical probability" during the Enlightenment, the intellectual process in which probabilistic thinking evolved into a crucial style of reasoning; the rise of statistics in the nineteenth century in astronomy, genetics, psychology, and sociology; and its influences on the modern form of governance by numbers in political, social, and commercial sectors. Some scholars have also examined the deliberations on the philosophical foundation of probability and their relevance to the epistemology of inductive inferences, the rise of statistical mechanics as a worldview in physical sciences, and the technical development of the stochastic process theory in mathematics.[7] In this scholarly context, noise is treated as a random

[6] For example, see Karin Bijsterveld, *Mechanical Sound: Technology, Culture, and Public Problems of Noise in the Twentieth Century* (Cambridge, MA: MIT Press, 2008); Mike Goldsmith, *Discord: The Story of Noise* (Oxford: Oxford University Press, 2012); Emily Thompson, *The Soundscape of Modernity: Architectural Acoustics and the Culture of Listening in America, 1900–1933* (Cambridge, MA: MIT Press, 2002); Roland Wittje, "Concepts and significance of noise in acoustics: Before and after the Great War," *Perspectives on Science*, 24:1 (2016), 7–28; and *The Age of Electroacoustics* (Cambridge, MA: MIT Press, 2016).

[7] For example, see Stephen Brush, *Statistical Physics and the Atomic Theory of Matter: From Boyle and Newton to Landau and Onsager* (Princeton, NJ: Princeton University Press, 1983); Gerd Gigerenzer, Zeno Swijyink, Theodore Porter, Lorraine Daston, John Beatty, and Lorenz Krüger, *The Empire of*

process. And the history of noise—such as that of Brownian motions or that of radioactive decay—is construed as an episode in the history of probabilistic and statistical mathematics.

The third body of literature inspects the appearance and popularity of information sciences in the mid-twentieth century and beyond. Scholars in this area have followed and analyzed the beginning of the technoscience on communication, control, computing, and signal processing in the context of military research circa World War II, and its further spread in life sciences, social sciences, policy making, business administration, and even countercultures afterwards. Manifested in different forms and specializations such as cybernetics, information theory, theory of computation, artificial neural networks, operations research, and system analysis, these information sciences en masse helped configure many domains of science and technology in the second half of the twentieth century, and opened new worldviews and new imaginations about living organisms, humans, society, and nature. Researchers have produced analyses on these information sciences' social, political, cultural, and intellectual significances to the world after World War II, while some thinkers have discussed the technosciences' broader philosophical implications. In these historical studies, characterizing and handling noise have emerged as issues in the development of information sciences.[8]

These three scholarships have had very limited interactions with one another, though. The sound-studies scholars who have paid attention to noise have concentrated exclusively on sonic noise, except for two works. Emily Thompson and Roland Wittje explored the beginning of electroacoustics in 1920s–30s America and Germany that turned sounds into electrical signals susceptible to the engineering-mathematical manipulations of filtering, equalization, modulation, and spectral analysis. In their investigations into electroacoustics, noise started to gain the sense of disturbances on electrical

Chance: How Probability Changed Science and Everyday Life (Cambridge, UK: University of Cambridge Press, 1997); Ian Hacking, *The Taming of Chance* (Cambridge, UK: Cambridge University Press, 1990); Stephen Stigler, *The History of Statistics: The Measurement of Uncertainty before 1900* (Cambridge, MA: Belknap Press of Harvard University Press, 1986); Jan von Plato, *Creating Modern Probability: Its Mathematics, Physics, and Philosophy in Historical Perspective* (Cambridge, UK: Cambridge University Press, 1998).

[8] For example, see William Aspray, "The scientific conceptualization of information: A survey," *IEEE Annals of the History of Computing*, 7:2 (1985), 117–140; Peter Galison, "The ontology of the enemy: Norbert Wiener and the cybernetic vision," *Critical Inquiry*, 21:1 (1994), 228–266; Steve Heims, *The Cybernetic Group* (Cambridge, MA: MIT Press, 1991); Ronald Kline, *The Cybernetic Moment: Or Why We Call Our Age the Information Age* (Baltimore, MD: Johns Hopkins University Press, 2015); David Mindell, *Between Human and Machine: Feedback, Control, and Computer before Cybernetics* (Baltimore, MD: Johns Hopkins University Press, 2002).

signals abstracted from sounds. Yet, the statistical treatments of noise in the following decades did not enter Thompson's or Wittje's historical narrative.[9]

The vast literatures on probability theory and statistics in social, life, physical, and mathematical sciences have largely overlooked the technological and engineering aspects, especially the sound technologies, from which the issues of noise emerged. Recently, Günter Dörfel, Dieter Hoffmann, Martin Niss, and I myself examined the research on Brownian motions in the 1900s–30s, and have placed it in—instead of its conventional scientific backgrounds of atomism, indeterminism, and visualization—the technological contexts of electronic noise in telecommunication and thermal fluctuations in precision measurement.[10] These works nonetheless rarely dealt with the further development of the stochastic theories of noise that levied a strong influence on the information and communication technologies of the 1930s–40s.

The researchers working on the rise of information sciences in the mid-twentieth century have indeed touched upon the issues of noise. But they have done so in ways in which the sonic contexts were mostly irrelevant, the probabilistic and statistical frameworks were treated as technical details on the backstage, and the dealing of noise was marginal to the forming of the major concepts. In Ronald Kline's, David Mindell's, Peter Galison's, and William Aspray's works, for example, the focus has been on the development of key notions such as information, redundancy, feedback control, intelligence, and computability; their applications to various engineering and technological tasks; and their extensions into novel worldviews and cultural imageries that in turn affected social or organizational relations.[11]

This book provides a historical study of noise that links all three distinct domains: sound, chance, and information. Looking into the transformation of noise from a sonic attribute to an informational notion through this integrated perspective unveils several primary features that are hidden when sound technologies, probabilistic theories, and information discourses are inspected separately. First, sound technologies loomed large throughout the abstraction

[9] Thompson, *Soundscape of Modernity* (2002); Wittje, *Age of Electroacoustics* (2016) and "Concepts and significance of noise" (2016).

[10] Günter Dörfel, "The early history of thermal noise: The long way to paradigm change," *Annalen der Physik*, 524 (2012), A117–A121; Günter Dörfel and Dieter Hoffmann, "Von Albert Einstein bis Norbert Wiener—frühe Ansichten und spate Einsichten zum Phänomen des elektronischen Rauschens." Preprint 301. Berlin: Max Planck Institute for the History of Science (2005); Martin Niss, "Brownian motion as a limit to physical measuring processes: A chapter in the history of noise from the physicists' point of view," *Perspectives on Science*, 24:1 (2016), 29–44; Chen-Pang Yeang, "Tubes, randomness, and Brownian motions: Or, how engineers learned to start worrying about electronic noise," *Archive for the History of Exact Sciences*, 65:4 (2011), 437–470.

[11] Aspray, "Scientific conceptualization of information" (1985); Galison, "Ontology of the enemy" (1994); Kline, *Cybernetic Moment* (2015); Mindell, *Between Human and Machine* (2002).

of noise and the corresponding development of the theories of random fluctuations. Even after the introduction of electroacoustics and the stable establishment of the concepts of signals and noise as entities not necessarily tied to sounds, various considerations with respect to sound reproduction—from the attempts to illustrate how FM radio combated statics to the performance evaluation of the pulse-coded modulation in telephone transmission—continued to influence the mathematical and physical modeling and understanding of informational noise. This heritage explains why today's theory of statistical noise (in stock markets, public health databases, gene expressions, climate records, and so on) is filled with the metaphors, vocabulary, and technicality of sound engineering—spectrum, filtering, signal-to-noise ratio, etc.—that were alien to the traditional statistics before World War II.

Second, the mathematics of Brownian motion, stochastic processes, and measure theory constituted the conceptual substrate for the modern theoretical treatments of random noise. This probabilistic machinery came to define what noise was, and to give it technical semantics. In light of this, the origin story of the information sciences warrants a reexamination. The dominant historical narratives about the beginning of cybernetics and information theory highlight the encompassing, generalizable, and nontechnical notions—e.g. information, redundancy, feedback, stability—marshalled by the areas' protagonists in their popular writings. Yet, findings from this book show that technical handling of random noise in the research program on stochastic fluctuations played a crucial part in the early development of information sciences. For instance, the debut of cybernetics lay in a theory of statistical prediction of signals embedded in random noise. Information theory became appealing to the engineering community only after its inventor introduced a method to specify the capacity of a channel that represented the effect of additive random noise on information transmission. Looking through the angle of technical handling of noise, we can make a much more salient historical connection between postwar information sciences on the one hand and prewar statistical mechanics and mathematical analysis on the other, as the new sciences' protagonists often referred to.

Third, the heavy presence of advanced mathematics and theoretical models in the stories does not imply a purely cognitive, mental, and ivory-tower exercise. Instead, the history of noise was configured with concrete problem situations, material cultures, and environmental factors. The sound technologies constituted the most important material conditions for the transformation of noise. To achieve high fidelity, phonographic technologists tried different materials for records, and came across the association between the cracking noise from the sound machine and the unevenness of a record's surface. To assist urban noise abatement, acoustic engineers devised schemes and instruments to quantify and measure din. To enhance the clarity of wired and wireless telephone conversations, physicists and industrial researchers pushed the limit of vacuum-tube amplifiers, experimented with electronic noise in amplifying circuits, and found its relationship with molecular thermal agitations or the shot effect. The impressive performance of FM radio prompted further thinking into the mathematical modeling of noise. Yet, the materiality of noise was not restricted only to sound reproduction. The theoretical studies of Brownian motions were embedded in the experimental systems of ultramicroscopic observations of suspended particles in a fluid, a galvanometer with a wobbling needle, and a clear liquid turning murky under a phase transition. During and after World War II, researchers on noise grappled with material conditions with a much bigger scale and more immediate uses in military arenas: radar, antiaircraft gunfire directing, encrypted telecommunication, anti-jamming for torpedo control, missile guidance, navigation and landing, and early warning systems for air defense. The highly mathematical deliberations in the theories of noise were almost always connected to these material conditions and pragmatic contexts.

Scope of the Book

Like the evolution of any of Williams's keywords, the historical transformation of noise was not the making of a single group, community, organization, or institution in a single culture or geographical region. Scholars have come up with original ways to write the histories of very broad subjects that either constitute the core human understanding of the world (à la Immanuel Kant's categories of synthetic a priori)—such as objectivity, time, and space—or form the basic conceptual elements characterizing human interventions of the material universe—such as efficiency and technology.[12] The temporal scale of these writings can span as wide as the whole history from Antiquity to the present or hundreds of years from early modern times to the twenty-first century. And the geographical or cultural scope can cover the entire Western civilization.

[12] See, e.g. Lorraine Daston and Peter Galison, *Objectivity* (Cambridge, MA: Zone Books, 2007); Stephen Kern, *The Culture of Time and Space, 1880–1918* (Cambridge, MA: Harvard University Press, 2003); Jennifer Karns Alexander, *The Mantra of Efficiency: From Waterwheel to Social Control* (Baltimore, MD: Johns Hopkins University Press, 2008); Eric Schatzberg, *Technology: Critical History of a Concept* (Chicago, IL: University of Chicago Press, 2018).

This kind of macroscopic or mesoscopic history inevitably involves the difficulty of selecting representative or informative cases and combining them into a meaningful narrative.

The historiographical challenge for this book may be more manageable. The transformation of noise occurred primarily within the first half of the twentieth century. The relevant geographical regions did not extend beyond North America and West and Central Europe. And the historical actors involved in the process were a narrowly defined social collective—specialists who received professional training and spoke the language of electrical engineering, mechanical technology, physics, and mathematics. Even so, the historical transformation of noise was a cross-cultural, transnational, and interdisciplinary process. It was too diverse and heterogeneous to subject to in-depth analysis and interpretation with respect to a single sociopolitical background, cultural underpinning, organizational setting, research community, school of thought, or a single scientific theory or technological product. In this book, rather, we follow specific conceptual and technical developments for the understanding and handling of noise across national, institutional, and disciplinary boundaries, and trace how these developments interacted with one another. Fortunately, some individuals in the story moved across regions and organizations, operated across disciplines, and brought ideas from one realm of study to another. These "translators" and the academic, corporate, and governmental research infrastructure surrounding their practice formed the continuing backbone of the narrative.

PART I

ABSTRACTION AND MATERIALIZATION OF NOISE

2
Discordance and Nuisance

Loud, annoying, and unpleasant sounds have been a sensory experience since the dawn of history. Noise is thc term to signify them. From the *Oxford English Dictionary* (*OED*), the earliest meanings of the word *noise* included:

- sound; aggregate of sounds occurring in a particular place or at a particular time;
- disturbance caused by sounds, discordancy;
- disturbance made by voices; shouting, outcry;
- a sound of any kind, especially a loud, harsh, or unpleasant one.

Be diverse as it might, noise possessed its most common connotation in the premodern and early modern world as loud and uncomfortable sounds. *Ancrene Riwle* circa 1250 stated "Hore meadlese nowse." In *Siege and Conquest of Jerusalem* in 1481, William, Archbishop of Tyre wrote "of the noyse that sourded emonge the hethen men discordyng in theyr lawe." According to the *OED*'s etymological analysis, the English word *noise*

> is perhaps from "sea-sickness", the literal sense of classical Latin *nausea*, "to upset, malaise" (compare figurative senses of classical Latin *nausea* s.v. NAUSEA n.), then to "disturbance, uproar", and thence to "noise, din" and "quarrel".

The *OED*'s claim connects unpleasant feelings and the aural senses arising from noisy sounds.[1]

This early meaning of noise was not unique to English. From *Dictionnaire de l'Académie française*, the French word *bruit*, which was originated from the verb *bruire* (roar) and initially meant "renowned" or "glow" as it appeared in the twelfth century, gradually became "the sounds of voice." Its more frequent

[1] John Simpson and Edmund Weiner (eds.), *The Oxford English Dictionary* (Oxford: Clarendon Press, 1989), the entry of "noise."

connotations later included "sound or aggregate of sounds that are produced against the regular harmony" and "turmoil and provocative movement."[2]

Two words in German denoted noise: *Geräusch* and *Lärm*. Derived from the verb *rauschen* (shout, roar), *Geräusch* referred to generic sounds during the medieval and early modern periods. According to Jacob and Wilhelm Grimm's *Deutsches Wörterbuch*, the primary meanings of *Geräusch* included "din, noise, sound of any form and strength," and "sound in opposite to the musical tone." *Lärm* derived from "the call to arms" or "alarm" in French and Italian around the fifteenth century. The meaning of the word broadened to "rebellion" and "turmoil" by the seventeenth century, and evolved into the sense of sonic noise by the eighteenth.[3]

Based on the early semantics of noise, *bruit*, *Geräusch*, *Lärm* (or *rumore* in Italian and *ruido* in Spanish for that matter), the core meaning of the term had to do with sound at large: sound of any form, aggregate of sounds, voice, cry, roar, etc. A noisy sound was voluminous, annoying, composite, or extraordinary. Under this sonic characterization, noise was often understood as something that possessed the power to disturb, disrupt, destabilize, and interfere. The social theorist Jacques Attali noted such significances. In *Noise: The Political Economy of Music*, he invoked a painting to illustrate the metaphoric implications of cacophony. Created in 1559 by the Dutch artist Pieter Bruegel the Elder, *The Fight between Carnival and Lent* depicted a vibrant street scene in Renaissance Europe: A man played a lute and another played a windbag. Beggars and lepers asked for money from passersby. Sitting on a barrel with meat on his head and a pierced swine head in his hand, a butcher was about to shout. A crowd in a tavern watched a farce. Another group danced around a circle. Vendors prepared food. Children toyed with tops. Contrasting this sonorous and colorful life was a solemn, dark, and quiet collective on the right of the canvas, where ladies and gentlemen finished their

[2] Maurice Druon, *Dictionnaire de l'Académie française*, 9th edition (Paris: L'Académie française, 2011), the entry of "bruit"; Émile Littré, *Dictionnaire de la langue française* (Paris: Librairie Hachette, 1883), vol. 1, A-C, 431.

[3] Jacob Grimm and Wilhelm Grimm, *Deutsches Wörterbuch* (Leipzig: Verlag Von S. Hirzel), vol. 4 (1878), 3583–3585, vol. 6 (1885), 202–205; Keith Spalding, *An Historical Dictionary of German Figurative Language* (Oxford: Basil Blackwell, 1974), 987, 1584. Karin Bijsterveld pointed out that by the twentieth century, *Lärm* and *Geräusch* had gained more definite and separate connotations in technical and social contexts: *Lärm* referred to noise as in clamor, whereas *Geräusch* meant noise as opposed to signal. Such a semantic distinction has not happened in English, French, or most other European languages [Bijsterveld, *Mechanical Sound* (2008), 105; also see Wittje, "Concepts and significance of noise" (2016), 9]. The contrast between noise as disturbing sound and noise as disturbance to signal is a major theme of this book. But this contrast did not begin to emerge until the turn of the twentieth century. For the time period being considered here, the distinction between *Lärm* and *Geräusch*, if any, was quite ambiguous.

mass and walked out of a church (Figure 2.1). Looking at the painting, we can easily imagine a noisy urban soundscape of early modern times.

But *The Fight between Carnival and Lent* was not a realistic portrait of a street filled with common noise. Instead, as Attali remarked, the artwork was an allegory of a historical moment when two "fundamental political strategies, two antagonistic cultural and ideological organizations" clashed with each other—the force of festivity, violence, indulgence, and chaos on the one hand, and the force of abstinence, harmony, norm, and order on the other. The bewildered and rampant crowds of Carnival embodied noise, whereas the disciplined and pious collective of Lent represented silence. Here noise was associated with disruption, aberration, and things being out of control. Noise interfered with churchgoers' regular proceedings, defied rules, engendered surprises, and disturbed order.[4]

What order did noise disturb? Two notions of noise were influential in the long history of civilization: as irregular and inharmonious sounds that went against the human senses, and as annoying sounds of the surroundings that invaded into public and private spaces and disrupted tranquility.

Figure 2.1 Pieter Bruegel the Elder, *The Fight between Carnival and Lent* (1559).
© KHM-Museumsverband.

[4] Jacques Attali, *Noise: The Political Economy of Music*, translated by Brian Massumi (Minneapolis: University of Minnesota Press, 1985), 21–24.

Although the idea of noise as discordance and that of noise as nuisance intertwined with each other, they came from different backgrounds. The concept of discordance tied to the theories of music since Antiquity, especially their preoccupation with harmonious tones and attempts to make sense of such tones with cosmic-numerological or (later on) psycho-physiological reasons. The concept of nuisance had a close relationship with the efforts by governments, local communities, and civic groups to control and "abate" din in urban or industrial settings. These two concepts became the dominant subjects of discussions on noise in technical literature and public discourse by the nineteenth century. In the context of discordance, noise was the opposite of musical tones. It defied the aesthetics, philosophical, and scientific principles of harmony, and appeared as an irregular, random, and chaotic sonic entity. In the context of nuisance, noise disrupted the communal norm of quietude, and disturbed the regular functioning of society.

This chapter provides a brief review of these two ideas of noise before the twentieth century. There is a massive literature in musicology on the history of harmonic, tonal, and consonant theories. Meanwhile, historians of technology and environmental historians have produced substantial works on noise abatement and noise control.[5] The aim here is not to make an original contribution to them, but to outline the understandings of noise before its major transformation in the twentieth century, when the term's notion of defying and disturbing the "signals" of conceptual and social orders was retained but its meaning extended to non-sonic fluctuations.

Discordance, Dissonance, and Disharmony

Noise as discordant and dissonant sounds creating unmusical, unaesthetic, and unpleasant feelings bore a relationship with a two-millennia-long tradition of harmonic theories in Western music. In the sixth century BCE, the Greek philosopher Pythagoras contemplated the nature of music as combinations of tones "in harmony" with one another. This notion of music as harmonies manifested in ratios of small numbers became popular among

[5] For an overview of the scholarship in musicology on the history of tonal, harmonic, and concordant theories, see Thomas Christensen (ed.), *The Cambridge History of Western Music Theory* (Cambridge: Cambridge University Press, 2002). Two representative historical works in recent years on noise abatement are Bijsterveld's *Mechanical Sound* (2008) and Thompson, *Soundscape of Modernity* (2002), chapter 4. A concise review of both senses of noise in their historical contexts can be found in Goldsmith, *Discord* (2012), chapters 1–8.

Greek and Roman savants, passed down to the Middle Ages, and formed a foundation of European musical theories.[6]

The Pythagorean notion of consonance centered on simple mathematical ratios. Two pitches were most concordant (i.e. mixed most "pleasantly") when they were in *unison* (identical). The ratio of the cord lengths or cord tensions generating two pitches in unison was 1:1. The next harmonious pitch-pair was an *octave* apart (e.g. from middle C to treble C in today's musical notation) produced by the ratio of 2:1. The less beautiful consonance than octave was *fifth* (e.g. from C to G) produced by 3:2. Next was *fourth* (e.g. from C to F) produced by 4:3. Further down the harmonic hierarchy was *tone*, the difference between fifth and fourth, produced by 9:8. Then came *semitone*, the difference between fourth and two tones, produced by 256:243.[7]

The consonances of unison, octave, fifth, fourth, tone, and semitone constituted the building blocks of European musical tones. Throughout the medieval and early modern periods, philosophers and music scholars used these consonances to construct various musical scales in mathematical manners, and to address crucial questions in performance. The assumption underlying musical harmonics was that mixing pitches of simple ratios was aesthetically more appealing. In contrast, mixing pitches with more complex or irregular relations was uglier, less pleasant, and more annoying. The former was music, the latter was noise.

Why were pitches of simple ratios musical whereas the other combinations were noisy? To answer this question, the Pythagoreans and Platonists resorted to the connection between cosmic order and arithmetic regularity. An example was to invoke tetractys—the sequence of the simplest numbers 1-2-3-4—to explain the harmonic hierarchy of unison, octave, fifth, and fourth. These tone-pairs corresponded to the ratios of number-pairs in the tetractys sequence with increasing complexity: 1:1, 2:1, 3:2, 4:3.[8]

In search of clues for explaining the beauty and pleasure of consonances, the numerological account of consonances gradually yielded to examinations

[6] Calvin Bower, "The transmission of ancient music theory into the Middle Ages," in Thomas Christensen (ed.), *The Cambridge History of Western Music Theory* (2002), 142–143.

[7] Jan Herlinger, "Medieval canonics," in Thomas Christensen (ed.), *The Cambridge History of Western Music Theory* (2002), 171–173. Retrospectively, the longstanding Pythagorean belief of the correspondence between the consonance and the ratio of the weights hung on the cords was effectively a statement that a tone was proportional to the tension of the cord. (The weight of the hammer caused the tension of the string.) This belief remained sound until Vincenzo Galilei, Galileo's father, challenged it in the sixteenth century. Vincenzo contended that the mathematical ratio corresponding to a consonance should be the *square root* of the weights. In today's language, the tone should be proportional to the square root of the tension. See Goldsmith, *Discord* (2012), 55–56.

[8] Catherine Nolan, "Music theory and mathematics," in Thomas Christensen (ed.), *The Cambridge History of Western Music Theory* (2002), 273–274.

of the physical characteristics of music and sounds. This change of direction aligned with the emergence of acoustics in the seventeenth century. Galileo was among the first to discuss consonances and dissonances in terms of acoustic vibrations and their effects on ears. In *Dialogues Concerning Two New Sciences* (1638), his avatar the wise man Salviati stated that:[9]

> the ratio of a musical interval is ... determined ... by the number of pulses of air waves which strike the tympanum of the ear, causing it also to vibrate with the same frequency. This fact established, we may possibly explain why certain pairs of notes, differing in pitch produce a pleasing sensation, others a less pleasant effect .... The unpleasant sensation produced by the latter arises, I think, from the discordant vibrations of two different tones which strike the ear out of time.

In the acoustic framework, Galileo contended that the ratios associated with different consonances were those of their vibrating frequencies. He explained the sensations about musical tones and noise in terms of a theory of "coincidence." Distinct tones were trains of pulses hitting an eardrum at different frequencies. When these frequencies were in simple mathematical ratios, the pulses coincided substantially, the tones produced synchronized mechanical effects on the ear, and the most harmonic sensation was incurred. When sounds did not oscillate at frequencies of simple ratios, the vibrating pulses were asynchronous to each other and "discordant and produce a harsh effect upon the recipient ear which interprets them as dissonances." This was noise.[10]

Galileo's discussion on consonances and dissonances marked a change in musical theory from numerological explanations to physical accounts. Yet, while the shift to acoustics might have altered the connotation of discordance, scholars still appealed to the quasi-aesthetic argument of simplicity and orderliness when they tried to explain why music sounded beautiful whereas noise sounded annoying.[11] The emergence of physiology of the human senses in the nineteenth century led to a further alteration in the scientific understanding of music, harmony, dissonance, and noise.

Enter the German scientist Hermann Helmholtz. In 1852, he began to work on the science of sound. Combining his familiarity with acoustics, hobby of piano performance, and interest in the physical foundation of aural senses, he

[9] Galileo Galilei, *Dialogues Concerning Two New Sciences*, translated by Henry Crew and Alfonso de Salvio (Buffalo, NY: Prometheus Books, 1991), 103–104.

[10] Ibid., 104.

[11] Burdette Green and David Butler, "From acoustics to *Tonpsychologie*," in Thomas Christensen (ed.), *The Cambridge History of Western Music Theory* (2002), 246–257.

produced a quantity of works on propagation of sound, harmonics, and tones in Bonn and Heidelberg. These works resulted in the publication of *On the Sensation of Tone as a Physiological Basis for the Theory of Music* in 1863.[12]

In *On the Sensation of Tone*, Helmholtz gave a formal definition of noise that became widely accepted by modern acousticians. He indicated "the first and principal difference" between various sounds experienced by human ears to be "that between *noises* and *musical tones*" (italics in the original). Specifically, "a noise is accompanied by a rapid alternation of different kinds of sensations of sound." To him, "the rattling of a carriage over granite paving stones," "the splashing or seething of a waterfall or of the waves of the sea," and "the rustling of leaves in a wood" were all noise, for they had "rapid, irregular, but distinctly perceptible alternations of various kinds of sounds." On the other hand, "a musical tone strikes the ear as a perfectly undisturbed, uniform sound which remains unaltered as long as it exists, and it presents no alternation of various kinds of constituents." In this definition, noise and musical tones differed in terms of their compositions. Noise comprised numerous components corresponding to different sensations of sound; their strengths varied with time; and they "are irregularly mixed up and as it were tumbled about in confusion." A music tone comprised a single or only a few of such components; and its (their) strength(s) did not vary with time. As he perceived, musical tones were more elemental, and one could produce noise by combining musical tones—for instance, striking all the keys of a pianoforte at the same time.[13]

Helmholtz's conceptualization of noise relied on a particular analysis that separated sound into its distinct components. Noise was the types of sound with numerous components changing over time and combining in irregular and complicated ways. This notion was familiar. As early as Galileo, natural philosophers, mathematicians, and music scholars had decomposed sounds into pendulum-like harmonic oscillations. In the 1820s, the French mathematician Joseph Fourier introduced a procedure to represent a continuous and smooth function of time in terms of sinusoidal waveforms at different frequencies. Later known as the Fourier-series expansion for periodic functions or the Fourier-integral expansion for duration-limited functions, the procedure became a powerful tool for analyzing the spectral elements of physical

[12] Hermann Helmholtz, *On the Sensation of Tone as a Physiological Basis for the Theory of Music*, translated by Alexander Ellis (New York: Dover, 1954). For Helmholtz's biography, see David Cahan, *Helmholtz: A Life in Science* (Chicago, IL: University of Chicago Press, 2018).

[13] Helmholtz, *On the Sensation of Tone* (1954), 7–8.

quantities. Helmholtz's components of noise or musical tones could be understood as their harmonic elements.[14]

Yet, Helmholtz argued that the components of noise and musical tones had a richer physical meaning than the outcomes of Fourier analysis. While it was possible to represent mathematically a sound as a sum of simple oscillations, ascertaining that these oscillations had incurred concrete sensations in the human ear and brain was a different matter. To him, the components of noise and musical tones were not only spectral terms from Fourier analysis, but also physical entities stimulating distinct sensations of sounds.[15] This assertion could be stated more explicitly with the so-called "Ohm's law," a principle proposed by the German physicist Georg Ohm in 1843:[16]

> Every motion of air ... is ... capable of being analysed into a sum of simple pendulum vibrations, and ... [to each of which] corresponds a simple tone, sensible to the ear, and having a pitch determined by the periodic time of the corresponding motion of the air.

How these spectral components of sounds became elements of sensations had to do with Helmholtz's physiological theory of hearing: A sound as vibrations of air propagated through the ear canal, tympanum, middle ear, and into the cochlea in the inner ear. The cochlea contained many nerve fibers of different lengths. These fibers functioned as resonators that responded to sound's distinct frequency components. The neural signal transmitted from each cochlear nerve to the brain corresponded to the sensation of a particular pitch. To Helmholtz, therefore, Fourier analysis of a sound was not only a mathematically viable representation but also a physically and physiologically meaningful decomposition, for the cochlea functioned as a spectral analyzer that separated distinct frequency components of a sound into distinct sensations.[17]

Helmholtz's auditory physiology explained why noise appeared featureless and atonal to human senses. As a combination of many frequency components varying over time, noise stimulated simultaneously a lot of nerve fibers—"resonators"—in the cochlea, which blanked the perception of any

[14] For the beginning of Fourier analysis, see Ivor Grattan-Guinness, *Joseph Fourier, 1768–1830: A Survey of His Life and Work* (Cambridge, MA: MIT Press, 1972); Morris Klein, *Mathematical Thought from Ancient to Modern Times*, Volume 3 (New York: Oxford University Press, 1972), 966–971.

[15] Helmholtz, *On the Sensation of Tone* (1954), 35.

[16] Ibid., 33.

[17] Ibid., 136–141.

recognizable pitches.[18] Similarly, why the pair of pitches appeared increasingly discordant along the canonic ladder of octave, fifth, fourth, and tone had a physiological basis. When the ratio of two pitches was not simple, the two tones interfered with each other to create a low-frequency beat, which caused a rough sonic sense and consequently unpleasant dissonance.[19]

Helmholtz's theory of sonic perception was influential. Although his strictly physiological explanations of musical sensations were challenged by psychological accounts toward the end of the nineteenth century,[20] his characterization of noise versus musical tones remained. In *The Theory of Sound* (1877), another classical scientific treatise on sound, Lord Rayleigh at Cambridge also laid out the major distinction in sonic experiences between musical notes and noise. Musical notes were perceived as smooth and continuous; noise was not. Whereas noise could be synthesized with musical notes such as by pressing all keys on a pianoforte, a musical note could not be a composite of noises.[21] This notion of noise as irregular sounds with no apparent periodicity and complicated spectra persisted into the twentieth century among physicists and engineers.[22]

Nuisance, Pollution, Public Problems

In addition to unpleasant and inharmonious sensory experiences, noise has also been associated with the disturbance of tranquility, damage to the quality of life, and hazards to the health of populations. Noise as discordance subscribes to an aesthetic, physiological, or psychological construal of human senses. The sounds contrary to harmony often subsume certain physical or mathematical forms. In contrast, noise as nuisance entails a disruption of social order. The sounds constituting threats to civil life may not have fixed acoustic characteristics. Rather, their threats come from their "improper" appearances at situations where they are not supposed to exist that disturb normal proceedings—any sounds can be annoying if they are present at the wrong occasions or times. For this reason, the historian Peter Bailey defined noise as "sound out of place."[23]

[18] Ibid., 150.

[19] Ibid., 185–196; Green and Butler, "From acoustics to *Tonpsychologie*" (2002), 260–261.

[20] Green and Butler, "From acoustics to *Tonpsychologie*" (2002), 262–266.

[21] John Strutt (Lord Rayleigh), *The Theory of Sound*, vol. 1 (London: Macmillan, 1877), 4.

[22] Robert Beyer, *Sounds of Our Times: Two Hundred Years of Acoustics* (New York: Springer-Verlag, 1999), 206.

[23] Peter Bailey, "Breaking the sound barrier: A historian listening to noise," *Body and Society*, 2:2 (1996), 50.

Restraining the invasion of "sound out of place" into quarters requiring quietude has been a longstanding endeavor. Legislation against noise dates back to Antiquity. Over the Middle Ages and early modern period, local regulations, municipal bylaws, and regional rules were passed to restrict the noise produced by various skilled trades and other common artifacts or occurrences in townsfolk's lives. By the seventeenth century, many European municipalities had one form of noise laws or another.[24] As these legal traces suggest, coping with noise as an intrusive plight to quotidian activities such as work, leisure, and sleep was a familiar experience to people in the premodern and early modern world.

The nature of this plight underwent a transformation around the nineteenth century. The accelerating urbanization made the city soundscape more and more unbearable to those whose preoccupations required quietude. The sounds from human activities associated with a conspicuous growth of population were so loud, heterogeneous, and varying that it became possible, as the physicist Mike Goldsmith points out, to talk about the overall noise level of the amalgamation of all the urban sounds instead of the individual sounds.[25]

Moreover, the new sounds from machines of the Industrial Revolution made the nineteenth century unprecedentedly noisier than before. The introduction of the steam engine, power loom, threshing machine, railway, hydraulic press, reverberatory furnace, and screw-cutting lathe elevated substantially the general noise levels in cities and even the countryside. Factories, machine shops, and manufacturing plants incessantly hummed and buzzed, whereas locomotives, steamboats, and trams roared across fields, water, and roads. The loud mechanical noise became a symbol of modern technology, as Hawthorne's 1844 encounter at Sleepy Hollow demonstrated. The industrial noise altered fundamentally the soundscape of that world. The new mechanical sounds were so numerous, deafening, congested, and encompassing that people had difficulties distinguishing different sounds. This amounted to what the composer Raymond Murray Schafer called a "lo-fi" soundscape.[26]

The overpopulation of urban areas and the mechanical sounds from new industrial technologies led to stronger calls for controlling noise. In the second half of the nineteenth century, restricting, containing, and eliminating noise had become a wide-reaching social movement. This period was marked by a significant rise of litigations and complaints against noise from streets,

[24] Raymond Murray Schafer, *The Soundscape: Our Sonic Environment and the Tuning of the World* (Rochester, VT: Destiny Books, 1977), 189–191; Bijsterveld, *Mechanical Sound* (2008), 55–57.
[25] Goldsmith, *Discord* (2012), 77–78.
[26] Schafer, *Soundscape* (1977), 71–73.

traffic, factories, manufacturers, and neighbors. Moreover, individuals and groups concerned with noise began to organize themselves and embark on concerted collective actions to lobby for noise-control legislations and raise the general awareness of noise's plight. With a clearer understanding of environmental hazards as byproducts of industrial technology, noise joined the other pollutants (garbage, exhausted air, contaminated water) and turned into an object of study and concern for medical and public health professionals. In this context, according to the technology studies scholar Karin Bijsterveld, noise was increasingly perceived and presented as a "public problem" that needed to be taken care of with policy measures, general education, research and standardization, and other collective actions.[27]

The spread of local civil actions against noise in Western societies in the second half of the nineteenth century resulted in much more conspicuous and broader noise abatement movements by the early twentieth century. In 1908, a philosophy professor Theodor Lessing in Hanover organized the German Association for the Protection from Noise. In New York, a physician Julia Barnett Rice created the Society for the Suppression of Unnecessary Noise in 1906. The London Street Noise Abatement Committee was established in 1908. These pioneer societies in noise abatement were followed by another surge of organizations in France, Germany, Britain, Austria, and the Netherlands. They published magazines for public education on noise's hazards, lobbied for policy changes and anti-noise legislations, and administered various activities to raise the general awareness of issues related to noise. Some of them achieved notable success. Lessing's Anti-Noise Society boasted more than a thousand members. The British Anti-Noise League staged a famous Noise Abatement Exhibition at the London Science Museum in 1935. Rice's Society for the Suppression of Unnecessary Noise pushed for the passing of a bill in Congress regulating boat whistles, and prompted the New York City authorities to enact a Noise Abatement Commission to study systematically the metropolis's situation of noise pollution. In 1930–32, the Commission issued a two-volume report *City Noise*, detailing the types of clamors and the results of 10,000 noise intensity measurements in about 90 distinct areas of the city.[28]

[27] Bijsterveld, *Mechanical Sound* (2008), 17–19, 27–31. For the concern of noise as a problem of mental health, see James Mansell, *The Age of Noise in Britain: Hearing Modernity* (Urbana: University of Illinois Press, 2017), 25–61.

[28] Bijsterveld, *Mechanical Sound* (2008), 101–102, 110–112; Mansell, *The Age of Noise in Britain* (2017), 49–53; Raymond Smilor, "Toward an environmental perspective: The anti-noise campaign, 1893–1932," in Martin Melosi (ed.), *Pollution and Reform in American Cities* (Austin, TX: University of Texas Press, 1980), 135–151.

In contrast to the notion of noise as discordance, the notion of noise as nuisance did not attach to specific sonic characteristics. Although music theorists, mathematicians, philosophers, and scientists had diverse views on what noise was, they attributed noise to sounds with no harmonic or rhythmic relationships, complicated spectral patterns, and irregular variations over time. The civil groups, concerned citizens, government officials, technical experts, and other stakeholders in the noise abatement movements at the turn of the twentieth century did not have such a consensus. In principle and in practice, sounds of any features could be considered noisy as long as the complainants perceived their presence to be improper and disturbing. Bijsterveld found that a lot of them dramatized the noise under dispute with the auditory topoi of "intrusive sounds" with the following characteristics: multitude of different sounds or a series of recurrent sounds, appearing to be close to and approaching the listener, and irregular or unpredictable in rhythm.[29] Schafer went further by trying to represent the waveforms of various environmental noises from mechanical or other sources.[30] In spite of these attempts, the temporal, spectral, or other acoustic attributes did not become the characteristics of noise for those involved in noise abatement. To them, noise was denoted by its sources (traffic, home, street, construction, trade, industry) and volume. And their major means of making it an object of technical control was to measure its intensity with various instruments.[31]

Anti-Music and Public Problems

Noise has been a longstanding human experience. Throughout history, noise was conceptualized and construed either as discordance, dissonance, and disharmony, or as nuisance, aural pollution, and public problems. Both concepts became more clearly articulated, developed, and specified in music, acoustics, physiology, urban planning, and environmental movements by the nineteenth century. Under these concepts, noise defied order, disturbed regularity, and obstructed the normal ways of beings: discordance was the opposite to regular musical tones; nuisance threatened social order. As we will see, these longstanding notions of discordance and nuisance affected the scientific studies and technical handling of noise in the twentieth century. The Helmholtzian characterization of entities with irregular spectra shaped engineers' and

[29] Bijsterveld, *Mechanical Sound* (2008), 44, 74.
[30] Schafer, *Soundscape* (1977), 79.
[31] Thompson, *Soundscape of Modernity* (2002), 144–157.

physicists' understanding of noise. The din abatement movements facilitated the introduction of techniques and instruments for quantitative measurements of noise. What unified these undertakings was their sonic nature: no matter whether noise was atonality or din from streets or workshops, it was still sound.

Yet, noise started to gain other meanings by the turn of the twentieth century. The invention of sound reproduction technologies played a crucial part in this new development. Telephone, phonograph, and radio transduced sounds into electric currents in wires and vacuum tubes, vibrations of mechanical membranes, grooves on metal or celluloid surfaces, and other physical forms. Sound became an entity that could be recorded, replayed, transmitted, amplified, filtered, and modified. This technical breakthrough transformed the connotation of noise: not only unpleasant sounds, but also the unwanted byproducts in the processes of sound reproduction, such as material defects of recording surfaces, interferences of atmospheric electricity with wireless communication, and random fluctuations of electric current in electronic tubes and resistors. In other words, the sound reproduction technologies enabled the materialization of noise.

3

Materializing Cacophony

Surface Noise on Phonographic Records

Noise and Sound Reproduction Technologies

The notion of sound violating the perception of harmony or breaking the tranquility of life dominated the understanding of noise from premodern times to the Industrial Revolution. Yet, the late nineteenth and early twentieth centuries witnessed the emergence of a new concept that treated noise as physical effects to be materialized, quantified, measured, mathematized, and manipulated in forms other than sound. Key to this transformation was a series of inventions and developments in the 1870s–1920s. The decades-long efforts to transmit human voices culminated in Alexander Graham Bell's electromagnetic diaphragmatic telephone in 1876. Thomas Edison's introduction of the tinfoil phonograph in 1877 opened the way for subsequent innovations in sound-recording machines in the 1880s–90s. Appropriation of newly found electric waves led to Guglielmo Marconi's establishment of wireless telegraphy in 1894, and radio telephony and broadcasting in the 1910s–20s. By the 1920s, the telephone, phonograph, and radio became major media around the world.[1]

Telephone, phonograph, and radio are what the media-studies scholar Jonathan Sterne called "technologies of sound reproduction." They "use devices called *transducers*, which turn sound into something else and that

[1] Much has been written on the technical, business, and social-cultural histories of the telephone, sound recording, and radio. Here is only a rudimentary and partial listing of the vast literature in the area. For the history of the telephone, see M.D. Fagen (ed.), *A History of Engineering and Science in the Bell System: The Early Years (1875–1925)*, vol. 1 (New York: Bell Telephone Laboratories, 1975); Claude Fischer, *America Calling: A Social History of Telephone up to 1940* (Berkeley: University of California Press, 1994). For the history of the phonograph/gramophone, see Roland Gelatt, *The Fabulous Phonograph: 1877-1977* (New York: Macmillian, 1977); Walter Welsh and Leah Brodbeck Stenzel Burt, *From Tinfoil to Stereo: The Acoustic Years of the Recording Industry, 1877–1929* (Gainesville: University of Florida Press, 1994). For the history of radio, see Hugh Aitken, *Syntony and Spark: The Origin of Radio* (Princeton, NJ: Princeton University Press, 1976); *The Continuous Wave: Technology and American Radio, 1900-1932* (Princeton, NJ: Princeton University Press, 1985); Susan Douglas, *Inventing American Broadcasting, 1899–1922* (Baltimore, MD: Johns Hopkins University Press, 1989).

something else back to sound" (italics in the original).[2] Telephone, sound recording, and radio all employed an eardrum-like *tympanic* mechanism that, via undulations on a carefully controlled membrane (the "diaphragm"), translated between sonic vibrations and oscillating electric currents, traces on designated surfaces, or modulated electric waves.[3]

The tympanic reproduction of sound had precedents well before Bell's telephone, Edison's phonograph, and Marconi's radio. Throughout the eighteenth and nineteenth centuries, scientists and inventors tried to convert sound into visible or analyzable effects: the German physicist Ernst Chladni's figures of acoustic vibrations on a sand-spread plate in the 1780s; Helmholtz's acoustic synthesizer (made of tuning forks) and analyzer (made of cavities) in the 1850s; the French inventor Leon Scott's phonautograph to transduce sound into curves via a human ear in 1857; the Parisian artisan Rudolph Koenig's manometric flame to visualize vowels; the French physiologist Étienne-Jules Marey's vocal polygraph to record a person's nasal, larynx, and lip movements in voice articulation in the 1870s; and the use of kymograph—originally introduced by the German physiologist Carl Ludwig to monitor blood pressure—to record voices by the 1890s.[4]

The rich technical traditions of visualizing and analyzing sounds and voices in the eighteenth and nineteenth centuries did not have an immediate impact on the understanding and treatment of noise. Scientists, engineers, and inventors did not pay much attention to the transduction, inscription, analysis, and abstraction of noise until the popularity of sound reproduction technologies at the turn of the twentieth century. The motivation for tackling noise along that line had to do with the historical condition of sound reproduction.

Industrial manufacturing, mass media, telecommunications, and popular cultures in the early twentieth century made reproduction a central issue in people's cultural, intellectual, and quotidian lives. In 1935, Walter Benjamin

[2] Jonathan Sterne, *The Audible Past: Cultural Origins of Sound Reproduction* (Durham, NC: Duke University Press, 2003), 22.

[3] Ibid., 34–35.

[4] For the Chladni figures, see Myles Jackson, *Harmonious Triads: Physicists, Musicians, and Instrument Makers in Nineteenth-Century Germany* (Cambridge, MA: MIT Press, 2006), 13–19. For Helmholtz's synthesizer and analyzer, see Cahan, *Helmholtz* (2018), 262–290. For Koenig's manometric flame, see David Pantalony, "Seeing a voice: Rudolph Koenig's instruments for studying vowel sounds," *The American Journal of Psychology*, 117:3 (2004), 435–440. For the phonautograph, see Sterne, *Audible Past* (2003), 31–34. For Marey's polygraph, see Robert Brain, "Standards and semiotics," in Timothy Lenoir (ed.), *Inscribing Science: Scientific Texts and the Materiality of Communication* (Stanford, CA: Stanford University Press, 1998), 262–263. For phoneticians' use of the kymograph, see Chen-Pang Yeang, "From modernizing the Chinese language to information science: Chao Yuen Ren's route to cybernetics," *Isis*, 108:3 (2017), 560–561. For an overview of the visual display of sound as well as other traces of physical effects in the nineteenth century, see V.J. Phillips, *Waveforms: A History of Early Oscillography* (Bristol: Adam Hilger, 1987).

wrote *The Work of Art in the Age of Mechanical Reproduction* to examine the sea change brought by printing, photography, and film to arts.[5] To Benjamin and his contemporaries, the most powerful aspect of mechanical reproduction (or "technical reproducibility," a more accurate translation of Benjamin's German term *technischen Reproduzierbarkeit*, as Wittje pointed out[6]) was the striking similarity between the original and the copy. The techniques of printing, photography, and motion picture made it difficult to distinguish between an artwork and its duplicate. The same sensibility also prevailed in the technical reproducibility of sounds. The witnesses of Bell's demonstration of the telephone at the Centennial Exposition in Philadelphia or the audiences of Edison's "tone test" concerts were deeply impressed by the uncanny sameness of the sounds and voices from machines and those from actual sources.[7] By the 1930s, the term "fidelity" was popular among the users, engineers, and businesses of the phonograph, gramophone, and wireless broadcasting.

Fidelity resulted from complicated sociocultural processes that required the forming of new institutions, aesthetics, and habitus.[8] It also intertwined with an epistemic transformation of noise. In the early years of sound reproduction technologies, inventors, engineers, and users experienced a variety of distracting sounds that degraded the quality of reproduction: a phonographic stylus sounded cracking under a bumpy recording surface; the telephone crosstalked; the radio suffered from static; hisses or snaps came out of an earphone or a loudspeaker. By the early twentieth century, it was more and more common among consumers and developers to understand such unruly sounds in terms of noise.

Yet, these new kinds of noise in sound reproduction technologies differed from the longstanding sounds of disharmony and nuisance. Unlike vendors' cries, roars of factory machines, or discordant utterances from musical instruments or vocal cords, telephone crosstalk, radio static, phonographic scratches, and other forms of noise in telecommunication and sound recording were associated with physical effects other than sounds: mechanical defects of record surfaces, electromagnetic interference from adjacent lines or stations, atmospheric discharge, or thermal fluctuations of electronic currents. A large

[5] Walter Benjamin, *The Work of Art in the Age of Mechanical Reproduction* (New York: Penguin, 2008).

[6] Wittje, *Age of Electroacoustics* (2016), 16.

[7] For Bell's demonstration of the telephone at the Centennial Exposition, see Robert Bruce, *Bell: Alexander Graham Bell and the Conquest of Solitude* (Ithaca, NY: Cornell University Press, 1990), 188–198. For Edison Phonograph's tone tests, see Emily Thompson, "Machines, music and the quest for fidelity: Marketing the Edison Phonograph in America, 1877–1925," *Musical Quarterly*, 79 (Spring 1995), 131–171.

[8] Sterne, *Audible Past* (2003), 215–286.

part of the technical improvement for acoustic fidelity in the first half of the twentieth century thus involved the examination, analysis, mitigation, and removal of these physical effects. At the same time, the very notion of noise extended to the physics, mathematics, and engineering of such physical effects. With the new theoretical, experimental, and technological developments to improve the fidelity of sound reproduction, noise was gaining more and more abstract and analytical connotations, and eventually became a concept referring to general physical effects.

Noise developed into an abstract and generalized entity of research and engineering along several principal lines. In this chapter, we examine the grappling with the so-called "surface noise" in phonographic records at the turn of the twentieth century. Surface noise referred to hissing, cracking, and scratching sounds in the background of mechanically reproduced music or speeches. It gained the name because most users and developers of early phonographs and gramophones believed that the annoying sounds were caused by records' uneven and defective surfaces. In the 1900s–30s, surface noise was one of phonographic engineers' and enthusiasts' major concerns regarding the quality of sound reproduction.

The very notion of *surface* noise marked a departure from conceiving noise only in terms of its sonic attributes, as the notion attached the unwanted sound in the mechanically reproduced music or speeches to records' physical conditions. Noise in this context associated not only with cacophony, but also with the damaged physical medium that caused it. Contrasting those of electroacoustics, atmospherics, and electronic noise, however, the early history of the surface noise did not involve a conversion between disturbing sounds and abstract signals associated with electrical communications. The phonograph and gramophone were primarily mechanical.

Moreover, the inventors, engineers, and enthusiasts of the phonograph rarely employed mathematical models, physical theories, or precision measurements to deal with surface noise. Instead, the discussions on surface noise's conditions, perception, and methods of reduction all concentrated on the materiality of sound-recording machines and records. To reduce surface noise, inventors and engineers experimented with various recipes of wax, rubber, shellac, and resin for phonographic cylinders or gramophonic discs, and undertook relentless trials of repetitive playing and breaking. Meanwhile, how smooth records looked and felt and how smooth they sounded became increasingly connected for enthusiasts. The noise of cracking, hissing, and clicking was perceived together with scratches and bubbles on record surfaces. The episode of surface noise thus testifies a specific route toward the generalization of noise from its original sonic attributes—a route that relied much less on mathematics, physical theories, and precision measurements than on the cultivation of a particular sensibility that linked defects of sound reproduction with material conditions of its medium.

Mechanical Reproduction, Sound Quality, and Surface Materials

The emergence of surface noise as a significant technical problem was embedded in the interplay between the technological and marketing developments of mechanical sound reproduction. Three inventions dominated the early history of sound recording. In 1877, Thomas Edison invented the first "talking machine," the tinfoil phonograph. Inspired by his working experience with telegram recorders, Edison found a way to inscribe and replay human voices. The phonograph was a cast-iron cylinder mounted on a screw and attached to two diaphragm-and-stylus units. The cylinder was wrapped by tinfoil with a fine spiral groove. The screw was connected to a handle that turned and shifted the cylinder (Figure 3.1).

When a person turned the handle and spoke into a diaphragm-and-stylus unit, the acoustic vibrations of the voices moved the diaphragm, which drove

Figure 3.1 Thomas Edison's tinfoil phonograph (1877). Thomas A. Edison Papers, Rutgers University, New Jersey.

the needle to indent varying depths (what Edison called "hill and dale") along the tinfoil's spiral groove. The pattern of this vertical cut recorded the voices. When the needle of the second diaphragm-and-stylus unit was placed against the groove while the cylinder was turned, it picked up mechanical vibrations from the cut, and the diaphragm amplified them into recognizable voices.[9]

Although Edison's tinfoil phonograph created an immediate sensation, it remained a mere curiosity until two competitors appeared. In 1885–86, Alexander Graham Bell's cousin Chichester Bell and a Massachusetts scientific instrument maker Charles Sumner Tainter at Alexander's Volta Laboratory in Washington, D.C. introduced a sequence of modifications to Edison's phonograph. Primarily, they replaced Edison's tinfoil-wrapped iron cylinder with a cardboard cylinder coated with wax. This yielded sharper and clearer sounds and permitted a thinner and denser groove over the recording surface, which increased the duration of a reproduced speech. In 1887, Bell and Tainter named their machine "graphophone."[10] The graphophone motivated Edison to further improve his phonograph. Working with staff at his newly established West Orange Laboratory, Edison followed Bell and Tainter's design to replace tinfoil iron cylinders with wax cardboard ones and introduce mechanical arrangements including a floating stylus and a foot treadle.[11]

Meanwhile, a German immigrant, inventor of telephone devices, and former engineer of Bell Telephone, Emile Berliner, in Washington experimented with mechanical sound reproduction via a different approach. Berliner also sought to record and replay sounds by cutting into a recording surface via a stylus-and-diaphragm unit. Different from the phonograph and graphophone, he employed a disc, not a cylinder. Following Leon Scott's phonautograph, moreover, the mechanical vibrations of sounds in Berliner's apparatus were not indented vertically but cut laterally on the record. To accentuate the stylus cut, he coated his zinc disc with beeswax and immersed it in a chromic acid solution to etch a deeper trace. Berliner named his machine "gramophone" and demonstrated its working in 1887–88.[12]

When Edison explored the commercial potential of the phonograph in the 1870s–80s, his major aim was to sell or rent it as a dictation machine for office use. This was also the first business model for the Bell–Tainter graphophone. This marketing strategy for early mechanical reproduction of sound

[9] Gelatt, *Fabulous Phonograph* (1977), 18–21; Welsh and Burt, *From Tinfoil to Stereo* (1994), 8–16.

[10] Gelatt, *Fabulous Phonograph* (1977), 33–35; Welsh and Burt, *From Tinfoil to Stereo* (1994), 19–23.

[11] Paul Israel, *Edison: A Life of Invention* (New York: Wiley, 1998), 280.

[12] Gelatt, *Fabulous Phonograph* (1977), 58–63; Welsh and Burt, *From Tinfoil to Stereo* (1994), 96–99; Emile Berliner, "The development of the talking machine," *Journal of the Franklin Institute*, 176:2 (August 1913), 189–200.

engendered specific considerations for sound quality. In an 1888 article to introduce Berliner's gramophone, the electrical inventor Edwin Houston stated that Edison's phonograph did not deliver practical applications, for it "failed to correctly reproduce articulate speech."[13] A focus of Edison's early phonographic research was to improve the machine so that it could produce louder voices with a clearer articulation of hissing consonants or sibilants.[14]

The marketing of mechanical sound reproduction shifted from dictation to entertainment in the following decades. By the 1890s, Edison Phonograph, Columbia Phonograph that held the patents for the graphophone, and various local companies all issued musical records. The gramophone firms Berliner established in Europe and Victor Talking Machine established by his former machinist Eldridge Johnson in Camden, New Jersey joined the force at the turn of the twentieth century. This reframing of sound reproduction toward music called for new considerations of sound quality.

While clarity of speech articulation was still central in this new listening culture, loudness and tone color became more important qualities for acceptable reproductions of music. Such a changing sonic sensibility was vividly reflected in the advertisements for talking machines in the late nineteenth century. In Edison Phonograph's house magazine *Phonoscope* in 1899, an advertisement for the records manufactured by Reed, Dawson, & Co. in New Jersey claimed that their products were "pure in tone, broad cut, loud and clear."[15] Another ad for Bettini Micro Phonograph described that their machine could produce "tone most natural in quality and musical"; it was "clear and louder than with other diaphragms, but with no metallic resonances."[16] Polyphone was proud for selling "the loudest talking machine in the world; twice as loud, twice as sweet and natural as any other talking machine and as loud and perfect as the human voice."[17] Columbia Phonograph remarked that their new Home Grand Graphophone had "marvelously loud and perfect reproductions of sound." It could offer "real music, actual voice, not a diminished copy or a far-away effect."[18] When advertising the Edison Concert Phonograph (Figure 3.2), Edison Phonograph stated:[19]

> Beginning with the early tin-foil machine Mr. Edison has developed the Phonograph step by step, until today the Edison Phonograph stands on the

[13] Edwin Houston, "The gramophone," *Journal of the Franklin Institute*, 125:1 (January 1888), 44.
[14] Israel, *Edison* (1998), 285.
[15] Reed, Dawson, & Co. advertisement, *Phonoscope*, 3:7 (July 1899), 5.
[16] Bettini Micro Phonograph ad, ibid., 6.
[17] Polyphone ad, *Phonoscope*, 3:8 (August 1899), 2.
[18] Home Grand Graphophone ad, *Phonoscope*, 3:11 (November 1899), 16.
[19] Edison Concert Phonograph ad, *Phonoscope*, 3:2 (February 1899), 5.

Pinnacle of Perfection. It perfectly reproduces the human voice; just as loud—just as clear—just as sweet. It duplicates instrumental music with pure-toned brilliancy and satisfying intensity. Used with Edison Concert Records, its reproductions are free from all mechanical noises; only the music or the voice is heard.

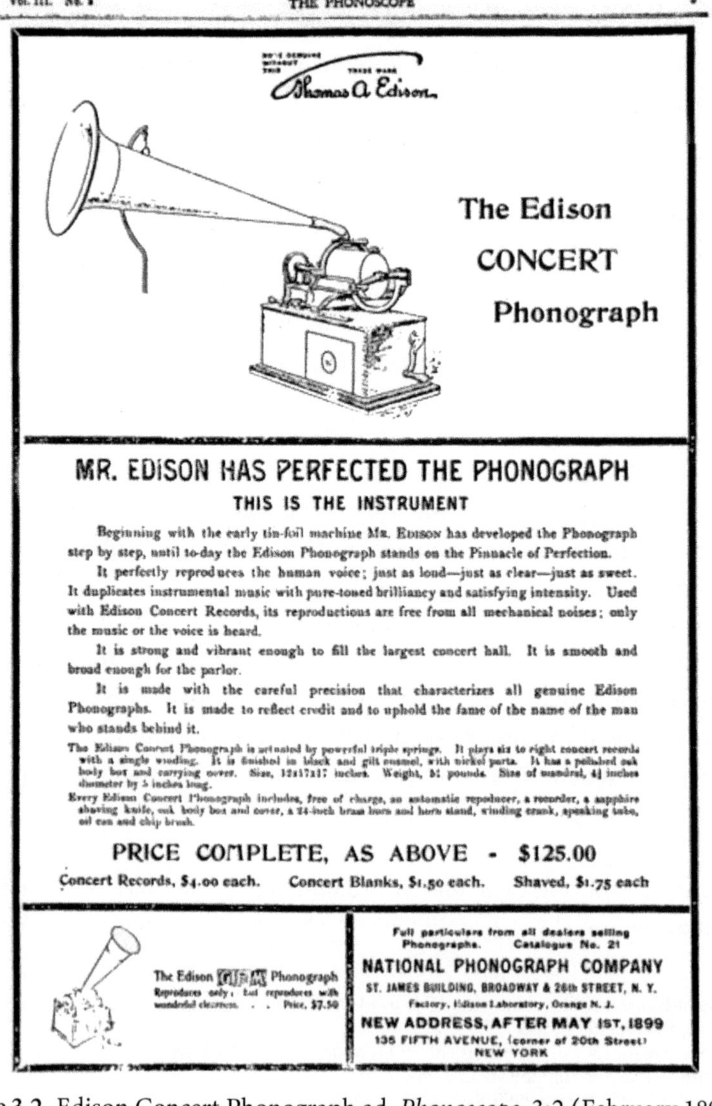

Figure 3.2 Edison Concert Phonograph ad, *Phonoscope*, 3:2 (February 1899), 5.

The message underlying these ads was consistent: Faint sounds and grotesque distortions were talking machines' most obvious shortcomings. Technical improvements that made sound reproduction louder or preserved more the tone quality of vocal or instrumental music could be good selling points for new phonographs, graphophones, or gramophones.

How to achieve this sound quality? In the early history of the phonograph, graphophone, and gramophone, there was a strong technical tradition to enhance sound quality via exploring and manipulating materials for making records. Bell and Tainter's replacement of Edison's tinfoil with waxed cardboard in 1887 marked the first major sonic improvement of talking machines. In the 1880s–1900s, technologists of mechanical sound reproduction devoted much effort looking for the right materials that were soft enough in the recording phase to carve out deep grooves but hard enough in the reproducing phase to generate loud and crisp sounds.[20] Inventors explored the use of celluloid in place of waxed cardboard for phonograph and graphophone cylinders.[21] To make gramophone discs, Berliner tried hard rubber and experimented with shellac, a resin secreted by insects. Being able to produce loud and crisp sounds, shellac became the standard material for gramophone discs until polyvinyl chloride (PVC, or "vinyl") was introduced in the 1920s.[22]

Working with materials also weighed heavily in record duplication. A generally conceived advantage of the gramophone over the phonograph or graphophone in the 1890s lay in the process of duplication. Whereas the latter involved mechanical indenting or engraving, the former electroplated a mold from a "mother" record and pressed the mold onto blank discs.[23] To compete with the gramophone, Edison and his associates Jonas Aylsworth, Walter Miller, and Charles Wurth utilized the comprehensive chemical substances and apparatuses at West Orange to invent a "gold sputtering" process in 1902 facilitating molding of cylinder records: depositing a thin layer of gold on wax or celluloid surfaces of blank records so that they were easily detachable from a mold.[24]

[20] Philip Mauro, "Recent development of the art of recording and reproducing sounds," *Phonoscope*, 3:9 (September 1899), 7–8.

[21] Welsh and Burt, *From Tinfoil to Stereo* (1994), 80–81.

[22] Berliner, "Development of the talking machine" (1913), 191–192.

[23] Gelatt, *Fabulous Phonograph* (1977), 58–62; Welsh and Burt, *From Tinfoil to Stereo* (1994), 96–98.

[24] Welsh and Burt, *From Tinfoil to Stereo* (1994), 81–82; Berliner, "Development of the talking machine" (1913), 195–196; Israel, *Edison* (1998), 400–401.

Surface Noise and the Entertainment Industry

By the turn of the twentieth century, the toy-like talking machine and the business-use dictaphone were transformed into a full-fledged instrument for music reproduction. While brass-band pieces, vaudeville, and folk songs were dominant genres for reproduced music, more "serious" classical music was expanding its share of the record market. Victor, Edison, and Columbia all employed professional vocalists and instrument players to make records of arias, opera segments, chamber music, and eventually symphonies. Some of these artists even had long-term collaborations with record companies—such as the Italian tenor Enrico Caruso and Victor—and became the "stars" of their brands. This broadening of musical genres for gramophone and phonograph records turned them from a medium of popular entertainment into a form of recreation for the upper middle-class.

The rise of the record market for classical music helped advance a new standard of sensibility for the quality of sound reproduction. Loudness, the primary concern for the first listeners of phonograph and gramophone, was no longer the dominant consideration to the connoisseurs and enthusiasts of musical records in the early twentieth century. Timbre, the length of music, and purity and clarity of sounds all attracted increasing attention. Under this new circumstance, mentioning of "surface noise" became more frequent in the discussions among users and developers of sound reproduction. Although surface noise had existed since Edison's tinfoil phonograph, it was increasingly perceived as a significant technical problem due to the new listening culture. The American music critic and editor Roland Gelatt depicted the sentiment on albums Caruso recorded for Victor Gramophone at the turn of the century:[25]

> Caruso's Milan series of March 1902 were the first completely satisfactory gramophone records to be made. Caruso's strong voice and slightly baritonal quality helped drown out the surface noise inherent in the early discs, and his vocal timbre seemed peculiarly attuned to the characteristics of the acoustic recording diaphragm.

As Victor was promoting the Caruso disc records, Edison Phonograph also marketed their cylinder products. In a letter to the editor in *The Phonogram* in November 1901, the author—presumably from the Edison firm—debunked the disc records and claimed that "all of the flat records I have ever heard have

[25] Gelatt, *Fabulous Phonograph* (1977), 115.

such a hissing, sissing, scratching sound, caused by a needle."[26] Similarly, when Columbia featured a new series of what they called the "Marconi Velvet Tone Record," they boasted that it was such a "flexible and unbreakable disc with so velvety a surface that the annoyance of the usual scratching sound is entirely eliminated."[27]

The increasing attention to surface noise aligned with the development of sound reproduction technology in the early twentieth century. In the 1900s–10s, the cylinder format represented by Edison and Columbia competed fiercely with the disc format represented by Victor. Both sides made significant engineering and marketing efforts to advance their technologies. The competition between the disc and cylinder records eventually ended with the victory of the disc.[28] But an important outcome of this competition was a considerable improvement in the quality of reproduced sounds. By the 1910s, Victor's "Red Seal," Edison's "Blue Amberol," and a few other series were well-acknowledged brands among music lovers for their exquisite volume, tone color, and clarity. Scratching, hissing, and clicking sounds became less and less tolerable to the consumers of the talking machines in the 1910s.

To illustrate how the developers and users of sound reproduction technology grappled with surface noise, we look into the technical treatments of phonographic surface noise at Edison's West Orange Laboratory in 1916–17, and general consumers' reports to Edison on surface noise they found from his company's records. Edison's laboratory epitomized the "high-tech" of the sound-recording industry at this time. The lab's dealing with surface noise was a good representation of the industry's technically advanced approach to noise. The comments, complaints, and praises he received from consumers were a fair indication of the general users' responses to surface noise.

The years 1916–17 were a critical period for Edison Phonograph. The cylinder records Edison had supported since 1877 had lost their war with gramophone discs a few years before, and the company had turned to the disc technology. To compete with Victor and Columbia, Edison and his associates devoted a tremendous amount of resource (estimated at $2–3 million) to develop a disc phonograph with higher sonic quality. The Edison Diamond Disc Phonograph was introduced in 1915. This new talking machine had a diamond stylus for recording and playing, which permitted much finer grooves on a record (meaning much longer play time) and created much louder resonances for reproduced sounds because of diamond's hardness.

[26] Ibid., 158.
[27] Ibid., 152.
[28] Welsh and Burt, *From Tinfoil to Stereo* (1994), 111–126.

Edison launched the diamond disc phonograph at the right moment. To prepare for the entrance of the US into the war in Europe, the American Navy contracted with Victor's factories in Camden to make underwater sounding equipment. This gave Edison Phonograph a chance to break Victor's dominance of the record market.[29] But the shortage of industrial materials due to the war also posed a challenge.

Tackling Surface Noise at Edison Phonograph, 1916–17

Improving the diamond disc phonograph was one of the major undertakings at Edison's West Orange Laboratory in 1916–17. A primary task of the laboratory's Phonographic Division was to reduce the surface noise of the records. Unlike the acoustic researchers since the mid-twentieth century,[30] Edison's staff did not employ quantitative measurements or statistical techniques to analyze surface noise. Rather, they dealt with surface noise via a program of relentlessly organized trials and testing. The core of this program was qualitative and empirical, not quantitative and theoretical. Its central focus was tinkering with and varying the material constitutions of the artifacts. This qualitative and empirical approach stressing the variation of materiality for grappling with noise in phonography aligned with the material-centered technical tradition in mechanical reproduction of sounds. This approach was also consistent with Edison's methodological hallmark to tackle technical problems, best represented by his team's yearlong search for a proper material for the filament of incandescent light.[31] While this kind of approach to technical problems might have been common in American industrial research at the turn of the twentieth century, the thoroughness of material testing and tinkering at West Orange was characteristically Edisonian.

A conspicuous part of Edison's program comprised a series of experiments to improve the surface quality of disc records. Here he and his staff conducted tests on many blank records produced by different methods or prepared with different ingredients. The materials they experimented with included resin, wood flour, various metallic compounds, china clay, and other types of clay. In addition, the West Orange researchers explored variations in the preparation of varnish, an oil coating on the surface of a record; the method of applying

[29] Gelatt, *Fabulous Phonograph* (1977), 208.

[30] D.H. Howling, "Surface and groove noise in disc recording media, I and II," *Journal of the Acoustical Society of America*, 31:11 (November 1959), 1463–1472; 31:12 (December 1959), 1626–1637.

[31] Israel, *Edison* (1998), 167–190.

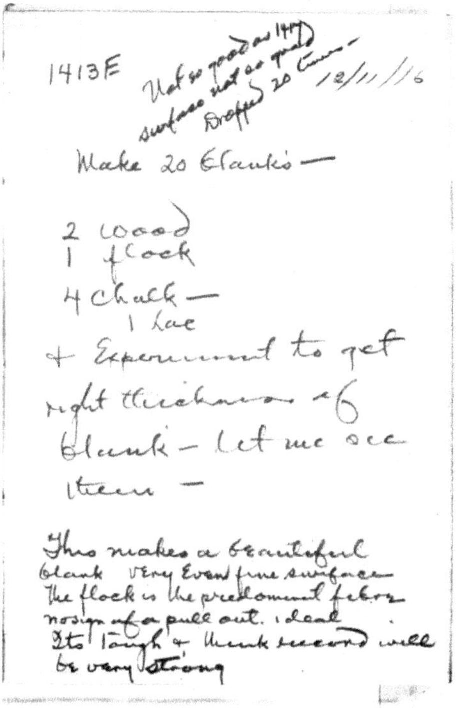

Figure 3.3 An example of the notes from Edison's laboratory on the trials for the material composition of phonographic records. Date: December 11, 1916. Microfilm 8, Notebook 16-11-30, Notebook Series: Notebooks by Edison and Other Experimenters, Reel 234 NA331, Thomas A. Edison Papers.

varnish; and the molding process for disc manufacturing. In the notebook Edison shared with his assistants William Dinwiddie and Archie Hoffman from November 1916 to March 1917, they documented the trials of numerous powder compositions for blank record manufacturing—even chalk and sesame were tested (Figure 3.3). They also experimented with different amounts of varnish on the record surfaces, how warm the surface was when the coating was employed, and how long to wait until the oil was dry.[32]

The ways Edison and his associates used to assess the effectiveness of their trials are worth noting. To determine whether the surfaces of the records they were experimenting with were satisfactory, they conducted both a listening test and a visual inspection. Some tested records were reported to have "quiet surfaces," or surfaces "not so quietful" as the others.[33] The report of the test conducted on May 4, 1918 exemplifies the thoroughness of the Edisonian program. Among the lot of 30 blank records under test, the experimenters documented that "25 scratch steady and quiet, 5 scratch steady but a little

[32] Notebook 16-11-30, Notebook Series: Notebooks by Edison and Other Experimenters, Reel 234 NA331, Thomas A. Edison Papers, Rutgers University, New Jersey, US.

[33] Microfilm 11, 19, and 21, Notebook 16-11-30, Reel 234 NA331, Thomas A. Edison Papers.

loud snaps." Eighteen were OK with snaps, but 12 had "quite a few" snapping sounds. Five had bad run out (abnormal disk spinning); 1 had slight run out; 24 were OK. Five had quite a few crackles; 25 were OK. Twenty-five were estimated to be commercially usable; 5 were not commercializable.[34] Edison paid close attention to this aural testing. When his hearing was not in a good condition (he suffered from a chronic deafening), he indicated his disability and relied on his assistants' report on the sonic quality of the tested surfaces.[35] Yet, auditory assessment was not the only way for the West Orange researchers to gauge the effectiveness of tested records. They relied equally on the examination of the surfaces' visual quality. From time to time, Edison, Dinwiddie, and Hoffman marked the tested surfaces as "soft," "very clotty," having "a lot of cracks" or "a lot of small holes," "very rough," and "beautiful and fine."[36] In another report, Edison documented that a record had "small granular crystals in sound groove causing scratching sound."[37] Inspecting record surfaces under a microscope was a common practice.[38] The historian Paul Israel found that Edison often determined what he considered to be sonic defects by visual examination of a record with a microscope.[39] Edison's preoccupation with the visuality of sounds contributed to the epistemic extension of surface noise.

In addition, the experimenters employed a "drop test" to assess the material strength of the blank records under examination: they dropped the records of a tested lot from a designated height (often five feet) to see how many broke apart.[40] Another similar form of testing was the "wear test." In March 1916, for instance, his staff tested 28 discs for wear. They ran each disc 250 times and inspected the surface under a microscope.[41] All these forms of assessment indicated a close relation between the handling of surface noise and the search for durable materials in the practice of sound-recording engineering at the time (Figure 3.4).

Edison's team at West Orange were not the only ones concerned with the surface noise of Edison Phonograph records. The inventor received a steady stream of letters from consumers, who expressed their opinions about the

[34] Microfilm 17, Notebook 16-11-30, Reel 234 NA331, Thomas A. Edison Papers.

[35] Microfilm 11, 16, Notebook 16-11-30, Reel 234 NA331, Thomas A. Edison Papers.

[36] Microfilm 4, 5, 7, 8, 19, Notebook 16-11-30, Reel 234 NA331, Thomas A. Edison Papers.

[37] Microfilm 3, Edison General Files Series, 1916 Phonograph—General E16-64, Reel 267 E1664A, Thomas A. Edison Papers.

[38] Microfilm 1, Edison General Files Series, 1916 Phonograph—General E16-64, Reel 267 E1664A, Thomas A. Edison Papers.

[39] Israel, *Edison* (1998), 436.

[40] Microfilm 10, 12, 15, Notebook 16-11-30, Reel 234 NA331, Thomas A. Edison Papers.

[41] Microfilm 0001, Edison General Files Series, 1916 Phonograph—General E16-64, Reel 267 E1664A-2, Thomas A. Edison Papers.

January 29th. 1916.

714

Mr. Edison.

The attached report covers two (2) weeks test. A great improvement in noisy motors has been found during the past week.

The motors in table marked "received noisy" all became quiet after running, showing that they had not been "run in" long enough, or has gotten by the inspectors in the Storage Battery Building.

A sound proof booth has just been finished in which all mechanisms will be tested for noise before sending to the saw tooth building.

"Drunken Governors" seem to be caused by poor adjustment, as after Mr. Halpin adjusted them no more trouble was experienced although everything was done to knock them out of adjustment.

"Noise developed." is the most serious trouble. This appears to be due to material, lubrication or adjustment. This trouble has decreased in the last week, and every effort is being made to locate the cause and correct it.

"Minor Defects" noted such as imperfect plating, or paint on horns, cabinet finish, or horn adjustment. These are immediately reported to the assembly department and inspection department.

The Amberola Machines seem to run more noisy than necessary, but are, however, commercial.

Detailed reports are on file of each machine tested, and also the results of special tests to determine cause of troubles, so that if you desire further information I can readily give it.

John H. Constable,

JPC:MBH Assistant Chief Engineer.

C.C. to: Messrs. Leeming, Maxwell, Ventres.

Figure 3.4 An example of the record testing results from Edison's laboratory. Date: January 29, 1916. Microfilm 16, Edison General Files Series 1916 Phonograph—General E16-64, Microfilm 16, Reel 267 E1664A, Thomas A. Edison Papers.

sonic quality of the company's discs. These users did not shy away from offering their own tips and tricks that they believed could alleviate the problem. On February 21, 1916, a J.C. Cross from Philadelphia indicated that some of the records he purchased from the Edison firm were "injured" on their surfaces. He greased these discs with a little cosmoline, a Vaseline-like product, and found that the coating "intensifies the whole record; second, it seems to bring out some of the softer over-tones; third, it obviates some of the scratch which is noticeable." Edison's reaction was unenthusiastic. He simply replied that grease and oil did not improve the record surface from their experiments.

He also pointed out that "the scratchy surface is the worry affecting life and I have been for the last few months working 18 hours daily to get rid of it and I shall certainly succeed."[42]

In another letter on September 7 of the same year, the author complained that a number of Edison records he had purchased recently were "very rough and noisy." The salesman had told him that this grinding noise would disappear over time as the diamond stylus polished the surface. But it did not happen. Edison's advice was to wipe the records with alcohol and to wait until they were dry. But he also expressed that this would only work when the records were dirty. He admitted that in the most recent batches of records he and his staff had tried to improve on the loudness and overtone by choosing a harder material. Yet, the side effect of a harder surface was more conspicuous surface noise. How to balance the overtone/loudness quality with the noise reduction quality was a major goal the company was working on.[43]

A letter from Dr. Adam William Coffing from the University of Chicago on October 14 shared the same frustration. Coffing purchased a number of Edison records early that year, and found the sonic quality very satisfactory. However, after a months-long trip to New England, he returned back home and found that the records had "marked increase of surface noise in the shape of a sharp hiss, characteristic of the new records just from the factory, in contrast with the velvety smoothness of the records I had purchased in early June, 1916." Edison's reply was frank: "the steamy sound on the record is the worry of my life. If I make everything dedicate so as to record all of the overtones then it will bring out the surface sounds, if I make it less sensitive to stop the steamy sound and lose overtones and hence quality." This dilemma was made worse by the shortage of supply for some materials due to the ongoing war in Europe. But he assured Coffing that "we are constantly at work at reducing [surface noise] and I shall never stop till I get rid of it altogether."[44]

Materializing Noise in Mechanical Reproduction

Tackling surface noise in mechanical reproduction of sounds at the turn of the twentieth century showcases a specific route toward the conceptual and

[42] Microfilm 0022, Edison General Files Series, 1916 Phonograph—General E16-64, Reel 267 E1664A, Thomas A. Edison Papers.

[43] Microfilm 0007, Edison General Files Series, 1916 Phonograph—General E16-64, Reel 267 E1664B, Thomas A. Edison Papers.

[44] Microfilm 0013-14, Edison General Files Series, 1916 Phonograph—General E16-64, Reel 267 E1664B, Thomas A. Edison Papers.

technical transformation of noise. Cracking and hissing sounds had existed since the invention of the phonograph but initially did not catch much attention. As the marketing of talking machines shifted from dictation apparatuses to music players and with the corresponding takeoff of the entertainment industry in the 1890s, demands for particular qualities of "high-fidelity" sound reproduction arose. Under this new listening culture, surface noise became a significant technical problem. Following the material-centered technical tradition in the business of mechanical sound reproduction and the Edisonian-style industrial research, phonographic technologists and engineers characterized the noise in terms of physical defects on the records, and suppressed it by experimenting with different materials for records. *Surface* noise thus connoted not only annoying sounds but also the mechanical disruptions of the recording medium. This materialization of noise marked the first step toward its full-fledged abstraction into quantifiable physical effects of disturbances in general, which we will see in the next chapter.

4

Measuring Noise

From Ear-Balance to Self-Registration

Metrology and Objectification of Noise

The emergence of sound reproduction technologies in the early twentieth century started to broaden the meaning of noise from a sonic attribute to more general physical effects. The telephone, phonograph, and radio introduced new aural experiences and opened up possibilities for associating unwanted sounds from such experiences with material, electrical, or geophysical processes involved in their production: scratches on record surfaces, atmospheric discharge, fluctuation of electric current, interference from adjacent lines, etc. This wider construal of noise owing to sound reproduction was the beginning of a much bigger transformation, which comprised the development of novel experimental techniques, instruments, theories, and mathematical approaches that turned noise into an object of scientific research and engineering manipulations. A central aspect of this transformation was the cultivation of the methods to measure noise.

Metrology started to become a hallmark of modern science and technology in the eighteenth century. During the Age of Enlightenment, as Thomas Kuhn, John Heilbron, and Ian Hacking showed, natural philosophers studying the qualitative "Baconian sciences" of heat, electricity, magnetism, and chemistry turned to quantitative and mathematical methods from the "exact sciences" of astronomy, optics, and mechanics. Meanwhile, the changing roles of the state called for numerical data from the fields of political economy, natural resource management, and demography. Transnationalization of trades and manufacturing demanded uniformity of metrics. By the nineteenth century, standardization of physical units shaped interactions between science and industry, and precision measurement characterized cutting-edge experimental science.[1]

[1] Thomas Kuhn, "The function of measurement in modern physical science," in *The Essential Tensions: Selected Studies in Scientific Tradition and Change* (Chicago, IL: University of Chicago Press, 1977), 178–224; John Heilbron, *Electricity in the 17th and 18th Centuries: A Study of Early Modern*

The measurements of sounds generated limited achievement along the line of precision metrology in the eighteenth and nineteenth centuries. From the early modern period to the Industrial Revolution, acoustics concerned either the physical nature of music or studies of sound wave propagation. Theoretical research of sound dominated over experimentation. Moreover, the primary focus of experimental research on sounds was the physiological and physical aspects of hearing. While scientists in the nineteenth century developed sophisticated research programs on sensory experiences, the quantitative determination of sounds remained challenging. And the phenomena of sounds were usually associated with hearing.

A breakthrough in sound measurement occurred in the first decades of the twentieth century. According to Wittje, acoustics at this time went through a fundamental change. The development of artillery ranging and underwater sonar for military uses during World War I and the rapid expansion of media technologies—telephone, talking machines, radio, sound motion pictures—turned acoustics from a bourgeois intellectual and aesthetic pursuit tied closely to music into a dynamic engineering field to serve industrial capitalism, mass media, and the state. As Wittje and Thompson indicated, a primary part of this change was the emergence of electroacoustics: the technology and instruments that transduced and integrated sounds into electric systems of microphones, loudspeakers, and vacuum tubes, and the conceptual tools of modeling hearing and other sonic phenomena in terms of such systems. Under electroacoustics, sounds became currents, voltages, or even abstract "signals" that could be represented, detected, and manipulated through electrical instruments and mathematical operations. Electroacoustics made it possible to measure the intensity, loudness, frequency, speed, and arrival time of sounds.[2]

Along with the development of sound-measuring apparatuses and techniques, the first attempts to quantify and measure noise emerged from electroacoustics. In the 1910s, radio engineers noted the effects of atmospheric statics in degrading wireless communication and tried to assess them. In the 1920s, American engineers and German technical physicists responded to the escalating noise abatement movement and introduced new instruments to measure the loudness of ambient noise in cities. As Mara Mills and Emily Thompson have shown, borrowing from the techniques of audiometry that

Physics (Berkeley: University of California Press, 1982), 9–97; Hacking, *The Taming of Chance* (1990), 60–63; Theodore Porter, *Trust in Numbers: The Pursuit of Objectivity in Science and Public Life* (Princeton, NJ: Princeton University Press, 1996). For a complete historical examination of measurement, see Norton Wise (ed.), *The Values of Precision* (Princeton, NJ: Princeton University Press, 1995).

[2] Wittje, *The Age of Electroacoustics* (2016); Thompson, *The Soundscape of Modernity* (2002), 229–293.

telephone engineers developed and physicians, public health experts, and educators used in the studies and assessments of deafness and hearing impairment, noise meters were employed as a means to quantify, measure, and domesticate urban din.[3] By the 1930s, audiometry spread into the applications for gauging various kinds of noise in sound reproduction, from statics in radio to vacuum-tube noise and electric power fluctuations in telephony.

Audiometry was built on an "ear-balancing" method: the apparatus comprised an electroacoustic transducer, an electronic amplifier, and a local oscillator that generated standard, monotonic reference signals. In measurement, an operator switched between listening to target noise and the sound from the local oscillator and adjusted the latter's intensity until the two appeared to have the same volume. Then the intensity of the local oscillating signal was treated as the measure of the noise volume. This was an effective method. The noise to be measured was often elusive, irregular, and fast-changing. Comparison with a much more stable and regular reference signal made noise considerably more discernible. But audiometry relied on human perceptions of hearing, which had become questionable by this time. The ear-balancing method measured the *loudness* of noise as perceived subjectively by the operator. This did not pose a problem when the noise to be measured was itself a sonic experience, like traffic noise or dins from factories. But when the measurable was a physical effect of which noise was a consequence (e.g. atmospheric static, tube current fluctuation, crosstalk), the measurement of the effect's *intensity* would be more relevant and useful. This consideration led to the introduction of "objective" methods of noise measurement, such as the oscillographic display and potentiometer. The eventual dominance of the "objective" methods over ear-balancing witnessed the power of the ideology of mechanical objectivity.[4] But it also testified to the subtle yet definite conceptual move of noise from its original aural contexts into a more generic physical domain. As noise was turned into an entity that did not have to be measured by listening, it appeared more like physical effects or processes, instead of annoying sensory experiences. Noise was objectified.

This chapter examines various attempts engineers and technologists made to turn noise into measurable quantities in the 1910s–30s. Several episodes in this metrological development are highlighted. They include American naval researchers' early exploration of recording the effects of atmospheric statics on wireless communication, AT&T engineers' development of the ear-balancing

[3] Mara Mills, "Deafening: Noise and the engineering of communication in the telephone system," *Grey Room*, 43 (2011), 119–142; Thompson, *The Soundscape of Modernity* (2002), 144–156.

[4] Daston and Galison, *Objectivity* (2007).

method and the 1-A Noise Meter for monitoring various telephone noise, the employment of the 1-A for gauging atmospheric noise in transoceanic radio links, AT&T's introduction of an "objective" noise meter in the 1930s, German researchers' exploration of measuring electronic noise, and the British Meteorological Office's use of the oscillograph in monitoring the waveforms of atmospheric noise after World War I. Although these episodes concentrated in the US and UK, they epitomized the general conceptual and technical developments in quantifying, measuring, and objectifying noise.

From the Metrology of Sound to the Metrology of Noise

Launching Sonic Measurements

The metrology of sound has a much shorter history than acoustics. Although scholars had investigated through theoretical work and observations into the physical and physiological properties of sound since the early modern period, attempts to measure it did not emerge until the second half of the nineteenth century. The first methods of acoustic measurements—including Helmholtz's cavity resonators and Koenig's manometric flame—relied on resonance. Helmholtz's device in the 1850s comprised a set of hollow metal spheres with sizes corresponding to different characteristic frequencies. When a sound had a component close to a sphere's characteristic frequency, an aurally recognizable resonance would be excited. The Helmholtz resonators could therefore identify specific frequencies in a complex sound. Koenig's manometric flame in the 1860s operated under a similar principle. Sound was directed through a speaking horn onto a diaphragm. The diaphragmatic vibrations created undulating air pressure to modulate the flame from a Bunsen lamp. Projecting the flame's image on a rotating mirror slowed down the visual effect of the vibrations. When the projected image illustrated a periodic pattern at a specific mirror rotating frequency, the sound had a harmonic component at the same frequency.[5]

Both Helmholtz's resonator and Koenig's manometric flame were more indicators than measuring instruments. They identified the presence of specific harmonic components in a sound, but showed qualitatively the components' amplitudes. These devices were used primarily in scientific studies of music

[5] Helmholtz, Hermann, *On the Sensation of Tone* (1954), 7, 43; David Pantalony, *Altered Sensations: Rudolph Koenig's Acoustical Workshop in Nineteenth Century Paris* (Dordrecht, Netherlands: Springer, 2009), 58–60.

to distinguish musical tones from a composite sound. They were unable to measure the quantity of the sound's overall strength or energy. An apparatus Lord Rayleigh conceived in the 1870s could arguably do that. This so-called Rayleigh disc comprised a lightweight, coin-sized disc vertically suspended in air via a string. The string was connected to a torsion meter. As a sound wave passed the apparatus, the corresponding air pressure twisted the disc. Although the sound incurred a back-and-forth air movement, the disc was driven into spinning along a single direction only, which accumulated torsion on the string. The measured torsion was interpreted as a measure of the sound intensity.[6] While the Rayleigh disc could measure the sound amplitude in principle, it was too fragile for most practical applications.[7]

The Helmholtz resonator, Koenig manometric flame, and Rayleigh disc were mechanical gadgets. Compared with contemporary scientific instruments in other areas of physical research, they had low precision, a limited range of applications, and a cumbersome operating procedure or a fragile build. Three subsequent developments led to a breakthrough of acoustic metrology in the early twentieth century. The first was the development of the microphone. In the 1870s, early telephone technologists including Edison, Berliner, and David Edward Hughes in London independently invented the carbon microphone, in which sound vibrations modulated the resistance of carbon granules between electrodes and were turned into a time-varying electric current. Afterward, the carbon microphone was integrated into expanding telephone systems. By the 1910s, telecommunication engineers improved the device and introduced new types, such as the condenser microphone (by modulating the device's capacitance) and the moving-coil microphone (by modulating the device's inductance). The microphone's function to convert sounds into electric signals altered the material means of engineering sounds and brought new theoretical tools for modeling them.[8]

Second, World War I opened a new page for acoustic instrumentation. The conflict in 1914–18 witnessed an unprecedented scale of mechanized and technological warfare. Physicists and engineers on both sides were brought into the research and development for employing sound technologies in various types of battlefields. They were particularly interested in the use of sound for detecting and locating enemy weapons and vessels: e.g. artillery ranging,

[6] John Strutt (Lord Rayleigh), *The Theory of Sound*, vol. 2 (London: Macmillan, 1878), 44–45.

[7] For a review of the early methods of sonic measurements, see Edward Elway Free, "Practical methods of noise measurement," *Journal of the Acoustical Society of America*, 2:1 (1930), 20–21.

[8] For a review of the early history of the microphone, see John Eargle, *The Microphone Book* (Oxford: Focal Press, 2005), 1–6.

locating aircraft, and underwater sounding. The idea was to "listen to" the sounds emitted from enemy targets to estimate their ranges, directions, and locations. Military researchers and engineers paid close attention to the precision measurement of sound, especially its arrival time, which formed the basis of ranging. To achieve accuracy, they resorted to electroacoustics. American, German, British, French, Austrian, and Italian scientists and engineers developed different types of microphones and hydrophones with more sensitive and selective responses to the sounds from battlefield targets than ordinary telephonic microphones. They also designed and improved electronic circuits for filtering and amplifying detected signals, and experimented with various schemes of automatic registration to display and record detected signals.[9]

Third, telecommunications engineers and physicians in the 1870s–1920s paid increasing attention to auditory measurements. The invention of the telephone had a close connection to the interest in assessing hearing conditions. Some telephone pioneers, including Bell and Hughes, actively explored the means of auditory evaluations and ways to help the aurally impaired. Continuing into the twentieth century, the relationship between telephony and hearing diagnosis or assistance was twofold. On the one hand, hearing loss and impairment became a notable public-health issue among medical and educational professionals, who sought to quantify the auditory disability of populations. Inventors, psychologists, and otologists viewed the telephone as a promising means to gauge the human subjects' hearing performance and repurposed the phone set into various types of electrical audiometers.[10] On the other hand, telephone engineers came up with "articulation tests"—asking human testers to listen to predesignated voices and checking their percentage of correct recognition—to adjudicate the quality of telephone communication. These articulation tests were applied to gauge the performance of phone circuits, the efficacy of signal transmission, and the impacts of interference on telecommunication networks.[11] Both lines of work centered on sonic measurements.

The invention of new electrical devices, the sounding technologies in World War I, the expansion of telephony, and the rise of the care for hearing disabilities transformed acoustics and advanced the metrology of sound. Underlying more elaborate techniques, an increasing precision, and a migration toward electrical instrumentation, we can find divergent views about the aim and

[9] Wittje, *The Age of Electroacoustics* (2016), 71–105.
[10] Mills, "Deafening" (2011), 125–127.
[11] Fagen, *A History of Engineering and Science in the Bell System: The Early Years (1875–1925)* (1975), 324–337.

nature of sonic measurements. To some, the metrology of sound was essentially a measure of sensory capabilities. That is, the measurements of sound were audiometry involving the participation and judgment of human operators. To others, the participation and judgment of human participants incurred too much "personal equation" (errors due to individual judgments) and "subjective" bias. It would be better if a large part of the human's role in those measurements could be replaced by more "objective" devices of automatic registration. For them, measuring sound was not identical to measuring hearing. Measurements of sounds should gauge their physical characteristics: energy, intensity, spectrum, times of arrival, speed, etc. This distinction between subjective and objective loomed large when the measurements of sound were further extended to the measurements of noise in the 1920s.

Audiometry and AT&T's Noise Measuring Sets

The first instrument used for systematic measurements of noise was the audiometer. The most popular audiometer, AT&T's 1-A Noise Measuring Set, was originally introduced for hearing tests (Figure 4.1). As early as the 1910s, the American telecommunication giant had initiated articulation tests to measure the proportion of speeches correctly reproduced through telephone transmission. In the early 1920s, Dr. Edmund Prince Fowler at Manhattan Eye, Ear, and Throat Hospital and Alexander Nicholson at Western Electric, AT&T's manufacturing branch in New York, conceived a device to record test sentences on a phonograph disc, and have a telephone transmitter pick up the replayed voices and send them to a headphone for test subjects to listen to. Harvey Fletcher, the company's authority of research on speech, hearing, and articulation, suggested that they replace the phonographed voices with an adjustable monotonic sound. In 1922, Fowler and Robert Wegel at Western Electric announced 1-A, the first commercial electronic audiometer.[12]

The 1-A Noise Measuring Set comprised an electronic oscillator, a control unit, and an earphone. The electronic oscillator included two No-215A vacuum tubes manufactured by Western Electric, and a passive electrical circuit of resistors, inductors, and condensers (capacitors). Like other typical electronic oscillators used in radio and telephone systems at the time, the 1-A audiometer's oscillator had a structure of a positive-feedback amplifier. The 215A vacuum tubes served as an electrical amplifier. Its output was passed

[12] Mills, "Deafening" (2011), 128–130.

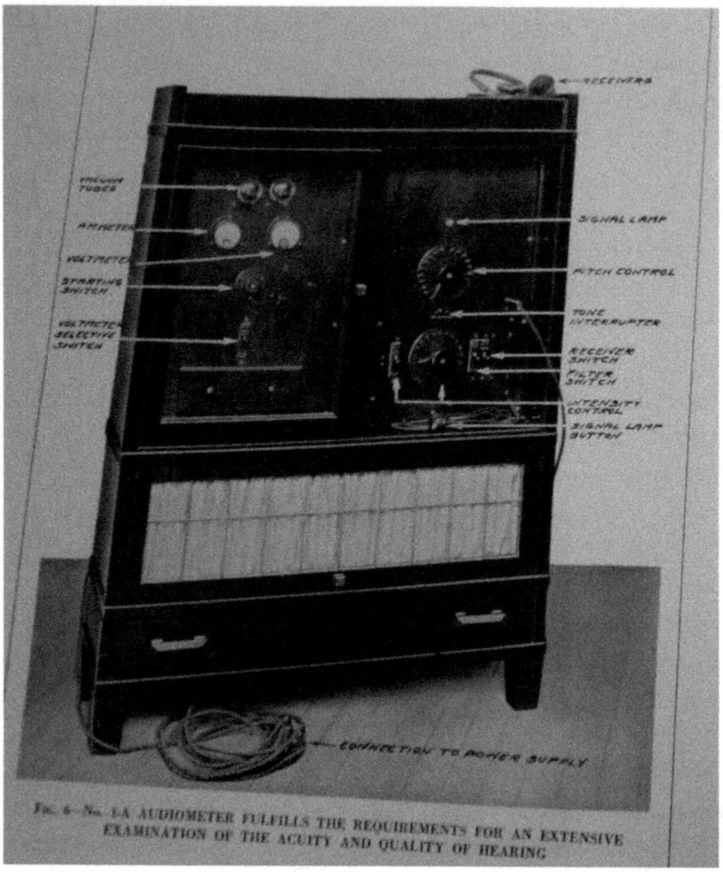

FIG. 6—NO. 1-A AUDIOMETER FULFILLS THE REQUIREMENTS FOR AN EXTENSIVE EXAMINATION OF THE ACUITY AND QUALITY OF HEARING

Figure 4.1 AT&T 1-A Audiometer. Western Electric Company, *The Audiometer: An Instrument for Measuring the Acuity and Quality of Hearing*, Western Electric Company Instruction Bulletin no. 145 (1924), 8, in bound volume #249-05, AT&T Archives, Warren, New Jersey, US.
Courtesy of AT&T Archives and History Center.

through the inductor-capacitance circuit for the selection of a particular oscillating frequency, and then was fed into the input for further amplification. The recursive process of such positive feedback built up a sinusoidal current at a single frequency determined by the feedback circuit's inductance and capacitance. The amplitude of the oscillating current was determined by the voltage bias to the vacuum tube and the resistors in the circuit. Thus, the frequency and intensity of the oscillating current could be adjusted through variable condensers and resistors. The range of the 1-A's oscillating frequency was from 32 to 16,384 double vibrations per second (64 Hz–32.768 kHz in today's expression), covering the whole human-audible spectrum. This oscillating current drove a receiving earphone that produced a monotone at the same frequency and proportional amplitude. The tone's pitch was varied in steps with a dial, and its amplitude was changed through another dial of intensity control. The electronic circuit was charged either with No. 6 dry batteries of 11.5 V or with a power outlet.[13]

The operating procedure for the 1-A audiometer followed its design mandate as an instrument to measure human hearing: The tester sat facing its control panel, while the subject, putting on an earphone, sat on the other side. Starting with a specific frequency, the tester changed the volume of the output tone until it became barely audible to the subject. The subject pressed a "signaling button" in the apparatus to indicate that they could still hear the sound. The tone volume was then recorded in a certain pre-calibrated and meter-inscribed "sensation unit" that engineers and hearing researchers developed in the 1920s to represent the perceived loudness of sound.

The common sensation unit among American technologists was the "decibel" proposed by the AT&T researcher Ralph Hartley in 1928: for a sound with mechanical pressure p, its strength in decibels is $20 \cdot \log_{10}(p/p_0)$, where p_0 is a base-level corresponding to the pressure of the faintest sound detectable by an average ear at the frequency to which human hearing is most sensitive (7×10^{-16} watts/m^2 at frequency 1000 cycles per second). Another similar sensation unit was the "phon" introduced by an engineering professor Heinrich Barkhausen at the Technische Hochschule at Dresden in the mid-1920s: the phon of a sound is the mechanical pressure level in logarithm of another sound at frequency 1000 cycles per second that appears to be as loud as the sound to be measured. These units referred to the logarithm of sound intensity to represent the nature of human perception—the Weber–Fechner law in nineteenth-century psychology had shown that exponentially increased stimulus would result in linearly increased sensation, hence the sensation's logarithmic dependence on stimulus.[14] The barely audible sound volume p_0 represented the subject's "threshold of audibility," for the sound level below this would not be recognizable to them. Similarly, the tester could increase the sound volume until the subject felt unable to bear it (again using the signaling button to indicate this). This volume represented the subject's "threshold

[13] Western Electric Company, *1-A Noise Measurement Set and 3-A Noise Shunt: Instructions for Operating*, Western Electric Company Instruction Bulletin no. 145 (1924), 1–4, 7–9, in bound volume #249-05, AT&T Archives, Warren, New Jersey, US; Western Electric Company, *Instructions for Operating 2-A Noise Analyzer* (1924), 1–4, in bound volume #249-05, AT&T Archives.

[14] Bijsterveld, *Mechanical Sound* (2008), 104–106; Free, "Practical methods of noise measurement" (1930); Mills, "Deafening" (2011), 130; Wittje, *The Age of Electroacoustics* (2016), 126.

of feeling," for the sound level above this would not be tolerable to them. After measuring the subject's thresholds of audibility and feeling, the tester switched to the next increment of frequency, and performed the same measurement. Repeating this procedure from the lowest to the highest tone, the tester could obtain a set of data for the subject's thresholds of audibility and feeling at various frequencies from 64 to 32,768 cycles per second. These data were then plotted on a chart named an "audiogram," with its horizontal axis representing frequency and its vertical axis representing the loudness in sensation units. The thresholds of audibility and of feeling were two curves on the audiogram.[15]

As good as it could get, the 1-A audiometer was nonetheless too bulky. A typical apparatus was the size of a cabin. This restricted the medical and public health applications of the instrument at places outside clinical settings. In 1924, Fowler and Wegel produced a portable and cheaper 2-A audiometer.[16] This model was the size of a small suitcase. It used one instead of two vacuum tubes. Its range of frequency was 64–8,192 double vibrations per second (128–16,384 Hz), and it covered 70% of the range of audible volume.[17]

The AT&T 1-A and 2-A audiometers were originally designed for hearing tests. As the apparatus's instruction bulletin stated:[18]

> Wherever hearing tests are required, one of the types of Audiometer can be used. Its combination of sustained pure tones automatically produced, and of pitch and intensity adjustable throughout a wide range, will be found of great usefulness in the delicate functional and diagnostic measurements of otologists. It may also replace to advantage many of the manually operated instruments now in use for conducting the standard hearing tests. In the field of physical examination, the Audiometer should prove extremely valuable in determining, so far as impaired hearing is concerned, the unfitness of applicants for insurance policies, automobile licenses, and for enlistment in the Army and Navy; also in the life protection tests of railroad and steamship companies; and in the health corrective examinations of schools, colleges, and gymnasiums.

The 1-A and 2-A measuring sets were used for large-scale hearing testing. Shortly after they became available, AT&T collaborated with the New York

[15] Western Electric Company, *1-A Noise Measurement Set and 3-A Noise Shunt* (1924), 5–7.

[16] Mills, "Deafening" (2011), 44.

[17] Western Electric Company, *1-A Noise Measurement Set and 3-A Noise Shunt* (1924), 7; Western Electric Company, *Instructions for Operating 2-A Noise Analyzer* (1924), 1–3.

[18] Western Electric Company, *1-A Noise Measurement Set and 3-A Noise Shunt* (1924), 2.

League for the Hard of Hearing and the American Federation of Organizations for the Hard of Hearing to deploy the audiometers in numerous tests that evaluated the hearing of thousands of schoolchildren and industrial workers.[19]

Hearing tests were not the audiometer's only applications. Soon after it became available, the instrument was used to measure the din from urban environments in the context of noise abatement. This is not surprising, since the anti-noise advocates, the activists working on hearing impairment, the medical, public-health, and educational professions, the investigators on hearing and speech, and the sound reproduction industrialists were all within proximate social circles in large cities such as New York and London. As "noise measurement sets," the 1-A and 2-A audiometers were designated to be sonic meters that could be repurposed for gauging not only sensory perceptions of tones but also sounds of various kinds. In 1925, H. Clyde Snook at Bell Telephone Laboratories employed a Western Electric audiometer to measure the automobile noise in New York City. In 1926, Edward Elway Free, the science editor of *Forum* magazine, used a similar audiometer to measure different types of street noise at distinct spots in New York City, resulting in what he claimed was "the first noise survey" of the city.[20]

As a semi-quantitative report on the noisiness of America's busiest metropolis, Free's article attracted wide attention among those concerned with the increasingly louder urban traffic and other street activities. A similar survey in Chicago was conducted. In 1929, the City of New York established a Noise Abatement Commission to address the city's escalating noise problems. The commission's first task was to map New York's noise landscape as a collection of data about the types and volumes of noise at different locations in different times. This task required a mobile yet sufficiently accurate noise measuring set. To secure such equipment, the commission sought Bell Labs' collaboration. The apparatus that AT&T engineers developed for mapping noise was a truck loaded with an audiometer and a sound-level meter that displayed the noise level on a potentiometer. The truck was staffed with technicians from Bell Labs, Johns-Manville Company, and New York's Department of Health. The team drove over 500 miles around the city and took 10,000 noise measurements at 138 locations. The Noise Abatement Commission published the results of this campaign in charts showing the noise levels in various circumstances. These charts also appeared in popular magazines.[21]

[19] Mills, "Deafening" (2011), 131–134; Thompson, *Soundscape of Modernity* (2002), 146–147.

[20] Free, "Practical methods of noise measurement" (1930), 21–23; Thompson, *Soundscape of Modernity* (2002), 148.

[21] Thompson, *The Soundscape of Modernity* (2002), 157–162.

The collaboration with the Noise Abatement Commission in New York marked AT&T's continual involvement with ambient noise measurements and the anti-noise movement. For instance, the firm was among the eight American companies that formed a National Noise Abatement Council and initiated a Noise Abatement Week on October 21–26, 1940.[22] Moreover, because the company's business was telephony, its major interest in noise measurements was to assess how different types of sonic disturbances degraded the quality of telephone communication. Thus, the noise levels in indoor environments were more relevant information to their undertaking than those in outdoor environments. While AT&T was assisting the Noise Abatement Commission's mapping of New York's noise landscape in 1929, it also collaborated with the National Electric Light Association to organize a "room noise survey" to measure the loudness of various types of sounds in indoor settings. In 1936–38, AT&T launched its own large-scale room noise survey. Four American cities were covered and measurements were taken at over 600 locations. Factories, stores, offices, residences, and other types of locations in "headquarter cities," small cities, small towns, and rural and suburban areas were chosen. In the survey, AT&T engineers used three kinds of measuring instruments: a sound-level meter made by the Electrical Research Product Company, a sound-level meter made by Western Electric, and the 2-A Measuring Set. The instruments were checked and calibrated at Bell Labs before their deployment in the field.[23]

How did researchers at Bell Labs transform the 1-A and 2-A audiometers from devices of hearing tests into instruments of ambient noise measurements? Key to this transformation was to modify the devices according to a concept of "ear-balancing." The origin of the ear-balancing method traced back to the rapid expansion of telephone networks in the 1900s–10s. As the envelope of telephony coverage was pushing to engulf the entirety of North America, engineers confronted an increasing challenge to long-distance signal transmission. To ensure the quality of communication, they needed a scheme to gauge the performance of telephone links. The earliest method was to get a relative gauge by comparing two links at the same time: testers just talked over two circuits alternatively and judged subjectively which one was better. Engineers used this means to compare copper with iron wire, open lines with cables, etc. But as telephone engineering became more sophisticated, a more rigorous approach was called for. By the early twentieth century,

[22] "Suggested plan to promote the observance of Noise Abatement Week" (1940), in Case No. 36,750-10, volume A, AT&T Archives.

[23] "Room Noise Data, Report #4; Detailed summary of data obtained during 1937-1938," in Case No. 36,750-10, volume A, AT&T Archives.

AT&T replaced the simple comparison between two circuits by a comparison between a targeted circuit and a "reference" circuit. The reference circuit comprised a telephone station and an artificial line with standard attenuation per unit length and a variable length. Testers could adjust the length of the reference circuit until its voice appeared the same as the targeted circuit's voice. That length, or equivalent overall attenuation, represented a gauge of the targeted circuit's signal intensity. In the 1910s, AT&T engineers introduced an adjustable local oscillator to replace the reference circuit and compared the voice from the tested telephone circuit with the sound from this local signal generator.[24]

In the original design of the audiometer, the monotone generated by the vacuum-tube oscillator was the only sound the subject listened to and responded to; and its loudness indicated the subject's hearing ability. In the setting for ambient noise measurement, however, this monotone became a reference signal for gauging the strength of external sounds. The listeners were no longer subjects under test, but operators of the measuring instrument. To measure the loudness of certain environmental noise, a subject alternated between listening to the external sound and the monotone on the earphone. They increased steadily the intensity of the monotone from the output of the audiometer's tube oscillator via adjusting an attenuator until the monotone appeared to "mask" the external sound and they were no longer able to recognize clearly the latter. The intensity of the monotone (the "local oscillator signal" as it is called now), which could be obtained from the audiometer's reading, was treated as a measure of the external sound. This was the method that Bell Labs researchers explored in their audiometric measurement of automobile noise in 1925. As Snook stated in a letter to the American Motor Body Corporation in 1926:[25]

> We have developed an apparatus for measuring noise in the air by the use of the "masking effect." The apparatus is known as our 3-A audiometer with offset receiver. It originally was designed to test the acuity of hearing; but has been adapted by the application of an offset receiver, to the measurement of noise in air. The offset receiver, standing a short distance away from the ear of the hearer, is in such a position that it permits sounds in the air to get

[24] Fagen, *A History of Engineering and Science in the Bell System: The Early Years (1875–1925)* (1975), 303–317.

[25] Letter from H. Clyde Snook to E.W. Templin, Engineer Motor Coach Division, American Motor Body Corporation, March 18, 1926, in "Development of audiometer," Box 444-05-01, Folder 08, AT&T Archives.

> to the ear of the hearer, along with the sounds that the receiver itself emits. That is to say, the observer's ear is by this simple expedient permitted to have both the sounds in the air and the sounds from the receiver to be admitted to it. If one of these sounds is louder than the other it will make the other one inaudible—that is it would drown it out.

The repurposed audiometer for noise measurement did not require some features in the original 1-A and 2-A sets for hearing tests. In particular, the noise measuring kit did not need a comprehensive frequency scan covering the entire audible spectrum. Rather, only the monotones at a very small number of specific frequencies were needed to serve the role of the masking sounds. Thus, the passive feedback circuit required redesign. After this modification, Western Electric featured a 3-A "Noise Shunt" for the measurement of environmental sounds.[26]

The 3-A audiometer was not the only noise measuring set using the ear-balancing method. In the 1920s, Barkhausen developed a similar noise meter based on the ear-balance between external sound and a monotone of 800 Hz generated by an electrical buzzer. In 1926, the Siemens-Halske Gesellschaft began to manufacture the Barkhausen audiometers, which became the main apparatus for measuring automobile and airplane noise in Germany, especially in the German aviation research sectors during the 1920s–30s. In another type of audiometer, the vacuum-tube oscillator or the electric buzzer was replaced with a phonograph to play a recorded monotone.[27]

The noise measuring sets developed in the 1920s provided a means to quantify the levels of various sonic disturbances in the ever busier and louder urban and industrial environments. Such apparatuses were an essential part of what Thompson called the "engineering of noise abatement."[28] Around the time when Western Electric 1-A, 2-A, 3-A, and comparable devices were widely used in numerous noise surveys and mapping, similar apparatuses gained an increasing traction in the handling of other kinds of disturbances affecting the quality of telecommunications technology. Although these disturbances appeared to be in one form of noise or another, they nonetheless represented broader physical processes than annoying sounds.

In Western Electric's plan to develop a portable indicating and measuring set for noise in telephone circuits ("line noise") during the early 1930s, for

[26] Western Electric Company, *1-A Noise Measurement Set and 3-A Noise Shunt* (1924).

[27] Wittje, *Age of Electroacoustics* (2016), 126; George William Clarkson Kaye, "The measurement of noise," *Proceedings of the Royal Institution of Great Britain*, 26 (1929–1931), 467–469.

[28] Thompson, *Soundscape of Modernity* (2002), 144–157.

instance, a memo from Bell Labs stated that the device would be designed to measure "four representative types of telephone line noise: tube noise, battery noise, commercial 110 volt D.C, and 60 cycles harmonics."[29] The tube noise denoted the electric current fluctuations due to the discretized current flow in a vacuum tube or electronic thermal agitation in a resistor (see chapter 6). The battery noise represented the variation of battery current. The D.C. (direct current) and 60 cycles noises were current fluctuations or excessive voltage from the electric power lines. Later, R.S. Tucker at Bell Labs provided a more thorough list of the quantities designated for measuring:[30]

- Metallic circuit noise
- Noise and crosstalk volume
- Noise to ground
- Noise in the receiver
- Program-circuit noise
- Speech volume
- Transmission loss, gain, and level
- Telephone interference factor
- Low-frequency potentials
- Acoustic noise and other sounds.

In fact, AT&T engineers had understood the notion of the "line noise" in telephone circuits in terms of any physical factors that impaired the transmission of telephoned voices, and had employed the articulation tests to assess the effects of such line noise on the performance of telephone systems.[31] This development was not restricted within AT&T. Neither was the extension of noise metrology confined to telephone engineering. A case in point was an instrument featured by the Tobe Company around the same time: a "radio noise and fault locator" that detected unwanted radio-frequency electric field in the form of "noise" affecting wireless broadcasting.[32]

[29] "Memorandum for File: Calibration tests of indicating line noise meter with various weighting characteristics, case 20267," March 14, 1930, in "Development of a portable, indicating line noise measuring set, 1931-36," 1–4, in File Case No. 34,494, vol. A, Reel No. FC4130, AT&T Archives.

[30] R.S. Tucker, "Memorandum: Portable apparatus for visual-indicating measurement of circuit's noise," January 18, 1935, in "Development of a portable, indicating line noise measuring set, 1931-36," 1–4, in File Case No. 34,494, vol. A, Reel No. FC4130, AT&T Archives.

[31] Joint Subcommittee on Development and Research, National Electric Light Association and Bell Telephone System, "Engineering method used by Bell System operating companies for evaluating effects of line noise on telephone transmission," Special Technical Report No. 7, October 25, 1930, 1–4, in File Case No. 34,494, vol. A, Reel No. FC4130, AT&T Archives.

[32] "Tobe Engineering Bulletin: Radio noise and fault locator model two-thirty-three," in "Development of a portable, indicating line noise measuring set, 1931-36," 1–4, in File Case No. 34,494, vol. A, Reel No. FC4130, AT&T Archives.

Note that the instabilities of a power supply, interfering radiation from the atmosphere or electrical appliances, current fluctuations in vacuum tubes or resistors, and crosstalks from adjacent phone lines were all labeled "noise" during this period. The boom of the phonograph, telephone, and radio by the early twentieth century had facilitated a broader construal of noise that included not only din on streets but also various physical disturbances to the quality of sound reproduction. The metrological development demonstrated that widening the connotation of noise did not occur merely at the semantic, lexicographic, or conceptual levels. The applications of audiometers and other din-measuring devices to the measurements of atmospheric noise, electronic noise, telephone line noise, metallic circuit noise, D.C. noise, and crosstalk suggested a convergence of the technical solutions for the quantitative determination of sonic noise on the one hand and those of other physical disturbances on the other. Atmospherics, electronic noise, line noise, and other types of disturbances in telecommunications not only bore the same name of noise as street din did. The science, technology, and instrumentation underlying the measurements of the former were profoundly shaped by the metrology for the latter.

This raises a question about the roles of aural perception in the measurement of noise in its broad sense. Atmospherics, thermal and shot noise, electromagnetic interference, crosstalk, and instabilities of power sources are electrical effects in telecommunications systems due to problematic technological design or functionality, geophysics, or statistical mechanics, not sound as such. While these physical effects do eventually manifest themselves as sonic disturbances at the terminal of sound reproduction, to scientists and engineers they are much better characterized as unwanted electromagnetic radiation or random fluctuations of electronic circuits' outputs. In principle, therefore, one does not have to grapple with such physical effects as sounds. The metrology of the broadly conceived noise in the 1910s–30s intertwined closely with the technologies of sound measurements, though. When the first instruments for measuring the generalized noise were developed, researchers and technologists employed similar methods and techniques to the sound-measuring sets such as audiometers and other noise meters. They discussed the pros and cons of "subjective" measurements using ear-balancing to determine the effective loudness of physical disturbances versus "objective" measurements using potentiometers, scopes, or other registration devices to determine the intensity or energy of physical disturbances.

In the rest of this chapter, we will focus on the metrology of two types of generalized noise: atmospherics and electronic noise. Atmospherics was electromagnetic interference from meteorological processes. Electronic noise corresponded to the current fluctuations in electronic circuits. These two types of noise caught more attention in scientific research and were generally viewed as factors due to the physical characteristics of the atmosphere or electrons, not faulty technological designs. The development of the schemes to measure atmospherics and electronic noise represented the efforts to determine empirically the extent of noise as an intrinsic uncertainty that affected the performance of sound reproduction.

Measuring Atmospherics

The First Attempt

Atmospheric noise has haunted radio technology since its beginning. In the 1890s, pioneers of wireless communications including Alexander Popov in Russia and Albert Turpain in France noted that thunderstorms usually accompanied hissing or cracking sound at the receiver earphone, which interfered with or even interrupted the reception of telegraphic signals.[33] But thunderstorms were not the exclusive factor for this background noise; it could appear without obvious weather disturbances. In 1902, the British radio experimenter Captain Henry Jackson found that the background noise was more severe during the daytime than at nighttime; it was also more serious in summer than in winter.[34] Thus, the noise seemed to have a close geophysical connection. By the 1910s, radio engineers, operators, and amateurs widely believed that fluctuations of this kind were due to erratic electromagnetic radiation from the atmosphere. They held a few rules of thumb about the effects of such atmospheric noise (also known as "static" or "stray wave"): winter was the best season, night was quieter, regions at higher latitudes suffered less noise, and the intensity of static decreased with radio frequency.

In the US, the first systematic observations of atmospherics took place during World War I. When the country entered the war in 1917, the transatlantic

[33] John Ambrose Fleming, *The Principles of Electric Wave Telegraphy and Telephony* (New York: Longmans & Green, 1916), 851–852. For the history of atmospheric measurements, see Chen-Pang Yeang, "The sound and shapes of noise: Measuring disturbances in early twentieth-century telephone and radio engineering," *ICON: Journal of the International Committee for the History of Technology*, 18 (2012), 63–85.

[34] Henry B. Jackson, "On the phenomena affecting the transmission of electric waves over the surface of the sea and the earth," *Proceedings of the Royal Society of London*, 70 (1902), 254–272.

wireless between the American Navy's high-power radio station in Arlington, Virginia and the French Army's counterpart on top of the Eiffel Tower in Paris became a critical communication method for both militaries. Atmospherics were a persistent problem for the transatlantic link. To find solutions to the problem, the Naval Radio-Telegraphic Laboratory led by physicist Louis Austin monitored atmospherics from August 1917 to May 1918 at the National Bureau of Standards' building in Washington, D.C.

The US Navy's observation campaign was foremost to explore atmospherics' qualitative features. Austin recognized that atmospherics incurred four kinds of sonic effects at a receiver earphone: (i) the most common was a rattling or grinding noise, (ii) a hissing tone, (iii) a sharp snap, and (iv) a noise somewhat like (i) but more crashing. These types, he pointed out, were associated with different meteorological processes: (ii) snow or rain, (iii) lightning flashes, (iv) thunderstorm.[35] Here he treated atmospherics as noise in its literal sense.

Austin did undertake quantitative measurement of atmospheric noise, though. He was interested in measuring the intensity of atmospherics at a receiver under different radio frequencies. This was not that simple, for static noise was too fluctuating and transient to measure directly with a radio detector. Austin's solution was to employ an indirect measuring procedure, using a so-called "audibility meter." The audibility meter was an electrolytic detector connected to a circuit of an earphone, a battery, and a variable resistor. A radio wave entering the receiver changed the detector's resistance, which in turn changed the electric current flowing through the earphone and created a sound on it. To measure the intensity of the radiation, an operator adjusted the variable resistor until the sound was "barely audible" and read the resistor's value. To Austin, the "bare audibility" of atmospherics was the condition "at which an average of three pulses of disturbances could be heard in telephone in ten seconds."[36] Following this procedure, Austin measured atmospherics at frequencies between 16.6 and 100 kHz. His results confirmed radio operators' folk wisdom that atmospherics intensity decreased with frequency.

Austin's audibility meter marked the first attempt to turn atmospheric noise into a measurable quantity. Although this meter appealed to engineers' preference for quantification, it had several problems. The method offered a relative instead of an absolute measure, as the reading was in the unit of resistance, not electromagnetic field strength. In addition, the proportionality between

[35] Louis W. Austin, "The reduction of atmospheric disturbances in radio reception," *Proceedings of the Institute of Radio Engineers*, 9:1 (1921), 41.

[36] Louis W. Austin, "The relation between atmospheric disturbances and wave length in radio reception," *Proceedings of the Institute of Radio Engineers*, 9:1 (1921), 29.

the reading at the variable resistor and the electromagnetic field intensity at the receiver was contingent upon the quantitative behavior of the electrolytic detector, which had not been totally determined at the time. Moreover, the condition of "three pulses in ten seconds" was arbitrary, and it implicitly conflated the category of amplitude with rate of repetition. Austin's method did not dominate the metrology of atmospheric noise.

Ear-Balancing Method

The more popular approach to measure atmospherics in the 1920s came from AT&T, as the company expanded its wireless telephonic services. Like the US Navy's audibility meter, AT&T's ear-balancing method also included human perception as part of its measuring instrumentation. And both sought to gauge the strength of noise. But in contrast to the audibility meter's reliance on the mediation of the detector's changing resistance, the ear-balancing scheme resorted directly to aural perception: a measurer balanced the sound of atmospherics and a local signal until the two seemed to have the same strength and took the local signal intensity to be the strength of the noise.

As early as the 1910s, AT&T employed the ear-balancing method in articulation tests and other adjudications of the quality of its telephone systems. As the company's business and activities became more plural by the early 1920s, this measuring method spread into other corporate undertakings. The development of the 1-A, 2-A, and 3-A audiometers in 1922–24 marked the extension of ear-balancing into otology and noise abatement. Around the same time, AT&T also adopted the ear-balancing method in the measurements of radio noise as the firm introduced its transatlantic wireless telephonic systems. In 1921, AT&T planned to launch a full-time commercial telephone service between America and Britain. The transmitters and receivers for this system were located at a radio station of the Radio Corporation of America (RCA) in Rocky Point, Long Island, and a branch office of Western Electric in Southgate near London. The transmitting power was 100 kW, and the operating frequency was 57 kHz. On January 15, 1923, 60 people from AT&T, RCA, Western Electric, and the British Post Office gathered in a testing room in London to listen to voices sent from AT&T headquarters in New York. The success of this test started a four-month measuring program.[37]

[37] Fagen, *A History of Engineering and Science in the Bell System: The Early Years (1875–1925)* (1975), 394, 400–401; H.D. Arnold and Lloyd Espenschied, "Transatlantic radio telephony," *Journal of the American Institute of Electrical Engineers*, 42:8 (1923), 815.

A central purpose of the measuring program was to acquire systematic, long-term data on transatlantic radio-wave propagation and atmospherics, particularly their diurnal and seasonal variations. Western Electric engineer Harold D. Arnold and AT&T engineer Lloyd Espenschied headed the program.[38]

Key to the program was an effective method to measure radio signals and noise. Here AT&T engineers encountered the same problem as Austin did: atmospherics were too irregular and fast changing to register at an off-the-shelf detector. To deal with this problem, Ralph Bown from AT&T and C.R. Englund and Harald T. Friis from Western Electric developed a measuring instrument in 1923 at Western Electric's receiver laboratory in Cliffwood, New Jersey. Their set was based on ear-balancing: a receiver with an earphone was connected to an antenna and a local signal generator via a switch. When the receiver switched to the antenna, radio signals or atmospherics were heard. When the receiver switched to the local signal generator, the local signal was heard. An operator switched between the antenna and the local signal generator and adjusted the local signal intensity until its volume at the earphone appeared identical to that of the radio signal or atmospherics. The local signal's intensity was taken to be the signal or noise intensity (Figure 4.2).

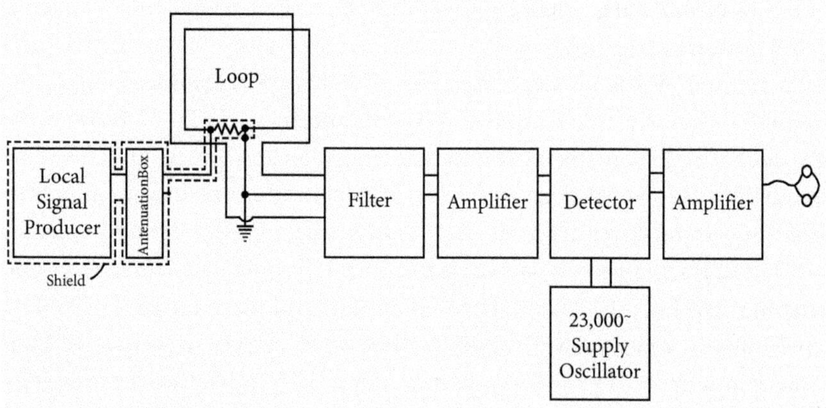

Figure 4.2 The block diagram of Bown et al.'s comparative method for radio signal measurement. Ralph Bown, Carl R. Englund, and Harald T. Friis, "Radio transmission measurements," *Proceedings of the Institute of Radio Engineers*, 11:2 (1923), 120, Figure 2.
Permission from IEEE.

[38] Arnold and Espenschied, "Transatlantic radio telephony" (1923), 815–816.

The core of this method was the local signal generator with which incoming radio input compared. What signals were comparable to the testing radio signals and atmospheric noise? For the testing radio signals, the answer was easy: since the signals were telegraphic dots and dashes with a monotonic carrier, choosing the local signals as monotonic waves would make the intensity comparison straightforward. The signal intensity obtained from this comparison, the AT&T engineers believed, would represent a good measure of the radio signal's real, physical strength.[39]

Measuring atmospheric noise was trickier, for it could not be properly emulated by single-tone local signals. But perhaps it was not relevant to measure the physical strength of noise, Bown et al. argued. Rather, they "approached noise from the telephone standpoint." That is, they were more concerned with the disturbing effect of noise on telephone users' hearing of normal speeches and voices than with the absolute intensity of noise. Consequently, they asserted, "a good way of measuring noise should be to express it in terms of a volume of speech upon which it produces a certain standard interfering effect." The engineers found a pragmatic way to circumvent the challenge of defining a standard speech for such measurements. Instead of trying to get a standard speech, they set the local signal for noise measurement to be an "artificial speech," which had a single audio-frequency tone changing uniformly between 600 and 1400 Hz at a rate of ten cycles per second. They justified this arrangement by noting that the audio tones in ordinary human voices followed that change on average. To measure atmospherics, an operator adjusted the volume of the local artificial speech until their ear was "just about to recognize its presence." That threshold volume was the *effective* strength of atmospheric noise. The measurements done in this manner were inevitably "subjective" and might suffer from inaccuracy, as Bown et al. recognized. But they contended that the data obtained from different unskilled observers using this method had only a variation less than 50%, which was smaller than actual variations of atmospherics. Thus, the method was stable enough to generate precise results.[40]

Bown et al.'s ear-balancing method well served Arnold and Espenschied's measuring program in 1923. The data they obtained exhibited patterns of atmospherics' diurnal and seasonal variations, which helped improve

[39] Arnold and Espenschied, "Transatlantic radio telephony" (1923), 822; Ralph Bown, Carl R. Englund, and Harald T. Friis, "Radio transmission measurements," *Proceedings of the Institute of Radio Engineers*, 11:2 (1923), 119–122, 144–148.

[40] Arnold and Espenschied, "Transatlantic radio telephony" (1923), 822; Bown et al., "Radio transmission measurements" (1923), 119–122, 144–148.

the noise-resisting performance of AT&T's transatlantic wireless telephony. Arnold and Espenschied presented the data in terms of not only signal and noise as such, but also *signal-to-noise ratio*. This was likely the first time that the term "signal-to-noise ratio" appeared in engineering literature.[41]

The ear-balancing method dominated AT&T's radio atmospherics measurements throughout the 1920s. In 1923, Arnold collaborated with C.N. Anderson and Austin Bailey to extend the transatlantic measuring program. Their new observations revealed the variation of atmospherics with radio frequency between 15 and 60 kHz, confirming Austin's earlier conclusion that atmospheric noise diminished with frequency, but reaching it with more accurate and abundant data. After the company switched focus to long-distance wireless telephony at shorter wavelengths, engineers employed the same method to measure atmospherics at frequencies above 1 MHz. In 1930–31, Ralph Potter employed the ear-balancing instrument to monitor the variation of atmospherics between 5 and 18.4 MHz.[42]

AT&T engineers' ear-balancing apparatuses for measuring atmospheric noise in radio communications resembled the company's audiometers used in hearing tests and measuring environmental noise. Both relied on an aural comparison between the "noise" to be measured and a locally generated sound. Arnold, Espenschied, and Bown's experimental transatlantic wireless telephony launched in 1923, after Fowler and Wegel invented the 1-A audiometer for hearing tests in 1922, but before Snook and Free adapted the instrument to measure urban noise in 1925. AT&T's work on the measurements of atmospherics and din was within the same technical culture of electroacoustic metrology, in which hearing, sounds, and signals were interchangeable.

A salient characteristic of this technical culture, and of the ear-balancing method which embodied it, lay in the central role of the human body in metrology of noise. The apparently non-anthropomorphic expression of atmospherics in terms of local signal intensity was a surrogate for a procedure involving a measurer's judgment and operation. In fact, the instrument did not measure the physical intensity of noise but gauged the noise's effect of masking the hearing of sound. In other words, what the ear-balancing method quantified was not atmospheric noise's magnitude of disturbance, but the degrees of perception the noise caused in specific communication systems. Thus, the ear-balancing device required a measurer's aural perception and discretion on the

[41] Arnold and Espenschied, "Transatlantic radio telephony" (1923), 822–826.

[42] Ralph K. Potter, "High-frequency atmospheric noise," *Proceedings of the Institute of Radio Engineers*, 19:10 (1931), 1731–1751.

equal volume between noise and a reference sound, and their manual intervention in tuning the reference sound to match the noise. As Graeme Gooday found, this dependence on human sensitivity and action has a long tradition in the history of metrology that can be traced back to electrical science and engineering in Victorian Britain.[43] In spite of their attempts in the 1920s to introduce fully automatic registration devices, AT&T researchers continued to utilize the ear-balancing method. Human perception was always considered part of measurement.

The ear-balancing method's reliance on aural perception made the choice of local buzzing sounds crucial. When the local signal was at the level comparable to that of the measurable, the former was supposed to "mask" the latter so that the listener could not (but just barely) hear the noise. Yet, the masking effect of the local oscillation was related not only to its intensity but also to its spectral features. A pure tone at a single frequency rarely sufficed for this purpose. To gauge the effect of atmospherics on wireless transmission of human voices, therefore, the measuring set in the transatlantic radio link had a local oscillator that generated an "artificial speech" with a tone undergoing a uniform frequency shift between 600 and 1400 Hz. Also, Free observed in 1930 that an audiometer gave the most accurate measurement of environmental noise when the reference sound from the local buzzer had a similar spectral pattern to the external sound. To increase accuracy, an operator sometimes took three noise measurements at three distinct frequencies of the buzzer sound—high pitch, medium pitch, and low pitch—and averaged the three data sets.[44]

The human-centered character of the ear-balancing method made sense, for it was designed for use in telephony and telegraphy. The impact of atmospherics or other types of noise on masking hearing telecommunicated sound was exactly what engineers at AT&T tried to gauge. Therefore, the method's dependence on aural perception did not introduce a distortion but constituted a *sine qua non* to measurement. Yet, problems ensued as researchers employed the same method to determine not the physiological effects of noise but its energy or electromotive force. As we will see, when German physicists around the same time used the ear-balancing method to measure electronic noise, ignorance of the method's physiological context in the initial experimental design led to a controversy that nearly terminated the first statistical theory of noise.

[43] Graeme Gooday, *The Morals of Measurement: Accuracy, Irony, and Trust in Late Victorian Electrical Practice* (Cambridge, UK: Cambridge University Press, 2004), 40–81.

[44] Free, "Practical methods of noise measurement" (1930), 23–24.

Measuring Electronic Noise with Ear-Balancing

Engineers and scientists had been familiar with fluctuations of electrical circuits and systems due to defective components. At the turn of the twentieth century, they started to worry about the more "fundamental" limit of electrical circuits due to fluctuations associated not with the imperfection of devices but with the properties of electrons. The statistical mechanics on Brownian motions in the 1900s provided a conceptual ground for this thinking about electronic noise (see chapter 5). The significant technological progress of vacuum-tube amplifiers in the 1910s, which enabled the amplification of very weak signals and hence were more vulnerable to electronic noise, gave researchers incentives to inspect its practical implications.

The German physicist Walter Schottky was one of the first individuals to pay attention to electronic noise. Director of Siemens-Halske Gesellschaft's communication electronics laboratory in Berlin during World War I, he had both the intellectual background and pragmatic motivation to study this issue. In 1918, he proposed a theory about the intrinsic electronic noise in a vacuum-tube circuit. Schottky acknowledged the presence of electric-current fluctuations due to electrons' thermal agitations. But he thought that in a vacuum tube, this "thermal noise" was much smaller than the fluctuations due to random shots of discrete electrons, known as "shot noise." He derived a mathematical formula for the average current intensity of shot noise (see chapter 6).

Schottky's shot noise remained a physical effect on paper for two years, as it was difficult to conduct scientific research in chaotic postwar Germany. In 1920, a Siemens researcher Carl A. Hartmann made the first attempt to empirically corroborate Schottky's theory. At the company's communication electronics laboratory, Hartmann conducted an experiment to measure a vacuum tube's shot noise.[45] His basic setup comprised a vacuum tube connected to a capacitor and an inductor. The goals of his experiment were to verify (i) whether this circuit indeed had a noisy tube current, and (ii) if so, whether the noise intensity was consistent with Schottky's prediction.

Hartmann proposed a procedure to address the second goal based on the fact that Schottky's quantitative prediction involved the charge of an electron e. He expressed Schottky's formula as a relation of e with other variables. This relation equated a ratio involving the measured shot current, the tube's bias

[45] Carl A. Hartmann, "Über die Bestimmung des elektrischen Elementarquantums aus dem Schroteffekt," *Annalen der Physik*, 65 (1921), 51–78.

current, and the circuit's resonance frequency to the electronic charge as a fundamental physical constant $e = 1.6 \times 10^{-19}$ coulomb.

Simple in concept, Hartmann's experiment was challenging in several senses. One challenge was to make precise measurements of shot current. Similarly to radio atmospherics, shot noise was irregular and difficult to measure with ordinary detectors. Hartmann adopted an ear-balancing method similar to AT&T's. In his design, the tube oscillator and a reference monotone generator tuned at the oscillator's resonating frequency were connected to a high-gain amplifier with a switch. He switched between the tube oscillator and the monotone generator and adjusted the latter until the intensity of both sounds appeared the same at the output earphone. The monotone intensity, easily measured by a galvanometer, was taken as the effective intensity of the shot noise.[46]

Hartmann's experimental results were qualitatively satisfactory yet quantitatively problematic. The apparatus indeed produced hissing tones at the output earphone, suggesting the presence of shot noise. He measured the effective shot-noise intensity for resonating frequencies from 238.73 to 2387.33 Hz, and calculated accordingly the charge of the electron *e*. To his disappointment, the results deviated significantly from Schottky's prediction. The value of *e* from the shot-noise measurements changed with frequency, instead of remaining a constant as it should. Hartmann's data were difficult to make sense of if Schottky's theory was right.

In the following years, physicists working on shot noise tried various approaches to figure out the nature of the discrepancy between Schottky's theory and Hartmann's experiment. Some corrected Schottky's mathematics, some blamed his physical assumptions, some redid the experiment with different setups (see chapter 6). Here we concentrate on a critique of Hartmann's measuring scheme. According to that critique, Hartmann made a serious mistake when using the ear-balancing method, for he ignored its dependence on human perception.

In 1922, a Czech physicist Reinhold Fürth at the University of Prague explained the variation of Hartmann's *e* value with frequency in terms of the "physiological" nature of the experimental method.[47] Fürth argued that one could not take the results of Hartmann's noise measurement via

[46] Ibid., 65–67.

[47] Reinhold Fürth, "Die Bestimmung der Elektronenladung aus dem Schroteffekt an Glühkathodenröhren," *Physikalische Zeitschrift*, 23 (1922), 354–362.

ear-balancing at their face value, for the method relied on the mediation of an experimenter's aural perception, a frequency-dependent feature. The problem with the German experimenter was that he did not choose the reference signals according to this aural perception (unlike the AT&T engineers' "artificial speech"), but naïvely set them to be *monotonic oscillations*. Consequently, the ear-balancing method required the experimenter to identify when the noise volume equaled the monotonic sound intensity and to take the monotonic current as the measure of the noise current. Yet, the two signals causing the same degree of aural perception had equal intensity if and only if the strength of human aural perception was proportional to the signal intensity (or, strength of stimulation) and both were at the same frequency.

Both assumptions were incorrect, as the Weber–Fechner law had proven the first assumption wrong, and the spectral character of the broad-band shot noise differed significantly from that of the monotonic reference signal. By removing these assumptions and making a more elaborate use of the Weber–Fechner law, Fürth reinterpreted Hartmann's data and produced a new estimate for the value of e with a slighter variation with frequency—his values deviated from 1.6×10^{-19} coulomb within 100%. Incorporating the physiological context saved Schottky's theory.

The ear-balancing method in the 1920s epitomized a particular route toward quantifying noise. Although both atmospherics and electronic noise were treated as physical effects, the method nonetheless sought to measure the degrees of human perceptive responses (in this case hearing) to those effects, not the effects themselves. The human measurer was thus a crucial part of the measuring instrument, and the whole metrological agenda did not separate from the technological contexts of telephone and radio. In addition, among all possible quantitative measurables of noise, the method only focused on its magnitude. Moreover, although the ear-balancing measuring set was not a spectral analyzer, engineers found it a handy instrument to determine the frequency response of noise.

This was not the only route, though. Around the time when American and German researchers used the ear-balancing method to measure noise, a few British researchers approached the same metrological problems with an alternative means. In contrast to the ear-balancing method's aural characteristics, the latter relied on visual display of noise waveforms.

The Visual Approach to Atmospherics

As Austin's team was measuring atmospheric noise in the 1910s, a group of British researchers was also examining radio disturbances from the atmosphere. Unlike Austin's team who were preoccupied with the implications of atmospherics for wireless communications, the British group was more concerned with studying atmospherics as meteorological effects and exploring their use in weather forecasting.

The head of the British group was Robert Watson Watt (later Watson-Watt), a researcher on radio probing of meteorological conditions and an early pioneer of radar technology. Trained in engineering at University College, Dundee, Scotland, he joined the Meteorological Office of the British Air Ministry (predecessor of the Royal Air Force) in 1915. A military engineering agency, the office aimed to provide weather forecasting to determine flight schedules. The weather phenomenon that brought the most serious risk to air fighters' safety was thunderstorm. Watson-Watt's first task was to predict the locations and times of thunderstorms. Knowing the connection between atmospheric noise and disturbing weather conditions, he proposed in 1915 to experiment on thunderstorm forecasting with radio direction finders, and led a group to conduct such experiments at the ministry's radio station in Aldershot, England.[48]

Throughout the second half of the 1910s, Watson-Watt and his colleagues used direction finders to triangulate the locations of atmospherics sources, and compared them with the locations of thunderstorms. They established twelve receiving stations representing distinct regions of the British Isles, obtained thousands of records from measurements, and conducted statistical analysis to find regular patterns that could be helpful to forecasting. Here, unwanted noise to radio engineers became wanted signals to meteorologists.[49]

At first, the Meteorological Office was interested only in thunderstorms. But Watson-Watt found that the thunderstorm was not the only meteorological cause of atmospherics; they correlated with other types of weather phenomena, too. After 1916, the meteorological events under consideration

[48] Robert A. Watson-Watt, *Three Steps to Victory: A Personal Account by Radar's Greatest Pioneer* (London: Odhams Press, 1957), 39–40.

[49] Ibid., 47–49; Robert A. Watson Watt, "Directional observations of atmospherics—1916-1920," *Philosophical Magazine*, 45:269 (1923), 1010–1026.; Robert A. Watson Watt, "Directional observations of atmospheric disturbances, 1920–21," *Proceedings of the Royal Society of London, A*, 102:717 (1923), 460–478.

included thunderstorms within 250 kilometers, thunderstorms within 1000 kilometers, hail showers within 200 kilometers, passing showers within a similar distance, squalls, and rainfall.[50] Watson-Watt compared atmospherics' directional data with weather reports in the British Isles, South and West Europe, and North Africa. The statistical data analysis obtained from April 1916 to April 1918 showed that the percentage of atmospherics geographically overlapping with thunderstorms was 36%, suggesting that thunderstorms were indeed a major source of atmospherics. Yet the analysis also showed that rain without thunder and lightning also had a strong correlation with atmospherics. The percentage of atmospherics sources located in rainy regions was more than 70%. Rainfall rather than lightning seemed the dominant factor to induce atmospheric electric disturbances at radio frequencies.[51]

Up to this point, Watson-Watt's program focused on measuring the locations of atmospheric sources, due to his preoccupation with weather forecasting. He broadened the scope of his atmospherics investigation after World War I, as the whole research program was transferred from the Meteorological Office to the newly established British Radio Research Board and moved from Aldershot to the board's research station in Ditton Park near Oxford.[52] Although Watson-Watt continued to study atmospherics' meteorological correlations via direction finding, under the new institutional affiliation he started to investigate the waveforms of radio noise using a visual display instrument: the cathode-ray oscilloscope.

As early as the eighteenth century, engineers had introduced apparatuses to trace the temporal changes of air temperature and pressure in steam engines. Throughout the nineteenth century, similar instruments were developed to represent the waveforms of measured results in graphic manners via electrical, magnetic, mechanical, or optical means. Many of these curve-tracing apparatuses had the function of recording displayed waveforms, which historian Anson Rabinbach dubbed as "devices of mechanical inscription." The

[50] Robert A. Watson Watt, "Atmospherics," *Proceedings of the Physical Society of London*, 37 (1924–1925), 26D.

[51] Robert A. Watson Watt, "The origin of atmospherics," *Nature*, 110 (1922), 680–681.

[52] Robert Naismith, "Early days at Ditton Park II, 1922-1927," *R.R.S. Newsletters*, 5 (September 15, 1961) and 10 (February 15, 1962), Box "Radio Research Board Committees, 1920-40s," Historical Records, Rutherford Appleton Laboratory; Subcommittee B on Atmospherics (Radio Research Board), "Programme and estimates for a general investigation of atmospherics," Sub-Committee B Paper No. 9, R. R. Board Paper No. 18, and "Programme of experiments M. O. Radio Station Aldershot," Sub-Committee B Paper No. 10, DSIR 36/4478, National Archives, London, UK.

curve-tracing apparatuses found extensive use in electrical engineering, acoustics, and physiology.[53]

Two inventions in the late nineteenth century dominated the art of curve tracers. William Duddell in England and André Eugène Blondel in France developed the oscillograph in the 1890s to track rapidly oscillating signals. In this apparatus, an input signal controlled the rotation of a mirror, which projected a fixed beam of light onto a photographic film. The trace of the light spot on the film indicated the temporal variation of the signal. Meanwhile, the cathode ray—an invisible beam of radiation in an electrically charged glass tube—discovered in the 1850s inspired the German Karl Ferdinand Braun's development of the oscilloscope in 1896, in which an input signal deflected a beam of cathode rays projected on a florescent screen to form a bright curve.[54]

Two points about the graphical apparatuses are worth noting. First, the visual character of the method fell into an "image" or "homomorphic" tradition (à la Galison) in experimental science that aimed at "representation of natural processes in all their fullness and complexity."[55] Under this tradition, visual evidence from the graphical method was presumably "direct," free of inference, and comprehensive. Second, compared with other visual instruments, the curve tracers were good at demonstrating the time relations of the measured effects. The waveform displayers were the right instruments to measure the elapsed times, periods of oscillation, or durations between events in observable physical or biological phenomena.

Watson-Watt first encountered the cathode-ray oscilloscope when he proposed to use it as a visual display of atmospherics direction-finding output in 1919. His proposal did not materialize, for the Meteorological Office lacked a cathode-ray tube fast enough to capture instantaneous atmospherics. As he broadened his focus to atmospherics' waveforms, a fast cathode-ray tube was needed even more. In November 1922, he knew from O. Webb of the Institution of Electrical Engineers that F. Gill from Western Electric had brought two newly developed fast cathode-ray tubes to the Institution. He rushed to the Institution at Savoy Place, London, and found that Edward Appleton at

[53] Anson Rabinbach, *The Human Motor: Energy, Fatigue, and the Origins of Modernity* (Berkeley: University of California Press, 1990), 96. For the early uses of the "graphical methods," see Phillips, *Waveforms* (1987); Sterne, *Audible Past* (2003); Frederic Holmes and Kathryn Olesko, "The images of precision: Helmholtz and the graphic method," in Norton Wise (ed.), *The Values of Precision* (Princeton, NJ: Princeton University Press, 1995), 198–221; and Norton Wise, *Aesthetics, Industry, and Science: Hermann von Helmholtz and the Berlin Physical Society* (Chicago, IL: University of Chicago Press, 2018).

[54] Phillips, *Waveforms* (1987), chapters 4, 5, 7.

[55] Peter Galison, *Image and Logic: A Material Culture of Microphysics* (Chicago, IL: University of Chicago Press, 1997), 19.

Cavendish Laboratory had borrowed one. He borrowed the other and started to collaborate with Appleton on the study of atmospherics' waveforms.[56]

Watson-Watt and Appleton developed a method to measure the noise's waveforms using high-speed cathode-ray tubes. The received atmospherics signal was coupled to a vacuum-tube amplifier and then to the cathode-ray oscilloscope's vertical electrodes. The oscilloscope's horizontal electrodes were connected to a triode oscillator. The oscilloscope displayed curves tracing the variation of the atmospherics' field intensity with time. The oscilloscope's refreshing period determined by its oscillating scan frequency 100–15,000 Hz was 66.6–10,000 microseconds, enough to accommodate an atmospheric pulse (~1000 microseconds)[57] (Figure 4.3).

Watson-Watt, Appleton, and their colleagues made a preliminary test on November 22, 1922, and conducted the actual experiment from December 22 to February 12, 1923. They collected hundreds of atmospherics waveform records and classified them into two types: the "aperiodic" (A) type in which

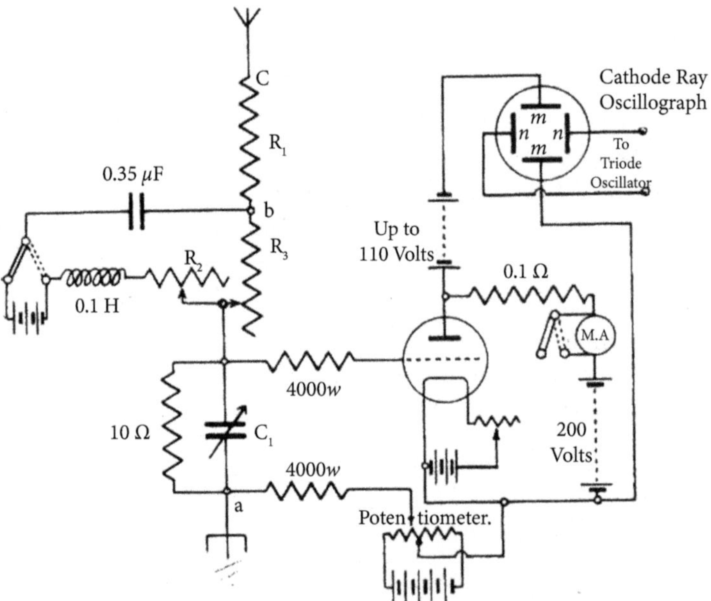

Figure 4.3 Watson-Watt and Appleton's oscilloscopic recorder. Robert A. Watson-Watt and Edward V. Appleton, "On the nature of atmospherics—I," *Proceedings of the Royal Society of London*, 103 (1923), 89, Figure 2.

[56] Watson-Watt, *Three Steps to Victory* (1957), 60–61.

[57] Robert A. Watson-Watt and Edward V. Appleton, "On the nature of atmospherics—I," *Proceedings of the Royal Society of London*, 103:720 (1923), 88–89.

the field change was in one direction only, and the "quasi-periodic" (Q) type in which the field change was first in one direction and then in the other. The A type was further classified into A+/A− groups in which the upper/lower plate of the condenser connected to the antenna became more positive at first, and the Q type was similarly classified into Q+/Q− groups. They also classified the atmospherics waveforms in terms of the rate of field change: those with convex waveforms were distinct from those with wedged concave waveforms.[58] They measured the principal constants of every atmospheric waveform, including the peak value(s), mean value, mean duration, and ratio of the positive duration to the negative duration. They performed statistical analysis for each waveform type and gave average values for the principal constants[59] (Figure 4.4).

Watson-Watt and Appleton's cathode-ray oscilloscope provided a new way to measure atmospherics. In contrast to the ear-balancing method that relied on a measurer's aural perception, the curves displayed on the screen were a visual representation of atmospheric radiation. The classes of atmospherics were associated with visual attributes of waveforms (periodic, aperiodic, positive, negative, round, peaked, and tilting angle) rather than aural attributes of noise (grinding, hissing, snap, and crashing).

Watson-Watt et al.'s cathode-ray oscilloscope was not yet a mechanical inscriber in the 1920s. The atmospheric waveforms displayed on the screen as florescent bright curves lasted only briefly and did not leave any record. The British researchers had to copy down those curves quickly on paper when they were still visible on the screen. This practice certainly introduced the measurer's perceptive factors and depended upon their craft skills of image recognition and drawing. But in the early 1930s, Watson-Watt and his collaborators succeeded in implementing fast automatic photography to record the waveforms. By the late 1930s, photography replaced drawing in Watson-Watt's atmospherics research.[60] With photography, the oscilloscope turned into a mechanical inscriber.

To contemporaries, the visual display had clear advantages over ear-balancing: directness, objectivity, and freedom from personal equations. Whereas scientists and engineers were aware of the perceptive factors of the ear-balancing scheme and discussed their effects on metrological precision,

[58] Watson-Watt, "Atmospherics" (1924–1925), 24D; Watson-Watt and Appleton, "On the nature of atmospherics—1" (1923), 91–96.

[59] Watson-Watt and Appleton, "On the nature of atmospherics—1" (1923), 99–100.

[60] Robert A. Watson-Watt, John F. Herd, and F.E. Lutkin, "On the nature of atmospherics—V," *Proceedings of the Royal Society of London*, 162:909 (1937), 267–291.

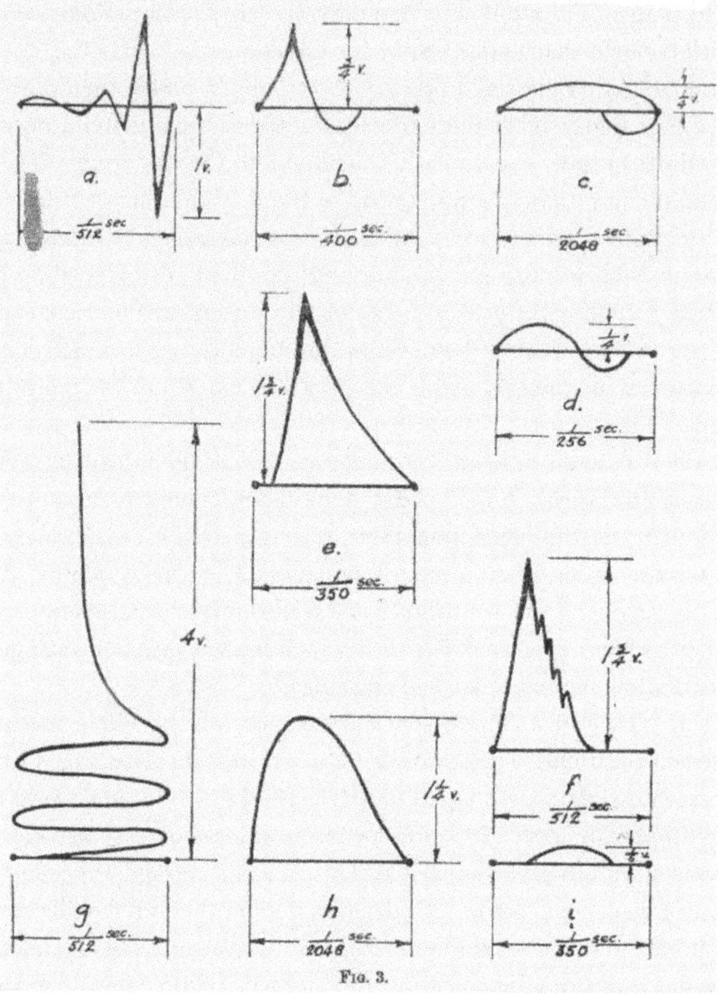

Figure 4.4 The waveforms obtained from Watson-Watt and Appleton's records. Robert A. Watson-Watt and Edward V. Appleton, "On the nature of atmospherics—I," *Proceedings of the Royal Society of London*, 103 (1923), 92, Figure 3.

they did not discuss possible physiological bias in oscilloscopic measurements. They did not find it necessary to calibrate the waveforms against optical illusions.

That does not mean that visual display was naturally a superior method to ear-balancing, since the former had a problem that the latter did not. The visual display produced more information than was necessary for noise measurement. Whereas the aural comparison only gave noise's average intensity, the oscilloscope demonstrated noise's entire waveform over time. Why should engineers or scientists know such details of noise as the amplitudes, durations, and shapes of its peaks, troughs, and plateaus? Although representing a measurable process in its fullness and complexity might be a virtue of the image tradition, those details seemed to have neither scientific nor practical implications to physicists or engineers. In radio and telephone engineering, measuring the average intensity of noise was adequate for gauging system performance and designing interference-robust systems. In atmospheric science and electronic physics, no theories had yet required any empirical information about noise other than its average intensity. Thus, why should researchers choose expensive and delicate oscilloscopes over the ear-balancing instruments?

Watson-Watt's answer to this question was twofold. First, the extra information from the visual approach had an implication for the physical studies of atmospherics. The waveforms of atmospherics pulses resembled those of electric discharge. Thus, we might be able to understand the discharge processes in thunderclouds or the upper atmosphere by inspecting atmospherics' waveforms. With the participation of John F. Herd, an old colleague of Watson-Watt since the days of the Meteorological Office, research on atmospheric waveforms continued to 1926. Watson-Watt, Appleton, and Herd obtained more measured data and explained some results with C.T.R. Wilson's theory of atmospheric discharge.[61]

This concentration on atmospheric waveforms evolved into a detailed examination of their time relations. Facilitated by photographic recording of oscilloscopic traces, Watson-Watt, Herd, and F.E. Lutkin collected data on atmospherics in the Middle East and North Africa during the 1930s. From these records, they identified two components in a typical atmospheric waveform: a faster pulse with higher oscillating frequencies followed by a slower pulse with lower oscillating frequencies. To Watson-Watt et al., the major quantitative information obtained from the waveforms was the difference between the two pulses' arrival times, their frequency components, and their amplitude ratio.[62]

Second, the oscilloscope was already a powerful tool for direction finding. The use of high-speed cathode-ray tubes enabled not only a visual display

[61] Edward V. Appleton, Robert A. Watson-Watt, and John F. Herd, "On the nature of atmospherics—II," *Proceedings of the Royal Society of London*, 111:759 (1926), 613–653; "On the nature of atmospherics—III," *Proceedings of the Royal Society of London*, 111:759 (1926), 654–677.

[62] Watson-Watt, Herd, and Lutkin, "On the nature of atmospherics—V" (1937), 267–291.

of atmospherics' waveforms but also a determination of their directions. In 1923, Watson-Watt and his colleagues succeeded in an experiment on oscilloscopic atmospherics direction finding in Aldershot. He connected two mutually orthogonal antennas (known as the "Belini-Tosi" direction finder) to the oscilloscope's horizontal and vertical electrodes. The incoming atmospherics incurred a tilted oscillatory waveform on the cathode-ray screen; the tilting angle indicated the atmospherics' direction of arrival.[63] Three years later, they applied the same principle, albeit with an oscillograph rather than an oscilloscope, to develop an automatic direction recorder for atmospheric noise.[64]

These considerations on the relevant information retrieved from visual display in noise measurement played an important part in the beginning of Britain's radar project. Anticipating the significant role of air battles in an upcoming war with Nazi Germany, the UK Air Ministry consulted Watson-Watt in 1935 about the possibility of producing a certain "death ray," strong electromagnetic pulses that could shoot down airplanes. Watson-Watt denied the possibility after performing calculations, but proposed instead that electromagnetic waves could be used to detect and locate enemy aircraft. He presented a memorandum to the Air Ministry on "radio ranging and detection." Watson-Watt's idea of radar was based on the principle of bouncing electromagnetic waves off a target. In the memorandum, Watson-Watt claimed that the oscilloscopic direction finder in his atmospherics research could be used to measure the direction of radar's echo pulse. The range of a target could be determined by gauging the returning time of the echo pulse. The target's angle of elevation could also be estimated via direction finding. Watson-Watt's proposed combination of pulse-echo sensing, direction, range, and elevation finding became the technological backbone of Chain Home, the UK's first working early-warning radar system that helped deter German air attacks during the Battle of Britain.[65]

The researchers on atmospherics were not the only ones exploring the implications of visual display for the studies of noise. In the 1920s, several instrument makers and engineers introduced various types of oscillographic, oscilloscopic, and other optical devices to record the waveforms of sounds. The "phonodeik" created by Professor Dayton Miller at the Case School of

[63] Robert A. Watson-Watt, John F. Herd, and L.H. Bainbridge-Bell, *Applications of the Cathode Ray Oscillograph in Radio Research* (London: H.M. Stationery Office, 1933), 126–132.

[64] Robert A. Watson-Watt, "The directional recording of atmospherics," *Journal of the Institution of Electrical Engineers*, 64 (1926), 596–610.

[65] Watson-Watt, *Three Steps to Victory* (1957), 84–87; Henry Guerlac, *Radar in World War II* (Washington, D.C.: American Institute of Physics, 1987), 127–145.

Applied Science was a well-known invention of this kind. The device used a horn to direct sound to a diaphragm and a spindle mechanism. The sound was transduced into motion that drove a mirror. A sharp beam of light was reflected by the mirror and projected on a photographic screen. As the mirror's location was modulated by the sound, the light spot on the screen undulated accordingly, which traced the waveform of the sound.[66] This waveform display was primarily applied in the demonstration and analysis of musical tones. But engineers were also exploring its uses in noise measurement. When Bell Labs were entrusted in 1926 by the musical instrument manufacturer Steinway to examine the acoustics of its studios, audience rooms, and pianos, Snook recommended to Steinway that a useful way to assess the amount of disturbing noise was visual display: retrieving room sounds with a microphone, electronically amplifying the signal, transducing it into a waveform with a General Electric oscillograph, and distinguishing graphically the waveform's frequency components with a harmonic analyzer.[67]

A Plethora of Means to Measure Noise

By the 1930s, multiple approaches to noise measurement were available to scientists and engineers. In a lecture given at the Royal Institution of London in 1931, George William Clarkson Kaye, superintendent of the Physical Department at the UK National Physical Laboratory, pointed out four primary operations of practical noise measurements:[68]

(1) The measurement of the "overall" power or energy-content of noise.
(2) The physical analysis of noise into its frequency components.
(3) The physical determination of noise's waveform.
(4) The aural measurement of noise's loudness perceived by a human ear.

Edward Elway Free in New York also made a similar observation about the plurality of noise-measuring apparatuses at this time. They included the audiometers using the ear-balancing method to gauge noise's loudness; oscillographic, oscilloscopic, or other optical-mechanical displays of noise waveforms; noise meters (also known as "acoustimeters") that indicated noise energy in a

[66] Dayton C. Miller, *The Science of Musical Sounds* (New York: Macmillian, 1922), 78–88.

[67] Letter from H. Clyde Snook to Paul Bilhuber, March 30, 1926, in "Development of audiometer," Box 444-05-01, Folder 08, AT&T Archives.

[68] Kaye, "The measurement of noise" (1929–31), 463.

potentiometer or galvanometer; spectrum analyzers that showed noise's frequency components; and tuning forks that measured noise's strength with the duration for which a fork was reverberating until its sound was masked by noise.[69]

These distinct types of apparatuses served different purposes. The tuning forks provided a handy, quick, simple, and portable tool for estimating noise levels, which was useful in preliminary or self-help surveys of street din. The audiometers incorporated human aural sensations and gauged perceived "noisiness" of noise. The acoustimeters boasted for being an "objective" instrument for measuring noise intensity. The spectrum analyzer and the oscillographic display promised to unveil the "footprints," frequency response, and other subtle characteristics of noise. Despite their differences, these approaches converged to a common trend that made noise quantifiable and measurable.

The most salient considerations in the development of noise metrology concerned its relationship with the aural contexts. The rise of electroacoustics in the 1910s–20s encouraged the treatments of sounds as electrical signals that could be manipulated with mathematical tools and physical devices. Scientists and engineers in the telecommunications industry began to expand the connotation of noise from annoying sounds to other phenomena such as atmospherics, stray radiations from appliances or power lines, and random fluctuations of electrical current. It makes sense to see a call for noise-measuring schemes that did not rely on aural perceptions or audible characteristics of the measurable.

A case in point is Western Electric's development of a portable indicating line-noise measuring set, known as RA-138. Unlike the ear-balancing devices that compared the measurable with a local reference tone, RA-138 converted environmental sound into an electrical signal with a microphone, directed the signal into a five-stage vacuum-tube amplifier, passed the amplifier output through a copper-oxide rectifier, and measured the current from the rectifier output with an indicator. In the measurement of electronic noise or atmospherics, the microphone was bypassed, and the set connected directly with an electronic circuit or radio receiver. The indicator had a potentiometer that controlled attenuation at the rectifier output and an ammeter or thermocouple taking readings of the signal intensity. The indicator was calibrated in decibel scales.[70]

[69] Free, "Practical methods of noise measurement" (1930), 21–29.

[70] Letter from W. Fondiller to K.W. Adams, April 3, 1935; "Proposed tentative standards for noise meters," March 26, 1935; "Electrical Research Product Company, Detailed Description of RA-138 Noise Meter"; all in Case No. 34,994, volume A, AT&T Archives.

Researchers at Bell Labs started to conceive this noise meter soon after Western Electric's 1-A measuring set was employed in the New York survey in 1929. The rising noise abatement movement and the expanding telephone networks anticipated a market for less bulky, handier, and quicker noise meters. The commercial need for this instrument diminished for a few years due to the Great Depression. The demand climbed up again by the mid-1930s, prompting researchers to complete design and prototyping.[71] But what were the strengths of a noise meter over an audiometer? A corporate technical report stated:[72]

> While it is not known that noise meters measure the effect of noise on telephone conversation any more accurately than the present ear-balance method, they possess a number of other advantages. By a meter method different observers will obtain practically the same reading on a given noise, especially if the noise is steady. Measurement of noise made at different times or in widely separated places will be more comparable than is the case at present. In addition, the time taken for noise measurement is reduced. Fluctuation of noise can be followed more readily.

Objectivity, not accuracy, was the major consideration when Western Electric engineers developed the RA-138 noise meter. Unlike the audiometer that relied on an operator's aural sensations and judgment about equal loudness, the acoustimeter presumably removed the roles of human discretion in the process of measurement. The direct reading of sound energy instead of "subjective" loudness guaranteed that measurements of the same noise by different operators or at different times and places gave the same results. Personal equations and perceptive biases would be reduced significantly, if not entirely. The noise meter marked the triumph of mechanical objectivity, and one further step toward the dissociation of noise from annoying sound.

This does not mean that the aural contexts became irrelevant to the metrology of noise. Neither did acoustimeters supersede audiometers by the 1930s. Although noise meters were promising, they had their own technical difficulties and challenges. One common problem was that their response time was not sufficiently fast, which made it hard to catch rapid and short noise pulses.

[71] Letter from J.J. Kahl to W.J. Shackelton, November 28, 1933, in Case No. 34,994, volume A, AT&T Archives.

[72] "Joint Subcommittee on Development and Research, National Electric Light Association and Bell Telephone System, Special Technical Report No. 16, Line Noise Meters," October 15, 1930, in Case No. 34,994, volume A, AT&T Archives.

Despite engineers' improvements on noise meters' integration time,[73] the ear-balancing apparatuses continued to be used in the measurements of noise and sound, especially the research on speech analysis and synthesis, phonetics, and psychoacoustics.[74]

Moreover, some researchers and engineers were reluctant to remove the aural and auditory characters from the metrology of noise. In a letter to Burgess Laboratories—a scientific instrument firm in Madison, Wisconsin—in 1926, Snook at Bell Labs stated:[75]

> You want a noise measuring instrument that is absolutely independent of the human ear. The term "noise" implies the hearing function. On this account, noise measurement must be related ultimately to some standard of noise that has been determined by the hearing function either of one individual or a group of individuals.

To Snook, measurement of noise was fundamentally about gauging the hearing effect it incurred, no matter whether the noise was a disturbing sound, or no matter whether the measuring device actually used an operator's aural perception. Under this rationale, Bell Labs researchers in the 1930s developed various shunt circuits for the RA-138 noise meter. Inserted between an acoustimeter's vacuum-tube amplifier and indicator, a shunt circuit functioned as a bandpass filter that simulated the frequency response of a human ear. Noise's energy components at distinct frequencies were weighted differently by the shunt circuit, presumably in the same way as the human aural perception did. In so doing, the indicator output was supposed to model the actual loudness of the noise as perceived by an ordinary human ear. The developers of RA-138 calibrated the shunt circuit and the indicator so that the noise meter's output was consistent with the data taken from audiometers with ordinary operators.[76]

[73] "Transmission observation at AT&T-BTL tie lines—Proposed experiment for evaluating the effect of room noise; Memorandum for file," November 7, 1938, in Case No. 36,750-10, volume A, AT&T Archives; A.H. Davis, "An objective noise-meter for the measurement of moderate and loud, steady and impulse noises," *Journal of the Institute of Electrical Engineers*, 83:500 (1938), 249–260.

[74] S. Millman (ed.), *A History of Engineering and Science in the Bell System*, Vol. 5, *Communications Science (1925-1980)* (New York: Bell Telephone Laboratories, 1984), 93–107.

[75] Letter from H. Clyde Snook to R.F. Norris, Burgess Laboratories, Wisconsin, Madison, March 25, 1926, in "Development of audiometer," Box 444-05-01, Folder 08, AT&T Archives.

[76] "Electrical Research Product Company, Detailed Description of RA-138 Noise Meter," in Case No. 34,994, volume A, AT&T Archives.

Between Subjective Perceptions and Physical Effects

The development of techniques and instruments in the 1910s–30s turned noise into quantifiable and measurable entities. This metrology epitomized two notions. Focusing on abating din and improving telecommunication audibility, audiologists and telephone technologists treated noise as disturbing sound and employed the ear-balancing method to measure the aural perceptions it incurred. Owing to electroacoustics, meanwhile, physicists and engineers began to view noise as disruption to abstract signals and devised "objective" schemes to gauge its effects without using human judgments. To researchers like Hartmann and Watson-Watt, noise was not only a plight to be subdued. It manifested physical processes such as electronic movements and atmospheric discharge that warranted investigations. As noise was materialized, abstracted, and quantified in sound reproduction technologies, it also became scientific objects.

Measurement and experimentation alone were insufficient for scientific research on noise, though. What were the causes of noise in its different manifestations? What was its behavior? How to represent it mathematically? While technologists, engineers, and experimenters in the early twentieth century coped with noise in sound reproduction through material manipulations and measurements, physicists and engineering scientists developed a theoretical understanding of noise. Their theory focused on a particular type of noise—irregular variations of current in electronic circuits. Contingent upon a breakthrough in statistical physics, they understood such noise as Brownian motion, random fluctuations of observables due to molecular agitations. How the studies of Brownian motion led to a statistical theory of noise is the subject of part II.

Between Subjective Perceptions and Physical Effects

PART II

BROWNIAN MOTION AND ELECTRONIC NOISE

5

Brownian Motion and the Origin of Stochastic Processes

From Probability and Statistics to Random Fluctuations

Today, the term "noise" refers not only to disturbing sound but also to random fluctuations of any kinds. Uncertainty, randomness, and irregularity had concerned savants, rulers, merchants, farmers, and artisans for millennia. Since Antiquity, people had reflected on unexpected outcomes under seemingly identical conditions in nature and society. Fortuna, the Roman goddess of luck and fate, and comparable deities in other cultures, metaphorically represented the general idea of uncertainty in natural and human domains as chance events controlled by blind, incomprehensible, and unpredictable forces of supreme origin.

Systematic and quantitative understanding and treatments of chance events did not appear until the introduction of probability theory in early modern times. From a received view, the first explicit theory of probability emerged as an arithmetic of gambling in the correspondence between Blaise Pascal and Pierre Fermat in 1654. During the Enlightenment, European mathematicians developed the theory into what the historian Lorraine Daston named "classical probability." It was a mathematical doctrine characterized by a preoccupation with chance events in economical and legal situations, modeling them with gambling-like scenarios with uniform a priori probabilities, estimating risks and gains, and deploying combinatorics. At the turn of the nineteenth century, mathematicians working on positional astronomy, including Pierre Simon Laplace, Adrien Marie Legendre, and Carl Friedrich Gauss, came up with a theory of errors to cope with random variations in astronomical data. Boosted by it, research on probability led to statistics as a data science to grapple with massive amounts of quantitative information in scientific, social, and economic realms. By the early twentieth century, probability theory and statistics penetrated major areas of the modern world and attained the status of what scholars called the "empire of chance."[1]

Historical studies of probability theory, statistics, and the "empire of chance" have pursued three directions. First, some have delved into the origin of statistical methods and techniques (least-squares methods, law of large numbers, central limit theorem, regression, etc.); their co-configuration with sociology, experimental psychology, evolutionary biology, genetics, and eugenics; and their uses in jurisprudence, insurance, clinical trials, public health, business management, and governmental administration. In the philosopher Ian Hacking's phrasing, this history testified to the "taming of chance" and building of a statistical regime that has governed many aspects of modern life.[2]

Second, researchers have examined the history of probability theory and statistics from the angle that emphasizes the development of philosophical interpretations, inductive logic, and mathematical systemization. They have focused on topics such as subjective and objective probabilities, foundation of statistical reasoning, Bayesian inferences, Andrey Kolmogorov's axiomatization, Richard von Mises' frequentist formulation and Brno de Finetti's empiricist formulation, hypothesis testing, and experimental design. Research along this line has sought to understand historically the logical and conceptual power of probability and statistics as a system of thought for analysis, decision-making, and prediction, and its philosophical grounds.[3]

Third, historians and philosophers have scrutinized the rise of the probabilistic and statistical approach to physical problems, and its ontological and epistemological implications. The subjects along this line of inquiry include James Clerk Maxwell's, Ludwig Boltzmann's, and Josiah Willard Gibbs' formulations of statistical physics, the problems with interpreting irreversibility associated with the second law of thermodynamics, the so-called "Maxwell's

[1] Lorraine Daston, *Classical Probability in the Enlightenment* (Princeton, NJ: Princeton University Press, 1988); Gigerenzer et al., *Empire of Chance* (1997).

[2] For example, see Stigler, *History of Statistics* (1986); Porter, *Trust in Numbers* (1996); Hacking, *Taming of Chance* (1990); Dan Bouk, *How Our Days Became Numbered: Risk and the Rise of Statistical Individuals* (Chicago, IL: University of Chicago Press, 2005). For the broader social implications of statistics as forces of control, power, and rationalization, see Michel Foucault, *Society Must Be Defended* (Paris: Picador, 2003), 242–243; Max Weber, *The Protestant Ethics and the Spirit of Capitalism* (New York: Schribner, 1958).

[3] For example, see Ian Hacking, *The Emergence of Probability: A Philosophical Study of Early Ideas about Probability Induction and Statistical Inference* (Cambridge: Cambridge University Press, 2006); von Plato, *Creating Modern Probability* (1998); Joseph Berkovitz, "The world according to de Finetti: On de Finetti's probability theory and its application to quantum mechanics," in Y. Ben Menachem and M. Hemmo (eds.), *Probability in Physics* (Dordrecht: Springer, 2012), 249–280; Allan Franklin, A.W.F. Edwards, Daniel J. Fairbanks, Daniel L. Hartl, and Teddy Seinfeld, *Ending the Mendel-Fisher Controversy* (Pittsburgh, PA: University of Pittsburgh Press, 2008).

demon" as a thought experiment connected to irreversibility, ergodicity, the beginning of quantum physics and its probabilistic interpretation, chaos, and nonlinear dynamics. In these studies, scholars' focus has been on the foundational questions in epistemology and ontology about the probabilistic-statistical approach in physical sciences: how to reconcile the approach to the Newtonian worldview, whether the success of the approach in statistical and quantum physics entails an indeterministic world, what is the limit of causal knowledge in the fundamentally probabilistic structures of statistical and quantum physics, etc.[4]

The conceptualization of noise in the early twentieth century opens a new historical perspective on thinking about uncertainty. Although theoretical studies of noise were connected to probability theory and statistics, they were less preoccupied with governance by numbers, biopolitics, personal equations, rational decisions, subjective and objective probabilities, the nature of inductive inferences, experimental design, determinism, or irreversibility. Rather, the focus was a new kind of randomness that had become conspicuous by the 1900s. Generally known as fluctuations, this new randomness concerned not deviations and dispersions of individual *data points*, but irregularities of natural or human phenomena embodied in *continuous trajectories* or *discrete sequences*. For example: agitation of small particles in a liquid, radioactivity, variation of current in an electronic circuit or that of chemical concentration in a solution, and a volatile stock market. While these phenomena did not alter the epistemological and ontological status of the probabilistic worldview, fluctuations nonetheless brought profound technical and conceptual changes that helped extend the statistical methods to grapple with the dynamics, stability, time-evolution, and spectral analysis of random events. In today's jargons, such fluctuations are conceived as "stochastic processes."

The early development of stochastic-processes theory came primarily from physics, economics, and engineering. Statisticians and "pure" mathematicians in the nineteenth and early twentieth centuries did contribute to this field. Francis Galton and Henry Watson's finding of a branching process in a statistical study of disappearing family names in the 1880s, Karl Pearson's and George Pólya's examinations of random walks, and Andrey Kolmogorov's inquiries

[4] Brush, *Statistical Physics and the Atomic Theory of Matter* (1983); Robert Purrington, *Physics in the Nineteenth Century* (New Brunswick, NJ: Rutgers University Press, 1997), 132–147; Lorenz Krüger, "Probability as a theoretical concept in physics," *Proceedings of the Biennial Meeting of the Philosophy of Science Association* (1986), 273–287; Mara Beller, *Quantum Dialogue: The Making of a Revolution* (Chicago, IL: University of Chicago Press, 1999); James Gleick, *Chaos: Making a New Science* (New York: Penguin Books, 1987).

into the Markov chains are cases in point.[5] Yet, these studies were overshadowed in quantity and visibility by a pioneering body of works on stochastic processes that emerged from theoretical studies of physical, economic, and engineering problems. A large part of random-process theory was originated from the explorations of Brownian-type fluctuations in physical quantities due to random molecular or electronic movements and in financial data associated with intrinsic irregularities of markets. Albert Einstein's and Marian Smoluchowski's theory of Brownian motion in 1905–06 attributed the macroscopic fluctuation to random molecular collisions. Louis Bachelier's characterization of the prices of financial products at Paris Bourse reached a similar conclusion. In the following decades, scientists delved further into this theory and extended their scope to other types of random fluctuations, including fast and irregular undulation of electric current in radio or telephone circuitry. Named "electronic noise" for its sonic effect on telephone and radio, this fluctuation was a subject of researchers who endeavored to improve communication technologies. Thus, noise gained a more abstract meaning as generic random disturbances from any wanted information or signals.

Several scholars have examined the early studies of random fluctuations. Roberto Maiocchi and Charlotte Bigg looked into the research on Brownian motion in the 1900s, stressing Einstein's and Smoluchowski's original contribution to focus on observing powder particles' mean displacement instead of mean velocity and Jean Perrin's ultramicroscopic emulsion experiment to verify their theory. Günter Dörfel and Dieter Hoffmann explored physicists' investigations into electronic noise in the 1910s–20s, highlighting Walter Schottky's discovery of shot noise and John B. Johnson and Harry Nyquist's finding of thermal noise. Jan von Plato surveyed the history of mathematics for stochastic processes. Leon Cohen overviewed the works on random fluctuations leading to the informational sense of noise in which engineers and scientists understand it today.[6]

[5] Francis Galton and Henry Watson, "On the probability of extinction of families," *Journal of the Royal Anthropological Institute*, 4 (1875), 138–144; Karl Pearson, "The problem of the random walk," *Nature*, 77 (1905), 294; George Pólya, "Über eine Aufgabe der Wahrscheinlichkeitrechnung betreffend die Irrfahrt im Strassennetz," *Mathematische Annalen*, 84 (1921), 149–160; David George Kendall, "Andrei Nikolaevich Kolmogorov," *Bulletin of the London Mathematical Society*, 22 (1990), 31–47.

[6] Robert Maiocchi, "The case of Brownian motion," *The British Journal for the History of Science*, 23 (1990), 257–283; Charlotte Bigg, "Evident atoms: Visuality in Jean Perrin's Brownian motion research," *Studies in the History and Philosophy of Science*, 39 (2008), 312–322; Günter Dörfel and Dieter Hoffmann, "Von Albert Einstein bis Norbert Wiener" (2005); von Plato, *Creating Modern Probability* (1998); Leon Cohen, "The history of noise: On the 100th anniversary of its birth," *IEEE Signal Processing Magazine*, 22 (2005), 20–45.

This literature on the early history of random fluctuations either concentrates exclusively on Brownian motion in its narrow sense (the movement of small particles in a fluid) or perceives the treatments of electronic noise as a natural and unproblematic extension of the theory dealing with agitation of powders. Yet, both views overlook the heterogeneity and diversity of the theoretical works on random fluctuations. In the following two chapters, we will see that the studies of suspended particles' Brownian motion and the exploration of randomly changing electronic current represented two distinct approaches to fluctuations in the 1900s–40s.

The first approach operated in the time domain. It grappled with random fluctuations as a bunch of particles' irregular trajectories. To its advocates, fluctuations were essentially consequences of dynamics. Like dynamics in general, the major aim of research in this approach was to predict and estimate future trajectories. In probabilistic calculus, this aim was equivalent to finding the temporal evolution of the trajectories' statistical attributes such as the mean, variance, and probability density function. Researchers probed these statistical properties by formulating the stochastic problems in terms of random walks, a diffusion-type differential equation for the probability density function, or an equation of motion with a random force term. Broadly, this approach reflected the research tradition of statistical physicists, who viewed fluctuations as instantiations of equipartition, ensemble averages, and Boltzmann distributions.

The second approach operated in the frequency domain. It conceived random fluctuations as erratic waveforms comprising different modes of harmonic oscillations. Black-body radiation or incoherent white light was a closer analogy than corpuscular dynamics, and Fourier analysis was the methodological cornerstone. To its advocates, the major aim of research was to find the amount of energy for a fluctuating waveform's components within a given spectral window. To achieve this aim, researchers often relied on the assumption that the waveform at a specific instant was uncorrelated to that before, which led to the conclusion that the energy of fluctuating noise remained uniform and constant over frequencies. In general, this approach reflected a research tradition in electrical engineering, as researchers treated fluctuations as random signals and undertook the canonical method of spectral analysis for electronic signal processing.

This book is not the first to pinpoint the distinction between the time-domain and frequency-domain approaches. In his correspondence to *Annalen der Physik* in 2012, Dörfel also observed these two different methods for treating random fluctuations in the 1920s.[7] Yet, while he has rightly associated the frequency-domain approach with those working on telecommunication engineering, his attribution of the time-domain approach to precision metrology was only part of the story. As we will see, the examinations of a random fluctuation's temporal variation were rooted in a deeper concern in statistical mechanics tracing back to Boltzmann. In addition, Dörfel's short exposé did not grapple with the full-fledged development of the theoretical treatments of random fluctuations.

In part II, we follow the origin of the time-domain approach (chapter 5) and the frequency-domain approach (chapter 6) to random fluctuations. Chapter 5 focuses on the inception of the stochastic theories of Brownian motion. First, we go through Einstein's paper in 1905 on Brownian motion, which is considered the foundational work on random fluctuations in physical systems. This work is placed in the context of the statistical-mechanical program. Then we examine Marian Smoluchowski's studies of random fluctuations and Paul Langevin's reformulation of Brownian motion, respectively. Finally, we look into Louis Bachelier's work on the modeling of stock market prices as Brownian motion.

The Statistical-Mechanical Program

The launch of theoretical investigations into Brownian motion was indebted to the statistical-mechanical program in the second half of the nineteenth century. In the 1730s–1850s, natural philosophers attempted to explain gas temperature, pressure, and volume in terms of movements of microscopic molecules.[8] In 1859, James Clerk Maxwell at King's College London came up with a statistical kinetic theory of gas. Maxwell treated gas as a collection of molecules undergoing random motions whose speeds could take any directions and magnitudes. From symmetry and uniform a priori probabilities, he deduced that at equilibrium the velocity amplitude v of gas molecules followed a distribution $Ae^{-v^2/\alpha}$. This distribution took the same form as the normal distribution of errors in astronomical data that the German mathematician Carl Friedrich Gauss had identified in 1809 and the bell-shaped curves of population distributions that the Belgian sociologist Adolph Quetelet had found in 1835. From statistical arguments, Maxwell also obtained what is now called

[7] Dörfel, "Early history of thermal noise" (2012), A117–A121.
[8] Purrington, *Physics in the Nineteenth Century* (1997), 132–147.

equipartition: the average kinetic energy of a molecule per degree of freedom is a constant proportional to temperature.[9]

Following Maxwell, the Austrian physicist Ludwig Boltzmann paved the foundation of statistical mechanics. When he was a Ph.D. student at the University of Vienna in the mid-1860s, Boltzmann was introduced to Maxwell's theory by his supervisor Josef Stefan. While moving back and forth between academic appointments in Vienna and Graz in 1868–77, Boltzmann published a series of articles on the statistical treatments of physical systems. He first extended Maxwell's distribution to the cases when molecules experienced an external force. Then he examined the cases when gas was moving and equilibrium had not been reached. He formulated a "transport equation" for the molecular density distribution. The Maxwell–Boltzmann distribution was a solution from this transport theory when the system converged to equilibrium. Next, he left transport dynamics and inspected gas configurations at which molecules occupied discrete energy levels. Assuming equally probable configurations and constant system energy and number of molecules, he used combinatorics to deduce his previous finding from the law of large numbers: When the gas had many molecules, its most likely energy distribution was the Maxwell–Boltzmann function; this normal distribution was much more likely than all others.[10]

Boltzmann's conceptual framework in the 1860s–70s was further developed at the turn of the twentieth century, as the physics professor Josiah Willard Gibbs at Yale University drew upon his studies of chemical reactions and gaseous kinetics and produced a formulation of statistical mechanics based on

[9] Elizabeth Garber, "Maxwell's kinetic theory 1859-70," in Raymond Flood, Mark McCartney, and Andrew Whitaker (eds.), *James Clerk Maxwell: Perspectives on His Life and Work* (Oxford: Oxford University Press, 2014), 139–153; James Clerk Maxwell, "Illustrations of the dynamical theory of gases," *Philosophical Magazine*, 19 (1860), 19–32, 20 (1860), 21–37; James Clerk Maxwell, "On the viscosity or internal friction of air and other gases," *Philosophical Transactions of the Royal Society*, 156 (1866), 249-268.

[10] John Blackmore (ed.), *Ludwig Boltzmann: His Later Life and Philosophy. Book I: A Documentary History* (Dordrecht: Kluwer, 2010), 1–4; Brush, *Statistical Physics and the Atomic Theory of Matter* (1983), 62–68; Ludwig Boltzmann, "Studien über das Gleichgewicht der lebendige Kraft zwischen bewegten materiellen Punkten," *Sitzungsberichte, Kaiserliche Akademie der Wissenschaften, Wien, Mathematisch-Naturwissenschaftliche Klasse*, 58 (1868), 517–560; "Einige allgemeine Sätze über Wärmegleichgewicht," *Sitzungsberichte, Kaiserliche Akademie der Wissenschaften, Wien, Mathematisch-Naturwissenschaftliche Klasse*, 63 (1871), 679–711; "Weitere Studien über das Wärmegleichgewicht unter Gasmolekülen," *Sitzungsberichte, Kaiserliche Akademie der Wissenschaften, Wien, Mathematisch-Naturwissenschaftliche Klasse*, 66 (1872), 275–370; "Über die Beziehung eines allgemeine mechanischen Satzes zum zweiten Hauptsatze der Wärmetheorie," *Sitzungsberichte, Kaiserliche Akademie der Wissenschaften, Wien, Mathematisch-Naturwissenschaftliche Klasse*, 75 (1877), 67–73; "Über die Beziehung zwischen des zweiten Hauptsatze der mechanischen Wärmetheorie und der Wahrscheinlichkeitsrechnung, respektive den Sätzen über das Wärmegleichgewicht," *Sitzungsberichte, Kaiserliche Akademie der Wissenschaften, Wien, Mathematisch-Naturwissenschaftliche Klasse*, 76 (1877), 373–435.

microcanonical ensembles, canonical ensembles, macrocanonical ensembles, and partition functions.[11]

The three founders of statistical mechanics and their peers were preoccupied with interpretations of thermodynamics. Inheriting the problematique from the kinetic theory of matters, they concentrated on theoretical understanding of phenomena in thermodynamics: the ideal gas law, van der Waal equation, viscosity, diffusion, heat conduction, chemical potential. A pressing issue was how to make sense of the second law of thermodynamics that Rudolph Clausius and William Thomson (Lord Kelvin) had proposed in the 1850s: Some processes in the material world cannot be reversed, or equivalently, the entropy of a closed physical system either remains the same or increases over time. While the nondecreasing entropy was believed valid in all thermodynamic phenomena, this trend was intuitively contradictory to a general principle in mechanics, according to which mechanical interactions should be reversible.

To highlight the puzzling character of irreversibility, Maxwell entertained a famous thought experiment about the "demon" in a letter to Edinburgh physicist Peter Tait in 1867: An intelligent genie able to recognize the speed of each gas molecule could, by simply controlling the gate on a wall dividing a container, make the gas hotter on one side and colder on the other, which contradicted the second law of thermodynamics.[12] Addressing the same issue, Boltzmann's longtime friend and colleague at the University of Vienna Johann Josef Loschmidt contended in 1876 that any thermodynamic system should be reversible. A thermodynamic process was a mechanical process of all constitutive molecules progressing from an initial state to a final state, Loschmidt maintained. The equations of motion for these molecules were reversible so that reversing all the molecular velocities at the final state should lead to their return to the initial state. Thus, the second law contradicted Newtonian mechanics. Boltzmann's response to Loschmidt's argument in 1877 was that the second law was statistical. The reverse counterparts of thermodynamically irreversible processes are not observable, not because they do not exist, but

[11] Brush, *Statistical Physics and the Atomic Theory of Matter* (1983), 72–78; J. Willard Gibbs, *Elementary Principles in Statistical Mechanics: Developed with Especial Reference to the Rational Foundation of Thermodynamics* (New York: Scribner, 1902).

[12] Andrew Whitaker, "Maxwell's famous (or infamous) demon," in Raymond Flood, Mark McCartney, and Andrew Whitaker (eds.), *James Clerk Maxwell: Perspectives on His Life and Work* (Oxford: Oxford University Press, 2014), 163–186. The philosophical-historical literature on Maxwell's demon is vast. A representative work is John Norton and John Earman, "Exorcist XIV: The wrath of Maxwell's demon. Part I: From Maxwell to Szilard," *Studies in the History and Philosophy of Modern Physics*, 29 (1998), 435–471; "Exorcist XIV: The wrath of Maxwell's demon. Part II: From Szilard to Landauer and beyond," *Studies in the History and Philosophy of Modern Physics*, 30 (1999), 1–40.

because their probability of occurrence is almost infinitely smaller than that of thermodynamically permissible processes. To make this argument, Boltzmann formulated a statistical interpretation of entropy $S = k \cdot \log W$ (S is entropy of a thermodynamic state, W is its number of possible microscopic configurations, and $k = 1.38 \times 10^{-23}$ m^2 kg s^{-2} K^{-1} is the Boltzmann constant).[13]

While Boltzmann's statistical understanding of the second law became the standard view among physicists by the late nineteenth century, challenges continued to emerge, and efforts were still made to reconcile mechanics with thermodynamics without statistics. In 1889, the mathematician Henri Poincaré in Paris found that for a closed mechanical system with a fixed total energy and a finite volume, its state in the phase space would return as close as possible to any initial state—a stipulation known as the "recurrence theorem." Seven years later, Max Planck's assistant Ernest Zermelo in Berlin employed the recurrence theorem to further an attack on the incompatibility between thermodynamics and mechanics despite Boltzmann's statistical interpretation. This move reraised the issue of reversibility and determinism.[14]

The statistical-mechanical program in the late nineteenth century was built upon the atomic theory of matter and probability calculus for molecular distributions. It aimed to establish a mechanical foundation of thermodynamics, account for the second law and irreversibility, and offer microscopic explanations of thermodynamic phenomena in gases and liquids at equilibrium or in the processes toward equilibrium. In the early twentieth century, however, statistical mechanics witnessed a fundamental change. Specifically, physicists broadened their scope to problem situations away from the statistical mean at thermodynamic equilibrium, and started to pay close attention to measurable effects associated with microscopic fluctuations.

This new focus on fluctuations had a conspicuous impact on modern statistical physics. Researchers identified various discernible effects of microscopic agitations. This included a fresh theoretical approach to phase transitions and critical phenomena, a general correlation between fluctuation and dissipation, and quantum noise. These studies showed that random movements of molecules, atoms, or electrons were not smoothed out in statistical average but

[13] Brush, *Statistical Physics and the Atomic Theory of Matter* (1983), 90–93; Josef Loschmidt, "Über den Zustand des Wärmegleichgewichtes eines Systems von Körpern mit Rucksicht auf die Schwerkraft," *Sitzungsberichte, Kaiserliche Akademie der Wissenschaften, Wien, Mathematisch-Naturwissenschaftliche Klasse*, 73 (1876), 128–142; Boltzmann, "Über die Beziehung zwischen des zweiten Hauptsatze der mechanischen Wärmetheorie und der Wahrscheinlichkeitsrechnung" (1877), 373–435.

[14] Henri Poincaré, "Sur le problème des trios corps et les équations de dynamique," *Acta Mathematica*, 13 (1889), 1–270; Ernst Zermelo, "Ueber einen Satz der Dynamik und die mechanische Wärmetheorie," *Annalen der Physik*, ser. 3, 57 (1896), 485–494.

were amplified to observable levels due to particular probabilistic structures of physical systems. This sea change of statistical physics began with research on an apparently minor phenomenon: the irregular movement of particles suspended in a fluid, or so-called Brownian motion.

Researching Brownian Motion before 1905

In 1827, the Scottish botanist Robert Brown—known for his expedition to Australia, keeping of a botanic collection at the British Museum, and naming of the cell nucleus—observed under microscope that pollen powders in water underwent irregular movements. Brown's finding attracted immediate speculations, involving renowned figures like Michael Faraday, David Brewster, and Charles Darwin. By the 1870s, the majority view among scientists was that Brownian motion was due to a fluid's internal physical movement such as capillarity, convection current, evaporation, interaction with light, or electrical force. Inspired by the contemporaneous kinetic theory of matters, the most popular candidate cause was irregular thermal collisions of liquid molecules on the particles. Yet, this molecular-kinetic theory encountered a difficulty in 1879, as the Swiss botanist Carl von Nägeli at the University of Munich gave a quantitative estimate of the motion. Nägeli was developing a theory of how sun-motes' growth was affected by bombardment from air molecules and thought it applicable to Brownian motion. In his calculation, the air molecular velocity at room temperature was 400–500 m/s. This velocity, imparted to a powder particle 300 million times heavier than a gas molecule, would not incur an observable movement. Three years later, the chemist William Ramsay at University College Bristol reached the same conclusion.[15]

Over eight decades, the studies of Brownian motion grew into a cottage industry. By Maiocchi's count, there were about forty publications on the subject between 1827 and 1903.[16] A series of works in the 1880s–90s turned Brownian motion into a hot spot for research in statistical physics. In 1888–89, Louis Georges Gouy, a physics professor at the University of Lyon, argued again that Brownian motion was caused by molecular collisions on suspended particles. Like Nägeli and Ramsay, Gouy knew the difficulty of fitting numerically the theory of molecular collisions with observation. But he believed that

[15] Stephen Brush, "A history of random processes: I. Brownian movement from Brown to Perrin," *Archives for History of Exact Sciences*, 5 (1968), 1–13; Maiocchi, "Case of Brownian motion" (1990), 257–259.

[16] Maiocchi, "Case of Brownian motion" (1990), 282–283.

the bombardment from fluid molecules could still incur a detectable effect on the particles through a collective and coordinated action. This conviction came from his observation that Brownian motion became more intense as temperature rose. Gouy did not produce a quantitative theory. Yet, he framed Brownian motion with the language of thermodynamics. This caught physicists' attention.[17]

While scientists did not settle down with a satisfactory explanation of Brownian motion, they explored its qualitative relationship with various physical factors including temperature, specific heat of the fluid, electric field, and magnetic field. These empirical investigations led to quantitative experimental studies of the effect that were built upon a material culture surrounding microscopy. In 1900, an Austrian physicist Felix Exner devised a means to measure the average speed of Brownian particles. A descendant of the most powerful scientist family in nineteenth-century Austria–Hungary and son of another prominent researcher on Brownian motion, Sigmund Exner, Felix was a doctoral student at the University of Vienna. Using the laboratory facilities of the university's physiological institute administered by his father, Felix prepared water doped with rubber particles of different sizes and put it under a microscope for observation. To perform measurement, he observed the particles through an eyepiece, drew their trajectories on blackened photographic plates, and optically enlarged the images. Since he timed the duration of each linear segment in a particle's zigzag trajectory, he could determine the particle's speed by dividing the length of a segment with the corresponding duration.

Following this procedure, Exner measured the velocity of Brownian motion at different temperatures and particle sizes. While his data showed an increase of velocity with temperature as the molecular-kinetic theory had predicted, their extrapolation gave zero velocity at a temperature well above 0°K, which contradicted the molecular-kinetic theory. Exner's own feeling for this result was that the molecular collisions were likely responsible for part of Brownian motion, but they might not be the only cause. Although Exner's experimental outcome did not provide a clear corroboration of the molecular-kinetic theory, his undertaking yielded the first systematic quantitative data on Brownian motion. Within two years, another Vienna-trained chemist Richard Adolf Zsigmondy joined the force. While working at the German optical manufacturer Carl Zeiss Allgemeine Gesellschaft, Zsigmondy developed an

[17] Louis Georges Gouy, "Note sur le mouvement brownien," *Journal de Physique*, 7 (1888), 561–564; Brush, "History of random processes" (1968), 12; Maiocchi, "Case of Brownian motion" (1990), 257–259.

"ultramicroscope" that observed tiny particles by immersing them in a solution, shining on it a beam of light, and receiving scattered instead of reflected light from the particles in the direction perpendicular to the beam. In 1902, Zsigmondy employed an ultramicroscope to observe a colloidal solution and noted the same Brownian motion—but the new instrument enabled a more precise measurement.[18]

Einstein's Work on Brownian Motion and Random Fluctuations

Einstein the Statistical Physicist

Albert Einstein tackled Brownian motion with minimum exposure to all these experimental and theoretical developments. Rather, he approached the problem as a stepping-stone for his research program in statistical mechanics. As John Stachel et al., Abraham Pais, and Jürgen Renn have pointed out, thermodynamics and statistical mechanics marked Einstein's point of entry to research in physics. His first two publications in 1901 and 1902, which were written immediately after he graduated from the Swiss Federal Polytechnic (ETH) in Zurich, concerned a theory of intermolecular forces applied to capillarity and electrolytic solutions. While serving at the Swiss Patent Office in Bern in 1902–04, he published three articles on the foundation of thermodynamics, dealing with the definition of temperature and entropy, irreversibility, and energy fluctuations.

In these works, Einstein extended Boltzmann's statistical interpretation of thermodynamics by invoking, without knowing Gibbs' similar work, the notion of canonical ensembles that treated a collection of physical systems with a fixed number of particles but variable energy due to their connection to a heat reservoir at a constant temperature. According to Renn, this reinterpretation of Boltzmann's statistical physics via canonical ensembles allowed Einstein to rederive the second law of thermodynamics and equipartition. Einstein's inquiry concerning canonical ensembles also led to an assertion that a system under thermodynamic equilibrium had an energy fluctuation $\Delta E^2 = kT^2 (d\langle E\rangle/dT)$.[19]

[18] Maiocchi, "Case of Brownian motion" (1990), 257, 262–263; Felix Exner, "Notiz zu Brown's Molecularbewegung," *Annalen der Physik*, 2 (1900), 843–847.

[19] John Stachel, David Cassidy, Jürgen Renn, and Robert Schulmann, "Einstein on the nature of molecular forces," and "Einstein on the foundations of statistical physics," in John Stachel, David

These articles configured Einstein's doctoral dissertation. In his thesis submitted to the University of Zurich in 1905, Einstein sought to determine the properties of atoms from physical-chemical properties of a solution with solute molecules larger than solvent molecules. Based on hydrodynamics and the kinetic theory of solutions, he derived two formulas connecting two microscopic parameters (Avogadro number, or number of molecules per mole, and the size of a solute molecule) to two macroscopic measurables (the solution's viscosity and the solute's diffusion coefficient). Einstein plugged experimental data on sugar solutions' viscosity and diffusion coefficient into these formulas and estimated the two microscopic parameters.[20]

Einstein's work in 1902–05 on statistical physics and the molecular-kinetic theory of sugar solutions led to his work on Brownian motion. While writing his dissertation, he noted a structural similarity between a solution containing solute molecules and one containing suspended grains of unsolvable substance, even though the size of the particles in the first system was much smaller than that in the second. Both systems followed the same statistical rules of kinetic thermal collisions. The murky liquid was not a new phenomenon to scientists. It was the effect that Brown discovered in 1827, and Nägeli, Ramsay, Gouy, Exner, and others tried to understand.

Einstein's 1905–06 Papers on Brownian Motion

In 1905, Einstein submitted a paper on the movements of small particles suspended in a liquid to *Annalen der Physik*. It was accepted on May 11, eleven

Cassidy, Jürgen Renn, and Robert Schulmann (eds.), *The Collected Papers of Albert Einstein*, Volume 2: *The Swiss Years, 1900–1909* (Princeton, NJ: Princeton University Press, 1989), 3–8, 41–55; Abraham Pais, *Subtle is the Lord: The Science and the Life of Albert Einstein* (New York: Oxford University Press, 2005), 55–92; Jürgen Renn, "Einstein's controversy with Drude and the origin of statistical mechanics: A new glimpse from the 'love letters,'" *Archives for History of Exact Sciences*, 51 (1997), 315–354; Albert Einstein, "Zur allgemeinen molekularen Theorie der Wärme," *Annalen der Physik*, 14 (1904), 354–362. To derive the formulation of the energy fluctuation, Einstein considered that under the canonical ensembles, the probability of the states falling between energy E and E + dE would be $dW = Ce^{-\frac{E}{kT}}\omega(E)\,dE$. Then the unitary requirement of overall probability resulted in $\int_0^\infty Ce^{-\frac{E}{kT}}\omega(E)\,dE = 1$, while the mean energy of the system would be $\int_0^\infty CEe^{-\frac{E}{kT}}\omega(E)\,dE = \langle E\rangle$. Combining both formulas led to $\int_0^\infty (E - \langle E\rangle)\,Ee^{-\frac{E}{kT}}\omega(E)\,dE = 0$. Taking the derivative of this equation with respect to T would give $\langle E^2\rangle - \langle E\rangle^2 = kT^2\,(d\langle E\rangle/dT)$.

[20] John Stachel, David Cassidy, Jürgen Renn, and Robert Schulmann, "Einstein's dissertation on the determination of molecular dimensions," in John Stachel, David Cassidy, Jürgen Renn, and Robert Schulmann (eds.), *The Collected Papers of Albert Einstein*, Volume 2: *The Swiss Years, 1900–1909* (Princeton, NJ: Princeton University Press, 1989), 170–182; Albert Einstein, *Eine neue Bestimmung der Moleküldimensionen*, Ph.D. dissertation, University of Zurich (1905), printed in John Stachel, David Cassidy, Jürgen Renn, and Robert Schulmann (eds.), *The Collected Papers of Albert Einstein*, Volume 2: *The Swiss Years, 1900–1909* (Princeton, NJ: Princeton University Press, 1989), 183–205.

days after he finished his dissertation. Retrospectively, this paper would be one of the three pillars constituting the works in Einstein's Annus Mirabilis that revolutionized modern physics: The one on electrodynamics under the moving frame of reference launched the theory of relativity, the one on light quanta pioneered quantum physics, and the one on Brownian motion vindicated the reality of atoms. Titled "Über die von der molekularkinetischen Theorie der Wärme geforderte Bewegung von in ruhenden Flüssigkeiten suspendierten Teilchen" ("On the movement of small particles suspended in stationary liquids required by the molecular-kinetic theory of heat"), the paper started with a modest position. As he stated:[21]

> It is possible that the motions to be discussed here are identical with the so-called "Brownian molecular motion"; however, the data available to me on the latter are so imprecise that I could not form a definite opinion on this matter.

Einstein's first question was whether the presence of small but insoluble particles in a liquid made a thermodynamic difference from a chemical substance dissolved in a solvent. Consider, Einstein suggested, a liquid of a fixed volume and within this a sub-volume confined by a membrane. The sub-volume contained either a substance dissolved in the liquid (like sugar in his dissertation) or tiny but visible particles suspended in the liquid (like pollen powders that Brown observed or rubber crumbs that Exner prepared), and the membrane was permeable to the liquid but not to the dissolved substance or particles. When the membrane confined the solution, the higher concentration of the liquid within the sub-volume resulted in an outward osmotic pressure. When the membrane confined the suspended particles, according to classical thermodynamics, there should be no such osmotic pressure on the membrane, since the system's free energy should not depend on the specific positions of the suspended particles and the membrane.[22]

[21] Albert Einstein, "On the movement of small particles suspended in stationary liquids required by the molecular-kinetic theory of heat," in John Stachel, David Cassidy, Jürgen Renn, and Robert Schulmann (eds.), *The Collected Papers of Albert Einstein*, Volume 2 : *The Swiss Years, 1900–1909* (Princeton, NJ: Princeton University Press, 1989), 123; article translated by Anna Becker from "Über die von der molekularkinetischen Theorie der Wärme geforderte Bewegung von in ruhenden Flüssigkeiten suspendierten Teilchen," *Annalen der Physik*, 17 (1905), 549–560.

[22] Ibid., 123–124. Einstein did not provide a further explanation on this claim. But his argument seemed to be like the following: The thermodynamic characteristics of this system depended only on the total mass and properties of the liquid, the membrane, and suspended particles, and temperature and pressure, not the particular positions of the particles and the membrane. Thus, the free energy of the liquid-filling sub-volume containing suspended particles should be equal to the free energy of the same liquid-filling sub-volume when all the particles were deposited in one corner. In the latter

But the conclusion would be utterly different from the perspective of the molecular-kinetic theory of heat, Einstein argued. According to that theory, the dissolved substance in the liquid was nothing but a bunch of molecules undergoing random motions due to their collisions with smaller solvent molecules. This picture was essentially the same as suspended particles undertaking random motions in the liquid due to their collisions with liquid molecules. The molecules of the dissolved substance and the suspended particles differed only in size—the latter were much larger than the former. Like thermal motions of solute molecules pushing the membrane outward, thermal motions of suspended particles would also exert an additional osmotic pressure on the sub-volume's wall. Based on the molecular-kinetic theory of heat, Einstein deduced a formula for the osmotic pressure due to suspended particles.

This derivation followed closely Boltzmann's statistical-mechanical program. Einstein treated the liquid containing suspended particles as a mechanical system fully characterized by the positions and momenta of all particles and solvent molecules, $\{p_1, p_2, \ldots, p_l\}$. According to Boltzmann's combinatoric interpretation of entropy, as Einstein had shown in his previous papers in 1902–03 on the foundation of thermodynamics, the system's free energy was proportional to the logarithm of the size of the microcanonical ensemble, which was an integral of the Maxwell–Boltzmann distribution over the state-space, $F = -kT\log\int e^{-E/kT}dp_1 \ldots dp_l \equiv -kT\log B$ (T was temperature, E was total energy, and B was the integral over the whole state-space). This integral was impossible to carry out directly. But Einstein managed to demonstrate (in a similar way to his establishment of the empirical formula for black-body radiation via entropy calculations using the hypothesis of light quanta) that when the liquid was homogeneous and the movements of particles were purely random, B was proportional to V^n, where V was the liquid volume and n was the number of suspended particles. This result gave the free energy's log-dependence on volume. By taking the partial derivative of the free energy with respect to volume, Einstein obtained an expression for the system's osmotic pressure, which was the same as the form of the osmotic pressure caused by a dissolved substance, $p = nkT/V$.[23]

case, the deposited particles obviously had no influence on the sub-volume's free energy. Therefore, the thermodynamic characteristics within and without the membrane should simply be those of the same liquid. Since the thermodynamic conditions on both sides of the membrane were identical, no additional osmotic pressure would be present on it. According to "classical thermodynamics," introducing suspended particles into the liquid was similar to putting rocks in it, which would not change its physico-chemical properties at all.

[23] Ibid., 125–127.

After establishing the equivalence between suspended particles and dissolved molecules through the consideration of osmotic pressure, Einstein ventured further into a theory of diffusion for the particles. The excessive osmotic pressure showed that the suspended particles in a liquid tended to spread out. To him, this tendency was diffusion. Meanwhile, because the liquid was viscous, a suspended particle in it would receive a friction proportional to its speed. And the viscosity tended to slow down and constrain particles' movements. In a stable condition, these two tendencies—diffusion and friction—formed a dynamic equilibrium, implying that the system's free energy should be invariant under a virtual displacement of particles. Using this equilibrium, Einstein deduced an expression for the particles' coefficient of diffusion when they were spheres: $D = kT/6\pi\kappa r$, where κ was the liquid's coefficient of viscosity and r was the particle's radius. He obtained this expression via the so-called "Stokes law": the friction a liquid imposed on a sphere of radius r and velocity v was $F = 6\pi\kappa rv$.[24]

While the first three sections of Einstein's paper were devoted to the thermodynamic properties of suspended particles, he examined the detailed characteristics of particles' motions in the next section. This was also the place where he engaged the notion of stochastic processes. The particles undertook random, diffusive, Brownian-like motions. While their trajectories were different from one another, they should have the same statistical properties. To characterize the trajectories of such motions, Einstein invoked the particles' number density $f(x, t)$: the number of particles per unit volume at location x (supposing one dimension without losing generality) and time t. To find the form of this distribution, he assumed that Brownian motions were random enough that (i) the movements of two distinct particles were uncorrelated to each other and (ii) the movements of a particle at two distinct instants were uncorrelated to each other if the interval between the two instants exceeded a very short correlation time. These assumptions enabled him to formulate the dynamic equation of $f(x, t)$ in terms of the numbers of particles kicking in and out of a small volume at consecutive time instants.[25]

Einstein considered a time interval τ much shorter than any observable duration but much longer than the correlation time. Then the displacements of particles after τ followed a distribution φ, so that for a total of n particles, the number of particles with displacements between Δ and $\Delta + d\Delta$ was $dn = n\varphi(\Delta)\, d\Delta$. This distribution φ should be symmetric; it should fall to zero

[24] Ibid., 128–130.
[25] Ibid., 130.

except for small $|\Delta|$, and its integral over the entire Δ axis should be one. With this, he expressed the number of particles between x and $x + dx$ at time $t + \tau$, $f(x, t + \tau)\, dx$, as the net number of particles "kicking into" the spatial window between t and $t + \tau$ (i.e. sum of the number of particles located at $x + \Delta$ at t and deviating between $-\Delta$ and $-\Delta + d\Delta$ at $t + \tau$):

$$f(x, t + \tau)\, dx = dx \cdot \int_{-\infty}^{\infty} f(x + \Delta, t)\, \varphi(\Delta)\, d\Delta. \tag{5.1}$$

To further reduce equation (5.1), Einstein expanded f with a Taylor series of τ on the left side and a Taylor series of Δ on the right side, and dropped the second- and higher-order terms of τ as well as the third- and higher-order terms of Δ (for their negligible magnitudes). In addition, he used the condition of normalization $\int_{-\infty}^{\infty} \varphi(\Delta)\, d\Delta = 1$ and defined $(1/\tau) \int_{-\infty}^{\infty} (\Delta^2/2)\, \varphi(\Delta)\, d\Delta \equiv D$. Consequently, equation (5.1) became[26]

$$\frac{\partial f}{\partial t} = D \frac{\partial^2 f}{\partial x^2}. \tag{5.2}$$

Equation (5.2) had the same form as the well-studied diffusion equation in heat conduction and the kinetic theory of gas. To make its solution even simpler, Einstein redefined $f(x, t)$ so that it represented not the number density of particles at location x and time t but the number density of particles with *net displacement* x from time zero to time t. Under this new definition, $f(x, t = 0)$ should behave like an impulse at $x = 0$, since at the beginning all particles should have zero displacement. This initial condition led to a ready-made solution of equation (5.2):

$$f(x, t) = \frac{n}{\sqrt{4\pi D}} \frac{\exp\left(-\frac{x^2}{4Dt}\right)}{\sqrt{t}}. \tag{5.3}$$

The number density in equation (5.3) followed the typical Gaussian distribution in statistics. Yet, its standard deviation was $\sqrt{2Dt}$, meaning that the mean displacement (in magnitude) of a Brownian particle increased with the square root of time. In other words, the distribution of the particles' displacement spread out over time. The mean displacement of the Brownian particles was $\bar{\Lambda} = \sqrt{t}\sqrt{(kT/3\pi\kappa r)}$.[27]

[26] Ibid., 130–131.
[27] Ibid., 132–133.

After publishing "On the movement of small particles" in 1905, Einstein continued to work on Brownian motion, along with his simultaneous pursuits of the quantum theory of black-body radiation and the theory of relativity. The once-anonymous Swiss patent officer now received feedback from the world of physics. The 1905 paper on thermal motions attracted the attention of Henry Siedentopf, Zsigmondy's colleague at Zeiss, co-developer of the ultramicroscope, and a researcher of Brownian motion. Siedentopf informed Einstein that he and Gouy "had become convinced that the so-called 'Brownian motion' is caused by the random thermal motion of the liquid's molecules." Notwithstanding the existing experimental data that Einstein himself considered "meager," the plan of the new star of physics was to further theoretical investigations of the colloidal phenomenon. In 1906, he published another article on the same subject.[28]

Based on the theory he developed in 1905, Einstein's 1906 paper took on a more general issue than offering a molecular-kinetic account of Brownian motion: In a physical system whose elements were undertaking random thermal motions, an observable parameter would experience spontaneous changes of value, even though the system was under a thermal equilibrium. The theory of Brownian motion could help determine the average intensity of such random fluctuations. Suppose the parameter under consideration was a, and the physical system was again fully characterized by the huge set of molecular state-variables $\{p_1, p_2, ..., p_l\}$. When the system had no other forces and thermal diffusion was the only physical process in action, the probability that the parameter's value fell between a and $a + da$ was $Ada = \int_{da} Ce^{-E/kT} dp_1...dp_l$, where $e^{-E/kT}$ was again the Maxwell–Boltzmann factor for canonical ensembles, C was a constant, and the integral was carried in the domain in which the state-variables made the parameter value fall between a and $a + da$. He also showed that in this condition, A was independent of a. If the physical system comprised particles suspended in a frictionless liquid with an infinite volume, the parameter a was the particle's displacement, no external forces existed, and the observation was made infinitely long after the initial setup, then the particles would diffuse out indefinitely and the parameter a would have equal likelihood to be present at any displacement. And the mean square value of a would be infinite, suggested by equation (5.3) when time went to infinity.[29]

[28] Albert Einstein, "On the theory of Brownian motion," in John Stachel, David Cassidy, Jürgen Renn, and Robert Schulmann (eds.), *The Collected Papers of Albert Einstein*, Volume 2 : *The Swiss Years, 1900–1909* (Princeton, NJ: Princeton University Press, 1989), 180–191; article translated by Anna Becker from "Zur Theorie der Brownschen Bewegung," *Annalen der Physik*, 19 (1906), 371–381.

[29] Ibid., 180–182.

In a realistic situation, the physical system was usually subjected to a constrained force. This force could be a hard wall confining the volume of the substance, hydrodynamic friction, gravity, or electrostatic or electrodynamic actions. When the observable parameter was between a and $a + da$, this force would correspond to a potential energy $\Phi(a)$. And the associated probability would be a typical integral of the Maxwell distribution over the microcanonical ensemble times a Boltzmann factor for the potential: $Wda = \int_{da} C' e^{-(E+\Phi)/kT} dp_1 \ldots dp_l = A' e^{-\Phi(a)/kT} da$. And the standard deviation, or mean square, of a would be

$$\langle a^2 \rangle = \frac{\int_{-\infty}^{\infty} A' a^2 e^{-\Phi(a)/kT} da}{\int_{-\infty}^{\infty} A' e^{-\Phi(a)/kT} da}. \tag{5.4}$$

Unlike the unbounded frictionless fluid with no external forces, the suspended particles in the liquid with a constrained force had a finite mean square for its observable parameter a. If a was a particle's displacement x and all particles were subjected to an elastic force $F = -Mx$ that tended to pull them back to their original points, then Einstein showed that $\langle x^2 \rangle = kT/M$, as a result of the balance between the outward thermal diffusion and the inward elastic pull. If a was x and all particles in the liquid were subjected to friction so that $F = -v/B$ (where v was the particle's speed and B was defined by Einstein as the coefficient of mobility), then Einstein demonstrated from equation (5.4) that $\langle x^2 \rangle = 2BkTt$, where t was the elapsed time. This was consistent with his results in 1905 that $\langle x^2 \rangle = 2Dt$ and $D = kT/6\pi\kappa r$, provided $B = 1/6\pi\kappa r$. Einstein went further to illustrate that this theory applied not only to suspended particles' mean square displacement but also to their mean square angular shift for rotational Brownian motion.[30]

Einstein's work on Brownian motion is often memorized as a milestone toward the eventual vindication of atoms' reality. In Walter Isaacson's best-selling biography of Einstein, the author commented on the scientist's 1905–06 research on thermal agitations:

> Within months, a German experimenter named Henry Siedentopf, using a powerful microscope, confirmed Einstein's predictions. For all practical purposes, the physical reality of atoms and molecules was now conclusively proven.

[30] Ibid., 183–190.

Isaacson also cited the contemporaneous physicist Max Born's remark:[31]

> At the time, atoms and molecules were still far from being regarded as real.... I think that these investigations have done more than any other work to convince physicists of the reality of atoms and molecules.

In his speech at the ceremony of the 1926 Nobel Prize honoring his 1908 experimental confirmation of Einstein's Brownian-motion theory, the French scientist Jean Perrin dedicated the subject to "discontinuous structure of matter."[32] In academic history of science, Maiocchi sketched the standard narrative about the development of Brownian-motion research from Brown's discovery and early explorations, through Einstein's theory to confirm molecules' existence in the early 1900s, toward Perrin's experimental corroboration in the late 1900s. Bigg also examined the experimental phase of this development. In this narrative, Einstein's theoretical breakthrough in 1905 was his identification of particles' net displacements as the objective of empirical measurement and statistical analysis. This avoided a discrepancy in previous works—which was a fictitious construction by later scholars as Maiocchi found—between theoretical prediction and experimental data on particles' velocities.[33]

This historiography has merits. Einstein himself admitted his plan of testing the molecular-kinetic theory of heat at the beginning of his 1905 paper. His focus on a particle's displacement instead of velocity has indeed been considered an important difference from, say, Nägeli's and Exner's treatments, and made easier experimental design for quantitative measurement. In the context of noise, however, Einstein's theory of Brownian motion was more significant as a debut of theoretical studies of random fluctuations, both as an observable effect and as a mathematical entity. Although nineteenth-century statistical physicists understood thermodynamic phenomena in terms of collective behavior of randomly moving molecules, they focused almost exclusively on stable occurrences under thermodynamic equilibrium when individual variations in the molecular-kinetic picture were "smoothed out"

[31] Walter Isaacson, *Einstein: His Life and Universe* (New York: Simon and Schuster, 2007), 106.

[32] Jean Baptiste Perrin, "Discontinuous structure of matter," Nobel Lecture in Physics, December 11, 1926, from the Nobel Prize website: https://www.nobelprize.org/prizes/physics/1926/perrin/lecture/ (accessed April 8, 2022).

[33] According to the incorrect standard narrative, the theoretical prediction of the particles' mean velocity was much larger than the experimental measurement. Retrospectively, such a measurement, which took velocity to be distance divided by time, did not give the real particle speed, since the actual particle trajectory was much more fractal than what was observed under microscope. For a review of this narrative and its problems, see Maiocchi, "Case of Brownian motion" (1990), 257–263. For Perrin's experimental research on Brownian motions, see Bigg, "Evident atoms" (2008), 312–322.

in average over many particles. While researchers on Brownian motion before the twentieth century did suggest the connection between the phenomenon and thermal molecular agitations, they did not produce a consistent theory from this thought. Einstein's work in 1905 demonstrated through statistical mechanics that random fluctuations at microscopic levels were not smoothed out in average over a large number, but rather led to observable and macroscopic effects. He also showcased a mathematical approach to grapple with the probabilistic characteristics of such fluctuations, which has been commonly marked as a beginning of stochastic-process theory.

Einstein was not the only individual to launch the studies of random fluctuations. As he was working on a theory of Brownian motion, another physicist Marian Smoluchowski in Austrian Poland independently examined the same problem. In contrast to the Swiss patent officer's heavy reliance on Boltzmannian statistical mechanics that treated fluctuations in terms of entropy, free energy, and transport process, the Polish professor dealt with Brownian motion in terms of random walks, Markov chains, conditional probability, and visible trajectories.

Smoluchowski on Brownian Motions and Random Fluctuations

A Statistical Physicist in Turn-of-the-Century Vienna

In a sense, Smoluchowki was Einstein's doppelganger, had the latter fared better in academic employment at the debut of his career, stayed in statistical physics, and refrained from venturing into disruptive relativistic spacetime and light quanta. These two individuals were similar to each other not only because they both subscribed to the statistical-mechanical program and were heavily influenced by Boltzmann, but also because they both explored phenomena that could not be explained by classical thermodynamics and thus launched the studies of random fluctuations.

Marian Smoluchowski was born into a Polish ruling-class family of Habsburg Vienna six years after the Austro-Prussian War. His father was a high official in Emperor Franz Joseph's privy council. He spent his formative years in Austria–Hungary's capital, then an epicenter of Europe's arts, literature, and science. In 1890, he matriculated at the University of Vienna to study physics under the supervision of Josef Stefan (Boltzmann's doctoral supervisor) and Franz Exner (Sigmund's brother and Felix's uncle), and he obtained a Ph.D. in

1895. He spent the following two years as a research fellow at Gabriel Lippmann's laboratory in Paris, Lord Kelvin's in Glasgow, and Emil Warburg's in Berlin. In 1899, he served as a Privatdozent (private lecturer) in Lwów (Lemberg in German), then a Polish-Lutheran city in Austrian Galicia and which is Lviv in the Republic of Ukraine today. A year later, Smoluchowski was appointed as a faculty member at the University of Lwów, where he developed a career in theoretical physics. In 1913, he moved to Jagellonian University in Kraków, then the largest city in Austrian Poland. He was elected the university's Rector (President) in 1917, right before his sudden death due to an acute disease.[34]

Smoluchowski was part of an influential intellectual community orbiting around *fin-de-siècle* Vienna that grappled with the issues of uncertainty, indeterminism, probability, and statistics. Physicist Elliott Montroll pointed out the presence of a "Vienna School of statistical thought" from the mid-nineteenth century to the mid-twentieth century. According to him, this "school" originated from Christian Doppler, the first director of the University of Vienna's Institute of Experimental Physics primarily known for his discovery of the Doppler effect about the frequency shift of waves emitted from a moving source. Doppler's protégés Josef Stefan and Johann Josef Loschmidt, Loschmidt's student Franz Exner, Stefan's student Ludwig Boltzmann, and Marian Smoluchowski were core members of this family tree. But it also included renowned Austrian figures of science, such as the positivistic philosopher Ernst Mach, founder of genetics Gregor Mendel, pioneer of quantum mechanics Erwin Schrödinger, and discoverer of nuclear fission Lise Meitner. As foreign students came to the university for studies and its graduates established academic careers abroad, the genealogy of this Vienna School spread from Austria–Hungary to countries like the Netherlands, US, and China. The international intellectual offspring of this network included, to name a few, Paul Ehrenfest, George Uhlenbeck, Samuel Goudsmit, Chen-Ning Yang, Tsung-Dao Lee, John Wheeler, Richard Feynman, and Kip Thorne.[35]

Philosopher Michael Stöltzner also identified a "Vienna indeterminism" as a distinct philosophical approach to the relationship between statistical

[34] Roman Smoluchowski, "Life of Marian Smoluchowski," and "Chronological table of Marian Smoluchowski's life," both in Subrahmanyan Chandrasekhar, Mark Kac, and Roman Smoluchowski (eds.), *Marian Smoluchowski: His Life and Scientific Work* (Warsaw: Polish Scientific Publishers, 1999), 9–14, 129–130; Arnold Sommerfeld, "Zum Andenken an Marian Smoluchowski," *Physikalische Zeitschrift*, 18 (1917), 533–539.

[35] Elliott Montroll, "On the Vienna School of statistical thought," *American Institute of Physics Conference Proceedings*, 109:1 (1984), 1–10.

properties, causality, and reality shared by Vienna-based thinkers—Mach, Boltzmann, and Franz Exner—at the end of the nineteenth century. While the three physicists differed in their views about the ontological status of atoms, according to Stöltzner, they held a common attitude against the then-popular Kantian position stipulating a definite connection between reality and a priori categories, and instead held a much more functional view about the relationship between empirical reality, observed data, and scientific theories. To them, the world was intrinsically uncertain, and the laws involving the basic constituents of nature were statistical.[36]

Historian Deborah Coen has contextualized the Vienna scientists' strong inclination to statistical and probabilistic thinking in terms of the imperial capital's political and social circumstances at the turn of the century. Focusing on the Exners, Austria–Hungary's most powerful academic family during this period, as representative of a broad sociocultural movement, Coen demonstrated that highlighting the centrality of statistics for achieving scientific understanding or decision-making in what was conceived as a highly uncertain world became Vienna progressive intellectuals' strategy to advance reform and modernize the empire. The liberalism these intellectuals embraced called for a rejection of religious dogmatism associated with the Catholic Church or regional nationalism, and their replacement with reason, science, and enlightenment. Skeptical of absolute certainty, the Austrian liberals used statistics and probabilistic calculus as a rational guide to deal with complex natural and social problems ranging from meteorology to census.[37]

As a member of this Vienna-centered community, Smoluchowski did not uphold an explicit political or philosophical agenda. His son Roman remarked that the insight he (Marian) gained through his father into the "intricacies and intrigues of domestic and foreign politics" gave him "a lifelong distaste for politics, even the politics of university life." Despite his university administrative positions, he was never as involved as the Exners in Austria–Hungary's academic or educational reforms. He was likely sympathetic to the liberal cause, but did not seem to tie its thinking to his scientific research.[38] Similarly, he did not develop a clear position or articulated reflections on the philosophical underpinnings of statistical approaches to understanding the material world.

[36] Michael Stöltzner, "Vienna indeterminism: Mach, Boltzmann, Exner," *Synthese*, 119:1 (1999), 85–111.

[37] Deborah Coen, *Vienna in the Age of Uncertainty: Science, Liberalism, and Private Life* (Chicago, IL: University of Chicago Press, 2007), in particular, 1–32, 183–226.

[38] Roman Smoluchowski, "Life of Marian Smoluchowski" (1999), 10.

In a tribute to Smoluchowski's contribution to statistical thought in physics, the Polish-American mathematician Mark Kac stated that:[39]

> Like Maxwell, Smoluchowski was a pragmatist and he was less concerned with *why* probability is introduced into kinetic theory than with *how* it can be *used* to explain known phenomena and to predict new ones. Unlike Boltzmann to whom probabilistic and statistical arguments were a line of defense against logical assaults on his theory, Smoluchowski, in the spirit of Maxwell, turned them into everyday working tools of physics.

Similarly, Einstein asserted in a eulogy to Smoluchowski that:[40]

> Smoluchowski's scientific effort concerned the molecular theory of heat. In particular, his interest was directed toward those consequences of molecular kinetics that could not be understood from the viewpoint of classical thermodynamics; because he felt that only from this perspective the strong resistance among the contemporaries of the end of the nineteenth century against the molecular theory would be overcome.

To Einstein, the significance of Smoluchowski's work was its affirmation of the atomic-molecular theory of matters through studies of phenomena that deviated from the predictions of classical thermodynamics, viz., random fluctuations, which was clearly within the technical domain of statistical physics at the turn of the century.

These assessments of Smoluchowski as a researcher of practicality and technicality in statistical physics nonetheless did not degrade the broader influence of his contribution. In fact, we may argue that he played an important role in the scientific and engineering approaches to noise precisely because his technical work opened up a new direction for representing and manipulating random fluctuations as stochastic processes.

Smoluchowski's diverse research in the first years of his career was shaped by where he was: acoustic properties of elastic soft materials and aerodynamic

[39] Mark Kac, "Marian Smoluchowski and the evolution of statistical thought in physics," in Subrahmanyan Chandrasekhar, Mark Kac, and Roman Smoluchowski (eds.), *Marian Smoluchowski* (1999), 16–17, italics in the original.

[40] Albert Einstein, "Marian Smoluchowski," *Die Naturwissenschaften*, 50 (1917), 107–108. Einstein's original paragraph is as follows:

"Smoluchowskis wissenschaftliches Ringen galt der Molekulartheorie der Wärme. Insbesondere war sein Interesse auf diejenigen Konsequenzen der Molekularkinetik gerichtet, welche vom Standpunkt der klassischen Thermodynamik aus nicht verstanden werden können; denn er fühlte, daβ nur von dieser Seite her der starke Widerstand zu überwinden war, den die Zeitgenossen am Ende des 19. Jahrhunderts der Molekulartheorie entgegenstellten."

properties of fluids (from his Ph.D. thesis at the University of Vienna and its extension), thermal radiation (from his work at Lippmann's laboratory in Paris), and the Roentgen ray and radioactivity (from his work at Kelvin's laboratory in Glasgow). He began to pay serious attention to thermodynamics when he studied thin air's heat conductivity at Warburg's laboratory in Berlin. In 1898, he published a theoretical account of a discontinuity in the temperature gradient of a gas separated into two parts by a partly insulated wall. In this theory, he employed the molecular-kinetic theory of gas and deduced a mathematical formulation to describe the different molecular motions on both sides of the wall that gave rise to such a discontinuity. This work marked the beginning of his lifelong commitment to the molecular-kinetic theory of matter, heat, and energy.[41]

From Density Fluctuation to Mean Free Path

Smoluchowski's exploration of the kinetic theory of matters at the beginning of the twentieth century led him to pay close attention to random fluctuations due to molecular statistics. Since Maxwell, followers of statistical mechanics had known that gas molecules at a non-zero temperature undertook motions with different speeds and along different directions, and the dynamic states of these molecules could only be characterized collectively with statistical distributions. Yet, Smoluchowski was among the first to note the macroscopic and irregular variations on observable quantities due to such molecular thermal motions. He first put this idea in writing in 1903 when he, along with several Austrian scientists, prepared a Festschrift to celebrate Boltzmann's sixtieth birthday.[42]

Published in February 1904, Smoluchowski's article in Boltzmann's Festschrift concerned the irregularities of gas density and their influences on the gas's entropy and state equation. His point of departure was that according to the kinetic theory of matters, the density of a gas might not be homogeneous or constant even under a thermal equilibrium. This assertion resulted from a simple rule of combinatorics: The physical system under investigation was a body of gas with N molecules confined within a volume V. Smoluchowski

[41] Marian Smoluchowski, "Über den Temperatursprung bei Wärmeleitung in Gasen," *Sitzungsberichte, Kaiserliche Akademie der Wissenschaften, Wien, Mathematisch-Naturwissenschaftliche Klasse,* 107: Abt. I a (1898), 304–329.

[42] Marian Smoluchowsi, "Über Unregelmäβigkeiten in der Verteilung von Gasmolekülen und deren Einfluβ auf Entropie und Zustandgleichung," in *Festschrift Ludwig Boltzmann* (Leipzig: Johann Ambrosius Barth, 1904), 626–641.

considered a tiny volume v within V. The probability that a molecule was found within v would be v/V. And the probability that n particular molecules were found within v while the other $N - n$ molecules were outside v would be $(v/V)^n(1 - v/V)^{N-n}$. Removing the constraint for *n particular* molecules and considering the probability that *n arbitrary* molecules were within v (which could be any possible combinations), he obtained from combinatorics the probability

$$W = \binom{N}{n}\left(\frac{v}{V}\right)^n\left(\frac{V-v}{V}\right)^{N-n} = \frac{N!}{n!\,(N-n)!}\left(\frac{v}{V}\right)^n\left(\frac{V-v}{V}\right)^{N-n}, \tag{5.5}$$

where $\binom{N}{n} = \frac{N!}{n!(N-n)!}$ represented the total number of combinations for selecting n members out of a set of size N, and $n! = n \cdot (n-1) \cdot \ldots. \cdot 1$ was the factorial of n. Employing the standard asymptotic approximation to the factorial $n! \cong \sqrt{2\pi n}(n/e)^n$ for large n (known as Sterling's formula, where $e = 2.71182281\ldots$ was the base of natural logarithm), Smoluchowski replaced (5.5) with a simpler form:

$$W = \frac{N!}{n!\,(N-n)!}\left(\frac{v}{V}\right)^n\left(\frac{V-v}{V}\right)^{N-n} \cong \left(\frac{\nu}{V}\right)^n \frac{e^{n-\nu}}{\sqrt{2\pi n}}, \tag{5.5'}$$

where $\nu = Nv/V$ was the average number of molecules within the tiny volume v.[43]

This result from the molecular-kinetic hypothesis differed clearly from classical thermodynamics, since the former indicated that the gas (molecular number) density within volume v could change randomly and (5.5') specified the probability distribution of this density fluctuation, whereas the latter stipulated that the gas density under a thermal equilibrium was a constant $N/V = \nu/v$ over the entire volume. Representing this deviation from average as $n = \nu(1 + \delta)$, Smoluchowski turned (5.5') into the probability for the deviation between δ and $\delta + d\delta$:

$$W(\delta)\,d\delta = \sqrt{\frac{\nu}{2\pi}}e^{-\nu\delta^2/2}\,d\delta. \tag{5.5''}$$

Equation (5.5") corresponded to a typical Bell curve familiar to nineteenth-century statisticians. From this distribution, he obtained the mean magnitude

[43] Ibid., 626–627.

of the percentage deviation δ to be $1/\sqrt{2\pi\nu}$. This implied that the percentage deviation of the number of gas molecules within the sub-volume v decreased with the average number of gas molecules within it. When v was about 1 cm^3, Smoluchowski noted, a rough estimate for the number of gas molecules within it from experimental data was $\nu \cong 6 \times 10^{19}$, which corresponded to a mean deviation of 0.5×10^{-8}%—a percentage way too small to be observed. But when v was about 0.2 μm^3, a considerably tinier volume that could only be visible under a microscope but was still much larger than the atomic scale, $\nu \cong 5 \times 10^5$ and the mean deviation was about 0.5%, which was much more discernible than 0.5×10^{-8}%. Thus, the density fluctuation was more visible on a smaller scale.[44]

In fact, when the sub-volume v was so tiny that the condition for large n did not apply and the volume contained only a few molecules, the probability in equation (5.5) could not be reduced asymptotically to the Gaussian distribution. Rather, the presence of molecules within v followed the so-called Poisson distribution: $W(n) = \nu^n e^{-n}/n!$ for $n = 1, 2, 3, \ldots$.[45]

Moreover, this density fluctuation implied a modification of certain central features in thermodynamics. The thermodynamicists in the late nineteenth century derived expressions for a gas's entropy in terms of its temperature and density. Since Smoluchowski predicted the gas density to be a random fluctuation with a probabilistic distribution, the classical formula for entropy had to be revised accordingly, which led to the appendage of more terms in the entropy expression. To him, the classical formula corresponded to "macroscopic entropy" when the scale of observation was large and the effect of gas density fluctuation was not discernible, whereas the additional terms corresponded to "microscopic entropy" when the scale of observation was small and the effect of density fluctuation was salient. Thus, Smoluchowski contended that the most updated equation of state in thermodynamics—the so-called van der Waal equation—had also to be revised according to the fluctuation of gas density.[46]

Smoluchowski's interest in the thermodynamic properties of gases was soon extended from density fluctuations to the mean free path. The mean free path referred to the average distance a particle (an atom, a molecule, or a bigger corpuscle) traversed freely between two consecutive collisions in a fluid. This notion was first proposed in 1857 by Rudolph Clausius in his kinetic theory of

[44] Ibid., 627.
[45] Ibid., 628–629.
[46] Ibid., 629–641.

gas. Maxwell revised Clausius's formula for the mean free path two years later and connected the new expression to the determination of a gas molecule's size. To Smoluchowski, the nineteenth-century kinetic theorists' numerical estimates of the mean free path were too inexact to compare with experimental data for fluids' viscosity, diffusion, and thermal conductivity. In 1906, he published a paper in *Bulletin of the Academy of Science in Kraków* on a statistical theory of the mean path in a gas and its relationship to diffusion.[47]

The physical system Smoluchowski tackled was once again a body of gas molecules undertaking thermal motions. Since molecular motions were random, the length of the free path differed from molecule to molecule and from time to time. Like the density fluctuation, the apparently irregular free path between collisions could be characterized in statistical ways. To do so, Smoluchowski considered three modeling scenarios from the simplest to the most complicated. First, he considered a one-dimensional case when all molecules had a constant speed c; molecular collisions were random and independent; and the chance of collision increased with elapsed time. Under this highly simplified idealization, the probability for a molecule to move without any collisions during a period t decayed exponentially with t, or explicitly, $p_0(t) = e^{-ct/\lambda}$, where λ was the gas's mean free path. The probability for a molecule to encounter one collision within t was the sum of probabilities when a collision occurred within the time window $[\theta, \theta+d\theta]$ for θ between 0 and t while no collision occurred between θ and t. This sum was $p_1(t) = \int_0^t \frac{c}{\lambda} e^{-c\theta/\lambda} p_0(t-\theta)\, d\theta = (ct/\lambda)\, e^{-ct/\lambda}$. Similarly, $p_2(t)$ could be computed from $p_1(t)$, and so on. From this iterative manner, the probability for a molecule to encounter n collisions was $p_n(t) = (1/n!)\,(ct/\lambda)^n e^{-ct/\lambda}$. This was the Poisson distribution for events occurring at random times of arrival. When t was on the scale of lab observation, the average number of collisions $N = ct/\lambda$ was large, and the Poisson distribution converged asymptotically via Sterling's formula to the normal distribution $p_n \cong 1/\sqrt{2\pi N} e^{-N\delta^2/2}$, where $N/n = 1 + \delta$.[48]

Second, each molecule had a constant free path λ but could go along any directions between two collisions in three dimensions with equal chances. Defining the origin as a molecule's location right after a collision, it would encounter another collision when it scattered onto the surface of a sphere with radius λ and center at the origin. The probability for the molecule to collide

[47] Marian Smoluchowski, "Sur le chemin moyen parcouru par les molécules d'un gaz et sur son rapport avec la théorie de la diffusion," *Bulletin de l'Académie des Sciences de Crocovie: Classe des Sciences Mathématiques et Naturelles* (1906), 202–213; reprinted in Wladysiaw Natanson and Jan Stock (eds.), *Pisma Mariana Smoluchowskiego* (*The Works of Marian Smoluchowski*) (Kraków: Academy of Sciences and Letters, 1924), 479–489.

[48] Ibid., 480–481.

anywhere on the sphere was the same. Projecting this three-dimensional scenario onto one dimension, Smoluchowski obtained that the probability for a molecule after leaving the origin to encounter its first collision between displacement x and $x + dx$ was $p_1(x)\,dx = dx/2\lambda$ for $-\lambda \le x \le \lambda$ and 0 otherwise. The molecule's probability to have its second collision within $[x, x+dx]$ was the sum of all probabilities when the first collision was at ξ between $x - \lambda$ and $x + \lambda$ while the second occurred after traveling from ξ to x. This gave $p_2(x)\,dx = (dx/2\lambda)\int_{x-\lambda}^{x+\lambda} p_1(\xi)\,d\xi$. Iteratively, he obtained the probability of nth collision $p_n(x)\,dx = (dx/2\lambda)\int_{x-\lambda}^{x+\lambda} p_{n-1}(\xi)\,d\xi$.[49] This form was more complex than the Poisson distribution in the first case. To compute, Smoluchowski employed a Fourier analysis and expressed $p_n(x) = \left(\frac{1}{\pi}\right)\int_0^\infty \left(\frac{\sin(q\lambda)}{q\lambda}\right)^n \cos(qx)\,dq$. When n was large, this expression converged asymptotically to the familiar normal distribution $p_n(x) \cong \sqrt{3/(2\pi n\lambda^2)}e^{-3x^2/2n\lambda^2}$. Accordingly, he obtained an expression for the molecule's mean displacement along the x-direction after n collisions, $r_n = \sqrt{8n/3\pi}\lambda$.[50]

Third, Smoluchowski considered a most realistic case in which a molecule could encounter a collision along any directions in three dimensions with an equal probability, but its free path between two collisions was a random variable that followed a Poisson distribution, not a constant. Following the same reasoning as in the first case, he obtained the probability that a molecule's first collision occurred within $[x, x+dx]$, $p_1(x)\,dx = (dx/2\pi)\int_x^\infty \left(e^{-\rho/\lambda}/\rho\right)d\rho$. The probability that the molecule's second collision occurred within the same window was the sum of probabilities when the first collision occurred at z while the second occurred after displacement $x - z$. This gave $p_2(x)\,dx = dx\int_{-\infty}^{\infty} p_1(z)\,p_1(x-z)\,dz$, and the general form $p_n(x)\,dx = dx\int_{-\infty}^{\infty} p_{n-1}(z)\,p_1(x-z)\,dz$, the probability of the nth collision.[51] In this expression, p_n was a "convolution" of p_{n-1} and p_1 from the linear-system theory in physics and engineering. A central result from this theory gave the relation $P_n = P_{n-1}{\cdot}P_1$, where P_1, P_{n-1}, and P_n were Fourier transforms of p_1, p_{n-1}, and p_n. Employing this relation recursively, Smoluchowski obtained $P_n = P_1{}^n$. Invoking Fourier analysis led to $p_n(x) = \left(\frac{1}{\pi}\right)\int_0^\infty \left[\frac{\varphi(q)}{q\lambda}\right]^n \cos(qx)\,dq$, where $\varphi(q) = \int_0^\infty \sin(q\alpha)\,e^{-\alpha/\lambda}/\alpha d\alpha$. When n became large, Smoluchowski showed again that the probability density converged asymptotically to a normal distribution $p_n(x) \cong \frac{1}{2\lambda}\sqrt{\frac{3}{n\pi}}e^{-\frac{3x^2}{4n\lambda^2}}$. Taking the large n as a measure of macroscopically elapsed time via the relation $t = n\lambda/c$, he represented a

[49] Ibid., 482–483.
[50] Ibid., 483–484.
[51] Ibid., 484–485.

molecule's probability to lie between x and $x + dx$ at time t as

$$p_t(x) \cong \frac{\beta}{\sqrt{\pi t}} e^{-\frac{\beta^2 x^2}{t}}, \tag{5.6}$$

where $\beta = \sqrt{3/4c\lambda}$. He further showed that a molecule's mean displacement was $\langle x \rangle = (1/2\beta)\sqrt{t/\pi}$, and the mean square of the distance between a molecule at time t and the origin was $\langle r^2 \rangle = 3t/2\beta^2$.[52]

Next, Smoluchowski applied this theory of the mean free path to modeling diffusion. The density function in (5.6) gave the probability of a molecule located at the origin at time 0 and displaced around x at time t. If the object under consideration was not a single molecule but a body of gas, then (5.6) depicted how the gas spread or, more precisely, the evolution of its density distribution. This described the process of diffusion. For a body of gas with an initial number-density distribution $f_0(x)$ at time 0, Smoluchowski utilized (5.6) to obtain its density distribution at a later time t:

$$f(X, t) = \frac{\beta}{\sqrt{\pi t}} \int_{-\infty}^{\infty} f_0(x)\, e^{-\frac{\beta^2 (X-x)^2}{t}} dx. \tag{5.7}$$

Equation (5.7) showed that the gas density spread out with time, and the way it spread followed a Gaussian distribution whose width increased with time. This matched the prediction from the diffusion theory. Also, (5.7) indicated that the diffusion coefficient was $D = 1/4\beta^2 = c\lambda/3$.[53]

Smoluchowski's works on density fluctuations and mean free paths set the tone for his scientific career and placed him deep in the research of statistical physics. They were preoccupied with fluctuation phenomena in gases and liquids that could not be accounted for by classical thermodynamics. His theoretical methods were built upon detailed examinations of a few kinetic-microscopic scenarios: molecules present in, entering, and leaving a tiny volume; a particle undergoing a zigzag motion due to kicks from other particles. The mathematical tools he employed to grapple with these scenarios had a strong Boltzmannian flavor, for they calculated the number of different combinations with equal chances and asymptotic approximations of probability into normal distributions. But he also came up with original mathematical techniques to facilitate such computations, such as the treatment of probability density functions with Fourier analysis. In contrast to focusing on the models

[52] Ibid., 486–487.
[53] Ibid., 487–488.

most relevant to physical reality, moreover, he often looked at different idealizations of the same physical problem, approached these idealizations with different mathematical means, and demonstrated the consistency between the results. These characters had profound influences on his later works on Brownian motion and other random fluctuations.

Looking into Brownian Motion

It is obvious to see the connection between Smoluchowski's studies of gaseous density fluctuations and the mean free path on the one side and his interest in Brownian motion on the other. Both density fluctuations and Brownian motion of small yet visible particles were random fluctuations that could not be explained adequately by thermodynamics but were better accounted for by the statistical molecular-kinetic theory of matters. The mean free path and diffusion in a gas described the movements of molecules as they collided with one another, while the Brownian motion of a pollen powder, a liquid droplet, or an air bubble was the similar movement of an object—albeit with a much larger size than the atomic scale—that resulted from molecular collisions.

Smoluchowski began to pay attention to the phenomenon in 1900 when he learned about Felix Exner's ongoing experiment to measure the speed of Brownian particles in a liquid at the University of Vienna. The hand-drawn diagrams Exner sent to Smoluchowski on the irregular displacement of a Brownian particle over time—known as the "Krix-Krax"—would become a visual highlight for the latter's demonstration of macroscopic fluctuations caused by microscopic randomness (Figure 5.1).[54] In Smoluchowski's own account, the conclusion of Einstein's papers on Brownian motion "completely agreed with the results that I had obtained a few years ago, following a totally different line of thought."[55] Smoluchowski's kinetic theory of gas molecules' mean free path and their diffusion took the same approach to his molecular-kinetic theory of the Brownian particles. Thus, it is quite likely that his work on Brownian motion was an outcome of an ongoing research program, and he might have come up with a kinetic-molecular theory of Brownian motion independent of and even earlier than Einstein's. But the professor at Lwów

[54] Kac, "Marian Smoluchowski and the evolution of statistical thought in physics" (1999), 18.

[55] "Die Ergebnisse derselben stimmen nun vollkommen mit einigen Resultaten überein, welche ich vor mehreren Jahren in Verfolgung eines ganz verschiedenen Gedankenganges erhalten hatte"; from Marian Smoluchowski, "Zur kinetischen Theorie der Brownschen Molekularbewegung und der Suspensionen," *Annalen der Physik*, 21 (1906), 756.

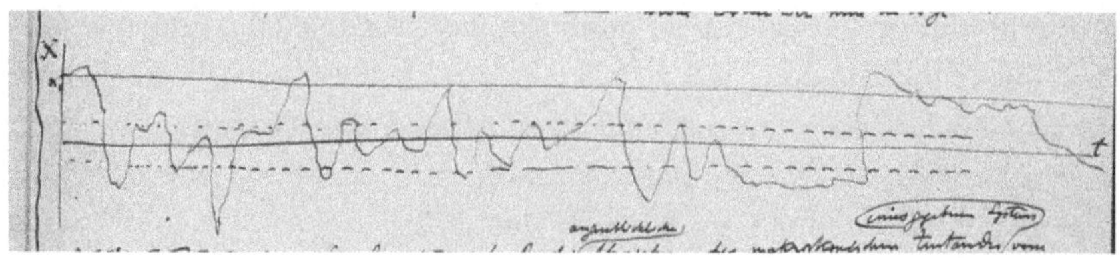

Figure 5.1 Smoluchowski's hand drawing of the "Krix-Krax," the irregular movement of a Brownian particle. Folder 9357, "Goettinger Referat. Gueltigkeitsgrenzen des zweiten Hauptsatzes des Waermetheorie," 13. Marian Smoluchowski Papers, Manuscript Collections, MS9397, Library of Jagellonian University, Kraków, Poland.

did not publish his work on this topic until he saw the Swiss patent officer's papers in 1905 and 1906. In July 1906, Smoluchowski submitted an article to the Krákow Academy of Science, which was published in *Annalen der Physik* in the same year. This paper laid out the ground for his kinetic-molecular theory of Brownian motion.

While both Einstein's and Smoluchowski's papers attributed the origin of suspended particles' irregular movements to molecules' random thermal motions, the two works exhibited starkly different styles of reasoning. Without engaging any specific experimental data, Einstein started with an intrinsic contradiction between classical thermodynamics and the kinetic-molecular theory in terms of their distinct theoretical implications for the effect of suspended particles on a liquid's osmotic pressure. And he followed strictly the canonical approach of Boltzmannian statistical mechanics: representing molecules' motions in high-dimensional phase space, lumping microdynamics into macroscopic partition-function-like quantities, coping with thermodynamic relations, and computing the probability density function. These fit squarely historian Suman Seth's characterization of Einstein's work as "physics of principles."[56]

In contrast, Smoluchowski demonstrated his close familiarity with the cutting-edge laboratory work on Brownian motion right at the beginning of his paper. He summarized the central experimental results from German, French, Austrian, Italian, and British scientists in the past thirty years. A core finding from these empirical studies, as he asserted, was the universality and stability of the effect. The suspended particles in a liquid undertook irregular and persistent agitations for a large variety of liquid substances, and no matter whether

[56] Suman Seth, *Crafting the Quantum: Arnold Sommerfeld and the Practice of Theory: 1890–1926* (Cambridge, MA: MIT Press, 2010), 13–46.

the particles were solid crumbles, liquid droplets, or gaseous bubbles. These motions did not stop or calm down. And they were not changed by numerous manipulations—illuminating the liquid with light of different colors and intensities or subjecting it to other forms of radiation, putting the liquid container in the dark, storing it for days or boiling it for hours, placing it in a vibration-free frame, controlling its temperature gradient to remove convection, or reduction of liquid layer thickness to a fraction of a millimeter, to name a few. The insensitivity of the phenomenon to these operations strongly suggested to Smoluchowski that Brownian motion was not caused by external factors, but rather by the liquid's inherent physical conditions.

What scientists did know, as he noted, was that the Brownian motions of suspended particles were dependent on the fluid's viscosity: the stickier the liquid, the slower the particles' zigzag movements. The size of the particles mattered, too. Smoluchowski cited Felix Exner's experimental data in 1901—which he deemed the only "absolute measurements" ("absolute Messungen") providing reliable quantitative results—and indicated that the suspended particles under microscope had higher speeds when the particles were smaller. At 23°C, Exner measured a speed of 0.00027 cm/sec when the particle's diameter was 0.00013 cm, 0.00033 cm/sec for a diameter of 0.00009 cm, and 0.00038 cm/sec for a diameter of 0.00004 cm. Moreover, the experimenters generally agreed on Brownian motion's dependence on the liquid temperature—the hotter the liquid, the speedier the suspended particles. Exner reported an increase of measured particle speed from 0.00032 cm/sec at 20°C to 0.00051 cm/sec at 71°C.[57]

Ruling out these external factors, Smoluchowski was convinced that the internal heat energy of the liquid was the true cause of Brownian motion, and the molecular-kinetic theory would offer a satisfactory explanation. If we observe the Brownian motions of suspended particles under a microscope, we will "get the immediate impression that the motions of the fluid molecules must just look like that."[58] This was a vivid depiction of Brownian motion as a visual manifestation of molecular thermal agitations. Smoluchowski acknowledged the fact that different versions of such a theory had been proposed in the nineteenth century, and they all encountered a challenge by Carl von Nägeli in 1879. From Nägeli's estimate, a water molecule at room temperature colliding with a particle of 10^{-4} cm diameter and unit density would impart a

[57] Smoluchowski, "Zur kinetischen Theorie der Brownschen Molekularbewegung und der Suspensionen" (1906), 757–761.

[58] "Wenn man die Brownsche Bewegung unter dem Mikroskop beobachtet, erhält man unmittelbar den Eindruck, daβ so die Bewegungen der Flüssigkeitsmoleküle aussehen müssen," ibid., 761.

velocity of 3×10^{-6} cm/sec on the particle. This speed was much smaller than the scale of Brownian motion and was certainly not discernible even under an ultramicroscope. Although successive impulses of this molecular order might add up to an observable effect, Nägeli contended that they should not accumulate additively to a large number, for liquid molecules collided into the much larger Brownian particle from all spatial directions with equal chances. Thus, the molecular collisions should cancel out on average, and the resultant effect of the collective collisions should not be markedly larger than that of a single collision.[59]

Smoluchowski's refutation of Nägeli's critique resorted to a statistical argument. In fact, the Polish physicist pointed out, Nägeli's reasoning had the same error as the claim that a player in a game of chance—such as tossing a coin—would not lose or gain more than a single stake on average. Rather, as Kac's account of Smoluchowski's comment had it, "good and bad luck do not cancel completely and the longer the game lasts, the greater is the average gain or loss."[60] To elaborate this point, Smoluchowski employed a probabilistic calculation. He considered the gambling of tossing a coin n times. In each toss, the player had a 50% chance of winning and a 50% chance of losing. Representing a winning as a move of a unit forward step and a losing as a move of a unit backward step, the time series of the gambling outcome was analogous to a one-dimensional random walk along a line. The player's net gain after n tosses could be represented as the displacement between the origin and the random walker's position after n steps. Since a single step could have two possibilities (+1 and −1) with equal probability, a sequence of n such steps would have 2^n possibilities. Smoluchowski supposed that among the n steps there were m forward moves and $n - m$ backward moves. Then there were $\binom{n}{m} = \frac{n!}{m!(n-m)!}$ combinations for this case. In other words, the probability for this case would be $\frac{1}{2^n}\binom{n}{m}$. And the case would correspond to a net displacement (or gain) of $m - (n - m) = 2m - n$. Summing all the cases with different m's, the mean value of the net displacement was $v = 2\sum_{m=n/2}^{n}\binom{n}{m}\frac{2m-n}{2^n} = \frac{n}{2^n}\binom{n}{n/2}$. When n became large, this mean value converged asymptotically to $\sqrt{2n/\pi}$. On average, the net gain (or loss) in a game of chance did not cancel out, but grew

[59] Ibid., 762.

[60] Ibid., 762; Kac, "Marian Smoluchowski and the evolution of statistical thought in physics" (1999), 18.

with the square root of the number of trials. This again was a manifestation of large-scale fluctuations due to small-scale randomness.[61]

The observation from this mathematical exercise gave Smoluchowski an opportunity to modify Nägeli's figure and bring it back to a more sensible level. In the Swiss-German botanist's estimate, a single molecule colliding with a particle of 10^{-4} cm diameter and unit density would impart a velocity of 3×10^{-6} cm/sec on it. Yet, when the particle was suspended in a liquid or gas, from Smoluchowski's count, it underwent not a single molecular collision per second, but rather 10^{20} collisions per second for a liquid and 10^{16} collisions for a gas. His results from the game of chance thus showed that even though most of the molecular collisions canceled out, the remaining collisions still caused detectable fluctuations. The velocity imparted on a particle due to molecular collisions should be amplified to $\sqrt{n} = 10^{10}$ times in a liquid or 10^8 times in a gas. This gave a Brownian particle's probable velocity at the order of 10^2 cm/sec in air and 10^4 cm/sec in water.[62]

This calculation of a Brownian particle's velocity was based on Nägeli's estimate for the transmission of velocity from a single molecule to the particle and the statistical computation of fluctuations. Nägeli's estimate was based on momentum conservation for a molecule colliding with the particle, which gave $C = mc/M$ (M and C were the mass and speed of the particle, and m and c were the mass and speed of the molecule). The actual situation was more complicated, argued Smoluchowski. Since a molecule's entire momentum might not be transmitted to the particle, Nägeli's formula was difficult to hold. To Smoluchowski, the more precise relation should be based on the principle that Maxwell and Boltzmann had established in statistical mechanics. According to what was later called equipartition theorem, at thermal equilibrium all the molecules in a gas should have the same average kinetic energy, which was proportional to the gas temperature. A suspended particle in the kinetic theory was nothing but a gigantic molecule, and thus had to follow equipartition, too. In other words, $MC^2/2 = mc^2/2$ on average, meaning that $C = c\sqrt{m/M}$. This modification brought down the estimated speed of a Brownian particle in water to 0.4 cm/sec, which was significantly lower than the previous estimate of 10^4 cm/sec. Yet, Exner's experimental data had the measured speed of Brownian particles in the range of 3×10^{-4} cm/sec, which was much smaller than the theoretical prediction of 0.4 cm/sec. The law of large numbers and the principle of statistical fluctuations elevated the estimated particle speed imparted

[61] Smoluchowski, "Zur kinetischen Theorie der Brownschen Molekularbewegung und der Suspensionen" (1906), 762–763.

[62] Ibid., 763.

from molecular collisions and thus saved the kinetic theory from Nägeli's critique. But the modification increased the estimated speed *too much* to fit the experimental data. How should this problem be resolved?[63]

To Smoluchowski, the problem laid in Exner's experimental data. The tiny particles moving at a speed of 0.4 cm/sec could not be meaningfully measured or sensibly detected even under an ultramicroscope. Exner thought that he obtained measurements of such a speed when he recorded visually a particle's locations at distinct instants and divided the distance between two consecutive locations by the time interval. But the actual trajectory of that particle was too fast for him to capture with his eyesight. What he obtained was a very crude set of samples of the trajectory, which had zigzag patterns too fine to be seen. A more proper way to understand Exner's measurements, according to Smoluchowski, was to construe them as particles' *displacements* from a certain point of reference at different instants. A molecular-kinetic theory of Brownian motion thus had to center on a particle's displacement, not its velocity.[64]

These discussions set the stage for Smoluchowski's kinetic theory of Brownian motion. By this point, he affirmed that a suspended particle behaved in the same way as liquid or gas molecules, and the particle could be conceived as nothing but an entity like other molecules colliding with one another and undergoing random thermal motions, albeit with a much larger size. This was the same conclusion Einstein reached in 1905. In contrast to Einstein's treatment of particles' collective trend as something similar to Boltzmann's transport phenomenon and direct tackling of particles' number density, however, Smoluchowski examined the trajectory of an individual particle, and its zigzag patterns owing to molecular collisions. This was a highly visual approach. His direct contact with the material conditions of the Brownian experiments and their experimenters—ultramicroscope, colloids, aerosols, Krix-Krax diagrams—might have played an important part in this choice. An equally likely reason for his concentration on individual trajectories was that he had already developed a theory of the molecular mean free path by this time. Tracing the trajectory of a suspended particle under molecular collisions was a straightforward extension from that of a molecule under the same kind of bombardments.

Smoluchowski first considered the change a single molecule's collision incurred on a suspended particle. The particle was so much bigger than the molecule that the actual effect of the collision on the former was minuscule.

[63] Ibid., 763–764.

[64] Ibid., 764. Maiocchi viewed this shift of focus from velocity to displacement as a crucial turn in the research on Brownian motion, see Maiocchi, "Case of Brownian motion" (1990), 263–269.

In addition, the molecule could move along any direction with respect to the particle's movement with equal probability. If the collision was elastic, this directional uniformity implied that the particle after collision would gain an average velocity $(3/4)(mc/M)$ perpendicular to the direction of its original motion. This meant that the particle would be kicked off from its original path with a tiny angle $\varepsilon = 3C/4c = 3m/4M$. Moreover, Smoluchowski took suspended particles' free paths as a constant to simplify calculations. A particle's trajectory was modeled as a disjointed collection of linear segments of equal length λ (mean free path). Each segment was tilted from its prior counterpart by an angle ε. Starting with O, the particle traversed along the segmented lines $OP_0, P_0P_1, P_1P_2, \ldots$ in three dimensions, where $OP_0 = P_0P_1 = P_1P_2 = \ldots = \lambda$, and the angle between OP_0 and P_0P_1 = the angle between P_0P_1 and $P_1P_2 = \ldots = \varepsilon$ (Figure 5.2). This assumption was identical to the second case (constant free path and directional uniformity) in Smoluchowski's 1904 work on the mean free path.[65]

Smoluchowski focused on a suspended particle's displacement instead of its speed. His aim in construing the model in Figure 5.2 was to obtain OP_n, the distance the particle traveled from origin O after n collisions. The computation of OP_n, denoted as Λ_n, could be decomposed into smaller parts involving OP_0, P_0P_1, P_1P_2, ... To do so, he projected these vectors onto Cartesian coordinate axes X, Y, and Z. If the angles between OP_0, P_0P_1, P_1P_2, ..., $P_{n-1}P_n$ and X were $\alpha_0, \alpha_1, \alpha_2, \ldots, \alpha_n$, then the X-coordinate of OP_n was $\lambda(\cos\alpha_0 + \cos\alpha_1 + \ldots + \cos\alpha_n)$. Similarly, he took the angles between

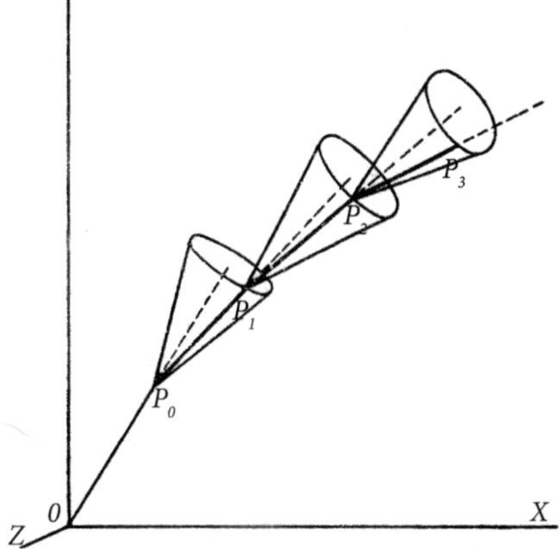

Figure 5.2 A suspended particle's trajectory under collisions. Marian Smoluchowski, "Zur kinetischen Theorie der Brownschen Molekularbewegung und der Suspensionen," *Annalen der Physik*, 21 (1906), 766.
Permission from John Wiley & Sons Books.

[65] Smoluchowski, "Zur kinetischen Theorie der Brownschen Molekularbewegung und der Suspensionen" (1906), 766.

the trajectory's linear segments and Y and Z to be $\beta_0, \beta_1, \beta_2, \ldots, \beta_n$ and $\gamma_0, \gamma_1, \gamma_2, \ldots, \gamma_n$, and obtained the Y- and Z-coordinates of OP_n to be $\lambda(\cos\beta_0 + \cos\beta_1 + \ldots + \cos\beta_n)$ and $\lambda(\cos\gamma_0 + \cos\gamma_1 + \ldots + \cos\gamma_n)$. Thus,

$$\Lambda_n^2 = \lambda^2\left[(\cos\alpha_0 + \cos\alpha_1 + \ldots + \cos\alpha_n)^2 + (\cos\beta_0 + \cos\beta_1 + \ldots + \cos\beta_n)^2 + (\cos\gamma_0 + \cos\gamma_1 + \ldots + \cos\gamma_n)^2\right]. \quad (5.8)$$

The randomness of molecular collisions prevented the exact determination of the displacement Λ_n. But its average could be obtained from the statistics of the directional angles $\alpha_0, \alpha_1, \alpha_2, \ldots, \alpha_n$, $\beta_0, \beta_1, \beta_2, \ldots, \beta_n$, and $\gamma_0, \gamma_1, \gamma_2, \ldots, \gamma_n$. Consider the relation between α_{m-1} and α_m. Smoluchowski supposed that $P_{m-1}P_m$ deviated from $P_{m-2}P_{m-1}$ with the angle ε. But since the particle moved in three rather than two dimensions, this constraint did not fix the direction of $P_{m-1}P_m$. Instead, it could deviate from the plane spanned by the X-axis and the previous segment $P_{m-2}P_{m-1}$ with an arbitrary angle φ_m. Smoluchowski showed from geometry that

$$\cos\alpha_m = \cos\alpha_{m-1}\cos\varepsilon + \sin\alpha_{m-1}\sin\varepsilon\cos\varphi_m. \quad (5.9a)$$

Via the similar consideration for the Y- and Z-components, he obtained

$$\cos\beta_m = \cos\beta_{m-1}\cos\varepsilon + \sin\beta_{m-1}\sin\varepsilon\cos\psi_m, \quad (5.9b)$$

$$\cos\gamma_m = \cos\gamma_{m-1}\cos\varepsilon + \sin\gamma_{m-1}\sin\varepsilon\cos\chi_m. \quad (5.9c)$$

Since the angles of deviation φ_m, ψ_m, and χ_m were due to molecular collision, they were random and had an equal probability between 0 and 2π. Taking the statistical average over φ_m, (5.9a–c) gave

$$\frac{1}{2\pi}\int_0^{2\pi}\cos\alpha_m d\varphi_m = \cos\alpha_{m-1}\cos\varepsilon. \quad (5.9a')$$

$$\frac{1}{2\pi}\int_0^{2\pi}\cos\beta_m d\psi_m = \cos\beta_{m-1}\cos\varepsilon. \quad (5.9b')$$

$$\frac{1}{2\pi}\int_0^{2\pi}\cos\gamma_m d\chi_m = \cos\gamma_{m-1}\cos\varepsilon. \quad (5.9c')$$

The statistical average of $\Lambda_n{}^2$ should be taken over φ_m, ψ_m, and χ_m for $m = 1, 2, \ldots, n$:[66]

$$\Lambda_n{}^2 = \frac{\lambda^2}{(2\pi)^n} \int d\varphi_1 \ldots d\varphi_n d\psi_1 \ldots d\psi_n d\chi_1 \ldots d\chi_n \left[(\cos\alpha_0 + \ldots + \cos\alpha_n)^2 + (\cos\beta_0 + \ldots + \cos\beta_n)^2 + (\cos\gamma_0 + \ldots + \cos\gamma_n)^2 \right]. \tag{5.8'}$$

Equations (5.9') reduced the average of α_n, β_n, and γ_n in terms of α_{n-1}, β_{n-1}, and γ_{n-1}. This relationship led Smoluchowski to express the statistical average of the square displacement $\Lambda_n{}^2$ in (5.8') via a recursive manner. Taking $J_n \equiv \Lambda_n{}^2/\lambda^2$, he represented J_n in terms of J_{n-1}: $J_n = J_{n-1} + 1 + 2\cos\varepsilon\frac{1-\cos^n\varepsilon}{1-\cos\varepsilon}$. Following an iterative procedure, he obtained

$$J_n = \frac{2n}{\delta} + 1 - n - 2\frac{(1-\delta)^2 - (1-\delta)^{n+2}}{\delta^2}, \tag{5.10}$$

where $\delta = 1 - \cos\varepsilon = 1 - \cos(3m/4M)$ was the small quantity representing the microscopic effect of molecular collisions. The mean square root displacement was $\bar{\Lambda} = \sqrt{\Lambda_n{}^2} = \lambda\sqrt{J_n}$.[67]

Equation (5.10) was the core formula in Smoluchowski's molecular-kinetic theory for Brownian motion. To explore this expression, he took a step highly consistent with the style of reasoning he had developed previously: looking at distinct cases in which the value $n\delta$ varied from small to large and discussing the numerical significance of the approximate results in these cases, even though some of them were implausible in actual physical situations. After this exercise, he tackled a more realistic case of a viscous fluid in which a suspended particle experienced not only the drive of outward diffusion due to molecular thermal motions but also a resistive force. This was the same condition that Einstein had dealt with in 1905. Based on the same Stokes formula for the friction coefficient that Einstein had used, Smoluchowski revised his model to include viscous resistance and employed (5.10) in calculating a Brownian particle's mean square root displacement. He obtained $\bar{\Lambda} = \sqrt{t}\left(8/9\sqrt{\pi}\right)\left(c\sqrt{m}/\sqrt{\kappa r}\right)$. Interpreting twice a molecule's kinetic energy mc^2 to be kT with equipartition in mind, he found that his relation for $\bar{\Lambda}$ equaled Einstein's $\bar{\Lambda} = \sqrt{t}\sqrt{kT/3\pi\kappa r}$ times a factor $\sqrt{64/27}$.[68]

[66] Ibid., 766–767.
[67] Ibid., 767.
[68] Ibid., 768–772.

Critical Opalescence, Recurrence, Markov Process

Smoluchowski's work on Brownian motion in 1906 set himself at the forefront of research on random fluctuations. In the following year, he directed his attention to another macroscopic manifestation of microscopic variations that had to do with a bizarre optical property of materials. His starting point was once again density fluctuations in a gas. In 1904, he established in Boltzmann's Festschrift that random molecular motions caused irregular deviations of an apparently homogeneous gas's concentration from its steady value. Specifically, he considered a body of gas at a thermal equilibrium with pressure p_0, temperature T_0, and volume V_0. This gas's molecular number density was supposed to be constant at the macroscopic scale. Thus, n_0 molecules should occupy a fixed volume v_0. At the microscopic scale, however, molecular motions shifted randomly the actual volume v of these n_0 molecules from its thermodynamic limit v_0. Such deviations could be indexed with a ratio $\gamma = (v - v_0)/v_0$. This concentration fluctuation was characterized with the probability $W(v)dv$ that the volume of n_0 molecules fell between v and $v + dv$. The Boltzmannian kinetic theory of fluid gave

$$W(v)\,dv = C\exp\left\{-\frac{1}{kT_0}\int_v^{v_0}\left[p\left(v'\right) - p_0\left(v'\right)\right]dv'\right\}dv, \tag{5.11}$$

where $\int_v^{v_0}\left[p\left(v'\right) - p_0\left(v'\right)\right]dv'$ represented the required amount of mechanical work to bring the gas isothermally and reversibly back to its "normal" state (p_0, v_0, T_0), and C was a coefficient of normalization. Smoluchowski called this formula the "Boltzmann-Einstein law."[69]

Smoluchowski visualized the fluctuations as "bubbles" embedded erratically in a uniform background gas with number density n_0/v_0. Since a bubble's density differed from that of the uniform gas, they had different refractive indices. When light was cast on the gas, a tiny bubble would scatter incoming electromagnetic waves to all directions. If there were a huge number of such bubbles and their refractive indices were sufficiently different from that of the uniform background, then a notable amount of light energy would be scattered, and the gas would stop being translucent and turn bright or murky. In the 1860s, the Irish physicist John Tyndall examined light scattering from dusts or

[69] Marian Smoluchowski, "Molekular-kinetische Theorie der Opaleszenz von Gasen im kritischen Zustande sowie einiger verwandter Erscheinungen," *Annalen der Physik*, 25 (1908), 210; Subrahmanyan Chandrasekhar, "Marian Smoluchowski as the founder of the physics of stochastic phenomena," in Subrahmanyan Chandrasekhar, Mark Kac, and Roman Smoluchowski (eds.), *Marian Smoluchowski* (1999), 26.

colloids suspended in a gas or liquid. The density fluctuations Smoluchowski considered would not yield a perceivable effect of the same kind in ordinary circumstances, since the deviation ratio γ due to a gas's molecular kinetics was typically much smaller than that due to externally introduced impurities.

The situation was different when the gas was at a particular state, though. It had long been well known that a substance transitioned from solid to liquid and then to gas with increasing temperature. Yet, when the pressure reached a certain critical value, there was no longer a phase transition from liquid to gas, and both states co-existed no matter how high the temperature was. Known as a "critical phenomenon," this effect was found by scientists in the early nineteenth century. By the 1860s, physicists noted that gradually moving a liquid or a gas toward its critical point would turn the substance from transparency to murkiness. This "critical opalescence" raised physicists' interest and inspired experimental studies.[70]

Smoluchowski believed that the density fluctuations due to molecular kinetics were the cause of critical opalescence. Whereas the volume/density deviation γ due to molecular kinetics was small for ordinary substances, it would be significant for a substance at the critical point. This difference became obvious when he expanded the mechanical work $\int_{v}^{v_0} [p(v') - p_0(v')]\, dv'$ in (5.11) with a Taylor series of γ. The first three terms in the expansion corresponded to $(\partial p_0/\partial v_0)\gamma^2$, $(\partial^2 p_0/\partial {v_0}^2)\gamma^3$, and $(\partial^3 p_0/\partial {v_0}^3)\gamma^4$. When the substance was in a gaseous state, the integral was dominated by the first term $(\partial p_0/\partial v_0)\gamma^2$, and the estimated mean square fluctuation $\langle\gamma^2\rangle$ from (5.11) was small. When the substance was at the critical state, however, its pressure was much less sensitive to volume change, and $(\partial p_0/\partial v_0) = (\partial^2 p_0/\partial {v_0}^2) = 0$. Thus, the integral was dominated by the third term $(\partial^3 p_0/\partial {v_0}^3)\gamma^4$. The estimated mean square fluctuation $\langle\gamma^2\rangle$ from (5.11) was considerably larger. This higher density fluctuation at the critical state led to stronger electromagnetic scattering that caused opalescence.[71]

To estimate light scattering from density fluctuations at a critical state, Smoluchowski invoked Lord Rayleigh's work in 1871 that dealt with electromagnetic scattering from particles much smaller than the optical wavelength. Rayleigh showed that the intensity of scattered light was inverse proportional to the fourth power of the wavelength. He used this result to explain why the sky was blue: the light we saw from the sky was scattered from tiny

[70] Smoluchowski, "Molekular-kinetische Theorie der Opaleszenz" (1908), 220.
[71] Ibid., 215; Chandrasekhar, "Marian Smoluchowski" (1999), 26–27.

colloids in the atmosphere; and the scattered light was much stronger at shorter wavelengths that skewed toward the blue side of the optical spectrum.

Smoluchowski believed that critical opalescence was similar to the blue-sky effect. In his expression of Rayleigh's formula, the ratio of the scattered light to the original light intensity was $h = \frac{32}{3}\pi^3 \frac{N_0 v_0^2}{\lambda^4}\left(\frac{\Delta\mu}{\mu}\right)^2$, where N_0 and v_0 were the concentration and volume of small scatterers created by density fluctuations, μ was the background refractive index, Δμ was the deviation of the scatterer's refractive index from μ, and λ was the optical wavelength. From a rough estimate of the refractive-index fluctuation $\Delta\mu/\mu$, he concluded that the effect of critical opalescence from his theory was in the same numerical order as experimental data.[72]

Like the case of Brownian motion, Smoluchowski was not the only one trying to crack the nut of critical opalescence. Once again, Einstein was working on the same problem. Einstein left the Swiss patent office for the University of Bern one year after Smoluchowski published a paper on critical opalescence in 1907, became a professor at the University of Zurich, and published his own work on the same subject in 1910. As usual, he started with the fundamental principles of Boltzmannian statistical physics and Maxwell's equations of electrodynamics, rederived the theory of Rayleigh scattering, and moved step by step toward the specific results of electromagnetic scattering due to density fluctuations. Compared with Smoluchowski's work, the major progress of Einstein's paper was an explicit formula for the intensity of Rayleigh scattering due to density fluctuations. Einstein's link between density fluctuations and refractive-index fluctuations was what he called the "Clausius-Mossoti-Lorenz" relation (or what is known as the "Lorenz" relation today): $\frac{1}{\rho} \cdot \frac{\mu^2-1}{\mu^2+2} =$ constant (ρ was the scatterer's density). Using this relation, Einstein came up with an explicit formula for the intensity of electromagnetic scattering due to density fluctuations at the critical state.[73]

Smoluchowski had been in touch with Einstein owing to their common research interest.[74] After Einstein published his article on critical opalescence, Smoluchowski wrote to him and praised his explicit formula for a fluid's electromagnetic scattering under a critical state as "absolutely right" ("vollständig

[72] Smoluchowski, "Molekular-kinetische Theorie der Opaleszenz" (1908), 216–219.

[73] Albert Einstein, "Theorie der Opaleszenz von homogenen Flüssigkeiten und Flüssigkeitsgemischen in der Nähe des kritischen Zustandes," *Annalen der Physik*, 33 (1910), 1275–1298.

[74] On June 11, 1908, for instance, Einstein sent Smoluchowski a short greeting card. See Folder 9414, "Letters to Smoluchowski, D-E," K189, Marian Smoluchowski Papers, Manuscript Collections, MS9397, Library of Jagellonian University, Krákow, Poland.

Recht").[75] In Smoluchowski's follow-up publication in 1911, he referred to his work on critical opalescence in 1907, and admitted that "I had drawn some quantitative conclusions therefrom, without however writing down the explicit final formula, for I considered it only as indicating roughly the order of magnitude." In contrast, "Einstein, however, has arrived in a very remarkable way by explicit calculation of the components of the dispersed waves at the same formula."[76] The theoretical prediction about critical opalescence in Einstein's explicit formula was corroborated by the Leyden-based physicist Willem Hendrik Keesom's spectroscopic measurement in 1911.[77]

The incomplete formulation of Smoluchowski's work on critical opalescence did not eclipse its contribution. He was the first to develop a statistical theory for critical opalescence, and to interpret it as an outcome of random fluctuations in molecular kinetics. The gist of his theory was that the density of a uniform gas nonetheless fluctuated due to molecules' random motions. This heterogeneity amounted to bubbles of different concentrations embedded in a homogeneous medium; and they scattered light owing to contrasting refractive indices. Under ordinary conditions, the quantities and heterogeneity of these bubbles were minuscule, and their light-scattering effect was not discernible. When the gas was at a critical state so that it coexisted with its liquid phase, the substance's heterogeneity became enormous. A huge number of bubbles emerged. And their light-scattering effect was conspicuous. From Smoluchowski's theory, critical opalescence was another manifestation of microscopic random variations amplified and turned into macroscopic and observable fluctuations.

To Smoluchowski, fluctuation phenomena included not only Brownian motion and critical opalescence. As suspended particles in a liquid underwent random Brownian motions, their concentration in any given region of the liquid also fluctuated randomly. Like the gas density fluctuation causing critical opalescence, the density fluctuation of Brownian particles in a liquid also led to observable characteristics. In the 1910s, Smoluchowski examined the statistical properties of Brownian particles' number density. His findings engaged the longstanding issue of irreversibility in thermodynamics.

Smoluchowski considered a small volume of a colloidal solution in which tiny but visible particles were suspended and moved freely. Putting the

[75] Smoluchowski to Einstein, December 12, 1911, Folder 9412, "Letters to Smoluchowski, D-E," 107, Smoluchowski Papers.

[76] Marian Smoluchowski, "On opalescence of gases in the critical state," *Philosophical Magazine*, 23:133 (1911), 168.

[77] Willem Hendrik Keesom, "Spektrophotometrische Untersuchung der Opaleszenz eines einkomponentigen Stoffers in der Nähe des kritischen Zustandes," *Annalen der Physik*, 35 (1911), 591–598.

solution under an ultramicroscope, one could observe particles within a well-defined volume and count their number. Repeating this count every few seconds, a sequence of measured values could be obtained. Since the particles underwent Brownian motions, they moved in and out of the volume erratically, and thus the measured number of particles within the volume should also fluctuate randomly. The question to Smolüchowski was, what were the statistical properties of such fluctuations?

To calculate the statistics of this number density fluctuation, Smoluchowski made two assumptions: that the particles' motions were independent of one another, and that a particle had equal a priori probability at all positions within the volume. These premises and results from his previous theoretical work led him to obtain the probability of finding n particles within the volume at any given instant: $W(n) = e^{-\nu}\nu^n/n!$, where ν was the average number of particles within the volume over a long sequence of measurements. This was the familiar Poisson distribution that characterized independently arriving random events.

Yet, Smoluchowski wanted to know not only the number density's probability distribution at any single instant, but also its probability distribution at any *two* consecutive instants. Specifically, he intended to calculate $W(n; m)$, the probability of finding n particles at one measurement, and finding m particles at the next measurement. This was a problem today's mathematicians call "Markov chains," in memory of the 1906 work of a Russian mathematician Andrey Markov at St. Petersburg University. Smoluchowski developed a technique for the calculation of $W(n; m)$ (later called transition probability) that would become a central result in the mathematical theory of Markov chains: to express $W(n; m)$ in terms of the conditional probabilities for i particles to escape from or enter the volume at the second instant if there were j particles at the first instant.

Smoluchowski's transition probability $W(n; m)$ was used to find the statistical average of another measurable quantity. If one observed n particles within the volume at an instant, then how long (i.e. how many steps) on average would it take for the particle count to be n again? Named "recurrence time," this quantity gauged the interval between two identical physical events (n particles within the volume). When n was close to the average number of particles within the volume, the recurrence time was reasonably short. When n was much larger than the average, the recurrence time became extremely long—in the order of $O(n^n)$. Recurrence of a large deviation from the statistical mean was extremely rare and took a long time.

To Smoluchowski, this finding provided an unexpected but welcome way to engage the longstanding debate on irreversibility in thermodynamics. At the beginning of this chapter, we encountered this problem that plagued nineteenth-century physicists—whereas the second law of thermodynamics stipulated irreversibility for entropy-increasing processes, mechanics predicted reversibility for all physical processes. To solve this problem, Boltzmann contended that the second law of thermodynamics held probabilistically and entropy-decreasing processes could actually occur but with an extremely low probability. Yet, Zermelo and Poincaré demonstrated that in a closed physical system, no matter how rare an event was, it would inevitably reoccur, albeit after a long time. Smoluchowski's finding offered a concrete problem situation for Zermelo's recurrence theorem. The number density fluctuation of Brownian particles in a liquid was precisely the kind of physical event Zermelo and Poincaré referred to. By observing this density fluctuation, one obtained a visualization of the core dilemma in thermodynamics.[78]

Between his publication in Boltzmann's Festschrift in 1904 and his premature death in 1917, Smoluchowski established an influential research program on random fluctuations. Along with Einstein, he examined phenomena that deviated from thermodynamics and were accountable only through the kinetic theory of matters and statistical mechanics: Brownian motion, density fluctuations, critical opalescence, etc. To his contemporaries, these phenomena provided compelling evidence for the reality of atoms and molecules, for they were interpreted as macroscopic effects of microscopic agitations that were not smoothed out in the averaging process but were rather amplified through peculiar statistical routes. Smoluchowski's research program also marked a fundamental transformation of statistical mechanics at the turn of the twentieth century from supplying a mechanical foundation of thermodynamics to studying the observable effects that were highly problematic in the thermodynamic framework.

The significance of Smoluchowski's work went beyond this conceptual sense. At the technical level, his studies of Brownian motion, density variations, mean free paths, critical opalescence, and coagulation helped enact a standard repertoire for physical-mathematical treatments of random fluctuations. To Einstein, the standard approach involved posing molecular equations of

[78] Marian Smoluchowski, "Drei Vorträge über Diffusion, Brownsche Molekularbewegung und Koagulation von Kolloidteilchen," *Physikalische Zeitschrift*, 17 (1916), 557–571, 587–599, reprinted in Natanson and Stock (eds.), *Pisma Mariana Smoluchowskiego* (1924), 530–594, especially 537–554; Chandrasekhar, "Marian Smoluchowski" (1999), 22–24.

motion in an extremely high-dimensional state-space, formulation of canonical ensembles, combinatoric expression of Boltzmannian entropy, and derivation of macroscopic fluctuations' probability density function. Smoluchowski added to this repertoire detailed considerations of elementary processes such as the trajectory of a molecule or a suspended particle, or the temporal variation of the number of colloids within a tiny volume. To him, an elementary process of this kind resembled a random walk. The core mathematical relation generating all statistical properties was its probability density function, which could be obtained via combinatoric and geometric calculations with respect to the elementary process. This probability density function could be used to compute the physical effects of macroscopic fluctuations: the concentration heterogeneity responsible for light scattering, the number of particles counted within a volume, the observed displacements of colloids, the mean free path, etc. Through the calculus of conditional probability and Bayesian reasoning, moreover, this time-variant probability density function could lead to higher-order statistical measures, such as the probability of finding *n* particles at one instant and *m* particles at the next. Because of this mathematical repertoire, the Indian-American physicist and trailblazer of stochastic astrophysics Subrahmanyan Chandrasekhar called Smoluchowski the "founder of the physics of stochastic phenomena."[79]

Yet, Smoluchowski did not treat random fluctuations as purely mathematical exercises. Rather, all of his theoretical works on random fluctuations were closely connected to concrete and tangible problem situations. They were all related to the experimental system of liquids or gases filled with suspended particles. Although his research was theoretical, he was familiar with the material culture of this colloidal experimental system marked by the preparation of emulsions and gases under critical or oversaturated states or close to phase transition, employment of an ultramicroscope, patient counts of particles over time under the scope, careful control of the temperature and pressure of the working substances, and measurements of thermal and optical properties of the substances. We have seen his correspondence with Felix Exner about the latter's experimental data on Brownian motion. In fact, Smoluchowski even considered performing an experiment himself to verify his predictions on Brownian motions.[80]

Smoluchowski's familiarity with the material culture of the colloidal experimental system was evident from the experimental details he engaged in his

[79] Chandrasekhar, "Marian Smoluchowski" (1999), 21.

[80] Smoluchowski, "Zur kinetischen Theorie der Brownschen Molekularbewegung und der Suspensionen" (1906), 756.

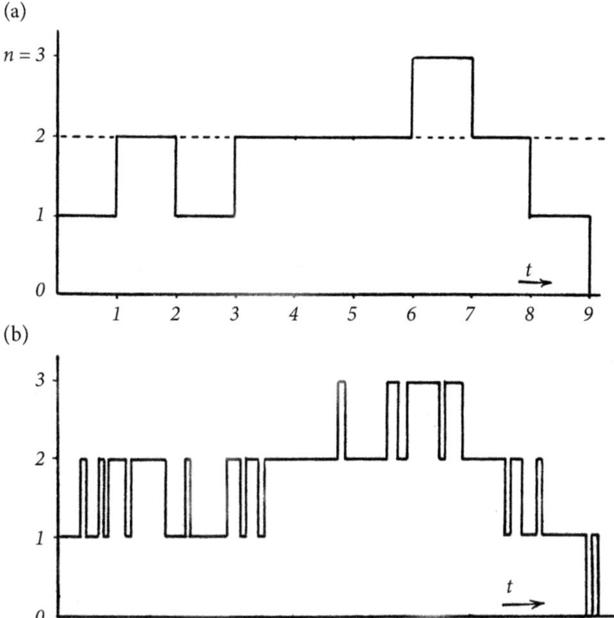

Figure 5.3 Smoluchowski's presentation of Svedberg's experimental data on the number of Brownian particles within a small volume at different times. Marian Smoluchowski, "Drei Vorträge über Diffusion, Brownsche Molekularbewegung und Koagulation von Kolloidteilchen," *Physikalische Zeitschrift*, 17 (1916), 565, Figure 1.

publications (which formed a sharp contrast with Einstein's works). In his examination of the recurrences for density fluctuations, he not only discussed the mathematical results of the expected recurrence time, but also presented the Swiss chemist Theodor Svedberg's experimental data in 1904. Svedberg prepared a solution of gold colloids and counted under an ultramicroscope the number of gold particles within a designated region every 1/39 minute, 519 times. In Smoluchowski's presentation, Svedberg's data clearly demonstrated frequent recurrences of number counts when they were sufficiently close to their average, and extremely rare recurrences when number counts deviated considerably from their average (Figure 5.3). This empirical measurement visualized Zermelo's and Poincaré's highly mathematical arguments about recurrences.[81]

[81] Smoluchowski, "Drei Vorträge über Diffusion, Brownsche Molekularbewegung und Koagulation von Kolloidteilchen" (1916), 538–549.

There were two other cases in point. In a letter to Perrin, Smoluchowski praised the French physicist's renowned experiment to corroborate Einstein's theory of Brownian motion by counting under an ultramicroscope the number of particles at distinct depths of a colloidal solution and determining Avogadro's number according to such data through Einstein's formula. Yet, Smoluchowski also pointed out issues with Perrin's experimental design that might affect the accuracy of the data, such as the convection current within the solution and the unequal sizes of the colloids.[82] In another instance, a researcher George Carse from Cavendish Laboratory at Cambridge wrote to Smoluchowski in 1906 and asked for help with a theoretical calculation to determine the two parameters in the van der Waal equation (the equation of states for a gas considering intermolecular forces) from experimental data. In the correspondence, the two individuals discussed specificities of the experiment with temperature-controlled gas in a glass chamber at Cavendish, reminding us of the Polish physicist's practical training at Kelvin's laboratory in Glasgow and Warburg's in Berlin.[83] Smoluchowski's theoretical works on random fluctuations had a clear underpinning of the colloidal experimental system.

Langevin's Approach through Stochastic Equations

A Physicist of the Modern Generation

At the turn of the twentieth century, Einstein and Smoluchowski launched the studies of random fluctuations. They introduced novel perspectives to examine Brownian motion, critical opalescence, density variations in gases and colloidal solutions, mean free paths, and recurrences through the framework of statistical mechanics. In so doing, they developed a mathematical repertoire for the treatments of random fluctuations. This repertoire represented the temporal evolution of a randomly varying physical quantity as a stochastic process x_t at instant t, found its probability density, and calculated its statistical average $\langle x_t \rangle$. In proper circumstances, the probability density function of a basic physical quantity (e.g. the colloidal concentration) was also used to compute

[82] Folder 9412, "Letters from Smoluchowski, 1897-1917," Smoluchowski to Perrin, undated, K124-125, Smoluchowski Papers.

[83] Folder 9413, "Letters to Smoluchowski, A-C," George Carse to Smoluchowski, October 16, 1906, and December 28, 1906, K134-136, Smoluchowski Papers.

the statistical average of another quantity (e.g. light-scattering power from a heterogeneous medium).

Despite its initial success, this mathematical repertoire nonetheless exhibited a notable limitation. When scientists began to show interest in knowing not only the statistical average of x_t but also its correlation at different times $\langle x_{t_1} \cdot x_{t_2} \rangle$ or even its higher-order statistics such as $\langle x_{t_1} \cdot x_{t_2} \cdot x_{t_3} \rangle$, this repertoire became inadequate, and new mathematical techniques were called for. An individual who explored a new mathematical technique to treat random fluctuations in the late 1900s was the French physicist Paul Langevin.

Paul Langevin was born in Paris at the dawn of the Third Republic. A veteran from the French Army, his father was a building inspector for the city government. Langevin showed a gift in science and mathematics from childhood. In 1888, he entered the City of Paris's Municipal School of Industrial Physics and Chemistry (l'École Municipale de Physique et de Chimie Industrielles de la Ville de Paris) for college education. Founded after the nation's defeat in the Franco-Prussian War, the Municipal School (ESPCI Paris now) was a *grande école* to train in and promote physics and chemistry for engineering and technology. The Municipal School hired Pierre Curie as an instructor and accommodated his and Marie Skłodowska Curie's experimental research on radioactivity and magnetism. In the 1880–90s, the Curies became the center of an influential social circle in the French natural sciences. Surrounding Pierre and Marie, this circle comprised young and ambitious physicists, chemists, and mathematicians in Paris, including Jean Perrin, George Urbain, and Andre Debierne (discoverers of rare earth elements), Emil Borel (set theorist and developer of the measure-theoretic interpretation of probability), Henri Becquerel (discoverer of radioactivity), and Pierre's brother Paul-Jacques (known for his collaboration with Pierre on piezoelectric research). Langevin came to know the Curies when he took courses with Pierre at the Municipal School, became an active member of the circle, and established a lifelong friendship with the Curie family.[84]

After graduating from the Municipal School, Langevin attended the École Normale Supérieure in 1894–97 to prepare for the agrégation exam in physical sciences, a civil-service qualification for teaching at high schools. Upon passing the agrégation in 1897, he received a fellowship from the City of Paris for studying abroad at Cavendish Laboratory, where he conducted research

[84] Frédérique Joliot-Curie, "Paul Langevin, 1872-1946," *Obituary Notices of Fellows of the Royal Society*, 7 (1951), 405; Pierre Biquard, *Paul Langevin: Scientifique, Éducateur, Citoyen* (Paris: Seghers, 1969), 35.

under the supervision of J.J. Thomson. Langevin returned to Paris in 1900 to start working as an assistant at the Sorbonne, obtained his Ph.D. in 1902, and succeeded Pierre Curie as a professor at the Municipal School in 1905 when his mentor moved to the Sorbonne. In the following decades, Langevin built up a prominent career and made himself a powerful figure of French science: professor at Collège de France in 1909, editor of *Journal de Physique* in 1920, director of the Municipal School in 1926, and chair of the Solvay Conference in 1928.[85]

Historians Helge Kragh and Richard Staley used the term "generation" to characterize the physicist community at the turn of the twentieth century. To them, the common mentality, practice, sense of destiny, and memory of these physicists en masse epitomized a major transformation of physics as a scientific discipline and physicists as a professional collective into the "modern" epoch.[86] Langevin's career and life witnessed this generational transformation into the "modern." Through the Solvay Conferences and journal editing, he was a core participant of the spectacular rise of quantum mechanics and relativity. He took part in military research on ultrasonic detection of German submarines during World War I, anticipating physicists' much more intense involvement in war technologies. His public opposition to Nazis and detention by the Vichy regime testified to political atrocity in "the age of extremes" (in Eric Hobsbawm's language). His reputation as a national hero and leadership in educational reform marked the social status physicists enjoyed as high expert-savants after World War II.

To the young Langevin of the 1890s–1900s, however, the most concrete and immediate impact of physics in the modern world had to do with atomism. Wilhelm Roentgen's discovery of X-ray, Becquerel's discovery of radioactivity, and J.J. Thomson's discovery of electrons at the end of the nineteenth century brought a profound ontological change to physics research. While chemists had gained significant success using the notions of atoms and molecules to account for chemical reactions, many physicists in the nineteenth century were still skeptical of the reality or relevance of microscopic entities. Roentgen's, Becquerel's, and Thomson's findings led to physicists' general conversion to an atomic worldview. Inspired by these new experiments, they launched theoretical research on the movements and interactions of electrons, ions, atoms,

[85] Joliot-Curie, "Paul Langevin" (1951), 405–407, 413–414.

[86] Helge Kragh, *Quantum Generations: A History of Physics in the Twentieth Century* (Princeton, NJ: Princeton University Press, 1999); Richard Staley, *Einstein's Generation: The Origin of the Relativity Revolution* (Chicago, IL: University of Chicago Press, 2009).

and molecules as primary mechanisms for phenomena from electrical polarization and heat conductivity to chemical bonds and spectral dispersion. Historians Jed Buchwald and Andrew Warwick referred to this development as "microphysics."[87]

From Microphysics to Statistical Physics

Microphysics was the guiding principle of Langevin's early studies. His first research at the École Normale, Cavendish Laboratory, and the Sorbonne was on the properties of Roentgen rays and ionization of air, trendy topics at cutting-edge physics laboratories in Europe at the time. Influenced by Thomson, Langevin was especially interested in the electrical properties of rarefied gases. Physicists had known that a body of gas under a high voltage would become electrically conductive. This was the basis for a wealth of gas-tube-related laboratory effects from neon illumination and cathode rays to X-rays. The atomic theory of matters provided a convincing explanation for this phenomenon: the neutral gas molecules in a strong electric field were decomposed into ions, which were positive and negative electrical charge carriers moving in opposite directions. These movements resulted in an electric current. When a positive ion and a negative ion collided, they recombined into a neutral molecule again. Langevin experimented with the conditions for decomposition of gas molecules into ions, the mobility for distinct types of ions, the rates of recombination and diffusion in an ionized gas, and its electrical conductivity. He also studied the potential applications of his research to the studies of ionized regions in the upper atmosphere. His doctoral thesis submitted to the Sorbonne in 1902 was on ionized gases.[88]

After the mid-1900s, Langevin's interest in microphysics extended beyond ionized gases. In 1910, he entertained a new account for the blue sky by interpreting it as an effect of Rayleigh scattering not from suspended aerosols or other Brownian-like corpuscles in the atmosphere but directly from free electrons in its gas molecules.[89] Yet, he made his fame for a novel theory of diamagnetism and paramagnetism. In 1905, Langevin explored the idea that the actions of electrons inside constitutive molecules of a material were responsible for its diamagnetic or paramagnetic properties. It had been known that

[87] Jed Buchwald and Andrew Warwick (eds.), *Histories of the Electron: The Birth of Microphysics* (Cambridge, MA: MIT Press, 2004).

[88] Joliot-Curie, "Paul Langevin" (1951), 409–410.

[89] Paul Langevin, "La théorie électromagnétique et le bleu du ciel," *Bulletin des Séances de la Société Française de Physique, Résumé des Communications*, (1911), 80–82.

near a strong magnet, some substances induced magnetism with an opposite polarity while some others induced weak magnetism with the same polarity as that of the magnetic source. The former effect was diamagnetism, and the latter paramagnetism. In his dissertation research in the 1890s, Pierre Curie experimented with the effects of temperature on the magnetic properties of various materials. He found that (i) magnetization of paramagnetic substances was inverse proportional to absolute temperature; (ii) magnetization of diamagnetic substances was independent of temperature; and (iii) a substance with innate magnetization (ferromagnet) was demagnetized above a critical temperature.[90]

As Curie's student in the early 1890s, Langevin was familiar with these experimental discoveries. His 1905 theory started with an atomic modeling of diamagnetism. Although Ernest Rutherford's planetary picture of the atom was still years ahead, by this time physicists had entertained hypothetical structures of the elementary constituents of matter. Langevin assumed that electrons undertook periodic circular motions within a molecule. These electronic orbital motions were equivalent to tiny electric current loops, which gave the molecule a minuscule magnetic moment. The modeling of a material's magnetism in terms of microscopic electric current loops dated back to André-Marie Ampère's supposition in the 1830s. The novelty of Langevin's theory was his dealing with the case when these electronic orbital motions were subjected to an external magnetic field. The external magnetic field exerted an additional force on the orbiting electrons, causing them to undertake precessions around the field's direction. This was equivalent to a change of the electrons' speeds, which corresponded to a change of their total magnetic moment bestowed on the molecule. Since the induced magnetic field had an opposite polarity to the external magnetic field, the change of the electrons' effective magnetization was negative with respect to the external magnetic field. This explained the material's diamagnetism. From Langevin's formula, the negative magnetization due to the change of an orbiting electron's magnetic moment in response to the external magnetic field did not depend on temperature.[91]

Langevin's theory of diamagnetism was praised for its simplicity and nice fit with experimental data.[92] After the publication of his 1905 paper, he received a letter from Rutherford—who had overlapped with Langevin at Cavendish in

[90] Pierre Curie, *Propriétés Magnétiques des Corps à Diverses Températures*, Ph.D. dissertation (Paris: University of Paris Sorbonne, 1895).

[91] Paul Langevin, "Sur la théorie du magnétisme," *Comptes Rendus des Séances de l'Académie des Sciences*, 139 (1905), 1204–1206; "Sur la théorie du magnétisme," *Journal de Physique Théorique et Appliquée*, 4:1 (1905), 678–693.

[92] Joliot-Curie, "Paul Langevin" (1951), 410–411.

1897–98 and was then a faculty member at McGill University in Montreal—on December 28, who said admiringly that "I read your article on magnetic theory with great interest. I think it is great and you certainly have the faculty of reducing your mathematics for a style readily intelligible."[93] Considering the electronic actions within a single molecule, he could give a good account of diamagnetism's empirically established rules, including its independence of temperature, proportionality to the square of the electron's charge and to the substance's density, and inverse proportionality to the substance's atomic number. The variations of individual molecules were not considered. According to him, this single-molecule model was valid because when there was no external magnetic field, the innate magnetic moments of all molecules due to their electronic orbital motions pointed to all possible directions and cancelled out one another; and the net magnetization was zero. When an external magnetic field was employed, the material's induced magnetization was due exclusively to the change of orbiting electrons' motions, which, from his theory, did not vary with individual molecules' conditions.[94] Hence, a statistical treatment of the microscopic entities was not needed.

Paramagnetism was different, though. To make sense of many materials' weak magnetization along the direction of an external magnetic field, Langevin argued that individual molecules' variations played a crucial part. In contrast to a diamagnetic material, the molecular thermal motions in a paramagnetic material affected the orientations of their innate magnetic moments due to orbiting electrons and prevented these tiny moments from cancelling out completely one another. As a result, the material had non-zero, albeit small, magnetization. When an external magnetic field was present, these molecules moved toward the state of lowest energy at which the molecular magnetic moments aligned with the external field. At a non-zero temperature, however, the molecules' random thermal motions rendered a perfect magnetic alignment impossible, and the molecules still possessed various orientations. Their deviations from the external magnetic field would thus be subjected to the laws of statistical mechanics and follow the familiar Boltzmann distribution. Since the spread of a Boltzmann distribution was controlled by temperature—the average kinetic energy—of the physical system, the paramagnetic material's magnetization was inverse proportional to temperature: the higher the

[93] Carlton 76, "Letters," Ernest Rutherford to Paul Langevin, December 28, 1905, L76/43, Paul Langevin Papers, Library of ESPCI Paris Tech, Paris, France.

[94] Langevin, "Sur la théorie du magnétisme" (1905), 1206.

temperature, the more disorderly the system, and the less molecular alignment with the external magnetic field.[95]

Langevin did not work out the technical details of the statistical treatment for paramagnetism in 1905. He possessed only a rough idea about the connection between the random variations of molecular magnetic moments and the temperature dependence of paramagnetism. He did not develop a full-fledged kinetic theory of paramagnetism utilizing the Boltzmann distribution until he gave a paper on this subject at the Solvay Conference in 1911.[96] Yet, the notions of statistical physics and random fluctuations became increasingly entrenched in his research from this point on. In 1905, he also published a kinetic theory of ions in a gas. Following Boltzmann's approach to gaseous diffusion and transport, Langevin developed a statistical formulation for collisions between ions, computed its corresponding rates of ionic recombination and mobility, and predicted an ionic gas's mean free path, electrical conductivity, and their dependence on temperature.[97] His microphysical inspection of magnetism and ionized gases led him step by step toward statistical physics and studies of random fluctuations.

Langevin Equation as an Alternative Representation of Brownian Motion

Although Langevin did not have direct research experience with colloidal solutions and Brownian motion, his friend Perrin did. Since the mid-1900s, the physical chemist at the Sorbonne had worked intensively on ultramicroscopic observations of emulsion, quantitative measurements of colloidal concentration and movements, and experimental testing of the Einstein–Smoluchowski theory of Brownian motion. Langevin interacted closely with the Perrin family during this period.[98] His familiarity with the studies of Brownian motions should not come as a surprise.

[95] Ibid., 1206–1207.

[96] Paul Langevin, *La Théorie Cinétique du Magnétisme et Les Magnétrons*, Rapport au Conseil Solvay (Paris: Gauthier-Villars, 1912); available in Carlton 133, "Electromagnetic theory and electron, magnetic theory and molecular orientations, relativity," L133/20, Langevin Papers.

[97] Paul Langevin, "Une formule fondamentale de théorie cinétique," *Annales de Chimie et de Physique*, 5 (1905), 245–288; available in Carlton 133, "Electromagnetic theory and electron, magnetic theory and molecular orientations, relativity," L133/04, Langevin Papers.

[98] For example, see Carlton 76, "Letters," Jean and Harriett Perrin to Paul Langevin, January 12, 1900, L76/21, and January 30, 1907, L76/22, Langevin Papers.

In March 1908, Langevin issued a research note, which was presented at the Académie des Sciences by Éleuthère Mascart, professor of Collège de France.[99] Titled "Sur la théorie du mouvement brownien," Langevin's paper began with a brief review of the major theoretical works on Brownian motions, from Gouy's hypothesis of molecular-thermal agitation in 1889, to Einstein's derivation of a suspended particle's mean square displacement from statistical mechanics in 1905, and on to Smoluchowski's "more direct" approach to the same problem in 1906 leading to the identical formula except for a factor of 64/27. Langevin assured that suspended particles' thermal agitation due to collisions with fluid molecules was the cause of Brownian motion. He claimed to show that a correct application of Smoluchowski's method would result exactly from Einstein's formula. Moreover, he declared that "il est facile de donner, par une méthode tout différente, une démonstration infiniment simple."[100]

So, what made up this method of Langevin's, which he boasted was "infinitely simpler" than and "entirely different" from Einstein's and Smoluchowski's? The French physicist's point of departure was, once again, the equipartition theorem. At equilibrium, Langevin insisted, a suspended Brownian particle in a fluid should have the same amount of average kinetic energy with respect to a given degree of freedom as a fluid molecule, despite the former's much larger size than the latter. From the kinetic theory of matters, this mean kinetic energy should be kT, where k was the Boltzmann constant and T the absolute temperature. Taking a suspended particle's one-dimensional displacement as x and its velocity dx/dt, equipartition thus determined the particle's mean square velocity via $m\langle(dx/dt)^2\rangle = kT$.[101]

Yet, a more pressing aim for the studies of Brownian motion was to find suspended particles' mean square displacement $\langle x^2\rangle$. Langevin's grappling with this problem showed a significant difference from Einstein's and Smoluchowski's approaches. Recall that Einstein sought to calculate particles' mean square displacement $\langle x^2\rangle$ by formulating a partial differential equation for the displacement's probability density function via a consideration similar to Boltzmann's treatment of continuity and transport phenomena, by solving the differential equation to get a Gaussian distribution $f(x, t)$ for probability density, and by calculating the statistical average using this distribution.

[99] Paul Langevin, "Sur la théorie du mouvement brownien," *Comptes Rendus des Séances de l'Académie des Sciences*, 146 (1908), 530–533; article translated by Anthony Gythiel and introduced by Don Lemons, "Paul Langevin's 1908 paper 'On the theory of Brownian Motion' ['Sur la théorie du mouvement brownien', C.R. Acad. Sci. (Paris), 146, 530–533 (1908)]," *American Journal of Physics*, 65:11 (1997), 1079–1081.

[100] Ibid., 531.

[101] Ibid., 531.

Smoluchowski resorted to an examination of random kicks along a particle's trajectory and undertook statistical and geometrical computations on these small stochastic components. In contrast, Langevin did not work directly on x's probability density function or its stochastic trajectory, but rather steered his attention to a Brownian particle's equation of motion. In the simplest scenario, a suspended particle in a fluid free of any external forces was subjected to a "resistance" (or friction) due to the fluid's viscosity; and this friction was proportional to the particle's speed. The Newtonian equation of motion for this particle was

$$m\frac{d^2x}{dt^2} = -\chi\frac{dx}{dt}, \tag{5.12}$$

where $\chi = 6\pi\kappa r$ according to Stokes's law of friction, κ was the friction coefficient, and r was the particle's radius. The simple equation of motion (5.12) suggested that no matter how fast a suspended particle moved initially, once it was in a viscous fluid, its kinetic energy would dissipate through resistive attrition and it would eventually come to a complete rest.

In a more realistic situation in which fine-grained microscopic actions were taken into account, Langevin argued that random collisions with molecules would continue to supply kinetic energy to the particle to withstand its tendency of falling stationary. This was why minuscule powders or colloids under a microscope appeared to undertake incessant motions. While both Einstein and Smoluchowski acknowledged the role of molecular thermal agitations in maintaining Brownian motion, Langevin differed from them in arguing that the effect of thermal molecular actions could be fully characterized with an additional force on a suspended particle. Considering the molecular actions, the particle's equation of motion (5.12) should be modified as follows:

$$m\frac{d^2x}{dt^2} = -\chi\frac{dx}{dt} + X, \tag{5.12'}$$

where X was a force due to random collisions of molecules on the particle. It would look like a dense train of (positive or negative) spikes of random amplitudes and arriving times. In Langevin's model, in other words, the simple, regular, and continuous aspects in connection to macroscopic classical physics and the complex, stochastic, and discontinuous aspects in connection to microscopic actions could be separated as different terms in the equation of motion; and the effects of thermal molecular agitations amounted to a random force.[102]

With the stochastic equation of motion at hand, Langevin then employed a number of mathematical maneuvers utilizing the assumed statistical properties of the particle's displacement x and the stochastic force X. X was unknown, which made it difficult to solve directly (5.12'). Langevin's tactic was to transform it into an equation for the statistical average $z \equiv \langle dx^2/dt \rangle$ that did not involve X, and he did so through the following algebraic manipulations. First, he multiplied both sides of equation (5.12') with x and rearranged several terms into total differentials:

$$\frac{m}{2}\frac{d^2x^2}{dt^2} - m\left(\frac{dx}{dt}\right)^2 = -\frac{\kappa}{2}\frac{dx^2}{dt} + xX. \tag{5.13}$$

Second, he employed the statistical average on both sides of equation (5.13). Third, he asserted from equipartition that $m\langle (dx/dt)^2 \rangle = kT$. In addition, he assumed that the stochastic force X should be "indifferently positive and negative," meaning that it should be symmetric with respect to the forward ($+x$) and backward ($-x$) directions in the statistical sense. Under this supposition, the statistical average $\langle xX \rangle$ should be zero. Plugging these two average values into equation (5.13), he obtained

$$\frac{m}{2}\frac{d}{dt}\left\langle \frac{dx^2}{dt} \right\rangle + \frac{\chi}{2}\left\langle \frac{dx^2}{dt} \right\rangle = kT. \tag{5.14}$$

Equation (5.14) was a first-order linear ordinary differential equation. Its solution was easily obtained: $z \equiv \langle dx^2/dt \rangle = 2kT/\chi + C\exp\left[-(\chi/m)\,t\right]$. The second exponential term in the solution could be neglected since it decayed quickly. By integrating z with time and by denoting the net displacement of a particle within duration τ to be $\Delta(\tau)$, Langevin obtained the solution $\langle \Delta(\tau)^2 \rangle = (2kT/\chi)\,\tau = (kT/3\pi\kappa r)\,\tau$.[103] This result was identical to Einstein's solution $\bar{\Lambda} = \sqrt{t}\sqrt{(kT/3\pi\kappa r)}$ from equation (5.3).

Langevin's short paper in 1908 on Brownian motion might look like a mere deployment of mathematical tricks to derive Einstein's and Smoluchowski's results via slightly different manipulations of symbols. But this deceptively simple outlook eclipsed the profound implications of the French physicist's work. A crucial novelty of Langevin's theory was his mathematical approach

[102] Ibid., 531.
[103] Ibid., 531–532.

to random fluctuations. Unlike Einstein and Smoluchowski, he bypassed the consideration and computation of a fluctuation's probability density function. He did not delve into the geometric details of a sequence of random walks as Smoluchowski did. Rather, Langevin represented mathematically a random process in terms of its equation of motion (5.12'). This equation of motion was easier to solve than the one describing the random process's probability function (such as (5.2)), since the former was a linear ordinary differential equation with constant coefficients, whereas the latter was a partial differential equation. To find the physical properties of random fluctuations, such as Brownian motion's mean square displacement $\langle x(t)^2\rangle$, Langevin did not have to obtain the full solution of x's probability density function. A direct grappling with the equation of motion would do.

By undertaking similar, albeit more complicated, algebraic manipulations with the equation of motion, in fact, one could in principle obtain x's higher-order statistics such as $\langle x(t)^m\rangle$ for $m > 2$, even though such calculations were often not pursued for their lack of empirical meaning. According to probability theory, this trait permitted the reconstruction of a fluctuation's probability density function through the so-called characteristic function.[104] From a mathematical perspective, therefore, Langevin's approach via the equation of motion provided a specification of a random fluctuation as complete as that which Einstein's and Smoluchowski's approaches did. For these reasons, the stochastic equation of motion—also known as the Langevin equation—has become a standard representation of random processes in modern mathematics.[105]

As Langevin framed the problem of Brownian motions in terms of the new mathematics of stochastic differential equations, he also introduced a novel empirical concept. That is, random fluctuations could be attributed to a stochastic force. While such a fluctuation acting upon an individual physical entity was due to a huge number of irregular microscopic interactions, Langevin's theory claimed that the collective effect of these interactions was reducible to a single random driving force. This was a highly visualized model. Under this picture, the motion of a suspended particle in a fluid could be

[104] The characteristic function $f(u)$ of a random variable X is the expected value of $\exp(iuX)$, i.e. $f(u) = <\exp(iuX)>$. It can be shown that $f(u)$ is the Fourier transform of X's probability density function. Moreover, the coefficients of $f(u)$'s Taylor-series expansion can be expressed in terms of X's moments; specifically, the nth coefficient is $i^n<X^n>/n!$. In principle, therefore, one can construct via the characteristic function X's probability density function from $<X>$, $<X^2>$, ..., $<X^n>$, ... See Anthanasios Papoulis, *Probability, Random Variables and Stochastic Processes* (New York: McGraw-Hill, 2002), 243–303.

[105] For example, see William Coffey, Yuri Kalmykov, and John Waldron, *The Langevin Equation: With Applications in Physics, Chemistry, and Electrical Engineering* (Singapore: World Scientific, 1996).

conceived as being driven by a series of random impetuses. The definite form of this stochastic force was difficult to know. But its statistical properties could be constructed via considerations of the physical system's symmetry, the law of large numbers, and common sense.

Langevin's visual grip of Brownian motion was not unique. Around the same time, his friend Perrin was performing a sequence of experiments toward a similar aim. He prepared a colloidal solution, observed it under an ultramicroscope, and counted the number of suspended particles within areas of the same size at different depths. His figures vividly showed that the "force" of thermal molecular agitations countered gravitation and resisted the tendency for particles to drop down to the bottom of the container. Taking into account this balance between the gravitational force and the stochastic force of thermal agitation, Perrin demonstrated how the difference of particle density at distinct depths could actually be used to determine Avogadro's number, affirming again the reality of atoms through an interpretation of Brownian motion. In another diagram, Perrin charted three suspended particles' trajectories under an ultramicroscope. These trajectories were clear outcomes of stochastic "kicks" from all possible directions with arbitrary magnitudes (Figure 5.4).[106] Perrin's visualization of Brownian motion as an effect driven by

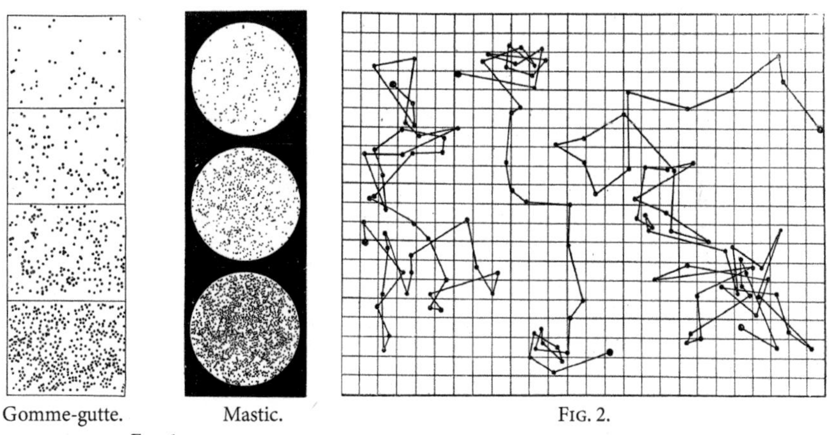

Figure 5.4 Perrin's observed Brownian particles under a microscope (left panel) and his trace of several particle trajectories (right panel). Jean Perrin, "Mouvement brownien et molécules," *Journal de Physique: Théorique et Appliquée et de Physique*, 9:1 (1910), 28 (Figure 1, 34 (Figure 2).

[106] Jean Perrin, "Mouvement brownien et réalité moléculaire," *Annales de Chimie et de Physique*, series 8, 18 (1909), 57, 81.

a stochastic force epitomized contemporary scientists' general view of random fluctuations.[107]

Be intuitive and sensible as it might, Langevin's model about the stochastic force was a bold assumption. It was far from straightforward to accept that all random fluctuations of a physical system could be concentrated and condensed into a clean force term, while the remaining parts were free from statistical variations. Chandrasekhar knew the significance of this point. As he stated in a review of the stochastic problems in physics and astronomy in 1943:[108]

> We can derive from Langevin's equation all the physically significant relations concerning the motions of the Brownian particles. But we should draw attention even at this stage to the very drastic nature of assumptions implicit in the very writing of an equation of the form (132) [i.e. Langevin's equation]. For we have in reality supposed that we can divide the phenomenon into two parts, one in which the discontinuity of the events taking place is essential while in the other it is trivial and can be ignored. In view of the discontinuities in all matters and all events, this is a prima facie, and *ad-hoc* assumption. They are however made with reliance on physical intuition and the *a posteriori* justification by the success of the hypothesis.

In his 1908 work, Langevin did not provide a justification for his, in Chandrasekhar's language, division of the phenomenon into the part of discontinuity represented by the stochastic force and the "trivial" part of Newtonian dynamics in a macroscopic and continuous system. But the French physicist indeed became preoccupied with the ontological issues of discontinuity, randomness, and determinism afterward.

In 1913, Langevin presented a paper in a conference at the French Physical Society titled "The physics of discontinuity." The subject of this report was the significance of the recently corroborated microscopic entities—molecules, atoms, and electrons—in physical sciences. These discrete and discontinuous units of nature led to the kinds of statistical distributions at equilibrium states that conformed to the laws of thermodynamics, as the nineteenth-century kinetic theorists had established. To Langevin, however, a more important implication of the microscopic entities was their irregular variations and deviations from their statistical mean, resulting in random fluctuations at the

[107] Bigg, "Evident atoms" (2008), 312–322.

[108] Subrahmanyan Chandrasekhar, "Stochastic problems in physics and astronomy," *Reviews of Modern Physics*, 15:1 (1943), 21.

macroscopic and observable level. To deal with the comparable problems in physics and chemistry such as the density fluctuations of a solution or a gas or the intensity fluctuations of radioactivity, he employed the typical chance-game model of roulette, in which the turnouts red and black represented two possible conditions of a microscopic object, such as the presence or absence of a molecule within a small volume, or the emission or silence of a radioactive particle within a temporal window.

Langevin also reiterated the familiar combinatorics to compute the probability distribution of the corresponding roulette events, such as the chance of having n_1 reds at the first designated slot, n_2 reds at the second designated slot, etc. The same model could even be used to compute higher-order statistics, such as the chance of having n_1 isolated reds, n_2 red pairs, n_3 red triplets, etc. In Langevin's view, the microscopic discontinuities gave rise to the intrinsic random fluctuations in nature, of which Brownian motion was an instantiation. The regularity of these random fluctuations could be captured with the theory of probability and statistical mathematics. He went even as far as to claim, à la Planck and Einstein, that certain incomprehensible distributions from recent physical measurements—e.g. the dispersion spectrum of black-body radiation or the variation of specific heat with temperature—could be adequately explained with the same statistics should more quantities, such as the energy of an oscillating electron, be assumed to possess discontinuous and quantized units.[109]

As Langevin advanced in his academic career, he showed an increasing interest in philosophical problems, especially the epistemological and ontological issues arising from the newly developed relativity and quantum mechanics. His thoughts on random fluctuations were extended from the technical aspects to the conceptual aspects, too. Atomism, statistics, and determinism—like space, time, causality, and observability inspired by relativistic and quantum theories—became subjects of his philosophical engagements with contemporary intellectuals from Henri Bergson and Gaston Bachelard to Max Born and Louis de Broglie.[110] Among the discussions he frequented were the occasions set up by the Centre International de Synthèse in Paris, a private

[109] Paul Langevin, "La physique du discontinu," *Conférence à la Société française de Physique*, 1913, published in *Les Progrès de la Physique Moléculaire* (Paris: Gauthier-Villars, 1914), 1–46; available in Carlton 132, "Thermodynamics and kinetic theory," L132/07, Langevin Papers.

[110] For Langevin's participation in philosophical discussions, see Bernadette Bensaude-Vincent, "When a physicist turns on philosophy: Paul Langevin (1911-39)," *Journal of the History of Ideas*, 49:2 (1988), 319–338.

foundation dedicated to interdisciplinary deliberations on philosophical problems. In June 1935, Langevin attended the center's conference on statistics.[111] Following Max Born, he gave a paper titled "Statistics and determinism." He began his presentation by noting the challenge brought by quantum physics to the notion of determinism, which was similar to the challenge brought by relativity to the notions of space and time. Yet, determinism became an issue before the introduction of quantum physics. In fact, atomism and the kinetic theory of matters had already opened a door for the discussions on determinism.

A major implication of atomism, Langevin continued, was that macroscopic fluctuations resulted from microscopic agitations. This was the case for Brownian motion, but also for critical opalescence and gas density fluctuations. A major scientific achievement at the turn of the century was to explain and represent these fluctuations with statistical theories of corpuscular interactions.[112] In these situations, according to Langevin, indeterminism was an indication of our lack of complete information about all the molecular configurations and dynamics. Yet, statistics could serve as a guide to save humans from ignorance and to achieve exactitude. He declared that "il y a un lien entre le déterminisme et la statistique qui est très profond puisque tout le déterminisme qu'on avait trouvé absolu prend un aspect statistique" [there is a very profound link between determinism and statistics, since the determinism that we have found absolute takes a statistical aspect].[113] From the specifics of ionized gases and colloidal solutions to the fundamentals of knowing and the material world, the road Langevin traversed marked the evolving conceptualization of fluctuations among the scientific community in the early twentieth century.

Louis Bachelier, Stochastic Processes, and Mathematical Finance

When random fluctuations became a subject of scientific studies in the 1900s, almost all the researchers tackled this topic from the perspective of statistical physics. The French mathematician Louis Bachelier was a notable exception. In a doctoral dissertation he defended at the University of Paris Sorbonne in

[111] Ibid., 333.

[112] Paul Langevin, "Statistique et déterminisme," Conférence au Centre International de Synthèse, June 8, 1935, 1–9; in Carlton 172, "Diverse scientific manuscripts," L172/14, Langevin Papers.

[113] Ibid., 10.

1900, he did not examine colloidal solutions, critical opalescence, phase transition, magnetization, or electronic noise, but looked into price volatility in a financial market. Bachelier's thesis modeled a bond's or stock's price over time as a sheer random and continuous "movement" that did not depend on its past records. He deduced its probability distribution and demonstrated the price's spread over time, derived a governing "equation of propagation" for this probability distribution, and identified important statistics of this "movement." The price of a financial product in Bachelier's treatment had the same mathematical structure as Brownian motion in Einstein's and Smoluchowski's works. Bachelier's findings echoed but preceded Einstein's and Smoluchowski's conclusions.

It is not easy to pin down Bachelier's position in the history of random fluctuation studies and the history of noise at large, though. Since he uncovered some major statistical properties of random fluctuations half a decade before Einstein and Smoluchowski, he is sometimes viewed as the originator of stochastic-process theory.[114] Moreover, because his work arguably marked the beginning of systematic investigations into financial markets via probability theory and stochastic analysis, he is often proposed as the founder of mathematical finance. He is especially esteemed as the forerunner who paved the way for Fischer Black, Myron Scholes, and Robert Merton's Nobel-Prize-winning research in the 1970s on the stochastic theory for pricing options.[115]

Yet, Bachelier had limited influence in the first half of the twentieth century, the crucial period when the central concepts, material cultures, research problem situations, and technical tools associated with the abstraction of noise as random fluctuations took shape. His financial theory fell into oblivion within the economist community until a few scholars on the US east coast "rediscovered" it in the 1950s. For decades, Bachelier's mathematical studies of stochastic movements never reached physicists and engineers working on random fluctuations in statistical mechanics and telecommunication technology. Neither did these studies have a clear linkage to the contemporaneous development of modern statistics as a mathematical science to process and analyze massive quantities of data from biometry, demography, and agriculture. His

[114] Cohen, "The history of noise" (2005), 32.

[115] Mark Davis and Alison Etheridge (translators and commentators), *Louis Bachelier's* Theory of Speculation: *The Origin of Modern Finance* (Princeton, NJ: Princeton University Press, 2006), 1–14, "Mathematics and finance"; Jean-Michel Courtault, Yuri Kabanov, Bernard Bru, Pierre Crépel, Isabelle Lebon, and Arnauld Le Marchand, "Louis Bachelier: On the centenary of *Théorie de la Spéculation*," *Mathematical Finance*, 10:3 (2000), 341–353; Edward Sullivan and Timothy Weithers, "Louis Bachelier: The father of modern option pricing theory," *The Journal of Economic Education*, 22:2 (1991), 165–171.

work was noticed and cited exclusively within the circle of mathematicians who endeavored to build an axiomatic, set-theoretic, or measure-theoretic system to characterize stochastic processes in functional spaces.[116] The significance of Bachelier's mathematical work to the studies of empirical random fluctuations did not become salient until this "pure" mathematics of stochastic processes engaged substantially engineering, economics, and physics after World War II.

In the year of the Franco-Prussian War, Louis Bachelier was born into a bourgeois family at Le Havre in Normandy. His father was a wine merchant, and his mother a banker's daughter. Soon after he graduated from high school, his parents passed away, forcing him to take over his family business and take care of his young siblings. Military service further delayed his college education until 1892, when he eventually entered the Sorbonne as an undergraduate and majored in mathematics. He obtained his B.S. in 1895, passed the agrégation in mathematical physics in 1897, and continued to pursue his Ph.D. In March 1900, he defended his thesis at the Sorbonne.[117]

As Bachelier's thesis committee members Henri Poincaré, Paul Appell, and Joseph Boussinesq remarked in their report, "the subject chosen by M. Bachelier is rather far away from those usually treated by our candidates."[118] Titled *Théorie de la Spéculation* (*A Theory of Speculation*), Bachelier's thesis employed probabilistic calculus to analyze the transactions at the Paris Stock Exchange (Paris Bourse). While this was an unusual, and perhaps even slightly questionable, topic for mathematicians at the time, Bachelier's family background in banking and business experience gave him sufficient knowledge about the financial market to undertake this project. The primary transactions he considered were two types of what we now call *derivative* financial products: forwards and options. At Paris Bourse, they were linked to bonds issued by the French government since the Ancien Régime, reformed by Napoleon, fueled by military campaigns, colonial expenditures, and public constructions throughout the nineteenth century, and turned into a prominent tool of investment at the beginning of the twentieth century.[119]

In a forward contract, a buyer purchased a quantity of bonds from a seller at their price on the date of the contract, but the transaction was completed on a designated future "liquidation date." That is, the buyer paid for the asset at

[116] Davis and Etheridge, *Bachelier's* Theory of Speculation (2006), 80–115, "From Bachelier to Kreps, Harrison and Pliska"; Courtault et al., "Louis Bachelier" (2000), 347.

[117] Courtault et al., "Louis Bachelier" (2000), 341–343.

[118] Paul Appell, Henri Poincaré, and Joseph Boussinesq, "Report on Bachelier's thesis (March 29, 1900)," translated and reprinted in Davis and Etheridge, *Bachelier's* Theory of Speculation (2006), 77.

[119] Ibid., 8–10.

its price on the contract date but owned the asset at its price on the liquidation date. In an option contract, a buyer bought a quantity of bonds from a seller in the same way as that in a forward contract but with a crucial difference: the buyer or the seller (the first party) had the option to terminate the transaction if the bonds' price on the liquidation date was lower (for the buyer) or higher (for the seller) than a pre-designated striking price. To compensate for the financial disparity in this contract, the first party paid a premium to the second. Bachelier devoted the first quarter of his thesis to the algebraic and geometric representation of forwards and options in terms of their profits and variations of an asset's market price.[120]

The amount of gain or loss for a buyer or a seller in a forward or option contract depended on the variation of the asset's price between the contract date and the liquidation date: a rising price would result in a profit for the buyer, whereas a falling price would result in a profit for the seller. Thus, knowing a bond's price movements was important for investors ("speculators" in Bachelier's term). Such information was probabilistic, for there were too many factors and stakeholders on the market. Bachelier distinguished two types of probabilities for asset prices: one that could be determined a priori like that in a game of chance (which he named "mathematical probability"), and the other that depended on the events during the market movement (which could not be foreseen). Speculators looked for information about the second type of probability: the government's fiscal performance, harvest, industrial output, war, etc. But only the first type of probability could be subjected to a mathematical analysis. To do so, Bachelier posed a basic assumption about the market: "At a given instant, the market believes neither in the rise nor in the fall of the true price."[121] In other words, for mathematical probability, the change of a bond's or stock's price was random at any given moment; it could be positive or negative, and did not depend on the product's past price record. This price "movement" resembled a random walk.

After introducing the basic assumption about the market, Bachelier established the form of mathematical probability governing an asset price. He considered the "relative price" of a bond or a stock as the price difference between the end and beginning of a specified duration. The probability that an asset's price at time t relative to that at time 0 fell between x and $x + dx$ was denoted as $p(x, t)dx$. He examined the situation when the relative price at time t_1 fell between x and $x + dx$ and the relative price at time $t_1 + t_2$ fell between z

[120] Ibid., 16–26.
[121] Ibid., 26.

and $z + dz$. The probability for this situation could be obtained by the "principle of compound probabilities," which was the Bayesian rule of conditional probability:

$$\begin{aligned}&\text{Prob}\{\text{price change} = [z, z + dz]\ \text{during}\ t_1 + t_2\ \&\ \text{price change}\\&= [x, x + dx]\ \text{during}\ t_1\}\\&= \text{Prob}\{\text{price change} = [z, z + dz]\ \text{during}\ t_1 + t_2 | \text{price change}\\&= [x, x + dx]\ \text{during}\ t_1\}\\&\times \text{Prob}\{\text{price change} = [x, x + dx]\ \text{during}\ t_1\}.\end{aligned} \tag{5.15}$$

In this expression, Prob{price change = $[x, x+dx]$ during t_1} = $p(x, t_1)dx$. Bachelier claimed that the conditional probability Prob{price change = $[z, z+dz]$ during $t_1 + t_2$ | price = $[x, x+dx]$ during t_1} = $p(z-x, t_2)dz$. This claim was true if the price change from $t = 0$ to t_1 was independent of that from $t = t_1$ to $t_1 + t_2$—a property of a Markov process as we know today. Bachelier did not articulate explicitly this "memoryless" assumption, but it was consistent with his random-walk model for the market. Under this supposition, he obtained that Prob{price change = $[z, z+dz]$ during $t_1 + t_2$ & price change = $[x, x+dx]$ during t_1} = $p(x, t_1)p(z-x, t_2)dxdz$. Moreover, the probability that the price change during $t_1 + t_2$ fell within $[z, z+dz]$ should be the sum of all the cases in which the price change during t_1 took any plausible value. In other words,

$$p(z, t_1 + t_2) = \int_{-\infty}^{\infty} p(z, t_1)\, p(z - x, t_2)\, dx. \tag{5.16}$$

This was Bachelier's core probabilistic law for the (relative) market price.[122]

From (5.16), Bachelier deduced several crucial properties for the market price's probability distribution $p(x, t)$. First, he demonstrated that a Gaussian distribution with its variance proportional to time was a solution of (5.16). This solution took the form

$$p(x, t) = \frac{1}{2\pi k_p \sqrt{t}} \exp\left[-\frac{x^2}{4\pi {k_p}^2 t}\right]. \tag{5.17}$$

Bachelier's solution took the same form as (5.3), Einstein's and Smoluchowski's solutions to the problem of Brownian motion. Like the displacement of a suspended particle in a solution, the price change of an asset on Bachelier's model

[122] This formulation was also a special case of what was later known as the Chapman–Kolmogorov equation that related a Markov process's joint probability distributions in different coordinate systems. In his mathematical examination of the Markov processes in 1931, Kolmogorov referred to Bachelier's formulation. See ibid., 29–30.

market undertook a random movement, followed a normal distribution, and spread to an ever broader range with the square root of time.[123]

Second, Bachelier derived (5.17) from an alternative approach that treated the variation of an asset price over time as a sequence of discrete movements with infinitesimal time and price steps. In this treatment, the variation of the price over time resembled a series of discrete one-dimensional random walks: at each time step, the price had an equal chance to increase or decrease by a fixed minuscule amount. Through combinatorics, he determined the number of possible combinations for reaching a given price after a certain number of time steps, and its ratio over the number of all possible combinations after the same number of time steps. Treating this ratio as the probability of having the given relative price after the fixed duration and taking an asymptotic approximation for a very large number of time steps, he obtained the same form of probability as (5.16). This again demonstrated that the Gaussian distribution with a spreading variance was consistent with the random-walk model.[124]

Third, from the probability distribution $p(x, t)$, he constructed the cumulative probability for the price change to be above a given value x, or $P(x, t) = \int_x^\infty p(x', t)\,dx'$. He explored a numerical method to compute the integral with a faster rate of convergence. Analogous to Fourier's treatment of heat flow, Bachelier considered the probability as a "flux," or a balanced quantity with the amount coming into and escaping out of a differential price window within a differential time slot. From this, he obtained the same diffusion equation for P as Fourier did for heat conduction, or as Einstein did for Brownian motion in (5.2):[125]

$$\frac{\partial P}{\partial t} = \frac{1}{c^2}\frac{\partial^2 P}{\partial x^2}. \tag{5.18}$$

This provided Bachelier with an opportunity to use the same mathematical apparatus in Fourier's theory to inspect the probability distribution for the market price.

Fourth, Bachelier considered the probability that a given price x would be attained or exceeded during an interval of time t. From the random-walk conceptualization of price variations, he invoked the French mathematician Désiré André's reflection principle in combinatorics to argue that this probability would be twice the probability that the price x was attained or exceeded

[123] Ibid., 29–33.
[124] Ibid., 33–36.
[125] Ibid., 37–42.

at the time instant t, which was by definition the cumulative probability $P(x, t)$.[126] With these mathematical results, Bachelier computed the probabilities for making different amounts of profits from forwards or options in various situations, the likely time periods for attaining profits, and compared the numerical outcomes with statistical data from Paris Bourse.[127]

Bachelier continued to work on probability theory after he obtained his Ph.D. in 1900. He published his thesis in *Annales Scientifiques de l'École Normale Supérieure* in the same year.[128] In the following four decades, he produced fourteen journal articles and five monographs on gambling, diffusion and collisions of particles, memoryless random processes, and continuous and discontinuous probabilities. Yet, his academic career did not proceed smoothly. In 1900–14, he did not have a formal academic position, and depended on fellowships and the family business for living. Although he taught mathematics at the Sorbonne, he was effectively a "free professor" and did not get paid by the university. When the University of Paris was about to make Bachelier's appointment formal in 1914, World War I broke out and he was drafted as a private in the French Army, where he served until 1918. In 1919, he obtained his first formal academic position as an assistant professor at the University of Besançon as a substitute for a permanent faculty member who was on leave. In the following decade, he moved around between the provincial universities at Besançon, Dijon, and Rennes as a substitute professor. When he was pursuing a permanent position at Dijon in 1926, he was rejected by the university's Council of the Faculty of Science largely because of a report from the mathematics professor at École Polytechnique, Paul Lévy. As a much more powerful figure in French academia and a founder of the measure theory of probability, Lévy in his report waged a strong critique of Bachelier's work on probability and random processes for what the former perceived to be conceptual mistakes. This infuriated Bachelier, who wrote a long letter of refutation. Bachelier eventually obtained a permanent professorship at Besançon in 1927 and made peace with Lévy later. But he never entered France's central mathematical circle in Paris throughout his career.[129]

Similarly, Bachelier's research on random fluctuations had mixed degrees of reception. In the first half of the twentieth century, his work was known

[126] Ibid., 62–66.
[127] Ibid., 45–62, 67–77.
[128] Louis Bachelier, "Théorie de la spéculation," *Annales Scientifiques de l'École Normale Supérieure*, series 3, 17 (1900), 21–86.
[129] Courtault, "Louis Bachelier" (2000), 344–346.

and cited exclusively by mathematicians developing the conceptual foundation and technical tools for stochastic analysis. They included Norbert Wiener at MIT, who tried to come up with a measure-theoretic characterization of Brownian motions in the 1920s; Andrey Kolmogorov at the Moscow State University, who examined Markov processes and explored an axiomatic definition of probability and random processes in the 1930s; and Joseph Doob at the University of Illinois Urbana-Champaign, who introduced a theory of martingales, a special kind of Markovian random processes. During this period, physicists or engineers studying random fluctuations in statistical mechanics or noise in sound reproduction systems were unaware of Bachelier's achievement. Bachelier's work on mathematical finance was simply ignored by the economist community. According to a famous legend in the field, economists' interest in the Bachelier-type modeling of asset prices as random processes only began in the 1950s when the Yale statistician Jimmy Savage sent a dozen or so postcards to economists around the US to bring to notice the French mathematician's work. Paul Samuelson at MIT was an inspired recipient who further developed this stochastic modeling for the financial market.[130]

Time-Domain Approach to Random Fluctuations

In the first two decades of the twentieth century, physicists and mathematicians started to pay close attention to random fluctuations in the empirical world: Brownian motions of suspended particles, variations of solution concentration, critical opalescence, and irregular changes of bond or stock prices on a financial market. While these researchers were heavily influenced by probability theory and statistical mechanics developed in the nineteenth century, the phenomena they examined posed new challenges and brought about new features: The atomic and molecular agitations did not smooth out on average, but were rather amplified into tangible fluctuations; the random changes of quantities occurred not in discrete steps, but over a continuous time or another parameter; etc. Einstein, Smoluchowski, Langevin, and Bachelier were among the first to explore a theoretical framework for the studies of such phenomena. They conceived fluctuations as time-evolution of random quantities, found their probability distributions as functions of time or constructed

[130] Paul Samuelson, "Foreword," and Mark Davis and Alison Etheridge, "Preface," in Davis and Etheridge, *Bachelier's* Theory of Speculation (2006), vii–xi, xiii–xv. For a more detailed historical account of Samuelson's "rediscovery" of Bachelier, see Donald MacKenzie, *An Engine, Not a Camera: How Financial Models Shaped Markets* (Cambridge, MA: MIT Press, 2006), 63–66.

their equations of motion that attributed the source of randomness to a certain stochastic force, and deduced their central statistical properties such as their mean or variance. To come up with such mathematical formulations, they resorted to equipartition in statistical physics, random walks, Bayesian probability calculus, and the implicit memoryless assumption. By the 1910s, this "time-domain" approach to random fluctuations had become familiar within the physicist community.

Shortly after the inception of the time-domain studies of random fluctuations, engineering scientists working on telecommunications noted similar agitations that gave rise to sonically noisy outputs in their technological systems. And they noted the relevance of statistical physics, probability theory, and stochastic process theory to their understanding of and coping with the noise they tried to tame. In contrast to statistical physicists and mathematicians, however, these engineering scientists followed the longstanding practice in electrical engineering to explore the spectral characteristics of random noise, i.e. its properties at distinct frequencies. This frequency-domain approach to random fluctuations in telecommunication systems will be the topic of chapter 6.

6

Electronic Noise in Telecommunications

Figuring out Random Fluctuations in Telephony and Radio

By the early twentieth century, noise started to gain a broader meaning than cacophony owing to the popularity of sound reproduction technologies. Surface noise in phonography, crosstalk in telephony, and statics in radio not only denoted disturbing sound but also embodied its mechanical, electrical, or atmospheric causes. The various methods and instruments devised by engineers and inventors—ear-balancing audiometer, oscillographic display, audiometer, acoustimeter—turned noise into measurable, quantifiable, and manipulable entities. As noise became an abstract concept and an object of quantification, there were increasing attempts to understand it theoretically: What physical processes generated noise? How to represent it mathematically? How to predict or estimate its effects on sound reproduction?

Physicists' and mathematicians' studies of random fluctuations in the 1900s provided the perspectives, methods, and tools for engineers to fill the theoretical void for noise. Probability theory and statistical mechanics became crucial intellectual resources for engineering researchers to grapple with noise in sound reproduction, especially that related to electrical telecommunication. Einstein's, Smoluchowski's, and Langevin's works on Brownian motion offered either conceptual inspirations or direct theoretical templates for treating noise. In the 1910s–20s, the physico-mathematical research on random fluctuations and the development in the quantification and metrology of noise in telephony and radio came to an unexpected conjugation.

Among numerous types of noise, engineering researchers in the early twentieth century paid particular attention to the one that they believed to be caused by random agitations of electrons in electrical devices and circuits—thermionic tubes and resistors in particular—of which telecommunication systems were made. The phenomena of interest involved the disturbances and fluctuations of electrical current due to the movements of microscopic charge carriers in electronic tubes and other circuit components. The "Schottky" (shot) noise and "Johnson" (thermal) noise were the best-known examples of such effects.

Dubbed "electronic noise," this kind of disturbance was not especially severe for sound reproduction. Electronic noise gave rise to a homogeneous cracking and hissing background at the output of a speaker or an earphone. This background was often weaker than much more conspicuous sonic disturbances such as atmospherics in wireless telegraphy, crosstalk in telephony, interference from stations at adjacent frequency bands in radio, or ambient dins. Electronic noise had a milder aural effect than some other types of sonic disturbances, which either incurred a more unpleasant sensation or caused a more acute disruption to voice communication. Yet, electronic noise bore certain theoretical significances with which surface noise, statics, crosstalk, interference, and dins could not compare. To researchers in the 1910s–20s, it embodied the same kind of random fluctuations as were responsible for Brownian motion, critical opalescence, irregular variations of concentration in a solution, and demagnetization above the Curie temperature.

The material culture of electronic noise was different from that appealing to Einstein, Smoluchowski, and Langevin. The former comprised tube amplifiers, resistors, inductors, capacitors, microphones, loudspeakers, cables, and antennas, while the latter involved ultramicroscopes, glassware, thermostats, emulsion, and magnets. Like Brownian motion of suspended particles in a liquid, however, electronic noise was the outcome of microscopic electric-charge carriers' fluctuations "amplified" into the macroscopic, observable level due to the statistical properties of random walk, combinatorics, and the partition function. In other words, electronic noise was considered intrinsic to the physical nature of telecommunication systems. Distinct from surface noise, crosstalk, interference, atmospherics, or din, electronic noise was not the result of a faulty engineering design, incorrect uses of technology, malfunctioning of systemic components, or intrusive factors from outside, but rather an outcome of atoms' fundamental properties. In contrast to many other types of noise, therefore, electronic noise could not be eliminated; it could only be controlled. Because of this feature, electronic noise was often viewed as the default disturbance of an electrical sound reproduction system. After correcting all design mistakes, stripping all operational failures, and removing all external interferences, electronic noise still stayed. It was an unavoidable and fundamental uncertainty of electrical sound reproduction.

Historians have noted the connection between random fluctuations and electronic noise. Günter Dörfel and Dieter Hoffmann examined the theoretical reasoning and experimental findings that led to the formulation of Schottky and Johnson noise, and situated these studies in the context of attempts to come to grips with the fundamental limit of measurement.[1] According to them, some physicists in the 1900s–20s considered electronic noise intrinsic fluctuations of electrical currents, similar to Brownian motion of pollen dust. From Einstein's schematic electrostatic electrometer for amplification of fluctuating electricity and Geertruida de Haas-Lorentz's analysis of thermal agitation's effect on metrology to Frits Zernike's and Gustaf Ising's examinations of a galvanometer's sensitivity, electronic noise seemed to impose an ultimate constraint on the accuracy of electrical measuring devices. The thermal agitations of galvanometers or electrometers at non-zero temperature reduced their sensitivity and restricted the minimum signal level they could measure. Before the uncertainty principle and the notion of quantum indeterminancy, Brownian noise was conceived as the fundamental barrier to the exactness of measurement.[2]

Electronic noise indeed posed challenges in the early twentieth century as scientists attempted to grapple with its implications for metrological accuracy and precision. While the discussions concerning these challenges constituted an important aspect of noise research during this period, another equally significant aspect was the specific engineering context within which such research proceeded. The purpose of this research, after all, was to improve the quality of thermionic tubes and electronic circuits that were just becoming popular in radio and telephony. After the invention of feedback amplifiers in the 1910s, vacuum-tube circuits were designed to amplify minute electrical signals with minor distortion.[3] As weaker and weaker signals could be detected, the previously inconsequential "background noise" became an increasing nuisance, and hence demanded a closer examination.

The studies of shot noise and thermal noise in the 1910s–30s concerned either the properties of thermionic tubes or the functionality of other electrical devices in tube circuits. Here engineering scientists were not necessarily concerned to single out one noise-producing factor over another as worthy of understanding from a fundamental standpoint. To them, no particular noise source was considered more "fundamental" than a plethora of other sources of uncertainties, for alleviating disturbances that corrupted signal transmission was their goal. In that context, the intrinsically stochastic character of Brownian motion was just one factor—albeit an important one—among those that

[1] Dörfel and Hoffmann, "Von Albert Einstein bis Norbert Wiener" (2005); Dörfel, "Early history of thermal noise" (2012).

[2] Dörfel and Hoffmann, "Von Albert Einstein bis Norbert Wiener" (2005), 12–17. Also see Beller, *Quantum Dialogue* (1999), 92–93.

[3] For the invention of feedback amplifiers, see David Mindell, "Opening Black's box: rethinking feedback's myth of origin," *Technology and Culture*, 41:3 (2000), 405–434.

generated electronic noise. Other, system-dependent factors, such as the aggregation of electric charge at different degrees of vacuum in thermionic tubes, resistors, and other electrical components, also played significant roles.

The engineering background of the early electron-noise studies led to a crucial theoretical development. The researchers on the Schottky and Johnson effects in the 1910s–20s were indebted to the theoretical legacies from the examinations of random fluctuations in the 1900s, and invoked intellectual resources from probability theory, stochastic processes, and statistical mechanics. In contrast to Einstein, Smoluchowski, Langevin, and their like, however, the engineering researchers working on electronic noise refrained from inquiring into the time-evolution of random electrical fluctuations. They did not attempt to formulate the probability density function of the agitating electrical current passing through a thermionic tube or a resistor, to seek the density function's governing equation, or to determine the time-dependence of the fluctuating current's statistics such as mean or variance. Rather, these researchers focused on the spectral properties of the fluctuating current. As Dörfel has indicated, this "frequency-domain approach," which differed from statistical physicists' "time-domain approach" in the studies of Brownian motions, was closely connected to the disciplinary matrix and practice of electrical and sound engineering.[4] There were important practical reasons for this focus on spectral properties: engineers needed to know the noise's energy distribution at distinct frequencies in order to design signals, filters, amplifiers, and electroacoustic transducers with pertinent frequency responses that could mitigate the impacts of noise. This preoccupation with spectral analysis in sound reproduction from Helmholtz, Edison, and Bell, to Marconi, Lee de Forest, and Harvey Fletcher configured the early research on electronic noise and prompted a new theoretical direction to inspect random fluctuations' frequency responses. This frequency-domain approach would eventually bring about important breakthroughs in the mathematical theory of random fluctuations and stochastic processes in the 1930s–40s.

The pragmatic mandate of electronic-noise research aligned with the agendas of the individuals and organizations that conducted it. Whereas Einstein, Smoluchowski, Langevin, and Bachelier were academics at universities (or at least aspired to be so), the early researchers on electronic noise—Walter Schottky, John B. Johnson, Harry Nyquist, Carl A. Hartmann—held a new kind of professional identity that barely existed before the twentieth century. Although they received advanced academic training at research universities

[4] Dörfel, "Early history of thermal noise" (2012), A117–A121.

and had doctoral degrees in physics, they worked at industrial corporations and devoted their efforts to the development and improvement of technologies. The establishments where they worked, e.g. AT&T Bell Telephone Laboratories and Siemens Research Laboratory, were not research facilities in academia or governments, but laboratories administered by manufacturing or infrastructure companies. The theoretical and experimental investigations on electronic noise therefore testified to the rapid rise of what historians call industrial or corporate research in the US and Germany in the 1870s–1930s.[5]

Walter Schottky's Route to the "Schroteffekt"

Engineers of sound reproduction had known about electrical circuits' fluctuating performance. Their systems and components were never perfect enough. Depending on quality control in manufacturing, the functioning of coils, condensers, resistors, and wires varied over time. Battery output was not steady. The new devices of radio-signal detectors, thermionic tubes, earphones, and microphones had even more uncertainties. The contemplation about the "fundamental" and "intrinsic" fluctuations in electrical circuits as a statistical property began with Brownian-motion studies in the 1900s. Einstein was the first individual to entertain this thought.

In 1907, two years after his proposal for a statistical theory of Brownian motion, Einstein suggested an analogy between irregular movements of suspended particles in a fluid due to molecules' random agitations and a fluctuating electrical current in a circuit due to electrons' stochastic motions. To make this analogy, he claimed, no detailed model of microscopic interactions was needed. Rather, only a general kinetic theory of matters was required. According to thermodynamics, a physical system under equilibrium was at a specific state whose physical parameters had fixed values (e.g. the system's parameter λ stayed at λ_0). These values corresponded to the state of maximum entropy. According to the molecular theory of heat, these parameter values were not deterministic, and there was a non-zero probability for a parameter to fall within a region surrounding its supposed value at thermodynamic equilibrium. That is, λ had a chance to be at other values than λ_0.

From statistical mechanics, the probability density for λ followed a Boltzmann distribution $dW = W(\lambda)\, d\lambda = C \cdot \exp[-(A/kT)] \cdot d\lambda$, where k was Boltzmann's constant, T was the system's temperature, and A was the amount

[5] For example, see Leonard Reich, *The Making of American Industrial Research: Science and Business at GE and Bell, 1876–1926* (Cambridge, UK: Cambridge University Press, 1985).

of mechanical work required to bring the system from its thermodynamic equilibrium at λ_0 to the state corresponding to λ along a reversible path. In other words, the system had an intrinsic thermal fluctuation over its parameter λ. When the system state under consideration was sufficiently close to the thermodynamic equilibrium and λ was close to λ_0 with a small deviation ε so that $\lambda = \lambda_0 + \varepsilon$, Einstein contended that the amount of required work was close to the second order of ε, $A = a\varepsilon^2$, for the energy at equilibrium should be minimum and its first order of ε vanished. Plugging this relation into λ's probability distribution and averaging over A, Einstein obtained the mean value of the work that brought the system from equilibrium to its current state: $\langle A \rangle = kT/2$.[6]

This mean work was also the average amount of energy corresponding to the non-equilibrium state at $\lambda = \lambda_0 + \varepsilon$. That was tantamount to claiming that the physical system could fluctuate away from its thermodynamic equilibrium with a Boltzmannian probability distribution and an average fluctuating energy $kT/2$. To Einstein, this claim applied not only to the familiar material systems in thermodynamics and statistical mechanics such as an oversaturated solution or a liquid under phase transition, but also to the components and instruments of electrical technology such as an electrical circuit.

He considered a short-circuited condenser with capacitance C, a pair of parallel conducting plates connecting to each other with a metal wire. From the theory of electromagnetism and electric circuits, nothing would happen on this device and no electrical voltage could be detected across the condenser, since the circuit was not connected to any source of power like a battery or an electrical generator. Yet, his theory of random agitations stipulated that this null condition was only the state at thermodynamic equilibrium. In reality, electrical energy due to microscopic fluctuations would be excited across the circuit, and one could detect a fluctuating voltage. Since the energy contained in a charged condenser followed $E = CV^2/2$ from electromagnetic theory (where V was the voltage across the plates), the fluctuating energy corresponded to a fluctuating electrical potential across the condenser $\langle E \rangle = C\langle V^2 \rangle/2 = kT$, or equivalently, $\sqrt{\langle V^2 \rangle} = \sqrt{2kT/C}$. Based on the value of capacitance for a typical on-the-shelf condenser (two interlocking sets of 30 plates each, adjacent-plates separation of 1 mm, and size of each plate as 100 cm^2), he estimated the average fluctuating voltage across the condenser at

[6] Albert Einstein, "Über die Gültigkeitsgrenze des Satzes von thermodynamischen Gleichgewicht und über die Möglichkeit einer neuen Bestimmung der Elementarquanta," *Annalen der Physik*, 22 (1907), 569–571.

room temperature to be in the range of 10^{-6} volts. This value could be amplified 10 or even 100 times via particular circuit arrangements. But to Einstein, the voltage was still too weak for the state-of-the-art measuring technology to detect. Before a substantial improvement of electrical metrology, Einstein's prediction of the fluctuating voltage remained a theoretical speculation.[7]

Einstein's idea about random fluctuations in electrical circuits was further developed by Geertruida de Haas-Lorentz. Daughter of the renowned Dutch physicist Hendrik Antoon Lorentz, Geertruida pursued a Ph.D. in physics at the University of Leyden in the early 1910s under her father's supervision. The subject of her thesis concerned Brownian motion, and what she conceived as "related phenomena." One of such phenomena was voltage and current fluctuations in electrical circuits. Comparable to Einstein's case, de Haas-Lorentz utilized the relation that the energy of a coil with inductance L and current I was $E = LI^2/2$, and obtained via solving the corresponding Langevin equation that an open-circuited inductor would have a fluctuating current determined by $\langle E \rangle = L\langle I^2 \rangle/2 = kT$, or equivalently, $\sqrt{\langle I^2 \rangle} = \sqrt{2kT/L}$. She also considered more complex electric circuits with multiple capacitors and inductors and calculated their fluctuating voltages and currents.[8]

In Einstein's and de Haas-Lorentz's works, electronic fluctuations were a theoretical entity. No one in the early twentieth century was able to observe its presence. If they existed, they were a weak physical effect that only sensitive scientific instruments could detect, not a type of recognizable disturbances that engineers of sound reproduction technologies should worry about. By World War I, the development of telephony and wireless telegraphy had made electronic fluctuations a non-negligible factor in system and device design. A researcher to explore their engineering implications was the German physicist Walter Schottky.

Schottky as an Industrial Physicist

Today, Walter Schottky is remembered for his achievements in electronic technology: exploration of the metal–semiconductor junction in solid-state devices and electric discharge in thermionic tubes; invention of a new electronic tube, microphone, and radio-receiver architecture. Although he

[7] Ibid., 571–572.

[8] Dörfel and Hoffmann, "Von Albert Einstein bis Norbert Wiener" (2005), 10–12; Geertruida de Haas-Lorentz, *Over de Theorie van de Brown'sche Beweging en Daarmede Verwante Verschijnselen*, Ph.D. dissertation (Leyden: University of Leyden, 1912), 79–99.

received training in theoretical physics at one of the field's central institutions and taught in academia for years, his expertise has often been associated with "industrial physics" or "applied science." This realm aligned with "technical physics" in German academia or British physicists' consultation with manufacturing or telecommunication industries, but did not become fully institutionalized until the launch of laboratories at large corporations in the US and Germany in the early twentieth century. Thus, it was individuals like Schottky, not Einstein or de Haas-Lorentz, who had a more acute sense of the technological relevance of electronic noise.

Son of a mathematician, Walter Hans Schottky was born in Zurich, Switzerland, and moved to Berlin when his father was appointed professor at the Friedrich-Wilhelm University in Berlin. He studied physics, took courses with Max Planck and Walter Nernst, and pursued a Ph.D. under Planck's supervision. His research was on the dynamics of electrons near the relativistic limit. Einstein's theory of relativity in 1905 predicted that the mass of an electron increased with speed. When an electron's speed was close to the light velocity, its dynamics would be significantly different. This velocity dependence of an electron's mass was soon observed experimentally by a Königsberg-based physicist, Walter Kaufmann. Although Kaufmann's interpretation of his experimental results differed from Einstein's prediction, Planck as a steadfast supporter of Einstein's theory held that the studies of electrons' behavior at high speeds opened an avenue for relativistic physics. Schottky's dissertation fit Planck's general interest in relativistic electrodynamics and relativistic theory of electrons. Schottky defended his thesis in 1912.[9]

After receiving his Ph.D. in Berlin, Schottky switched to experimental physics. As he recalled later, "I had finished a lengthy theoretical work. Now—as remedy for the pure abstraction—an experimental interlude should follow connected with a change of location. After the long years of my studies in Berlin this could be granted to me." He decided to move to Jena and work with Max Wien, cousin of the Nobel Laureate and pioneer of black-body radiation and quantum physics Wilhelm Wien and researcher on electrical metrology, wireless telegraphy, and electronic tubes. In Jena, Schottky participated in Wien's graduate seminar for experimental physics, with the thought that "it should

[9] Otfried Madelung, "Walter Schottky (1886-1976)," *Festkörperprobleme*, 26 (1986), 1–2; available in Box 062, Walter Schottky Papers (NL 100), Deutsches Museum Archives, Munich, Germany; Reinhard Serchinger, "Walter Schottky und die Forschung bei Siemens," in Ivo Schneider, Helmuth Trischler, and Ulrich Wengenroth (eds.), *Oszillationen: Naturwissenschaftler und Ingenieure zwischen Forschung und Markt* (Munich: R. Oldenbourg Verlag and Deutsches Museum, 2000), 172. For the details of Kaufmann's experiment and the subsequent development of the relativistic theory of electrons, see Staley, *Einstein's Generation* (2008), 219–293.

be left to fate if I would find something of interest for further deeper investigation."[10] He did, and entered applied physics. His point of contact was once again an examination of electrons, but this time in the context of devices for telecommunications, not relativistic electrodynamics.

Schottky's first task as Wien's assistant was to measure electrical currents on light-illuminated metal surfaces. In 1887, when Heinrich Hertz in Karlsruhe was experimenting with spark-gap discharge that would lead to the discovery of radio waves, he found that the intensity of induced sparks changed with ultraviolet light influx.[11] Dubbed the photoelectric effect, this phenomenon was closely studied in the following decade. By the early twentieth century, scientists generally believed that light illuminated on metal provided its electrons sufficient energy to escape into space and form an electric current. An empirical rule from this study was the so-called Stoletov's law (à la Russian physicist Aleksandr Stoletov) that the strength of the photoelectric current was proportional to light intensity.[12]

At Wien's laboratory, Schottky found that the measured current remained weak regardless of light intensity, which contradicted Stoletov's law. This result implied that electrons were prevented from leaving a metal surface. Schottky conjectured that space charge, the electric charge in air surrounding the metal surface, created a potential barrier against further electronic motions. Based on this hypothesis, he derived a "$U^{3/2}$ law"—the space-charge-limited current I was proportional to the 3/2 power of the voltage U across two metallic plates and to the inverse square of the plates' separation d, i.e. $I \propto U^{3/2}/d^2$. He discovered that the space-charge effect did not impose a limit on photoelectric current at room temperature. But it was the dominant factor for current emitted from heated cathodes, meaning that the $U^{3/2}$ law was applicable to electronic tubes. This motivated him to investigate vacuum tubes. He measured tube current and voltage with different cathode materials and filament temperatures. The voltage-current curves from his measurements were consistent with the $U^{3/2}$ law, verifying his theory of the space-charge effect. Unfortunately, Schottky was not the first to discover this. Irving Langmuir at General Electric and Clement Child at Cornell University had independently

[10] Quoted in Madelung, "Schottky" (1986), 4.

[11] Heinrich Hertz, "On an effect of ultraviolet light upon the electric discharge," in Hertz, *Electric Waves* (New York: Dover, 1962), 63–79.

[12] Aleksandr Stoletov, "Sur une sorte de courants electriques provoques par les rayons ultraviolets," *Comptes Rendus de l'Académie des Sciences*, 106 (1888), 1149.

found and published the same law in 1913–14. Schottky's paper came months later.[13]

Schottky's experience with the $U^{3/2}$ law shaped his future career on the studies of electronic tubes, semiconductors, and physical chemistry. In 1914, he continued electronic-tube experiments at the University of Berlin. He presented his work on the $U^{3/2}$ law in 1915 at the university's physics colloquium. The presentation impressed a Swedish physicist, Ragnar Holm, who was an attaché from Siemens & Halske Aktiengesellschaft—Germany's largest electrical manufacturers. (Siemens sent research staff to scientific conferences and college colloquia to build connections with academia.) Holm invited Schottky to conduct research for Siemens. Schottky was eager to help the nation win the ongoing World War with more "useful" work. Working for Siemens was preferable to pure scholarly research. Though still at the University of Berlin, he received funding from the firm to build a laboratory for vacuum-tube experiments. Reckoning however on the increasing demand for military technology, he left the university in 1916 to work directly at Siemens Research and Development. In 1917, he became head of "K Laboratory" of Siemens's general research establishment (Werner-Werks für Bauelemente).[14]

While the corporate-sponsored scientific research in Germany had prevailed in chemical and optical industries by the second half of the nineteenth century, the advancement of electrical technology opened a new page. As a major German manufacturer for electrical power and telecommunications systems and devices, Siemens & Halske Gesellschaft integrated in 1905 its existing small laboratories at production facilities in Berlin into a research laboratory. Werner Bolton, a chemist who invented a tantalum filament for Siemens' incandescent light bulbs, assumed directorship of the laboratory. After Bolton's death in 1912, a physicist Hans Gerdiens succeeded his position and turned the establishment into a more formal Laboratory for Physics and Chemistry, with an ambitious plan for a new building housing seven

[13] Madelung, "Schottky" (1986), 4–5; Walter Schottky, "Jena 1912 und das $U^{3/2}$-Gesetz: Eine Reminiszenz aus der Vorzeit der Elektronenröhren," in Siemens & Halske Aktiengesellschaft, *50 Jahre Entwicklung und Fertigung von Elektronenröhren im Hause Siemens*, reprinted from *Siemens-Zeitschrift*, 36:2 (1962), 22–24; available in Box 009, Folder 4.1.1, Teil 2, Schottky Papers. The $U^{3/2}$-law was also known as the Langmuir-Child equation.

[14] Serchinger, "Walter Schottky" (2000), 174–176; Schottky, "Lebenslauf," (undated) and "Übersicht über meine wissenschaftlichen und technischen Untersuchungen," September 1948, in Box 009, Folder 4.1.1, Teil 1, Schottky Papers.

laboratories at what was later called *Siemensstadt* in the northern part of Berlin.[15]

Thermionic tubes were a primary focus of Siemens' corporate research undertaking. Also known as "vacuum tubes," this device was an evacuated glass bulb with two electrodes (anode and cathode) and a filament. Unlike resistors, capacitors, or inductors that responded passively to external electromagnetic energy, a thermionic tube produced its own electrical current. When a tube was in operation, its filament was powered with a battery. Electrons in the filament gained sufficient energy and escaped into the space within the tube. This provided a steady source of electrical charge, which formed a current across the tube. The voltage on the tube's electrodes could control the magnitude and direction of this electrical current. The first thermionic tube was patented in 1904 by John Ambrose Fleming, professor of electrical engineering at University College London and longtime consultant of Edison's and Marconi's companies. Dubbed the "valve," Fleming employed the diode tube as a rectifier that permitted electrical current only along one direction and used it to detect signals in wireless telegraphy. Three years later, the American inventor-entrepreneur Lee de Forest added a grid between the cathode and the anode and used this triode tube—also known as an "audion"—to amplify weak signals. By the early 1910s, vacuum tubes became a cutting-edge device technology for the telephone and wireless.[16]

The German industries were fully aware of vacuum tubes' strategic value to the telecommunications business. Their entry point for the fabrication and research of tubes was patents held by a Vienna-based physicist, Robert von Lieben. In the late 1900s, von Lieben explored a series of methods for grid control of tube current and their applications to signal amplification. In 1912, Siemens, Allgemeine Elektricitäts-Gesellschaft (AEG), and Telefunken (Siemens's and AEG's joint venture enacted in 1903 to resolve the two companies' disputes over patents on wireless telegraphy) formed a certain "Lieben-Konsortium" to mutually license one another for the making and uses of vacuum tubes covered by von Lieben's patents. After this legal proceeding, Siemens started to pursue the research, development, manufacturing, and

[15] "Research for Progress: The Development of the Central Research Laboratory," Siemens website, https://new.siemens.com/global/en/company/about/history/stories/research-laboratory.html (accessed April 8, 2022).

[16] For the history of Fleming's and de Forest's inventions of the thermionic tubes, see Sungook Hong, *Wireless: From Marconi's Black-Box to the Audion* (Cambridge, MA: MIT Press, 2010), 119–190.

applications of electronic tubes. The outbreak of World War I further accelerated the company's work in response to the need in military communications. In 1915, Siemens featured its first thermionic tube. Named the "type-A tube" ("A-Röhre"), the device was a triode with flat metal plates as anode, cathode, and grid. The type-A tube constituted the core device for Siemens's multi-stage electronic amplifiers. During the war, Siemens's A-tube amplifiers were deployed in the long-distance military telephone connection between the Belgian town of Charleroi on the Western Front and Bucharest on the Eastern Front.[17]

Schottky's wartime work at Siemens was closely embedded in the firm's development and manufacturing of thermionic tubes for electrical telecommunications. Located at Charlottenburg in suburban Berlin, K Laboratory was Siemens's main facility for the research and development of electrical devices used in signal transmission such as loading coils, vacuum tubes, and amplifiers. (K denoted *Kabel*, the German word for *cable*.) At K Laboratory, Schottky and his associates introduced a way to improve current-control efficiency of a thermionic tube's grid and to prevent emitted electrons' unwanted bouncing from the grid by wrapping it with a net of metal sheeting. This led to the invention of the so-called "screen-grid" tubes, which marked Siemens's significant improvement in fabricating thermionic devices. In addition, Schottky conceived a method for wireless reception by mixing radio-frequency signals with a waveform generated from a local oscillator with a slightly higher frequency. Named "superheterodyne," this idea was explored independently by Edwin Armstrong in the US and Lucien Lévy in France around the same time and would become a dominant architecture for radio systems after the war. Screen-grid tubes and superheterodyne were critical improvements to German military communications.[18]

The motivations for developing screen-grid tubes and heterodyne reception were to increase amplifier gain and to reduce external noise. Both concerns led Schottky to a question: Was there a limit of amplification due not to individual device characteristics but to fundamental physical principles governing the operations of all electronic devices? In Schottky's own reminiscence, the solution to this pragmatic question came from a theoretical inspiration. In 1916, he attended a lecture given by Einstein, then a professor at the University

[17] Franz Michel and Oskar Pfetscher, "Siemens-Röhren haben Tradition: Vorgeschichte und Werden der Siemens-Röhrenfabrik," in Siemens & Halske Aktiengesellschaft, *50 Jahre Entwicklung und Fertigung von Elektronenröhren im Hause Siemens*, reprinted from *Siemens-Zeitschrift*, 36:2 (1962), 9–11; available in Box 009, Folder 4.1.1, Teil 2, Schottky Papers.

[18] Serchinger, "Walter Schottky" (2000), 176–177; Madelung, "Schottky" (1986), 6–7.

of Berlin, on statistical mechanics. Einstein, who had introduced the statistical theory of Brownian motion and entertained the concept of electronic noise due to random thermal fluctuations, stimulated Schottky's thinking of the issues he dealt with in the handling of thermionic tubes.[19] In 1918, Schottky proposed a theory for electronic noise.[20]

Shot Noise in a Resonating Tube Circuit

Schottky began by noting that two kinds of noise due to electrons' fundamental properties posed limits to tube amplification: (i) electrons moved randomly at any non-zero temperature; (ii) electrons were discrete. The first feature caused a random fluctuation in a supposedly constant current—"thermal noise" (*Wärmeeffekt*). The second feature implied that a current was a discrete series of surges, not a continuous flow, and thus also caused a fluctuation in an apparently constant current—"shot noise" (*Schroteffekt*).[21] Schottky focused on shot noise. An electron flow's discreteness was most conspicuous in a high-vacuum diode tube in which all electrons emitted from the cathode arrived at the anode. He modeled the number of emitted electrons within a duration τ as a random process $n(\tau)$. When the tube current was apparently constant, the average number of electrons per unit time was a constant N. The electron-number deviation from its average was a random fluctuation $\Delta n(\tau) = n(\tau) - N\tau$. He assumed that the emitted electrons were uncorrelated with one another. Thus, $\Delta n(\tau)$ followed a Poisson process with zero mean and standard deviation $N\tau$: $<\Delta n(\tau)> = 0$, $<[\Delta n(\tau)]^2> = N\tau$.

The number $n(\tau)$ determined the tube's electric current via $i(\tau) = en(\tau)/\tau$ ($e = 1.6 \times 10^{-19}$ coulomb was the charge of an electron). Thus $<[\Delta i(\tau)]^2> = e^2N/\tau$. The current's instantaneous value fluctuated. An operator connecting an earphone to an operating vacuum tube would hear a hissing tone caused by the fluctuation—this was shot noise. The strength of shot noise was measured by the current's standard deviation $<[\Delta i(\tau)]^2>$. For a direct bias current $i_0 = eN$, the effective "shot current" was $i_S \equiv \sqrt{\langle[\Delta i(\tau)]^2\rangle} = i_0\sqrt{e/i_0\tau}$.[22]

[19] Walter Schottky, "Übersicht" (1948), Schottky Papers.

[20] Walter Schottky, "Über spontane Stromschwankungen in verschiedenen Elektrizitätsleitern," *Annalen der Physik*, (Ser. 4) 57 (1918), 541–567.

[21] Ibid., 541–547.

[22] Ibid., 548–553.

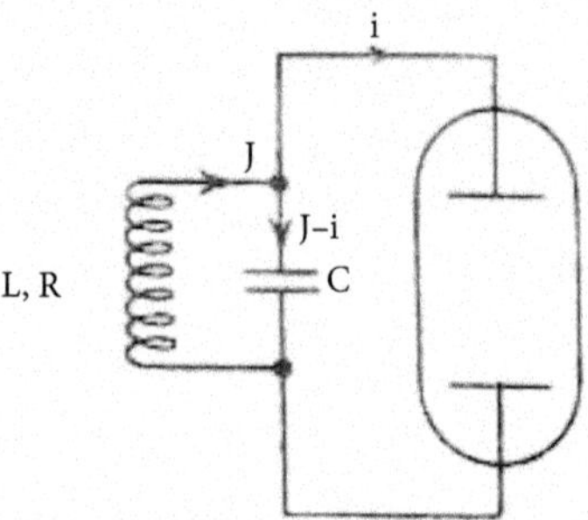

Figure 6.1 Schottky's tube circuit for evaluating the shot effect. Walter Schottky, "Über spontane Stromschwankungen in verschiedenen Elektrizitätsleitern," *Annalen der Physik*, (Ser. 4) 57, (1918), 554, Figure 6.
Permission from John Wiley & Sons Books.

The above formula expressed the shot current of a diode thermionic tube alone. In reality, a tube was rarely used without being connected to other circuit elements. As we will see below, shot noise in Schottky's theory had a flat, frequency-independent "white" power spectrum, which led to an unphysical outcome of infinite signal power if the bandwidth of the tube circuit was not limited. Therefore, he considered the shot effect of an oscillating circuit with a diode tube in parallel with a capacitor (capacitance C) and a resistive inductor (inductance L and resistance R). The RLC circuit formed a resonator that selected the tube's shot current at a specific range of frequencies (Figure 6.1).

To evaluate the circuit's noise current intensity, Schottky used a method based on Fourier analysis. This method resembled a technique his mentor Max Planck adopted in treating black-body radiation in 1897–98. Its core was averaging over frequency to distinguish the slowly from the quickly varying elements of the Fourier series and attributing physical meanings only to the slowly varying elements.[23]

A Fourier Analysis of Shot Noise

Schottky wrote down the equation linking the current i into the tube with the current J out of the resistive inductor:

$$\frac{d^2J}{dt^2} - \rho\frac{dJ}{dt} + {\omega_0}^2 J = {\omega_0}^2 i,$$

[23] Ibid., 554; Thomas Kuhn, *Black-Body Theory and the Quantum Discontinuity, 1894-1912* (New York: Oxford University Press, 1978), 72–84.

where $\omega_0^2 = 1/LC$ and $\rho = R/L$. If the tube current i was time harmonic with a single frequency, $i = I\sin(\omega t + \varphi)$, then the solution led to the circuit's time-average oscillating energy $E_S = (I^2L/2)/[(1 - x^2)^2 + r^2x^2]$ ($x = \omega/\omega_0$ and $r = \rho/\omega_0$). In reality, the tube current i was a randomly varying shot current. That means, the above circuit equation resembled Langevin's equation for Brownian motion, and the fluctuating tube current i was analogous to the stochastic force driving a suspended particle. Schottky used a typical way of treating finite-duration, time-varying functions and expressed i as a Fourier series sampled over a long period T:

$$i = \sum_{k=0}^{\infty} i_k = \sum_{k=0}^{\infty} I_k \sin\left(\omega_k t + \varphi_k\right),$$

where $\omega_k = 2\pi k/T$. The amplitude I_k and phase φ_k of each oscillating mode varied randomly. The circuit's oscillating energy for the shot current was the sum over all oscillating modes:

$$E_S = \sum_{k=0}^{\infty} E_k = L\sum_{k=0}^{\infty} \frac{I_k^2}{2} \frac{1}{\left(1 - x_k^2\right)^2 + r^2 x_k^2}, \tag{6.1}$$

where $x_k = \omega_k/\omega_0 = 2\pi k/(\omega_0 T)$.

The value of E_S in equation (6.1) depended on the values of all I_k^2. From Fourier analysis,

$$I_k = \frac{2}{T}\int_0^T i\,(t) \sin\left(\omega_k t + \varphi_k\right) dt.$$

But since the current's average vanished after the integral,

$$I_k = \frac{2}{T}\int_0^T \Delta i\,(t) \sin\left(\omega_k t + \varphi_k\right) dt.$$

Plugging this expression of I_k into equation (6.1) helped little with computing the oscillating energy E_S, for $\Delta i(t)$ was random. To facilitate this computation, Schottky needed a certain average.

Schottky's procedure of calculating average E_S differed from today's standard practice. Students of modern probability theory will base their computation on tools like ensemble average and correlation function. To them, the average oscillating energy $<E_S>$ is an ensemble average with respect to random configurations of $\{I_k^2\}$. They will move the ensemble average into the sum in equation (6.1) and turn it into $\langle E_S \rangle = L \sum_{k=0}^{\infty} \frac{\langle I_k^2 \rangle}{2} \frac{1}{(1-x_k^2)^2+r^2 x_k^2}$, and will express $<I_k^2>$ as,

$$\left\langle \left[\frac{2}{T} \int_0^T \Delta i(t) \sin(\omega_k t + \varphi_k)\, dt \right]^2 \right\rangle = \frac{4}{T^2} \left\langle \int_0^T \int_0^T \Delta i(t) \Delta i(t') \sin(\omega_k t + \varphi_k) \sin(\omega_k t' + \varphi_k)\, dt dt' \right\rangle,$$

exchange the order of ensemble average and integrals, replace $<\Delta i(t)\Delta i(t')>$ with a known correlation function of $\Delta i(t)$ (a Dirac delta function $I\delta(t - t')$ for shot noise), and perform the integrals and summation. This was *not* what Schottky did. In contrast to modern probability theory, his averaging procedure was carried out in the frequency domain. Schottky's apparently cumbersome procedure bore a striking similarity to his mentor Max Planck's approach to the black-body radiation problem in 1897–98. To understand better Planck's intellectual legacy to Schottky's 1918 paper, it is worth sketching the mentor's early treatment of black-body radiation, which the philosopher of science Thomas Kuhn reconstructed in his 1978 monograph.

Techniques from Planck's Work on Black-Body Radiation

Planck's overarching agenda in his research during 1897–98 was to demonstrate irreversibility—the central character of thermodynamics—in black-body radiation. To do so, he considered a model with a spherical cavity (representing a black-body radiator) and a resonator sitting at the center of the cavity. The resonator served both as a portion of the black-body model and a measuring device. It had a resonating frequency ν_0 and a damping constant ρ. Like other resonators, this device had a bandpass frequency response whose center frequency was ν_0 and bandwidth proportional to ρ. In Planck's choice, ρ was large enough so that the resonator's bandwidth corresponded to the range of average over frequency, which was unavoidable for an actual measuring instrument. Thus, Planck's resonator was a device that generated measurable physical quantities. For this reason, he called it an "analyzing resonator."[24]

When electromagnetic waves entered and reverberated in the cavity, part of their energy was absorbed by the resonator, which then reradiated secondary

[24] Kuhn, *Black-Body Theory* (1978), 80.

waves. Planck expressed the axial component near the resonator of the total electric field via a Fourier integral:

$$E = \int_0^\infty d\nu C_\nu \cos(2\pi\nu t - \theta_\nu).$$

Then he showed that the radiation intensity J_0 (the amount of energy radiated from the entire cavity per unit time) was

$$J_0 = \int d\mu \left[A_\mu^0 \sin(2\pi\mu t) + B_\mu^0 \cos(2\pi\nu t)\right],$$

with

$$A_\mu^0 = \frac{2}{\rho\nu_0}\int d\nu C_{\nu+\mu} C_\nu \sin^2\delta_\nu \sin\left(\theta_{\mu+\nu} - \theta_\nu\right),$$

$$B_\mu^0 = \frac{2}{\rho\nu_0}\int d\nu C_{\nu+\mu} C_\nu \sin^2\delta_\nu \cos\left(\theta_{\mu+\nu} - \theta_\nu\right),$$

and δ_ν was defined by $\cot\delta = \pi\left(\nu_0{}^2 - \nu^2\right)/\rho\nu_0\nu$. Here $\sin^2\delta_\nu$ represented the frequency response of the analyzing resonator, as it took its maximum at $\nu = \nu_0$ and diminished quickly beyond the bandwidth 2ρ. Moreover, the energy U_0 of the resonator was

$$U_0 = \int d\mu \left[a_\mu \sin(2\pi\mu t) + b_\mu \cos(2\pi\nu t)\right],$$

with

$$a_\mu = \frac{3c^3}{16\pi^2\rho\nu_0{}^3}\int d\nu C_{\nu+\mu} C_\nu \sin\delta_{\nu+\mu} \sin\delta_\nu \sin\left(\theta_{\mu+\nu} - \theta_\nu\right),$$

$$b_\mu = \frac{3c^3}{16\pi^2\rho\nu_0{}^3}\int d\nu C_{\nu+\mu} C_\nu \sin\delta_{\nu+\mu} \sin\delta_\nu \cos\left(\theta_{\mu+\nu} - \theta_\nu\right).$$

According to Planck[25], the Fourier components $A_\mu{}^0$, $B_\mu{}^0$, a_μ, b_μ in the equations for J_0 and U_0 corresponded to measurable physical quantities,

[25] Ibid., 79–81.

as they were discernible by the analyzing resonator. However, they were expressed in terms of $C_{\nu+\mu}C_\nu\sin(\theta_{\nu+\mu} - \theta_\nu)$ and $C_{\nu+\mu}C_\nu\cos(\theta_{\nu+\mu} - \theta_\nu)$ determined by the amplitudes C_ν and phases θ_ν of the electric field's Fourier components, which did *not* correspond to measurable physical quantities since they fluctuated too rapidly over frequency. As Kuhn commented, the electric field's Fourier components represented by C_ν and θ_ν in Planck's model resembled the non-measurable microstates in Boltzmann's kinetic theory of gas, like the positions and momenta of individual molecules, while the radiation intensity's and resonator energy's Fourier components $A_\mu{}^0$, $B_\mu{}^0$, a_μ, b_μ resembled the measurable macrostates in Boltzmann's theory.

Planck wanted to eliminate the "unphysical" C_ν and θ_ν entirely and represented the measurable quantities $A_\mu{}^0$, $B_\mu{}^0$, a_μ, b_μ in terms of one another. To do so, he approximated the fast-varying $C_{\nu+\mu}C_\nu\sin(\theta_{\nu+\mu} - \theta_\nu)$ and $C_{\nu+\mu}C_\nu\cos(\theta_{\nu+\mu} - \theta_\nu)$ with the slowly varying $A_\mu{}^0$, $B_\mu{}^0$:

$$C_{\nu+\mu}C_\nu \sin\left(\theta_{\mu+\nu} - \theta_\nu\right) = A_\mu^0 + \varepsilon\left(\mu, \nu\right),$$

$$C_{\nu+\mu}C_\nu \cos\left(\theta_{\mu+\nu} - \theta_\nu\right) = B_\mu^0 + \eta\left(\mu, \nu\right).$$

The functions ε and η captured the fast-varying parts of $C_{\nu+\mu}C_\nu\sin(\theta_{\nu+\mu} - \theta_\nu)$ and $C_{\nu+\mu}C_\nu\cos(\theta_{\nu+\mu} - \theta_\nu)$. However, within the sufficiently large bandwidth of the resonator's frequency response $\sin^2\delta_\nu$, ε and η changed so unsystematically that their contributions from different frequency components cancelled out one another. That is, $\int\varepsilon\sin^2\delta_\nu d\nu = \int\eta\sin^2\delta_\nu d\nu = 0$. This condition, coined "natural radiation" by Planck, guaranteed that the slowly varying physical measurables $A_\mu{}^0$ and $B_\mu{}^0$ (obtained by averaging, with the resonator's frequency response, over neighboring Fourier components of the actual field) could well approximate the fast-varying quantities $C_{\nu+\mu}C_\nu\sin(\theta_{\nu+\mu} - \theta_\nu)$ and $C_{\nu+\mu}C_\nu\cos(\theta_{\nu+\mu} - \theta_\nu)$ corresponding to the actual field's Fourier components. By replacing $C_{\nu+\mu}C_\nu\sin(\theta_{\nu+\mu} - \theta_\nu)$ and $C_{\nu+\mu}C_\nu\cos(\theta_{\nu+\mu} - \theta_\nu)$ with $A_\mu{}^0$ and $B_\mu{}^0$ in the expressions for a_μ and b_μ and employing the law of energy conservation, Planck established a differential equation relating J_0 and U_0, $dU_0/dt + K\nu_0 U_0 = (P/\nu_0)J_0$, which produced evolutions of U_0 irreversible with time.[26]

While *fin-de-siècle* physicists' (and Kuhn's) focus with respect to Planck's 1897–98 papers was on the relationship between his natural radiation and Boltzmann's molecular disorder as well as Planck's clandestine introduction

[26] Ibid., 80–84.

of a statistical argument, his early work on black-body radiation nonetheless conveyed a more generic lesson. Instead of treating directly the rapidly varying Fourier components of a complicated physical quantity, we should work whenever we can on their averages over distinct windows of frequencies. In so doing, we smooth out the quantity's fine-scale fluctuations and transform spectral analysis into scales that are much coarser and hence more "physical" and much easier to handle.

Replacing Ensemble with Spectral Average

Schottky did not cite Planck in his 1918 paper (there was no citation in the article), but the Siemens researcher's approach followed closely the aforementioned lesson. Returning to the calculation of the shot current's energy E_S in the resonating circuit, the sum in equation (6.1) represented a Fourier analysis. The energy spectrum in this expression contained a quickly changing factor (with respect to frequency index k, not time) $I_k{}^2$ and a slowly changing factor $L/2[(1-x_k{}^2)^2+r^2x_k{}^2]$. Following Planck's generic lesson, Schottky treated these two factors separately by first averaging $I_k{}^2$ over finer windows and then performing the summation over a coarser scale. Specifically he partitioned the set of non-negative integers $k = 0$ to ∞ into disjoint and consecutive windows $[k_1(1),k_2(1)]$, $[k_1(2),k_2(2)]$, ..., $[k_1(n),k_2(n)]$, ... In the nth window, the factor $L/2[(1-x_k{}^2)^2+r^2x_k{}^2]$ varied so little with k that it could be approximated by $L/2[(1-x_{k0(n)}{}^2)^2+r^2x_{k0(n)}{}^2]$, where $k0(n)$ was an integer between $k_1(n)$ and $k_2(n)$. (A natural partition of this kind would be $[0,K-1]$, $[K,2K-1]$, ..., $[(n-1)K, nK-1]$, ..., while $k0(1)$, $k0(2)$, ..., $k0(n)$, ... could be the midpoints of the intervals.) Consequently, equation (6.1) was approximated as follows:[27]

$$E_S = \frac{L}{2}\sum_{k=0}^{\infty}\frac{I_k{}^2}{2}\frac{1}{(1-x_k{}^2)^2+r^2x_k{}^2} = \frac{L}{2}\sum_{n=1}^{\infty}\sum_{k=k_1(n)}^{k_2(n)}\frac{I_k{}^2}{(1-x_k{}^2)^2+r^2x_k{}^2}$$

$$\cong \frac{L}{2}\sum_{n=1}^{\infty}\frac{1}{\left(1-x_{k_0(n)}{}^2\right)^2+r^2x_{k_0(n)}{}^2}\sum_{k=k_1(n)}^{k_2(n)} I_k{}^2 .$$

The sum $\sum_{k=k_1(n)}^{k_2(n)} I_k{}^2$ in this expression averaged out the rapid fluctuations of $I_k{}^2$. For this to be effective, the size of the window should be large enough to contain sufficient terms, i.e. $k_2(n)-k_1(n) \gg 1$. Meanwhile, the window should

[27] Schottky, "Über spontane Stromschwankungen" (1918), 556.

be small enough—$[k_2(n) - k_1(n)]/x_{k_0(n)} \ll 1$—so that $1/[(1 - x_k{}^2)^2 + r^2 x_k{}^2]$ varied little from $k = k_1(n)$ to $k = k_2(n)$.

To evaluate the sum $\sum_{k=k_1(n)}^{k_2(n)} I_k{}^2$, Schottky plugged in the relation $I_k = \frac{2}{T}\int_0^T \Delta i(t)\sin(\omega_k t + \varphi_k)\,dt$ and obtained:

$$\sum_{k=k_1}^{k_2} I_k{}^2 = \frac{4}{T^2}\sum_{k=k_1}^{k_2}\int_0^T\int_0^T \Delta i(t)\,\Delta i(t')\sin(\omega_k t + \varphi_k)\sin(\omega_k t' + \varphi_k)\,dt dt'.$$

Then he approximated the double integral with a double Riemann sum with short time increments Δt and Δt'. To justify this approximation, he argued that the resonance circuit's frequency response $L/2[(1 - x_k{}^2)^2 + r^2 x_k{}^2]$ suppressed high-frequency contributions from the double integrand, while the integrand's low-frequency components behaved like constants within the time windows Δt and Δt', as long as they were much shorter than the circuit's resonating period. Thus,[28]

$$\int_0^T\int_0^T \Delta i(t)\,\Delta i(t')\sin(\omega_k t + \varphi_k)\sin(\omega_k t' + \varphi_k)\,dt dt'$$
$$\cong \sum_{t=0}^{T}\sum_{t'=0}^{T} \Delta i(t)\,\Delta t \Delta i(t')\,\Delta t' \sin(\omega_k t + \varphi_k)\sin(\omega_k t' + \varphi_k),$$

where t and t' in the Riemann sum took discrete values between 0 and T.

To evaluate the double Riemann sum, Schottky grouped its terms into two clusters: those with identical time indices ($t = t$') and those with different time indices ($t \neq t$'). The first cluster was

$$\sum_{t=0}^{T} [\Delta i(t)\,\Delta t]^2 \sin^2(\omega_k t + \varphi_k).$$

The modern approach to evaluate this sum is to calculate the ensemble average of the shot current's square to obtain $<[\Delta i(t)\Delta t]^2> = e i_0 \Delta t$, and to plug the value back into the sum. Schottky did not do this. Instead of averaging over distinct configurations, he worked on a single time series extended over a very long period T and performed average over time based on an implicit ergodic assumption. Schottky observed that in the sum the factor $\sin^2(\omega_k t + \varphi_k)$ was periodic in time. Thus, within a long period T, each distinct value of $\sin^2(\omega_k t + \varphi_k)$ occurred many times in the sum. Grouping the terms with an equal value of $\sin^2(\omega_k t + \varphi_k)$ (say, $= \sin^2(\omega_k t_0 + \varphi_k)$) together, it led to $\sin^2(\omega_k t_0 + \varphi_k) \times$

[28] Ibid., 557.

$\{[\Delta i(t_0)]^2 + [\Delta i(t_1)]^2 + [\Delta i(t_2)]^2 + ...\}\Delta t^2$, where t_0, t_1, t_2 were the instants at which $\sin^2(\omega_k t + \varphi_k)$ had the same value. With a coarser resolution, the sum $\{[\Delta i(t_0)]^2 + [\Delta i(t_1)]^2 + [\Delta i(t_2)]^2 + ...\}\Delta t^2$ was time average in a sense. Under the ergodic assumption, it equaled $<[\Delta i(t)\Delta t]^2> = ei_0\Delta t$ for shot current. The first cluster of the Riemann sum became

$$\sum_{t=0}^{T} [\Delta i(t)\Delta t]^2 \sin^2(\omega_k t + \varphi_k) = ei_0\Delta t \sum_{t=0}^{T} \sin^2(\omega_k t + \varphi_k) = ei_0 T/2.$$

Since this result was independent of k, its contribution to $\sum_{k=k_1(n)}^{k_2(n)} I_k^2$ was $(2/T)(k_2 - k_1)ei_0$.[29]

The Riemann sums' second cluster was

$$\sum_{t=0}^{T} \sum_{\substack{t'=0 \\ t' \neq t}}^{T} \Delta i(t)\Delta t \Delta i(t')\Delta t' \sin(\omega_k t + \varphi_k)\sin(\omega_k t' + \varphi_k).$$

Since t and t' represented separate time intervals, Schottky asserted that the shot current $\Delta i(t)$ and $\Delta i(t')$ were uncorrelated to each other. Again, the modern approach would plug the correlation $<\Delta i(t)\Delta i(t')> = 0$ into the sum and obtain zero result. Although Schottky also aimed to get this result, his reasoning, without resorting to the direct ensemble average, was more strenuous. He tackled the problem from the observation that unlike the terms $[\Delta i(t)\Delta t]^2\sin^2(\omega_k t + \varphi_k)$ in the first cluster, the terms $\Delta i(t)\Delta t\Delta i(t')\Delta t'\sin(\omega_k t + \varphi_k)\sin(\omega_k t' + \varphi_k)$ in the second cluster were incoherent and did not exhibit any clear order. In fact, Schottky argued, because $\Delta i(t)$ and $\Delta i(t')$ were independent of each other, the factor $\Delta i(t)\Delta t\Delta i(t')\Delta t'$ varied randomly in such a way that it could be equally often positive and negative ("ebenso oft positiv wie negativ sein können"). The sum of such terms also randomly varied with the same frequency to be positive and negative, and the multiplication of the factor $\sin(\omega_k t + \varphi_k)\sin(\omega_k t' + \varphi_k)$ would not change this property.

Moreover, he claimed, it was "well known" ("bekanntlich") that for p quantities of this kind with order of magnitude a, the order of magnitude for the sum of these p quantities was $\sqrt{pa}$.[30] Since the order of magnitude for $\Delta i(t)\Delta t\Delta i(t')\Delta t'$ was less than $<[\Delta i(t)\Delta t]^2> = ei_0\Delta t$, it followed from the stated theorem that the order of magnitude for the second cluster was less than

[29] Ibid., 558.

[30] Ibid., 558–559. Schottky did not prove nor provide citation information for this theorem. But it seemingly was an intermediate result in the proof of the law of large numbers. Suppose there are p identical and independent random variables $x_1, x_2, ..., x_p$ with zero mean and variance $<x^2> = a^2$. Define $s = x_1 + x_2 + ... + x_p$. Then it can be shown that s has zero mean and variance $<s^2> = pa^2$. The root mean square of s, $[<s^2>]^{1/2} = p^{1/2}a$, can be interpreted as the "order of magnitude" of s.

$[(T/\Delta t)(T/\Delta t')]^{1/2} ei_0 \Delta t = Tei_0$. This double sum was also a zero-mean, randomly varying quantity over k. Hence Schottky could apply the same theorem to the sum of the second cluster over k from k_1 to k_2, and get an upper bound for the order of magnitude for the second cluster's contribution to $\sum_{k=k_1(n)}^{k_2(n)} I_k^2$, which was $(4/T)(k_2 - k_1)^{1/2} ei_0$. Since $k_2 - k_1 >> 1$, the contribution from the second cluster at $t \neq t'$, $(4/T)(k_2 - k_1)^{1/2} ei_0$, was much smaller than that from the first cluster at $t = t'$, $(2/T)(k_2 - k_1) ei_0$. Consequently, $\sum_{k=k_1(n)}^{k_2(n)} I_k^2 \cong (2/T)(k_2 - k_1) ei_0$. This relation suggested that the shot current's average power around frequency ω_k was independent of frequency and corresponded to a "white" spectrum: $\langle I_k^2 \rangle = 2ei_0/T$.[31]

The Result

Schottky substituted shot current's white spectrum into equation (6.1) to calculate the circuit's overall oscillating energy. He approximated the discrete sum with an integral and obtained:

$$E_S \cong \frac{ei_0 L}{T} \cdot \frac{\omega_0 T}{2\pi} \int_0^\infty \frac{dx}{(1 - x^2)^2 + r^2 x^2}. \tag{6.2}$$

He gave the value of the integral $\int_0^\infty \frac{dx}{(1-x^2)^2 + r^2x^2} = \frac{2\pi}{r^2}$, leading to the total oscillating energy $E_S = \omega_0^3 Lei_0/\rho^2$. To transform this expression into a form directly comparable to the results from measurements, he argued that the energy $E_S = \omega_0^3 Lei_0/\rho^2$ was equivalent to the energy of the tube oscillator when the current i of the RLC circuit had amplitude $\sqrt{2} i_S$ and frequency ω_0, viz., $i = \sqrt{2} i_S \sin(\omega_0 t + \varphi)$. The circuit's energy at its resonance frequency was $E_S = (i^2 L/2)/r^2$. Comparing both expressions, he obtained the measurable effective shot current

$$i_S = \sqrt{\frac{2\pi e i_0}{\tau}}, \tag{6.3}$$

[31] Why Schottky did not use zero correlation $\langle \Delta i(t) \Delta i(t') \rangle = 0$ to get rid of the second cluster was not obvious in his text. The most likely reason was that he did not want to rule out the possibility that $\langle \Delta i(t) \Delta i(t') \rangle \neq 0$ as $t - t'$ was small. Even though zero correlation did not hold in that case, he might have been convinced that the order of magnitude for the second cluster was much smaller than that for the first cluster.

where $\tau = 2\pi/\omega_0$ was the circuit's resonance period. He also estimated the numerical scale of the shot-noise energy compared to that of the thermal-noise energy at room temperature and found that shot noise was usually much stronger than thermal noise.[32]

Schottky's theory of shot noise proposed an operable experimental condition to verify its underlying hypothesis and gave a quantitative prediction for a measurable entity. Yet, his work was not followed up on immediately. Schottky's article was published in June 1918, when the war had consumed Germany and an unconditional capitulation seemed inevitable. Before the end of World War I in November 1918, vacuum-tube research was a top military secret not open to public discussions. After the war, Germany's political turmoil and economic depression interrupted normal academic activities and impeded the exchange of scholarly information with other countries. For these reasons, Schottky's work did not reach the international scientific and engineering communities and even the German academia outside Siemens for years. Schottky himself did not continue the research on shot noise, either. As the war ended and a scientist's duty to serve his nation ended, he left Siemens in 1919 for the University of Würzburg, joined Wilhelm Wien's group, and got habilitation (qualification to teach at universities).[33]

Shot Noise after Schottky

Despite the mathematical tricks in ensemble and frequency averages, Schottky's theory conveyed a simple idea: the random flow of discrete electrons in a thermionic tube gave rise to a kind of fundamental noise in electronic circuits. However, the situation became notably messier when the theory of shot noise was put into empirical testing or more realistic modeling. In the experimental and engineering work on shot noise during the 1920s, more complex factors had to be taken into account, and whether the shot noise posed a fundamental performance limit for electronic circuits was far from clear.

Experiments with Schottky Noise

The Siemens researcher Carl A. Hartmann made the first attempt to empirically corroborate Schottky's theory of shot noise. In 1920, Hartmann

[32] Schottky, "Über spontane Stromschwankungen" (1918), 555–562.
[33] Madelung, "Schottky" (1986), 7.

conducted an experiment at K Laboratory to measure a vacuum tube's shot noise.[34] His basic experimental setup was the same as Schottky's model in Figure 6.1: a vacuum tube connected in parallel to a capacitor and an inductor. The goals of his experiment were to verify (i) whether this circuit indeed had a noisy tube current; and (ii) if so, whether the noise intensity was consistent with Schottky's prediction in equation (6.3).

Hartmann proposed a procedure to meet the second goal from the fact that Schottky's quantitative prediction involved the charge of an electron, e. He expressed equation (6.3) as a relation of e with other variables, $e = i_S{}^2/(i_0\omega_0)$. This relation indicated that the ratio involving the measured shot current i_S, the tube's bias current i_0, and the circuit's resonance frequency ω_0 should be a fundamental physical constant $e = 1.6 \times 10^{-19}$ coulomb.

Though simple in concept, Hartmann's experiment was challenging in several senses. First, the tube had to be at a high vacuum to satisfy Schottky's assumption that all electrons leaving the cathode arrived at the anode. To fulfill this requirement, Hartmann designed a tube continually evacuated by a pump. Second, since the shot noise was tiny, the resonance circuit's quality factor should be very high to prevent signal dissipation. Therefore, he reduced the circuit's resistance so that $\omega_0 L/R > 150$. Third, the tiny shot noise also meant that signals from the oscillating circuit should be considerably amplified. Here he used Siemens's high-gain multistage wideband electronic amplifiers.

Making precise measurements of shot current was critical. Like radio atmospherics, shot noise was irregular and difficult to measure by common galvanometers. Hartmann adopted an ear-balancing method similar to AT&T engineers' in atmospherics measurements. In his design, the vacuum-tube oscillator and a reference monotone generator tuned at the oscillator's resonance frequency were connected to a high-gain amplifier with a switch (Figure 6.2). Like those in atmospherics measurements, Hartmann switched between the tube oscillator and the monotonic generator and adjusted the latter until the intensity from both appeared identical at the output earphone. Then the monotonic signal intensity, easily measured by a galvanometer, was taken as the effective intensity of the shot noise (also see chapter 4).[35]

Hartmann used this setup to experiment with shot noise. His results were qualitatively satisfactory yet quantitatively problematic. The instrument indeed produced hissing tones at the output earphone. Shot noise existed. He measured the effective shot-noise intensity for various resonance frequencies ω_0 from 238.73 to 2387.33 Hz, with bias currents i_0 at 2 and 20 milliamps, and

[34] Hartmann, "Über die Bestimmung" (1921), 51–78.

[35] Ibid., 65–67.

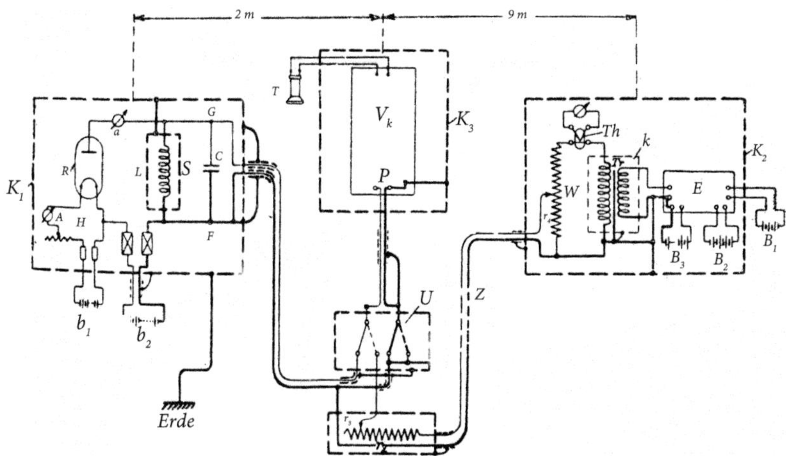

Figure 6.2 Hartmann's experimental setup. Carl A. Hartmann, "Über die Bestimmung des elektrischen Elementarquantums aus dem Schroteffekt," *Annalen der Physik*, 65 (1921), 65, Figure 8. The block on the left hand is the vacuum-tube oscillator and the block on the right hand is the monotone generator. The block "V_k" is the high-gain amplifier. "*U*" is the switch connecting the three blocks. Permission from John Wiley & Sons Books.

used the measured i_S to calculate the charge of an electron *e*. To his disappointment, however, the experimental results deviated significantly from Schottky's prediction. The problem was twofold. The values of *e* from the shot-noise measurements were all in the order of 10^{-22} coulomb, 1000 times smaller than its commonly known value 1.6×10^{-19} coulomb from Robert Millikan and Harvey Fletcher's oil-drop experiment in 1909. Also, the measured *e* varied with frequency; it did not remain a constant as Schottky's theory had predicted (Figure 6.3).

Hartmann's experimental results were difficult to understand if Schottky's theory was right. The experimenter's own explanation was that the emission of electrons from the cathode was not really a Poisson process. When an electron was emitted, a part of energy was taken away and the cathode was cooled down for a period. The lower temperature reduced the cathode's ability to further emit electrons. As a result, the actual number of emitted electrons was smaller than that from a Poisson process. The measured shot current (and hence the estimated value of *e*) was smaller than Schottky's prediction.[36] Yet, no further research supported this physical picture.

[36] Ibid., 74–76.

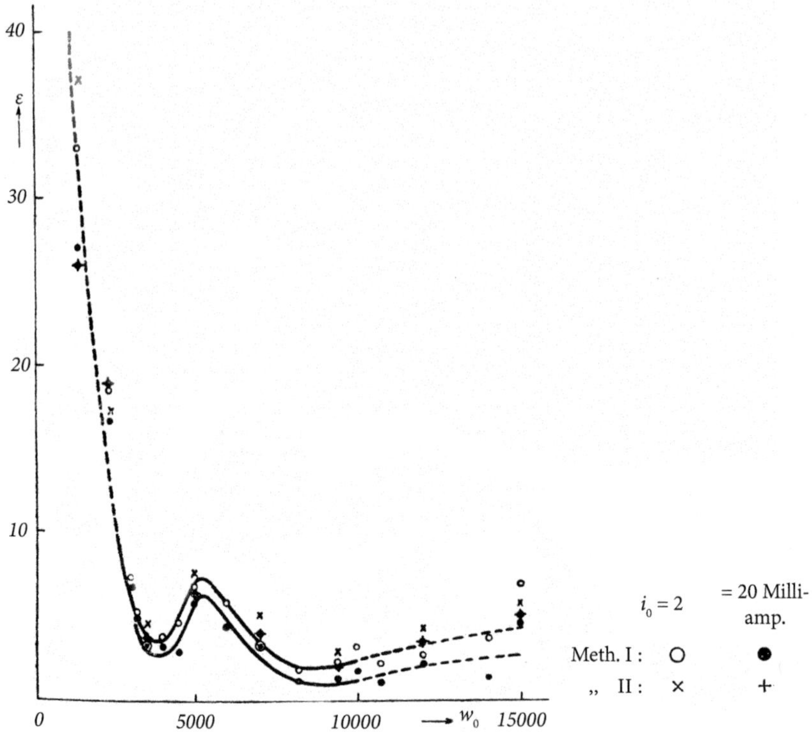

Figure 6.3 Hartmann's experimental results. The abscissa is the resonance frequency ω_0 and the ordinate is the value of e from shot-noise measurements (unit 10^{-22} coulomb). The two curves correspond to $i_0 = 2$ mA and 20 mA. Hartmann,"Über die Bestimmung" (1921), 71, Figure 9.
Permission from John Wiley & Sons Books.

The American physicist John B. Johnson found a more commonly accepted reason for the huge discrepancy between Hartmann's experimental data and Schottky's theoretical prediction: a mathematical error. Born in Sweden, John Bertrand Johnson immigrated to the US when he was a teenager, received a Ph.D. in physics from Yale University in 1917, and joined Western Electric's Engineering Department in New York. Johnson was among the first generation of Bell-System researchers with advanced training in physics. His job at Western Electric was to develop components for AT&T's transcontinental telephony. This brought him to the problems of noise.[37] In 1920, Johnson read Schottky's 1918 article in *Annalen der Physik* (the 1918 issue did not reach

[37] Anonymous, "John B. Johnson, recent Sarnoff Award winner" (obituary), *IEEE Spectrum* 8:1 (1971), 107.

America until 1920, owing to Germany's postwar postal delay). He was skeptical of Schottky's integral $\int_0^\infty dx/\left[\left(1-x^2\right)^2+r^2x^2\right]=2\pi/r^2$ for equation (6.2). He tried to calculate the integral himself but could not find a solution from the table of integrals. To solve the problem, he consulted a mathematician L.A. MacColl, who "suggested splitting the Schottky equation [the integrand] into four complex factors, integrating them separately and then recombining them [...] MacColl again looked at the equation and said this was a case for the method of poles and residues and, without putting pencil to paper, read off the correct result."[38] With MacColl's help, Johnson obtained the correct value of the integral $\int_0^\infty dx/\left[\left(1-x^2\right)^2+r^2x^2\right]=\pi/2r$. He immediately wrote to Schottky and published this new result.[39]

Johnson's mathematical exercise bridged the gap between theory and experiment. After receiving Johnson's letter, Schottky recalculated the integral with the help of his mathematician father and found Johnson right. Then he took Johnson's correction to recalculate the shot-noise energy E_S and the effective shot current i_S:

$$i_S=\sqrt{\frac{ei_0R}{2L}}. \tag{6.4}$$

Consequently, $e=2Li_S{}^2/(i_0R)$ instead of $i_S{}^2/(i_0\omega_0)$. For the same i_S, the new result $e=2Li_S{}^2/(i_0R)$ gave an estimate of the e value approximately 1000 times larger than the calculation from the old result $e=i_S{}^2/(i_0\omega_0)$. After correction, therefore, Hartmann's shot-noise data led to an estimate of e much closer to its actual value—the order of magnitude was right.[40]

The problem of e's frequency dependency still remained, though. The corrected data still yielded e values that varied with ω_0. In 1922, a Czech physicist Reinhold Fürth at the University of Prague, specialist on Brownian motion,[41] explained this anomaly in terms of the "physiological" (or cognitive) nature of the experimental method.[42] Fürth argued that one could not take the results of noise measurements from the ear-balancing method at their face value, for the method relied on the mediation of experimenters' aural cognition, a

[38] John B. Johnson, "Electronic noise: the first two decades," *IEEE Spectrum* 8:1 (1971), 42.

[39] John B. Johnson, "Bemerkung zur Bestimmung des elektrischen Elementarquantums aus dem Schroteffekt," *Annalen der Physik*, (Ser. 4) 67 (1922), 154–156.

[40] Walter Schottky, "Zur Berechnung und Beurteilung des Schroteffektes," *Annalen der Physik*, (Ser. 4) 68 (1922), 157–176.

[41] P. Weinstein (ed.), *J. C. Poggendorffs Biographisch-Literarisches Handwörterbuch*, Band 5 (Berlin: Verlag Chemie, 1922), 403.

[42] Fürth, "Die Bestimmung der Elektronenladung" (1922), 354–362.

frequency-dependent feature. The ear-balancing method required an experimenter to identify when the noise sound intensity equaled the monotonic sound intensity and to take the monotonic current as the measure of the noise current. Yet the two signals causing the same degree of aural perception in human ears had equal intensity if and only if the strength of human aural perception was proportional to the signal intensity (or, strength of stimulation) and the signals were at the same frequency. Both conditions were absent in Hartmann's experiment.

The Weber–Fechner law in physiology states that the strength of (aural) perception s is in a logarithmic rather than a linear relation with the strength of stimulation V (signal intensity): $s = 2c \cdot \log(V/V_0)$, where V_0 is a perceptive threshold (note $s = 0$ when $V < V_0$). While the reference signal in the ear-balancing method was monotonic with a definite intensity, the noise to be measured had an extended spectrum with a random intensity. Accordingly, Fürth revised the Weber–Fechner law for noise in a statistical form:

$$s_1 = \frac{2c \int_{V_0}^{\infty} dV \cdot \log(V/V_0)\, W(V)}{\int_{V_0}^{\infty} dV \cdot W(V)},$$

where $W(V)dV$ was the probability that the noise intensity was between V and $V + dV$. Under Fürth's assumption, the random noise intensity had a Rayleigh distribution $W(V) = A \cdot \exp(-V^2/2\langle V^2\rangle)$. Meanwhile, the aural perception s_2 stimulated by a monotonic reference V' followed the original law $s_2 = 2c\log(V'/V_0)$. In the ear-balancing method, experimenters established that $s_1 = s_2$ by equalizing noise and reference. But unlike what Hartmann had assumed, $s_1 = s_2$ did not entail that the mean noise intensity equaled the reference signal intensity $[\langle V^2\rangle]^{1/2} = V'$. Instead, Fürth's formulae for s_1 and s_2 showed that $[\langle V^2\rangle]^{1/2}/V'$ was a function of V'/V_0. Moreover, the threshold V_0 was not a constant, either—it was a function of frequency depending on the frequency responses of both the human aural perception and the amplifying telephonic circuit.

Fürth obtained empirical values of V_0 at different frequencies from Hartmann's setup, and computed and plotted $[\langle V^2\rangle]^{1/2}/V'$ as a function of V'/V_0. From both, he constructed a relation between $[\langle V^2\rangle]^{1/2}$ and V', which he used to retrieve the noise intensity $[\langle V^2\rangle]^{1/2}$ from Hartmann's data for the reference signal intensity V'. Reinterpreting Hartmann's data led to new estimates for the value of e with a much slighter variation with frequency than the original estimates'—Fürth's values deviated from 1.6×10^{-19} coulomb within 100% (Figure 6.4).

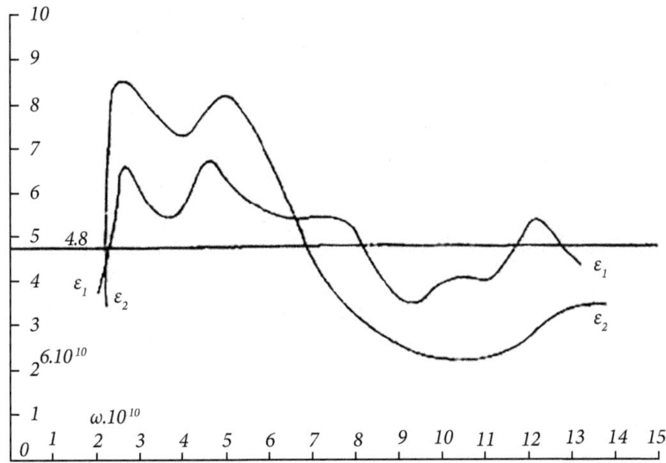

Figure 6.4 Fürth's correction of Hartmann's experimental results. Reinhold Fürth, "Die Bestimmung der Elektronenladung aus dem Schroteffekt an Glühkathodenröhren," *Physikalische Zeitschrift*, 23 (1922), 361, Figure 6.

Albert W. Hull and N.H. Williams at the Schenectady Research Laboratory of General Electric (GE) in New York further improved the accuracy of the shot-noise experiment in 1924. As GE's expert on vacuum-tube circuits, Hull had developed multistage high-gain amplifiers using shielded-grid tubes in the early 1920s.[43] These devices soon played an important part in the work of shot noise, in which he and his assistant Williams became interested. They attacked the shot-noise problem from the perspective of precision experiment, especially the determination of *e* from accurate noise measurements. With the high-gain amplifiers, they could get rid of Hartmann's ear-balancing method. They connected the amplified shot-noise current—now strong enough to be measured with more common means—directly to a crystal detector and a current meter that measured the mean square value of the shot current.[44]

Hull and Williams used the new experimental setup to determine the value of *e* from the shot-current measurements. Working on a single frequency at 725 kHz, they obtained experimental data leading to a value of *e* extremely close to its canonically recognized value. Their estimated *e* at i_0 = 1–5 mA had an error range of less than 3% with respect to 1.6×10^{-19} coulomb (Figure 6.5).

[43] Two years later, Hull's vacuum-tube amplifiers would achieve a gain of 2 000 000 below 1 MHz. See Albert W. Hull, "Measurements of high frequency amplification with shielded-grid pliotrons," *Physical Review*, (Ser. 2) 27 (1926), 439–454.

[44] Albert W. Hull and N. H. Williams, "Determination of elementary charge *e* from measurements of shot-effect," *Physical Review*, (Ser. 2) 25 (1925), 148–150.

Shot-effect of temperature-limited electron current i_0 in α U. V. 199 radiotron.

i_0 (m-amp.)	γ (m-amp.)	l (cm)	v_1 (μ-volts)	R (ohms)	F	v_0^2 (μ-volts)	J^2 (μ-amp)	e (coulombs)
1	.443	9.0	65.0	3.045	.776	73.8	.204	1.541×10^{-19}
2	.925	6.0	89.4	3.37	.763	102.2	.282	1.640
2	.602	9.0	88.3	3.37	.763	101.0	.279	1.603
3	1.05	6.0	102.7	3.65	.750	118.5	.327	1.595
3	.700	9.0	102.7	3.65	.750	118.5	.327	1.595
4	.580	12.0	113.4	3.85	.740	131.8	.364	1.570
4	.780	9.0	114.4	3.85	.740	133.1	.367	1.595
5	.835	9.0	122.5	4.06	.727	143.8	.367	1.566
5	.625	12.0	122.2	4.06	.727	143.5	.396	1.556
							mean	1.586×10^{-19}

Figure 6.5 Hull and Williams's determination of the value of *e* from shot-noise measurements. Reprinted Table 2 with permission from Albert W. Hull and N.H. Williams, "Determination of elementary charge *e* from measurements of shot-effect," *Physical Review*, (Ser. 2) 25 (1925), 166.

Copyright (1925) by the American Physical Society. The last column represents the estimates of *e* from measurements.

Hull and Williams did more than improve the experiment to come closer to the known value of *e*, however. They found a new factor that qualified Schottky's theory of shot noise: the space-charge effect. W.L. Carlson at GE's Radio Department had noted that a tube's noise intensity fell as the tube was more strongly charged. This phenomenon caught Hull and Williams's attention. At first, they believed that it was an artifact due to the reducing tube resistance. But they changed their mind when they still observed low noise values after stabilizing the tube resistance. Now they thought that the reduction of the shot noise was caused by the accumulated space charges between the cathode and the anode; such space charges created a potential barrier against the flow of electrons. If the space charges inside a tube reached saturation after a long time (the "space-charge limited" case), then all the relative motions between electrons inside the tube disappeared and they moved regularly in a uniform stream. In this case, the lack of randomness and discontinuity eliminated shot noise. Schottky's theory of shot noise was valid when a tube's electronic current was limited only by its filament temperature—the filament's ability to emit electrons (the "temperature limited" case). When the current was limited by the tube's space charges, shot noise was greatly reduced and Schottky's theory no longer applied.[45]

[45] Ibid., 166–170.

Theoretical Revision

As Hartmann et al. worked on shot-noise experiments, efforts were made to refine Schottky's theory. Schottky's theory of shot noise was built on statistical reasoning. Translated into mathematics, his problem was to solve a differential equation (the circuit equation) with a random source (shot noise) whose statistics were partially known—a problem similar to Langevin's. Instead of transforming the stochastic differential equation into another differential equation for the mean or variance of the statistical quantity as Langevin had done, however, Schottky solved the problem via a Fourier analysis. His approach had shortcomings: calculations were complicated, each Fourier component's physical meaning was unclear, the assumption underlying frequency average was questionable, and his spectrum of random noise lacked a rigorous definition. In the 1920s, physicists and engineers attempted to solve the shot-noise problem without resorting to Fourier analysis.

The Dutch physicists Leonard Ornstein and Herman Carel Burger at Ryks University in Utrecht were the earliest to revise Schottky's approach. In 1923, they claimed to solve the random differential equation in a more direct way. Key to their solution was to conflate statistical average with time average. In so doing, they simplified the differential equation using integration by parts and obtained the mean square of the current J out of the resistive inductor: $<J^2> = e^2n^2 + e^2n\omega_0{}^2/\rho$. In this expression, the first term corresponded to the constant direct current and the second term corresponded to the current fluctuation, which was equivalent to Schottky's formula of E_S.[46]

The British physicist Norman Campbell at General Electric London made another attempt to revise the formulation. Aiming to explain both thermionic and photoelectric emissions, he formulated a generalized principle that considered a series of randomly occurring events. An instrument measured the effect of these events. Suppose $\theta = f(t)$ gave the instrument reading at time t after an event occurred at time 0, $<\Delta n^2>$ gave the standard deviation of the number of events in a unit time. Then the standard deviation of θ was $\langle\Delta\theta^2\rangle = \langle\Delta n^2\rangle\int_0^\infty f^2(t)\,dt$. This principle implied that the overall shot effect could be decomposed into a factor describing the statistics of the *collection* of electrons and a factor related to the waveform excited by an *individual* electron.[47]

[46] Leonard Ornstein and Herman Carel Burger, "Zur Theorie des Schroteffektes," *Annalen der Physik*, (Ser. 4) 70 (1923), 622–624.

[47] Norman Campbell, "The theory of the 'Schrot-effect,'" *Philosophical Magazine*, 50 (1925), 81–86.

Thornton C. Fry at AT&T furthered Ornstein and Campbell's reasoning. A mathematician, Fry had taught at the University of Wisconsin and MIT before joining AT&T. In 1925, he assumed directorship of Bell Labs' Mathematical Research Department, where he built a staff of computing and mathematics to serve the firm's industrial research. A specialist in engineering probability,[48] Fry thought that the major problem of Schottky's theory was the lack of a mathematically rigorous treatment of probability and randomness. He was prepared to offer one of his own.

Like Campbell, Fry began by considering a general case in which a measuring device detected current I and voltage E excited by independently arriving electrons in a long period T. The device's mean instantaneous power was $\langle EI\rangle = \sum_{n=0}^{\infty} p(n) \langle (EI)_n \rangle$, where $p(n)$ was the probability for n electrons to arrive within T that followed a Poisson distribution with the average rate υ, and $<(EI)_n>$ was their mean instantaneous power. $<(EI)_n>$ could be calculated from the assumption that the current or voltage excitations from distinct electrons were additive. The additive formula could be expressed with an iterative relation between $<(EI)_n>$ and $<(EI)_{n-1}>$. Repeating the iteration n times led to $<(EI)_n>=<(EI)_1> + n(n-1)<E_1><I_1>$. Substituting $<(EI)_n>$ and $p(n)$ into $<EI>$, Fry obtained $<EI>=(\upsilon T)<(EI)_1> + (\upsilon T)^2<E_1><I_1>$. He interpreted $<EI>$ as the system's overall power, $(\upsilon T)<(EI)_1>$ as the power of the shot noise, and $(\upsilon T)^2<E_1><I_1>$ as the power of the direct current. This implied shot-noise energy $E_S = \upsilon<w_1>$.

The quantity $<w_1>$ was the average energy generated in the measuring device by an electron traveling from the tube's cathode to its anode. To facilitate its calculation, Fry modeled the tube as a parallel-plate condenser. A unit-charge particle's movement from one plate to another induced a time-variant voltage $v(t)$ and current $i(t)$ across the plates. Consequently, $\langle w_1\rangle = e^2 \int_0^{\infty} v(t)\, i(t)\, dt$. The quantities $v(t)$ and $i(t)$ depended exclusively on the circuit. In Schottky's theory, the circuit was the tube connected to a resonance network (Figure 6.1). Fry modeled the tube as a capacitor, obtained $v(t)$ and $i(t)$, and substituted them into his formula for E_S. He found that Schottky's prediction for the shot-noise energy should be modified by a multiplicative factor $[1 + R^2(C+C_{tube})/L]$, where R, L, C were the resonance circuit's resistance, inductance, and capacitance, and C_{tube} was the tube's effective capacitance.[49]

[48] Mindell, *Between Human and Machine* (2002), 191.

[49] Thornton C. Fry, "The theory of the Schroteffekt," *Journal of the Franklin Institute*, 199 (1925), 203–220.

Flicker Noise

Hull and Williams's experiment and Fry's theory inspired Johnson to research further into shot noise. In 1925, Johnson performed an experiment at Bell Labs to measure shot noise in Schottky's circuit using Hull and Williams's method (high-gain amplifier, crystal detector, and direct current meter). He compared his experimental results with Fry's prediction. His goal, relevant to AT&T in particular and to the electronics industry in general, was to find vacuum-tube circuits' performance limit imposed by electronic noise. Johnson measured the noise strength of triode tubes from about 100 commercial electronic amplifiers. His results confirmed Hull and Williams's finding: the vacuum tubes of amplifying circuits operated at the space-charge-limited condition, so their shot-noise intensity was much lower than Schottky's and Fry's predictions.[50]

In addition to that, Johnson discovered a new phenomenon. To test the general applicability of Schottky's or Fry's theory, Johnson measured the noise of diode tubes operating at the temperature-limited condition (i.e. without space charges). He changed the circuit's resonance frequency from 0 to 10 kHz to observe the variation of the measured noise intensity. This arrangement led to a novel effect: at high frequencies, the measured noise intensity was close to Fry's prediction; but the noise intensity increased rapidly with decreasing frequency when it was low. This low-frequency deviation changed with the filament material. Tungsten had much higher noise intensity than oxide coating. With a tungsten filament, the ratio of e obtained from the noise data to its canonical value of 1.6×10^{-19} coulomb was 0.7 for frequencies above 200 Hz, but increased to 50 at 10 Hz. With an oxide-coated filament, the ratio increased from 1 at 5 kHz to 100 at 100 Hz (Figure 6.6). It seemed that at low frequencies certain much stronger noise *of a different kind* superseded ordinary shot noise. Johnson named this new kind of noise the "flicker effect" for its special sound pattern.[51]

Flicker noise, Johnson argued, differed fundamentally from shot noise. Its much higher magnitude implied that it was not due to random and independent emissions of discrete electrons. Since the flicker effect depended upon the filament material, its cause was related to the activities at the filament. A cathode surface changed continually with evaporation, diffusion, chemical actions, structural rearrangements, and ion bombardment. These factors altered the

[50] John B. Johnson, "The Schottky effect in low frequency circuit," *Physical Review*, 26 (1925), 81–83.
[51] Ibid., 76–80.

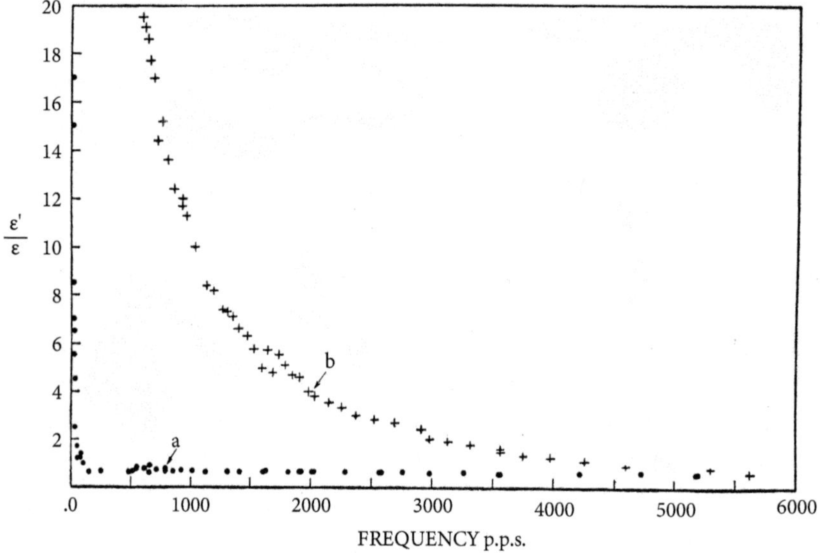

Figure 6.6 Johnson's flicker noise. "a": tungsten filaments, "b": oxide-coated filaments. Reprinted Figure 7 with permission from John B. Johnson, "The Schottky effect in low frequency circuit," *Physical Review*, 26 (1925), 79.

filament's rate of electronic emission, which caused current fluctuations, the source of this noise.[52]

Fry's revision of Schottky's theory, Hull and Williams's findings of the space-charge effect, and Johnson's discovery of the flicker effect at low frequencies opened new directions for shot-noise research in the rest of the 1920s. Following Fry's approach, Stuart Ballantine of the Radio Frequency Laboratory in Boonton (New Jersey) calculated the voltage and current excited by an electron between two parallel plates to determine shot noise's power spectrum in the temperature-limited case.[53] After leaving GE for the Physics Department at the University of Michigan, Williams collaborated with his colleagues to improve the methods of noise measurements. In the space-charge-limited condition, they also used experimental techniques to measure the charge of a positive thermion and to examine the filament material's effect on charge emission.[54]

[52] Ibid., 85.

[53] Stuart Ballantine, "Schrot-effect in high-frequency circuits," *Journal of the Franklin Institute*, 206 (1928), 159–167. This approach culminated in a classic textbook on random noise written in 1958 by MIT Lincoln Laboratory researchers Wilbur Davenport and William Root, *An Introduction to the Theory of Random Signals and Noise* (New York: McGraw-Hill, 1958), 112–144.

[54] N.H. Williams and H.B. Vincent, "Determination of electronic charge from measurements of shot-effect in aperiodic circuits," *Physical Review*, (Ser. 2), 28 (1926), 1250–1264; N.H. Williams and S.

Schottky developed a theory to explain the flicker effect's frequency dependence. He assumed that the large noise intensity at low frequencies was caused by fluctuation of the filament's electron-emitting capability owing to the coming and going of "foreign" atoms on the filament surface. A foreign atom, once it reached the filament surface, did not depart immediately; instead, it stayed there for a short period of time t_a. Within this short period, the numbers of foreign atoms on the filament surface at two distinct instants were close to each other; but such numbers fluctuated significantly with respect to each other if the separation between the two instants far exceeded t_a. In other words, the correlation between the numbers of foreign atoms (which determined the correlation of the filament's electron-emitting capability) at two instants separated by time t was not an impulse at $t = 0$ and zero otherwise, like shot noise. Rather, it had a flat top within t_a and fell off to zero (quickly or slowly, depending on the nature of atomic attachment and detachment) outside that duration. This nontrivial correlation led to flicker noise's frequency dependence: $<I_{\text{flicker}}^2> \propto f^{-n}$ (n was between 1 and 2).[55]

These diverse studies revealed that noise was much more complicated than Schottky's original theory in 1918 had expected. A vacuum tube was a microcosm with various physical mechanisms. Random emissions of discrete electrons could only partially explain the microcosm's fluctuations. Schottky's theory failed in the space-charge-limited case at which tube amplifiers operated. In the temperature-limited case, the flicker effect dominated over the shot effect at low frequencies. Emission of ions, generation or recombination of ions and electrons, and lumping of charged particles all affected tube noise. Schottky's theory was a proper concept to start grasping electronic noise but was by no means a useful framework to determine the fundamental limit of vacuum-tube circuits' amplification that engineers wanted.

Then what was that limit?

Thermal Noise in Resistors

While the engineering researchers at Siemens, GE, and smaller firms were inquiring into shot noise in thermionic tubes, studies of another type of electronic noise were being actively pursued at AT&T. Like GE and Dupont,

Huxford, "Determination of the charge of positive thermions from measurements of shot effect," *Physical Review*, (Ser. 2), 33 (1929), 773–788; H.N. Kozanowski and N.H. Williams, "Shot effect of the emission from oxide cathodes," *Physical Review*, (Ser. 2), 36 (1930), 1314–1329; John S. Donal, Jr., "Abnormal shot effect of ions of tungstous and tungstic oxide," *Physical Review*, (Ser. 2), 36 (1930), 1172–1189.

[55] John B. Johnson, "Thermal agitation of electricity in conductors," *Physical Review*, (Ser. 2) 32 (1928), 98–101.

AT&T was a major explorer of industrial research that facilitated American technological prowess in the early twentieth century. In-house development and research had been part of the telephone cartel's corporate culture. In 1891, AT&T convened an Engineering Department that absorbed the company's existing small laboratories and testing facilities dating back to its Boston days. A similar department was formed shortly afterwards at Western Electric, AT&T's manufacturing wing. In 1911, AT&T's chief engineer John Carty built a Research Branch within Western Electric's Engineering Department. AT&T's own Engineering Department also created a branch for Engineering and Research in 1919. By the late 1910s, AT&T and Western Electric hosted numerous research laboratories and facilities—most of which were in New York City—with hundreds of staff members. In 1924, the company merged all these R&D capacities into a single organization, Bell Telephone Laboratories, which would become one of the most famous corporate laboratories in the twentieth century.[56]

Similar to what happened at Siemens' Berlin laboratory and GE's Schenectady laboratory, AT&T's industrial research focused not on equipment testing, direct technical support, or solutions to immediate engineering problems, but rather on the development of new components, systems, or procedures and on studies of the sciences associated with new technologies. Moreover, AT&T's research establishment employed a significant number of individuals with advanced degrees in physics, mathematics, electrical engineering, or chemistry from leading American universities: Frank Jewett, Harold Arnold, Edwin Coplitts, John Carson, John B. Johnson, Harry Nyquist, and Ralph Hartley, to name a few.

A preoccupation of AT&T's research divisions in the 1910s was thermionic tubes. When de Forest's audion was brought to the attention of Jewett, Arnold, and Colpitts in 1912, they immediately recognized its potential to the company's business. At this time, AT&T endeavored to build the first transcontinental telephone network from New York to San Francisco. The telephone signals as weak electrical current suffered severe decay, distortion, and dispersion along telephone cables before they could reach even a small fraction of the coast-to-coast distance. To enable the transcontinental connection, AT&T engineers had to install many "repeaters" along the phone network to transduce electromechanically telephone signals back to sounds and to convert

[56] Fagen, *A History of Engineering and Science in the Bell System: The Early Years (1875–1925)* (1975), 37–58; Reich, *The Making of American Industrial Research* (1985), 151–184.

them again into electrical currents with their power boosted through electromagnetic relays or other equivalent components. Since the de Forest audion could amplify weak electrical signals, it could serve as an effective and compact alternative solution to the bulky, complex, and power-consuming relays. Moreover, thermionic tubes could detect and amplify signals for wireless or radio telegraphy and telephony which AT&T engineers planned to experiment with for transatlantic and airborne communications. For these business incentives, AT&T devoted considerable resource and effort to the development, improvement, and production of thermionic tubes. This work not only enabled AT&T's New York–San Francisco telephone network in 1915 and its successful transatlantic wireless experiments in the early 1920s, but also turned the company into a major supplier of vacuum tubes for American military communications in World War I.[57]

Johnson's investigation into electronic noise was part of Bell Labs' research and development on thermionic tubes and related circuitry for AT&T's telephone and radio systems. In this applied research, he nonetheless found something with profound theoretical significances. When Johnson conducted the shot-noise experiment in 1925 to find the fundamental limit of tube amplification, he measured the noise intensity and the gain of about 100 commercially available triode tube amplifiers. He plotted the results in a Cartesian coordinates system (the ordinate was the gain and the abscissa was the noise strength) and found an interesting pattern (Figure 6.7). The points representing various amplifiers' gains and noise levels distributed to the right of a sloped line passing the origin. That is, for a fixed amplification, the noise intensity was always higher than the value given by the sloped line, meaning that the line represented the lower bound of the amplifiers' noise intensity. This minimum noise intensity was the residual tube noise when disturbing factors were suppressed.[58]

Johnson's Inquiry

What was the physical significance of the minimum noise represented by the sloped line in Figure 6.7? At first, Johnson thought it was Schottky's shot noise without the space-charge effect and the flicker effect. But he changed his mind

[57] Reich, *The Making of American Industrial Research* (1985), 160–184. For AT&T's transcontinental telephone network and its connection to the research on electronic tube amplifiers, see Mindell, "Opening Black's Box" (2000). For AT&T's early experiments on transatlantic wireless communications, see chapter 4.

[58] Johnson, "The Schottky effect" (1925), 83–85.

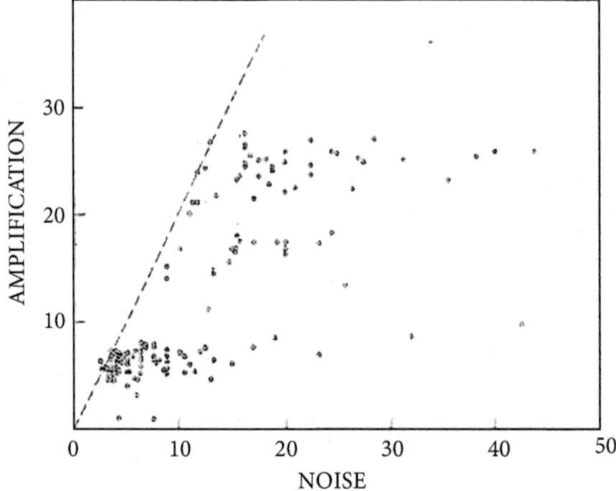

Figure 6.7 Amplification as a function of noise in triode tubes. The dots represent measurements taken from about 100 commercial tube amplifiers. Reproduced Figure 13 from John B. Johnson, "The Schottky effect in low frequency circuit," *Physical Review*, 26 (1925), 84.

after noting a fact: since a sloped line passing the origin represented a proportional relation, the minimum tube output noise strength in Figure 6.7 was proportional to the tube amplifier's gain. This implied that a tube's input noise strength was a *constant* when its output noise was a minimum. If so, then the minimum residual noise had nothing to do with the vacuum-tube amplifier itself. Rather, it was determined only by the amplifier's input condition.

To testify this hypothesis, he put a resistor at the input of an amplifying circuit and changed its resistance to observe whether the output noise strength varied accordingly. It did, and the measured noise intensity was proportional to the input resistance. This resistor-dependent noise reminded Johnson of a less discussed aspect of Schottky's 1918 article: thermal noise. Was the residual noise a result of electrons' thermal agitation at the amplifier's input resistor? Informed by this assumption, Johnson measured the noise of the least noisy amplifiers with input resistors different in temperature, size, or material. The measured noise intensity was proportional to the input resistor's temperature only, not size or material. This strongly suggested that the minimum residual noise was a fundamental physical effect related to the input resistor's thermal agitation.[59]

[59] Johnson, "Electronic noise" (1971), 43–44.

In 1926, Johnson performed a systematic experiment on thermal noise at Bell Labs with his assistant J.H. Rohrbaugh. His focus now shifted from tubes to resistors. His circuit comprised a resistor connected to a six-stage tube amplifier with the output coupled to a thermocouple for current measurements (Figure 6.8). The high-gain amplifier suppressed shot noise, flicker noise, and other fluctuations to the maximum extent, and was shielded against external electric, magnetic, acoustic, and mechanical shocks. The input resistor was in a constant-temperature bath. The aim of the experiment was to determine the mean square voltage $<V_m{}^2>$ of the thermal noise at the input resistor from the mean square noisy current $<I^2>$ measured at the amplifier's output via the relation $<I^2>=Y^2<V_m{}^2>$. (Y was the ratio of the amplifier's output current to its input voltage.)[60]

Johnson's experiments in 1926 reached the same but a more expressive conclusion: The ratio of thermal noise's mean square voltage to the input resistance $<V_m{}^2>/R$ was independent of the resistor's shape and material. Johnson tried different materials including metal wires, graphite, thin metal films, films of drawing ink, and electrolytes such as NaCl, $CuSO_4$, K_2CrO_4, and $Ca(NO_3)_2$. All yielded the same measured values. Also, the ratio $<V_m{}^2>/R$ was proportional to the resistor's absolute temperature. At room temperature, a 5000-Ω resistor had $<V_m{}^2>/R \cong 10^{-18}$ watts. He obtained thereby an empirical formula for the thermal noise: $<V_m{}^2>/R = <I^2>/(Y^2R) = KT$. ($K$ was a proportionality constant.)[61]

As Johnson continued the experiment, he was obliged to revise the empirical formula for thermal noise $<I^2> = KTRY^2$, since both the bandpass amplification Y and the input resistance R were functions of frequency. To incorporate

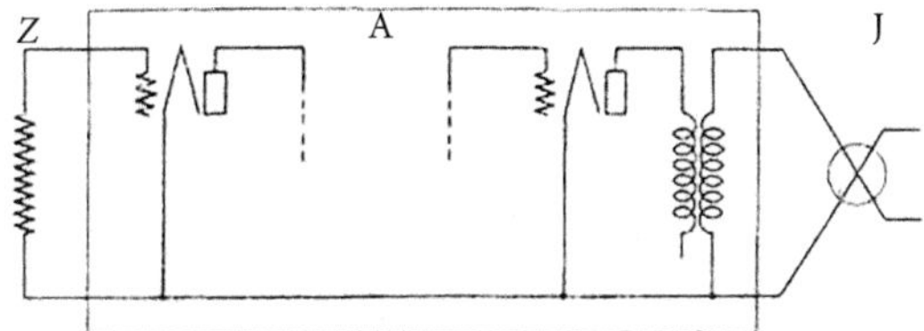

Figure 6.8 Johnson's experimental apparatus on thermal noise. Reproduced Figure 1 from John B. Johnson, "Thermal agitation of electricity in conductors," *Physical Review*, (Ser. 2) 32 (1928), 98.

[60] Johnson, "Thermal agitation" (1928), 98–101.
[61] John B. Johnson, "Thermal agitation of electricity in conductors," *Physical Review*, 29 (1927), 50–51, 367–368.

the frequency dependence, he suggested that $I^2 = KT\int_0^\infty d\omega R(\omega)\lfloor Y(\omega)\rfloor^2$. Then he compared the measured data with the empirical formula. He determined the amplification $|Y(\omega)|^2$ by measuring the amplifier's input-output characteristics. The input resistance $R(\omega)$ was the real part of the amplifier's input impedance, which was modeled as a resistor R_0 connected in parallel to a capacitor C. By substituting the measured $|Y(\omega)|^2$ and the modeled $R(\omega)$ into the empirical formula, he obtained an estimate of $<I^2>$ from numerical integration. The measured $<I^2>$ varied in the same pattern as the estimated $<I^2>$ had predicted. Johnson's formula seemed right.[62]

Nyquist's Theory

The formula's empirical adequacy did not satisfy Johnson; he was curious about the theoretical foundation of his formula. In 1927, he brought this issue to his colleague Harry Nyquist. Nyquist was another Swede who immigrated to the US during his teenage years. He received a B.S. and M.S. in electrical engineering from the University of North Dakota, and a Ph.D. in physics from Yale. Nyquist began to work at AT&T Development and Research in 1917. When the Bell Telephone Laboratories were formed in 1924, his department was merged into this new establishment.[63] To explain Johnson's formula, Nyquist spent a month developing a theory based on electric circuit analysis, thermodynamics, and statistical mechanics.[64]

According to Schottky, thermal noise was fluctuations generated by electrons' random thermal motions in a conductor. Its simple relation with temperature indicated that it was a thermodynamic phenomenon. Thus, Nyquist's theory began with a simple scenario at thermal equilibrium: two identical conductors (I and II), each with resistance R, connected together by perfectly conducting wires and at temperature T. Electrons in the two conductors underwent continual thermal agitation as long as T was not zero. The thermal agitation in conductor I induced an electromotive force on the entire circuit, including conductor II. The electromotive force yielded a current around the circuit. The current heated conductor II, and hence transferred power from conductor I to conductor II. In the same manner, power was transferred from

[62] Johnson, "Thermal agitation" (1928), 102–103.
[63] Mindell, *Between Human and Machine* (2002), 105–110.
[64] Harry Nyquist, "Thermal agitation in conductors," *Physical Review*, (Ser. 2) 29 (1927), 614.

conductor II to conductor I. At thermal equilibrium, the power transferred from I to II equaled that from II to I.[65]

Nyquist contended that at thermal equilibrium, not only the total power but also the power at any frequency exchanged between conductors I and II equaled each other.[66] This showed that the thermally agitated electromotive force was a universal function of frequency, temperature, and resistance—like other thermodynamic functions. Its form did not change with physical setups not involving frequency, temperature, and resistance. Nyquist considered a different setup with the two conductors connected by a long non-dissipative transmission line with length l (Figure 6.9). The transmission line had inductance L and capacitance C per unit length so that its characteristic impedance $(L/C)^{1/2}$ equaled R, meaning no reflection at both ends. At thermal equilibrium, two trains of energy traversed the transmission line: one from left to right being the power of the thermal agitation from I to II and the other from right to left being the power of the thermal agitation from II to I.

To calculate the transferred power, Nyquist isolated the transmission line from the conductors at $t = 0$ by short-circuiting its two ends, trapping the energy on the line. The transmission line containing the energy transferred

Figure 6.9 Nyquist's scenario for thermal-noise calculation. Reprinted Figure 3 from Harry Nyquist, "Thermal agitation of electric charge in conductors," *Physical Review*, (Ser. 2) 32 (1928), 111.

[65] Harry Nyquist, "Thermal agitation of electric charge in conductors," *Physical Review*, (Ser. 2) 32 (1928), 110.

[66] If not, a contradiction ensued. Suppose that conductor I sent more power to conductor II than it received from conductor II in a frequency band. Then it sent less power to II than it received from II in the rest of the frequency spectrum, for the total power from I to II equaled that from II to I. When a resonance circuit blocking the energy transfer in the frequency band was set between the two conductors, the total power transferred from I to II equaled the power from I to II without the resonance circuit less the power in the frequency band. Similarly, the total power from II to I equaled the power from II to I without the resonance circuit less the power in the frequency band. With the resonance circuit, therefore, the total power from I to II was less than that from II to I, a contradiction to thermal equilibrium. So the assumption that the power transfer in the frequency band was unbalanced was wrong.

from both conductors resembled Planck's black-body radiator. It was a resonator with modes of vibration corresponding to stationary waves. The modes of vibration had frequencies $nv/(2l)$ (v was the speed of the energy transfer and n was a positive integer). Since l was long, the number of vibrating modes (degrees of freedom) was huge and could be represented as an approximately continuous function of frequency: the number of modes between f and $f + df$ was $2ldf/v$. For so many vibrating modes, the energy distribution followed Boltzmann's equipartition law: each degree of freedom had the mean energy kT, where k was the Boltzmann constant. Therefore, the transmission line's mean energy between f and $f + df$ was $2lkTdf/v$. This was the mean energy within $(f, f+df)$ transmitted from the two conductors to the line during the transit time l/v. This implied that the mean power each conductor transferred within $(f, f + df)$ was $P(f)df = (2lkTdf/v)/(2l/v) = kTdf$.

From the power spectrum $P(f) = kT$, Nyquist obtained the electromotive-force spectrum $E(f)$. The electromotive force $E(f)df$ generated a current $I(f)df = [E(f)/2R]df$ in the circuit comprising conductors I and II. The power spectrum was therefore $P(f) = I^2(f)R = [E^2(f)/4R]$. This relation, along with $P(f) = kT$, implied $E^2(f) = 4RkT$, which expressed the electromotive force induced by the thermal agitation of a resistor with resistance R and temperature T. Nyquist extended this relation by considering the thermal effect of a passive network with a frequency-dependent impedance $Z(f) = R(f) + iX(f)$. From circuit theory, he showed that $E^2(f) = 4R(f)kT$.

In Johnson's experiment, $E^2(f)df$ was the mean square thermal voltage between f and $f + df$ at the amplifier's input. That is, the mean square thermal current between f and $f + df$ at the amplifier's output was $E_f^2(f)|Y(f)|^2df$ ($Y(f)$ was the ratio of the amplifier's output current to its input voltage at frequency f) and the overall output mean-square current was:[67]

$$\langle I^2 \rangle = \int_0^\infty dfE^2(f)\lfloor Y(f) \rfloor^2 = \frac{2kT}{\pi}\int_0^\infty d\omega R(\omega)\lfloor Y(\omega) \rfloor^2. \qquad (6.5)$$

Equation (6.5) was identical to Johnson's empirical formula $I^2 = KT\int_0^\infty d\omega\, R(\omega)\lfloor Y(\omega) \rfloor^2$, except that the proportionality constant K was now replaced by a fundamental constant $2k/\pi$.

Nyquist tested his formula with Johnson's experimental results. Key to his test was the Boltzmann constant k. Johnson had data for $R(\omega)$, $Y(\omega)$, T, and $\langle I^2 \rangle$. Nyquist calculated k from these data using (6.5) and compared the

[67] Nyquist, "Thermal agitation of electric charge in conductors" (1928), 113.

results with its standard value $k = 1.372 \times 10^{-24}$ joule/°K. The outcome was satisfactory: the estimated value of k from the thermal-noise measurements was only about 7.5% below the standard value. Johnson attributed this discrepancy to the inaccuracy of the amplifiers' gain.[68]

Within two decades, the physicist community would view Johnson's finding of thermal noise and Nyquist's theorization of it as the first illustration of a generalized theory of random fluctuations in all physical systems.[69] To Johnson and Nyquist in the 1920s, however, their scientific inquiries into thermal noise were motivated by the considerations for electronic circuit design. Equation (6.5) showed that the thermal noise increased with the input resistance, temperature, and the amplifier's bandwidth. Thus, Johnson argued, there were three practical means to suppress an amplifier's thermal noise: to reduce the device's temperature, to restrict the amplifier's input resistance, and to make the amplifier's bandwidth no greater than needed.[70]

Engineering Treatments of Electronic Noise

By the 1930s, radio engineers had known two kinds of electronic noise: thermionic tubes' shot noise (including the flicker effect) and resistors' thermal noise. Although device defect or malfunction still mattered, such electronic noise became more and more critical in determining the performance limit of electronic devices as the device quality improved. Also, the originally tiny shot noise and thermal noise were considerably magnified as tube amplifiers' gains increased significantly. For cutting-edge radio receivers free from most quality problems, shot and thermal noise seemed to pose the "ultimate" limit of signal amplification. As a Bell Labs researcher G.L. Pearson asserted:[71]

> It is well known that the noise inherent in the first stage of a high gain amplifier is a barrier to the amplification of indefinitely small signals. Even when fluctuations in battery voltage, induction, microphonic effects, poor insulations, and other obvious causes are entirely eliminated, there are two sources of noise which remain, namely, thermal agitation of electricity in the circuits

[68] Johnson, "Thermal agitation" (1928), 104–105.

[69] Herbert Callen and Theodore Welton, "Irreversibility and generalized noise," *Physical Review*, 83:1 (1948), 34–40.

[70] Johnson, "Thermal agitation" (1928), 106–107.

[71] G.L. Pearson, "Fluctuation noise in vacuum tubes," 2, manuscript submitted on April 30, 1934, in Legacy No. 622-08-01, Folder 05, AT&T Archives.

and voltage fluctuations arising from conditions within the vacuum tubes of the amplifier.

Schottky's, Hartmann's, Hull and Williams's, Johnson's, Nyquist's, and others' scientific research into shot and thermal noise provided engineers with the means to push the envelope of the "ultimate" performance limit of electronic devices. Shot noise and flicker noise occurred only when a thermionic tube had zero or little space charge. As the tube's filament temperature was high enough to saturate itself with space charge, the tube current became coherent and the random emission of discrete electrons did not cause current fluctuations. Thermal noise was the result of the thermal agitation in an amplifier's input resistor. It was saliently suppressed if the amplifier's input stage had a low resistance and was maintained at a low temperature.

Yet, engineers found the guidelines from the scientific theories of noise inadequate. According to the theories, a tube amplifier with zero input resistance and high filament temperature had zero output noise. In 1930, Frederick Llewellyn at Bell Labs measured the noise of tube amplifiers with zero input resistance and high filament temperature. Under this almost perfect condition, however, the measured output noise intensity was still conspicuous. The discrepancy between theory and experiment made Llewellyn think about the *real* limit of amplification. He proposed three possible causes for the output noise with zero input resistance and space-charge-limited tubes. First, shot effect still existed at space-charge saturation, since the space charge only diminished rather than eliminated the emitting fluctuation's influence. Second, even at zero input resistance, electrons traversing a tube still had thermal agitation. Their noisy effect could be represented by an internal resistance at the tube's plate. Third, the traversing electrons ionized air in the tube, and the secondary charged particles from ionization bombarded the tube's electrodes to create some output current fluctuation. Unfortunately, the effects of shot noise at the space-charge-limited condition, thermal noise of the plate resistance, and noise from ionization were difficult to analyze theoretically or quantitatively.[72]

Llewellyn's discovery showed that electronic noise was more complex than Schottky et al.'s theories could capture. The shot-noise and thermal-noise models derived from electrodynamics, statistical mechanics, and thermodynamics were replaced by pictures of tangled interactions between charged particles

[72] Frederick Llewellyn, "A study of noise in vacuum tubes and attached circuits," *Proceedings of the Institute of Radio Engineers*, 18:2 (1930), 243–265.

and electronic tubes' physical setups. Research into the field in the 1930s indicated that a more accurate understanding of noise could only be achieved by studying the complicated details of electronic physics in thermionic tubes.

Engineers could not wait until physicists confidently grasped the phenomena. Even though physicists could obtain a theoretical understanding of the noise, as Johnson and Llewellyn remarked in 1935, "the greater part of the noise in practical tubes is caused by things that have not been included in the theory and that are still in a state of flux."[73] More than a fundamental limit of noise in ideal tubes, engineers needed to know the noise strength in real tubes for design and operation. And they needed a practical, systematic method to obtain such information. Measurement was perhaps the only feasible way. Working along the line of Johnson and Llewellyn's investigation, engineers at Bell Labs developed a scheme to characterize experimentally the tube noise strength. In a theoretical and experimental study at Bell Labs in 1934, Pearson identified four types of electronic noise:

- Thermal agitation in the internal plate resistance of the tube.
- Shot effect and flicker effect from space current in the presence of space charge.
- Shot effect from electrons produced by collision ionization and secondary emission.
- Space charge fluctuations due to positive ions.

To characterize the quantitative behavior of these distinct types of noise, Pearson designed a scheme of measurement. He employed a 60-ohm vacuum thermocouple and a micro-ammeter to detect and record the noise level. He took a measurement of the noise intensity, inserted a variable resistor into the input stage (the grid) of the vacuum tube, and then recorded the noise level while increasing the resistance of this inserted resistor at the input until the measured noise level was double the original one without the resistor. Pearson treated this value of inserted resistance as an equivalent indication of the tube's input electronic noise. This work showed not only AT&T engineers' grasping of electronic noise's complex physical nature, but also their attempts to integrate their development of metrology for ambient noise (see chapter 4) with their studies of the fluctuations due to vacuum tubes and electrical circuitry.[74]

[73] John B. Johnson and Frederick Llewellyn, "Limits to amplification," *The Bell System Technical Journal*, 14 (1935), 92.

[74] Pearson, "Fluctuation noise in vacuum tubes" (1934), 1.

In an even more popular metrological scheme, engineers measured a tube's noise intensity under various operating conditions and, most importantly, when the tube's grid (input) was short-circuited to ground. Under this condition, the tube's input resistance was zero and the measured values were treated as the tube's intrinsic noise levels. This measure, called the tube's "noise figure," was represented by the thermal noise of an effective resistor at the tube's input. The noise-figure scheme offered a systematic method to rate off-the-shelf thermionic-tube devices. Throughout the 1930s, American, British, and German radio engineers studied, measured, rated, and published the noise figures of commercial electronic tubes.[75] Systematic measurements inspired by, but independent of, noise physics became the canonical engineering treatment of electronic noise.

Moreover, engineering researchers found that the electronic noise was not a standalone effect of scientific curiosity, but was something that should rather be treated along with other factors that generated cacophony and disturbance in sound reproduction and voice communication. For instance, C.B. Aitken at Bell Labs in 1934 investigated the effect of thermal noise as a background in radio broadcasting when interferences from other channels were present. Aitken was interested in the condition when unwanted radio-broadcasting signals at the same or adjacent frequency channels entered a receiver and caused interferences. Such interferences were often weak and thus under reasonable control after the governments in the US and Western Europe enacted laws of spectrum allocation in the 1920s–30s. When some background noise like the thermal noise appeared in the receiver, however, the radio interference would interact with the electronic noise to form quite unpleasant and growling sounds called "flutter" at a loudspeaker or earphone. When the interfering carrier was swinging in and out of phase with the radio program signals, the flutter intensity turned high and low, which made the noise even more unbearable. Aitken conducted an experiment to gauge the flutter effect and claimed that the impact of background thermal noise on the reception of radio broadcasting could be represented in terms of percentage of modulation with the carrier frequency.[76]

Around the same time, another Bell Labs researcher W.A. MacNair inquired into the loudness level of random noise as a distinct measure. MacNair invoked the famous results from his colleagues Harvey Fletcher and Wilden Munson

[75] Johnson and Llewellyn, "Limits to amplification" (1935), 92–94; Johnson, "Electronic noise" (1971), 46.

[76] C.B. Aitken, "The effect of background noise in shared channel broadcasting," May–July 1934, in Legacy No. 622-08-01, Folder 01, AT&T Archives.

that had established the difference between a tone's energy level (intensity) and the subjective feeling about its strength in human hearing (loudness), as well as the quantitative relationship between the two at various frequencies. For the sake of designing and improving electrical reproduction of speech and music, MacNair was convinced of the importance of finding not only the relationship between loudness and intensity for homogeneous tones but also that for noise. As he indicated:

> the type of noise considered here is that which has a uniform frequency distribution of intensity. Many noises encountered in amplifying and reproducing equipment are of this general type. An example of a noise which is strictly of this type is that due to the thermal agitation of electricity in conductors.

To obtain a simple but reasonably precise estimate of the thermal noise's loudness–intensity relationship, MacNair modeled the wideband noise in terms of a discrete number of harmonic tones at uniformly sampled frequencies. The loudness–intensity relationship for each of such tones could be obtained from the Fletcher–Munson curves. Then the results from all these sampling frequencies were added to obtain the overall relationship. MacNair's conclusion was that the difference in loudness for thermal noise was often more salient than its difference in intensity. This meant that the wideband electronic noise, after passing through tube amplifiers and speakers into a room or an auditorium, could indeed create conspicuous disturbance to listeners.[77] Here the electronic noise was not a subtle laboratory effect that Einstein and de Haas-Lorentz endeavored to discover, but rather a notable part of general listeners' experience in the age of technical reproducibility of sounds and voices.

Physicists' research into electronic noises contributed to engineers' endeavors to tame uncertainties in electronic circuit designs and operations, but not necessarily in the way they had anticipated. The discoveries of the shot, flicker, and resistive thermal effects helped engineers to understand noise's fundamental nature. The theories of these effects provided noise-reduction guidelines (increasing the filament temperature, reducing the input resistance, etc.) that engineers often used. Yet Schottky's and Johnson's original hope to find the ultimate limit of amplification posed by noise diminished after more complex physical processes inside electronic tubes were found. The theories of shot noise and thermal noise could not predict electronic noise. To obtain

[77] W.A. MacNair, "The loudness level of random noise," presented at the Acoustical Society of America, December 4, 1934, in Legacy No. 622-08-01, Folder 05, AT&T Archives.

the electronic-noise data useful to radio engineering, systematic measurement was still the only viable method in the 1930s. This does not mean, however, that noise physics was irrelevant as engineers chose to characterize electronic noise with measurements, a practical art, not with predictions. Rather, the theories shaped the measuring methods. The noise-figure scheme, for example, was built on the theory of thermal noise at both the operational level (short-circuiting the input resistor) and the representational level (representing the noise with effective resistance).

The electronic-noise studies sought a new kind of engineering knowledge—a technology's fundamental limitation—with the assistance of physics. To grapple with electronic noise, one of the most common uncertainties in electrical devices, the German and American physicists and engineers aimed to characterize the *upper bound* of *ideal* devices' performances, not the actual performances of *real* devices. Conceptually, they focused on ideal-type devices free from "human factors"—flaws, defects, unsatisfied qualities—so that the performance imperfection was caused by the devices' fundamental working principles and hence posed an intrinsic limit to real devices of the same kind. Using statistical mechanics, thermodynamics, circuit theory, and electrodynamics, they constructed theories to predict such an upper bound. Underneath these theories was the conviction that random Brownian-like electronic motions generated fluctuations of performance in electrical devices. Therefore, the physical quantities (such as voltage and current) and the abstract engineering entities (such as signals) were represented by stochastic variables and the measurable noise effect was the result of their statistical behavior. In so doing, engineers and physicists converted technological systems into models of stochastic systems.

The noise studies of the 1920–30s did not fulfill their goal. Experiments showed that neither shot noise nor thermal noise posed the ultimate limit to vacuum-tube amplifiers. Other kinds of noise were present even when the Schottky effect and the Johnson effect were null. A thermionic tube, a complex microcosm with many types of electron–material interactions, did not achieve its optimum state simply by eliminating shot noise and the input resistor's thermal noise. Failing to fulfill their goal, however, did not make the theories of noise useless to engineers. The theories might not be adequate in explaining or predicting. But they gave engineers a *language* to talk about noise. From the theories, engineers gained intellectual tools to make sense of, to quantify, and to define electronic noise.

This scientific language of electronic noise was characterized with its frequency-domain approach. Schottky's theory of shot noise inspected the effect of discrete electrons' movements in a thermionic tube on different frequency components in an amplifier circuit, making an analogy between these components and Planck's spectrum of black-body radiation. Similarly, Nyquist treated electrons' thermal agitations as a source to stimulate energy exchange at different frequencies. The further experimental investigations on shot noise, the discovery of the flicker effect, and the engineering studies of electronic noise's specific impacts on sound reproduction all focused on spectral analysis. Although physicists and engineers at Siemens and AT&T invoked statistical physics and the theory of Brownian motions in their research on electronic noise, unlike Einstein, Smoluchowski, and Langevin, they were not really concerned with the time-evolution of random fluctuations in the electronic systems they studied. Rather, they treated electronic noise as a disturbance to sound reproduction with "a uniform frequency distribution of intensity." This view resonated with Helmholtz's claim in 1863 that "a noise is accompanied by a rapid alternation of different kinds of sensations of sound."

Whereas Helmholtz had a quasi-musical model of sonic noise as a physical entity stimulating aural sensations at distinct tones, the physical picture in the 1910s–30s had noise as a wideband signal with a flat and continuous spectrum at all frequencies. Moreover, the physicists and engineers in the early twentieth century had developed a set of mathematical concepts and tools—random processes, equipartition theorem, the Langevin equation, spectral analysis of stochastic signals—to characterize such noise as random fluctuations. In part III, we will see how these concepts and tools for electronic noise intertwined with two other lines of research: one pursued by physicists to explore Brownian motion and other types of random fluctuations, and the other initiated by mathematicians to give a more rigorous underpinning of random processes.

PART III
BUILDING A THEORETICAL REGIME

7

Dynamics of Brownian Motion

Three Theoretical Approaches to Noise during the Interwar Period

The studies of suspended particles' Brownian motions and electronic systems' disturbances in the 1900s–20s taught a central lesson. Noise, or at least some important types of it, could be construed as random fluctuations that exhibited certain statistical regularities. Einstein's, Smoluchowski's, and Langevin's work showed that molecules' thermal motions gave rise to stochastic agitations of observable quantities—e.g. granules' positions and gas density—that were not smoothed out through averaging over a large number. Such agitations followed a normal distribution that expanded asymptotically over time. Schottky's, Nyquist's, Johnson's, and other engineering researchers' work indicated that shot noise and thermal noise in electronic circuits of radio and telephony had the same physical origin as Brownian motion. Because of this origin, electronic noise spread its energy uniformly over frequencies. The Brownian-type fluctuations epitomized what were believed to be the most "fundamental" noise in sound reproduction. Thus, while Edison Phonograph technologists coped with surface noise via trying different recording materials and AT&T engineers quantified ambient din and telephone distortion with measurements, researchers at Bell Labs and Siemens theorized electronic noise with the mathematical tools of statistical mechanics, random processes, and spectrum analysis.

The research in the early twentieth century by no means completed the theoretical development of noise. Einstein, Smoluchowski, Langevin, and Bachelier launched probabilistic theorization of Brownian motion. Schottky, Johnson, and Nyquist inquired into the conceptual connections between electronic noise and Brownian particles. But such inquiries raised crucial questions that these pioneers did not answer: What was the temporal behavior of Brownian particles before they settled down to the ever-spreading Gaussian distribution? What would happen to these particles if they were subjected to an external force or a particular boundary condition? What were noise's implications for the precision of measurements? What would be the effect of random noise if it went along with a transmitted signal through a telephone or radio receiver? How was it possible to analyze the spectrum of noise that did not have a well-defined Fourier series or integral? Mathematically, what did it mean to be a random process?

In part III, we examine the theoretical development of random noise in the 1920s–30s that addressed these questions. By the end of World War I, scientists and technologists had ideas about the ubiquity, physical causes, and some quantitative behavior of Brownian fluctuations, but did not know much more than that. At the dawn of World War II, they possessed a more fully developed technical language to talk about noise, a computational platform to calculate its effects in physical or engineering circumstances, and a framework to understand the structure of random fluctuations. The interwar period witnessed the building of a theoretical regime for noise.

Being mathematical, formal, and quantitative, this regime was nonetheless neither homogeneous nor uniform. Rather, it comprised three distinct approaches. In chapters 5 and 6, we have seen the origin of two of them. In this part, we will see their further development in the 1920s–30s. The first approach operated in the time domain. It grappled with random fluctuations in a physical system as an outcome of microscopic particles' irregular trajectories. Here, fluctuations were phenomena of dynamics, and the major aim of research was to determine the temporal evolution of the trajectories' statistical attributes. This approach came from the research tradition of statistical mechanics. In the 1900s–10s, Einstein, Smoluchowski, and Langevin established the basic framework of this approach. During the interwar period, this approach was advanced further. Scientists sought a more comprehensive governing equation—known as the Fokker–Planck equation—for the probability density of Brownian particles affected by external forces or peculiar boundary conditions. They examined fine-scale temporal evolution of random fluctuations before it settled down asymptotically. And they applied their theories to metrological problems. The Dutch-American physicist George Uhlenbeck played a central part in this research program.

The second approach operated in the frequency domain. It conceived random fluctuations as waveforms comprising different modes of oscillations. Fourier analysis was its methodological cornerstone. The major aim of research was to find the energy for a fluctuating waveform's components within a frequency window. To achieve this aim, researchers often relied on the assumption of noise waveforms' narrow self-correlation, which resulted in a flat, "white" spectrum. This approach came from the R&D of electrical sound reproduction. Schottky, Johnson, and Nyquist paved the foundation for this approach in their work on shot and thermal noise. The interwar engineers explored further how electronic noise and atmospherics affected the performance measure of telecommunication systems as both signals and noise passed through electrical circuits in the radio and telephone. John Carson at Bell Labs represented this approach.[1]

The third approach focused on a rigorous characterization of the central concepts for random fluctuations. Under this approach, a Brownian motion was configured in a functional space, while a new formulation of spectral analysis was developed for the fluctuations to which conventional Fourier analysis could not apply. The mathematicians taking this approach were more concerned about building a solid logical foundation and conceptual structure for random fluctuations than about statistical physics or telecommunication engineering. Their work was inspired by traditions in modern mathematics: set theory, measure theory, and axiomatization of probability. The American mathematician Norbert Wiener at MIT was a leader in this development.[2]

The disunity of the interwar theoretical works on noise is a familiar theme in historical and social studies of science and technology. The anthropologist Karin Knorr Cetina coined the term "epistemic cultures" to signify "amalgams of arrangements and mechanisms—bonded through affinity, necessity, and historical coincidence—which, in a given field, make up *how we know what we know*" (italics in the original). To her, "epistemic cultures are cultures that create and warrant knowledge,"[3] pointing to different standards of knowledge making and validation and their implications for institutions and practice along disciplinary boundaries—e.g. high-energy particle physics versus molecular biology. The philosopher Evelyn Fox Keller introduced a similar notion of "epistemological cultures" surrounding what constituted legitimate understandings and explanations. She maintained that we could see different styles of reasoning and norms for valid knowledge within the same scientific

[1] Dörfel and Hoffmann, "Von Albert Einstein bis Norbert Wiener" (2005); Dörfel, "Early history of thermal noise" (2012); Chen-Pang Yeang, "Two mathematical approaches to random fluctuations," *Perspectives on Science*, 24:1 (2016), 45–72.

[2] For the rise of set theory, measure theory, and axiomatization in the late nineteenth century and their applications to probability theory and stochastic processes, see von Plato, *Creating Modern Probability* (1994); Morris Klein, *Mathematical Thought from Ancient to Modern Times*, Volume 3 (1972), 1023–1039. For a historical review of the development of stochastic process theory, see Davis and Etheridge (translators and commentators), *Louis Bachelier's* Theory of Speculation (2006), 81–115; Robert Jarrow and Philip Protter, "A short history of stochastic integration and mathematical finance: The early years, 1880–1970," *Lecture Notes—Monograph Series, Institute of Mathematical Statistics: A Festschrift for Herman Robin*, 45 (2004), 75–91.

[3] Karin Knorr Cetina, *Epistemic Cultures: How the Sciences Make Knowledge* (Cambridge, MA: Harvard University Press, 1999), 1.

discipline or sub-discipline such as evolutionary developmental biology.[4] Elsewhere I also pointed out an essential difference between mathematicians', engineers', and experimental physicists' theoretical studies of long-distance radio-wave propagation in the early twentieth century. I argued that these theories possessed different epistemic status for their different goals: to represent, to explain, or to compute.[5]

The three theoretical approaches to noise during the interwar period epitomized diverse aims, standards, and cultures. Statistical physicists' goal was to determine the behavior of dynamic systems under the influence of random fluctuations. Electrical engineers' purpose was to gauge the effects of noise on the performance of sound reproduction. Mathematicians' agenda was to develop what they held to be a rigorous formulation and structure of the theoretical knowledge about noise. While all of them utilized probability calculus, statistics of Gaussian distributions, and intuitive or axiomatic concepts of stochastic processes, their epistemic status was different.

Moreover, these three approaches were not a mere exercise with abstract symbols and formulas. They were associated with distinct material cultures. Statistical physicists' time-domain approach was closely related to ultramicroscopy, suspended particles in a fluid, optics of polarized materials, and galvanometry. Electrical engineers' frequency-domain approach was embedded in the design and testing of electronic systems for radio and telephony. Even mathematicians' "pure" theoretical examinations of random fluctuations traced material underpinnings in hydrodynamic turbulence, incoherent light, the meteorological record, and electric power grids.

The three interwar theoretical approaches to noise were not three separate island empires. Despite their conceptual and technical differences, the researchers taking each of these approaches were not bounded tightly by their own epistemic culture. Instead, they did not shy away from appropriating elements of other theoretical approaches when opportunities came up. For instance, when Uhlenbeck examined Brownian motion of a galvanometer that affected metrological precision, he employed a harmonic analysis to characterize the meter's dynamics. Norbert Wiener's generalized harmonic analysis was inspired by his collaboration with the engineering faculty at MIT for tackling frequency responses in electrical systems.

[4] Evelyn Fox Keller, *Making Sense of Life: Explaining Biological Development with Models, Metaphors, and Machines* (Cambridge, MA: Harvard University Press, 2002), 1–10.

[5] Chen-Pang Yeang, *Probing the Sky with Radio Waves: From Wireless Technology to the Development of Atmospheric Science* (Chicago, IL: University of Chicago Press, 2013).

Dynamics of Random Fluctuations

By the 1910s, physicists had a good grasp of Brownian motion. From Einstein's, Smoluchowski's, and Langevin's works, they were aware that Brownian motions of suspended particles and other similar fluctuations resulted from molecular thermal agitations amplified into observable phenomena. They knew how to represent the effects with random walks, an equation for probability density, or an equation of motion with a random force. They also obtained some important features of the Brownian-type fluctuations from these mathematical representations.

Physicists' theoretical research on Brownian motion did not stop here, though. Experimental, disciplinary, and conceptual factors provided fuels for its continuation in the 1920s–30s. While Perrin's ultramicroscopic observation in 1909 corroborated Einstein's theory, further experimental studies painted a more complex picture. Ultramicroscopic measurements of colloidal solutions from the Swedish chemist Theodor Svedberg in Uppsala, his disciple Arne Westgren, and Einstein's Prague colleague Reinhold Fürth showed that the statistical variance of particles' displacements changed before it settled down with a steady increase with time as Einstein had predicted.[6] The fine-scale temporal variation of a Brownian motion's statistical attributes became a meaningful research topic.

Moreover, explorers of the emerging quantum physics made implicit or explicit connections between the thermal fluctuations underlying Brownian motion and the newly found quantum fluctuations. Although their physical natures were utterly different, they were nonetheless both chance phenomena subjected to probabilistic calculus. In relation to this analogy, the measurement problem imposed by Werner Heisenberg's recently introduced uncertainty principle underlying quantum fluctuations was sometimes juxtaposed with the limit that thermal fluctuations incurred on the precision of measuring instruments. At the discourse level, both types of fluctuations were construed as "fundamental" constraints a probabilistic world posed to the exact empirical determination of physical quantities.

In this chapter, we examine statistical physicists' further theoretical research on Brownian motions in the 1910s–30s: the development of the so-called "Fokker–Planck equation" that represented the probability density of Brownian entities as an outcome both of more general considerations than Einstein

[6] For a brief overview of Svedberg's, Westgren's, and Fürth's experimental work, see Chandrasekhar, "Stochastic problems in physics and astronomy" (1943), 43–50.

had stipulated and of inspirations from quantum physics; and Uhlenbeck and his collaborators' work on the detailed time-evolution of Brownian motions' statistical attributes as well as the limit they imposed on the precision of electrical measurements.

The Fokker–Planck Equation

If we treat a large number of particles undergoing random thermal motions as a dynamic system, then an inevitable question concerns the form of its governing equation. As we have seen in chapter 5, the physicists working on Brownian motion in the 1900s came up with two answers. One, as Langevin proposed in 1908, was to write down the equation of motion for an individual particle and to model the effect of irregular collisions from other particles with a random force (cf. (5.12')). The other, developed by Einstein in 1905–06, was to formulate a partial differential equation that governed the particles' probability density function (cf. (5.2)).

Einstein's work in 1905–06 did not set the final tone for inquiries into the governing equation for random thermal motions' probability density function. A series of studies in the 1910s resulted in a more complicated form of partial differential equation than Einstein's diffusion equation, later known as the "Fokker–Planck equation."

Fokker's Initiative

The first extension of Einstein's diffusion-type equation was made by the Dutch physicist Adriaan Daniël Fokker. Fokker was born into a prominent family in the Dutch East Indies during the Netherlands' rapid colonial expansion into today's Indonesia in the late nineteenth century. He was a son of the president of the Netherlands Trading Society at Batavia (today's Jakarta), and a cousin of Anthony Fokker, the well-known airplane designer–manufacturer. Adriaan's family moved back to Europe when he was seven. He studied mining engineering at the Technical University of Delft, but switched to physics and moved to the University of Leyden to pursue a Ph.D. with Hendrik Antoon Lorentz.[7] When Fokker entered academia in the late 1900s, science in the

[7] Hendrik B.G. Casimir and Sybren R. de Groot, "Levensbericht van Adriaan Daniël Fokker," *Jaarboek, Huygens Institute—Royal Netherlands Academy of Arts and Sciences (KNAW)* (1972), 114–118; "Adriaan Daniël Fokker," in Huygens-Fokker Foundation, Centre for Microtonal Music website, http://www.huygens-fokker.org/whoswho/fokker.html (accessed April 8, 2022).

Netherlands was on a booming rise. The period of the 1870s–1920s is sometimes referred to as "the Second Golden Age of Dutch Science." State-initiated educational reform; transformation of the centuries-old universities at Leyden, Utrecht, and Groningen into German-style research universities; and enactment of new universities helped make a few Dutch institutions research centers in physics, chemistry, and physiology.[8] Lorentz was a pivotal figure in this development. His longtime career in Leyden and nationwide influence facilitated the country's strong research activities in electrodynamics, microphysics, relativity, and quantum mechanics.

The leading Dutch physicists in Leyden, Utrecht, and Amsterdam during this period paid close attention to the problems of matter–radiation interactions and what is now called "condensed matter": Johannes van der Waal's study of the equation of state, Pieter Zeeman's discovery of the split of gaseous spectral lines under a magnetic field, Lorentz's theory of electrons to interpret the Zeeman effect and other electrical properties, Heike Kamerlingh-Onnes's experiments with low-temperature substances, and the Vienna-trained immigrant Paul Ehrenfest's formulation of quantum mechanics with many bodies. This intellectual milieu was encouraging to statistical physics of random fluctuations in the 1900s–20s. Lorentz's own daughter Geertruida de Haas-Lorentz produced a thesis in 1912, under her father's tutelage in Leyden, on electrons' Brownian motions and their impact on metrology (cf. chapter 6). Fokker's Ph.D. thesis in 1913 was another project on Brownian motions supervised by Lorentz.

Fokker's thesis was inspired not only by Einstein's work on Brownian motion in 1905–06, but also by Planck's investigation of black-body radiation circa 1900. Planck's theory treated thermal radiation from a substance as electromagnetic waves generated by many molecular radiators. The object of Fokker's study was not particles moving in a liquid, but molecules of a matter that irradiated energy. Since these microscopic radiators also underwent thermal agitations, random fluctuations also characterized thermal radiation from a substance.[9]

After graduating from Leyden in 1913, Fokker moved to Zurich to further his study of Brownian motion with Einstein, who was then teaching at the ETH. In 1914, Fokker extended his thesis work on thermal radiation and published it in *Annalen der Physik*. Titled "The mean energy of rotational

[8] Bastiaan Willink, "Origin of the Second Golden Age of Dutch Science after 1860: Intended and unintended consequences of educational reform," *Social Studies of Science*, 21:3 (1991), 503–526.

[9] Adriaan Fokker, *Over Brown'sche Bewegingen in het Stralingsveld, en Waarschijnlijkheids-Beschouwingen in de Stralingstheorie*, Ph.D. dissertation (Leyden: University of Leyden, 1912).

electric dipoles in a radiation field," Fokker's article considered a scenario, in which an electromagnetic field was deployed in a region that contained a large number of polarized molecules functioning like tiny electric dipoles. To simplify the treatment, he supposed that these dipoles could rotate only along a single direction. Thus, each of these dipoles possessed an angular momentum. Like in other molecular systems at non-zero temperature, these angular momenta changed irregularly. When a dipole's angular momentum increased (or decreased), it absorbed (or emitted) energy from (or to) the background radiation. Fokker's first goal was to find an equation that governed the probability density function for these irregular molecular angular momenta.[10]

To achieve this goal, he considered a parameter q that represented a molecular dipole's angular momentum. This parameter took random values with a certain statistical distribution. For a system of N dipoles, the number of dipoles whose angular momenta fell between q and $q + dq$ was $N \cdot W(q)\, dq$ (dq being a differential quantity). Here $W(q)$ represented the probability density function for the dipoles' angular momenta. To obtain the governing equation for $W(q)$, Fokker also needed the average loss of a dipole's angular momentum over a unit time interval $f(q)$, the change R of a dipole's angular momentum within a very short time interval τ, and its mean $\langle R \rangle$ and mean square $\langle R^2 \rangle$. Fokker argued that when the system achieved a stationary state so that the probability density $W(q)$ before and after τ remained unchanged and the terms at second and higher orders of τ were ignored, $W(q)$ was governed by the following differential equation:

$$W(q)f(q)\,\tau - W(q)\,\langle R \rangle + \frac{1}{2}\frac{\partial}{\partial q}\left\{W(q)\,\langle R^2 \rangle\right\} = 0. \tag{7.1}$$

To Fokker, this equation could be applied to represent problem situations other than molecular dipoles under thermal radiation. For instance, one could also interpret q as the height of a Brownian particle suspended in a fluid, $f(q)$ as the average speed of falling for the Brownian particles, R as the change of a particle's height within τ, and $W(q)$ as the probability density function of the Brownian particles. Equation (7.1) could describe that physical situation, too.[11]

[10] Adriaan D. Fokker, "Die mittlere Energie rotierender elektrischer Dipole im Strahlungsfeld," *Annalen der Physik*, 43 (1914), 810–811.

[11] Ibid., 811–813.

Fokker made an explicit analogy between radiated molecular dipoles and Brownian particles in a liquid. And he contended that equation (7.1) could describe various physical situations exhibiting thermal fluctuations. Yet, he did not explain how he came up with this equation. Its derivation was missing in Fokker's 1914 paper. Not until three years later did another researcher return back to the governing equation for thermal fluctuations. He was Max Planck.

Planck's Follow-up

As a founder of quantum physics, advocate for relativity, and chair professor at the University of Berlin, Planck was at the forefront of physics at the turn of the twentieth century. His groundbreaking work circa 1900 on black-body radiation introduced a discrete-energy hypothesis that debuted quantum theory. To him, the problem situation for black-body radiation resembled those in statistical physics, for the huge number of microscopic electromagnetic oscillators in thermal radiation resembled randomly moving gas molecules in a volume or agitated electrons in a conductor. From this perspective, Brownian motion and thermal radiation had a structural similarity. While Planck ventured into other realms of research in the 1900s–10s, he held a continual interest in thermal radiation. When Niels Bohr presented a successful quantization of the hydrogen atom in 1913, the subject of thermal radiation became relevant again. Planck, along with a few other physicists at the time, endeavored to extend Bohr's one-particle quantization to a quantum theory of multiple entities. A reconsideration of some problems in statistical physics, especially those related to Brownian motion connected to thermal radiation, might provide an avenue to resolving the issues of quantization.[12]

This was the context for Planck's revisit to the statistical theory of thermal radiation. In 1917, he published an article titled "About a law of statistical dynamics and its extension to quantum theory" in the annual proceedings of the Prussian Academy of Science.[13] In this paper, Planck started by acknowledging Einstein's and Fokker's contributions toward understanding thermal fluctuations of physical systems in which a large number of entities underwent Brownian-like motions. Like Fokker, Planck's formulation for such systems

[12] Max Born, "Max Karl Ernst Ludwig Planck, 1858-1947," *Obituary Notices of Fellows of the Royal Society*, 6:17 (1948), 175.

[13] Max Planck, "Über einen Satz der statistischen Dynamik und seine Erweiterung in der Quantentheorie," *Sitzungsberichte der Königlich Preussischen Akademie der Wissenschaften*, 24 (1917), 324–341.

was generic. Each entity in the system possessed a physical state parametrized by a quantity q, which could be a molecular dipole's angular momentum, a suspended particle's displacement, a gas molecule's velocity, etc. Again, his aim was to obtain the governing equation for the probability density function $W(q)$: for a system of N entities, the number of entities between the states q and $q + dq$ was $N \cdot W(q)\, dq$. As probability density, the integral of $W(q)$ over its entire sample space was unity: $\int_{-\infty}^{\infty} W(q)\, dq = 1$.

In contrast to Fokker in 1914, here Planck presented a derivation of his equation. To do so, he introduced another probability function $\phi_q(r)$, where $\phi_q(r)\, dr$ represented the probability for a particle at a state between q and $q + dq$ to shift between r and $r + dr$ within a short duration τ. Within τ, the magnitude of the shift $|r|$ could not exceed R_0. Thus the sample space for $\phi_q(r)$ was the domain for r to be between $-R_0$ and R_0, and the unitary condition became $\int_{-R_0}^{R_0} \phi_q(r)\, dr = 1$.[14]

To formulate $W(q)$, Planck considered the continuity of density flow—a strategy that Einstein had adopted in 1905. He argued that if a particle's movements (i.e. changes of q) were independent of each other at different times, then the number of particles within [q', $q' + dq'$] at the beginning of τ and which moved into [q, $q+dq$] at the end of τ would be $N \cdot W(q')\, dq' \phi_{q'}(q - q')\, dq$. The total number of particles within [q, $q+dq$] at the end of τ would then be the integral of the states from all possible positions q' at the start of τ that entered [q, $q+dq$] at the end of τ. This was the integral of $N \cdot W(q')\, dq' \phi_{q'}(q - q')\, dq$ over the valid domain of $\phi_{q'}(q - q')$, which was $q' \in [q - R_0, q + R_0]$. That is,

$$N \cdot dq \cdot \int_{q-R_0}^{q+R_0} W(q')\, \phi_{q'}(q - q')\, dq'.$$

With a change of variable $r = q - q'$, the integral became

$$N \cdot dq \cdot \int_{-R_0}^{R_0} W(q - r)\, \phi_{q-r}(r)\, dr.$$

Planck already knew that the number of particles [q, $q+dq$] at the beginning of τ equaled $N \cdot W(q)\, dq$. Thus, the change of the number of particles within [q, $q+dq$] during τ was $N \cdot dq \cdot \int_{-R_0}^{R_0} W(q - r)\, \phi_{q-r}(r)\, dr - N \cdot W(q)\, dq$. This change of particle number could also be written in terms of the rate of change for the

[14] Ibid., 325–327.

probability density function as $N \cdot dq \cdot (\partial W/\partial t) \cdot \tau$. Equating both quantities, he obtained[15]

$$\frac{\partial W}{\partial t} \cdot \tau = \int_{-R_0}^{R_0} W(q-r)\,\phi_{q-r}(r)\,dr - W(q). \tag{7.2}$$

To simplify (7.2), Planck treated $q - r$ as a small perturbation from q (since $|r| < R_0 \ll q$). This made possible a Taylor series expansion of the integrand with respect to q. Keeping the terms to the second order of r, Planck obtained a partial differential equation for W:

$$\frac{\partial W}{\partial t} \cdot \tau = -\frac{\partial}{\partial q}\left[W(q) \cdot \langle r\rangle\right] + \frac{1}{2}\frac{\partial^2}{\partial q^2}\left[W(q) \cdot r^2\right], \tag{7.3}$$

where $\langle r\rangle \equiv \int_{-R_0}^{R_0} r\phi_q(r)\,dr$ and $r^2 \equiv \int_{-R_0}^{R_0} r^2\phi_q(r)\,dr$.[16]

Two things about (7.3) are worth noting. First, although its form looks different from Fokker's (7.1), we can deduce (7.1) from (7.3), if we replace r in (7.3) with R in (7.1), integrate over q on both sides of equality in (7.3), acknowledge $dq/dt = -f(q)$, employ the relation $\int dq\,(\partial W/\partial t)\cdot\tau = \int dq\,(\partial W/\partial q)\cdot(dq/dt)\cdot \tau \cong -f(q)\cdot\tau\cdot\int dq\,(\partial W/\partial q) = -f(q)\cdot\tau\cdot W(q)$, and set the integral constant to zero. This shows the equivalence between Fokker's and Planck's formulations, and hence justifies the name "Fokker–Planck equation."

Second, in spirit, if not in precise calculus, Planck's derivation of (7.3) was similar to Einstein's derivation of (5.2), and Planck's $\phi_{q-r}(r)$ resembled Einstein's $\varphi(\Delta)$ in (5.1). Yet, Planck's (7.3) was not Einstein's diffusion equation (5.2)—the former had one more term: $-\frac{\partial}{\partial q}\left[W(q)\cdot\langle r\rangle\right]$. Why the difference? When Einstein derived his governing equation for the probability density of Brownian particles' displacements, he did not assume any physical force on the particles. The only factor that drove the particles into motion was thermal agitation, which diffused them uniformly along all directions. In Einstein's one-dimensional case in 1905, the displacement probability function $\varphi(\Delta)$ was thus symmetrical with Δ. When he took a Taylor series expansion to the second order of Δ in (5.1), therefore, the integral of the first-order term $\int_{-\infty}^{\infty}\Delta\varphi(\Delta)\,d\Delta$ became zero. The zeroth-order term on both sides canceled each other, while the second-order term contributed to the $\partial^2/\partial x^2$ part of Einstein's diffusion equation. Planck's theory did not entail such a symmetric condition, however. His parameter $\langle r\rangle = \int_{-R_0}^{R_0} r\phi_q(r)\,dr$ did not vanish as Einstein's $\int_{-\infty}^{\infty}\Delta\varphi(\Delta)\,d\Delta$

[15] Ibid., 328.
[16] Ibid., 328.

did. Consequently, his equation contained a term corresponding to the first-order partial derivative $\partial/\partial q$. Physically, this indicated that Planck did not exclude forces other than diffusion tendency. When a force was present, the particles could be driven toward a particular direction. The shift probability density $\phi_q(r)$ would thus be skewed, meaning that $\langle r \rangle = \int_{-R_0}^{R_0} r\phi_q(r)\,dr$ would not become zero. In other words, Planck's theory of Brownian motion was more general than Einstein's.

Smoluchowski's "Intuitive" Take

Planck might not have known that he was not the only researcher after Einstein and Fokker to work on the governing equation for the probability density of entities under thermal agitations. In fact, someone else had already obtained a similar formulation to the "Fokker–Planck equation" through a different route. In 1915, Marian Smoluchowski published a paper in *Annalen der Physik* that tackled this issue.[17] His subject was once again Brownian motion. Differently from many prior works, he focused on the case when an external force was imposed on the particles. Instead of resorting to probability flow as Einstein and Planck did, Smoluchowski's way of proceeding was to reverse the direction of reasoning and to infer the form of the governing equation from the probability density function obtained via another means.

Without an external force, the particles diffused uniformly, their trajectories were statistically independent of one another, and each trajectory at different instants was also independent. This core condition of Brownian motion led Smoluchowski to an integral expression in the form of a Markov chain. He set $W(x_0, x, t)$ as the probability density function of particles with initial position x_0 at the beginning of a time interval t, and final position x at the end of the interval. Then the mutual independence between the particles yielded the relation

$$W(x_0, x, t) = \int W(x_0, \alpha, \vartheta) \cdot W(\alpha, x, t - \vartheta)\, d\alpha. \tag{7.4}$$

The integrand represented the probability density of particles at the initial position x_0 at the beginning of t, at the position ϑ at an intermediate time ϑ, and at the final position x at the end of t, owing to mutual independence.

[17] Marian Smoluchowski, "Über Brownsche Molekularbewegung unter Einwirkung äußerer Kräfte und deren Zusammenhang mit der verallgemeinerten Diffusionsgleichung," *Annalen der Physik*, 48 (1915), 1103–1112.

The overall probability density $W(x_0, x, t)$ was the integral of the scenarios at all possible intermediate ϑ. Smoluchowski showed that this Markovian relation gave a solution of $W(x_0, x, t)$ as a Gaussian distribution whose variance increased with time:[18]

$$W(x_0, x, t) = \frac{1}{2\sqrt{\pi D\vartheta}} e^{-\frac{(x-x_0)^2}{4D\vartheta}}.$$

When an external force $g(x)$ was imposed on particles, the form of the probability density function was slightly modified. Smoluchowski claimed that he had derived the form of the probability density if the external force was elastic $g(x) = -ax$. But the form of the function could retain for a general force that depended on particle position x, as long as the duration ϑ was short:

$$W(x_0, x, t) = \frac{1}{2\sqrt{\pi D\vartheta}} e^{-\frac{[x-x_0-\beta\vartheta g(x)]^2}{4D\vartheta}}. \tag{7.5}$$

In this expression, β represented Brownian particles' mobility. Since the particles were suspended in a fluid, the exerted force would eventually balance with the fluid's resistance. On average, therefore, the particles would not accelerate indefinitely by g, but rather reached a constant drift velocity that equaled $\beta f(x)$. Within a short duration ϑ, this drift velocity would cause a mean displacement of $\beta\vartheta f(x)$. This additional displacement due to the external force explained the additional term $\beta\vartheta f(x)$ in (7.5).[19]

Without providing details, Smoluchowski asserted further that he could calculate from (7.5) the number of particles moving across the position x within a short duration Δt divided by the total number of particles in the fluid. This quantity, which we may call the Brownian particles' "probability current," was

$$\left[-D\frac{\partial W(x)}{\partial t} + \beta W(x)\, g(x)\right]\Delta t.$$

He noted that this probability current contained two terms. One was $-D\frac{\partial W(x)}{\partial t}$. He named it "diffusion current" (der Diffusionsströmung) and associated it with the particles' spreading due to molecular thermal fluctuations. The other was $\beta W(x)\, g(x)$. He named it "convection particle current" (der konvektiven Teilchenströmung) and associated it with the particles' movements driven

[18] Ibid., 1104.
[19] Ibid., 1104–1105.

by the external force. The number of particles between x and $x + \Delta x$ during Δt was

$$N\left\{D\frac{\partial^2 W(x)}{\partial x^2} - \beta\frac{\partial}{\partial x}\left[W(x)g(x)\right]\right\}\Delta x\Delta t.$$

This number of accumulated particles within the temporal and spatial windows was the same as the changing rate of the number of particles $N\frac{\partial W(x)}{\partial t}$ around x and t times the widths of the differential windows Δt and Δx. That gave the equation

$$\frac{\partial W}{\partial t} = D\frac{\partial^2 W}{\partial x^2} - \beta\frac{\partial}{\partial x}\left[Wg(x)\right]. \tag{7.6}$$

Smoluchowski's governing equation (7.6) for the Brownian particles' probability density function was identical to the Fokker–Planck equation (7.3). They were both linear parabolic partial differential equations of the diffusion type. Instead of parametrizing the equation with the statistical attributes of particles' random shift as Planck did, Smoluchowski parametrized the equation with macroscopic and observable quantities, including the fluid's diffusion coefficient D, the particles' mobility β in the fluid, and the external force $g(x)$.[20]

Burger's Perturbation Analysis

Smoluchowski's examination of the Fokker–Planck equation for Brownian motions stimulated further studies in the Netherlands. The Polish physicist posed a new question about the behavior of Brownian particles with an external force. But his somehow intuitive approach did not seem convincing to a few contemporary researchers. In 1917, two Utrecht-based scientists, Herman Carel Burger and Leonard Ornstein, separately presented their own deductions of the governing equation for the probability density of Brownian particles with an external force.

Both individuals were products of the "Second Golden Age" of Dutch science. A native of Utrecht, Leonard Ornstein studied physics at the University of Leyden, and pursued a Ph.D. under Lorentz's supervision. Inspired by Lorentz's interest in statistical physics, Ornstein examined Josiah Willard Gibbs's formulation of statistical mechanics and its connections to the molecular kinetic theory of matter. He obtained his doctoral degree in 1908 and was

[20] Ibid., 1105.

appointed professor of physics at the University of Utrecht in 1914. Herman Carel Burger was another native of Utrecht. He entered the university of his hometown to study physics. Under Ornstein's supervision, Burger presented a dissertation on the growth and dissolution of crystals and obtained a Ph.D. in 1918. Both scientists built their careers in Utrecht. Their most recognized research during the interwar period consisted of experimental studies and precision measurements of spectral lines from materials' thermal radiation. This was a subject preoccupying numerous contemporary Dutch physicists from Pieter Zeeman and Lorentz to Peter Debye due to its connection to the emerging quantum theory of atoms and molecules. Because of the practical value of precision metrology, both individuals maintained a collaborative relationship with the Dutch electrical conglomerate Philips.[21] In the late 1910s, their common interest was Brownian motion, a recent branch from statistical physics.

At a Royal Netherlands Academy of Arts and Sciences (KNAW) meeting on April 27, 1917, Burger presented a paper on the theory of Brownian motion and its comparison with certain recent experimental results. The Utrecht Ph.D. student was well aware of the major results Einstein and Smoluchowski had obtained in 1905–06: the diffusion equation, the Gaussian distribution whose variance spread with time, etc. He also knew about Smoluchowski's examination of Brownian motion under the influence of an external force in 1915–16. Burger's motivation to revisit the problem was to rederive these results with what he considered a more rigorous method, and to compare his theoretical predictions with recent experimental data on colloidal solutions. His point of entry was Smoluchowski's Markovian relation (7.4). Burger took the probability of particles at displacements between x and $x + dx$ during a time interval t to be $f(x, t)dx$, and expressed (7.4) as[22]

$$\int_{-\infty}^{\infty} f(x, t_1) f(a - x, t_2)\, dx = f(a, t_1 + t_2). \tag{7.4'}$$

Unlike Einstein's and Planck's detailed count of particles entering and leaving a differential spatial window at adjacent instants, Burger tackled the mathematical implications of (7.4'). When the Brownian particles did not have an external

[21] Hendrik Anthony Kramers, "Levensbericht van L.S. Ornstein," *Jaarboek, Huygens Institute—Royal Netherlands Academy of Arts and Sciences (KNAW)* (1940–1941), 225–231; J.G. van Cittert-Eymers, "Burger, Herman Carel," in Charles Coulston Gillispie and Frederic Lawrence Holmes (eds.), *Dictionary of Scientific Biography*, 2 (New York: Scribner, 1970), 600–601.

[22] Herman Carel Burger, "On the theory of the Brownian movement and the experiments of Brillouin," *Proceedings of the Huygens Institute—Royal Netherlands Academy of Arts and Sciences (KNAW)*, 20:1 (1918), 643.

force, he expanded $f(a - x, t_2)$ on the left side of (7.4') in a Taylor series of powers of x, and $f(a, t_1 + t_2)$ on the right side of (7.4') in another Taylor series of powers of t_1. Assuming tiny t_1 and (without justification) that the contributions from high-order powers of x were small, Burger kept the Taylor series on the left side of (7.4') up to the order of x^2 and the Taylor series on the right side of (7.4') up to the order of t_1. Utilizing the relations $\int_{-\infty}^{\infty} f(x,t)\,dx = 1$ and $\int_{-\infty}^{\infty} xf(x,t)\,dx = 0$ (since $f(x,t)$ was an even function of x, given the symmetry of thermal agitation), he reduced (7.4') into

$$\frac{1}{2}\frac{\partial^2 f(a,t_2)}{\partial a^2}\int_{-\infty}^{\infty} x^2 f(x,t_1)\,dx = t_1\frac{\partial f(a,t_2)}{\partial a}.$$

In this equation, the only t_1-dependent part on the left side was $\int_{-\infty}^{\infty} x^2 f(x,t_1)\,dx$, while the only t_1-dependent part on the right side was t_1. These two terms must equal each other, aside from a constant coefficient. Burger thus defined that $\int_{-\infty}^{\infty} x^2 f(x,t_1)\,dx = 2Dt_1$, where D was a constant. This led to the familiar diffusion equation for $f(a,t_2)$:[23]

$$\frac{\partial f(a,t_2)}{\partial t_2} = D\frac{\partial^2 f(a,t_2)}{\partial a^2}.$$

With this new derivation of Einstein's main result, Burger went further to deduce the governing equation for Brownian particles when an external force was present. Again, the Markovian relation (7.4') played a central role. Burger changed slightly the form of (7.4'), so that the probability density f now had three arguments $f(x,a,t)$, where x represented the particle's initial position at the beginning of the time interval t, and a represented the particle's final position at the end of the interval. Then (7.4') became

$$\int_{-\infty}^{\infty} f(x,p,t_1)f(p,a,t_2)\,dp = f(x,a,t_1+t_2). \qquad (7.4'')$$

To examine the effect of an external force, Burger considered the situation when t_2 in (7.4'') was small. Then within this short duration a particle moved only a small amount. That meant the probability density $f(p,a,t_2)$ was close to that of a Brownian particle undergoing a displacement of $a - p$ during t_2. If there was no external force, then thermal agitation would be the particle's only drive, and such a probability density would be an even function of $a - p$,

[23] Ibid., 645–646.

$f(p, a, t_2) = \varphi(a - p, t_2)$. But it was not the case. In addition to thermal agitation, the particle was also driven by the external force. Burger supposed that the particle's displacement due to the external force within t_2 was δ. Thus, only the displacement component $a - p - \delta$ was caused by thermal agitation; and only the probability density associated with $a - p - \delta$ was an even function. That is, $f(p, a, t_2) = \varphi(a - p - \delta, t_2)$. As mentioned, the homogeneity of Brownian motion gave $\int_{-\infty}^{\infty} \varphi(\xi, t)\, d\xi = 1$, $\int_{-\infty}^{\infty} \xi \varphi(\xi, t)\, d\xi = 0$, and $\int_{-\infty}^{\infty} \xi^2 \varphi(\xi, t)\, d\xi = 2Dt$.[24]

To advance, Burger needed to know the analytical form of δ, and its relationship with the external force. Here he made a similar claim as Smoluchowski's in 1915–16: When the duration t_2 was short, the velocity caused by a position-dependent force $\psi(p)$ was $\beta\psi(p)$ (β was the particle's mobility in the fluid), and the displacement caused by the same force was $\delta = \beta t_2 \psi(p)$. Given that $\varphi(x, t)$ was an even function of x, Burger asserted $f(p, a, t_2) = \varphi(a - p - \delta, t_2) = \varphi(p + \delta - a, t_2) = \varphi(p + \beta t_2 \psi(p) - a, t_2)$. Plugging this relation into (7.4"), he obtained

$$\int_{-\infty}^{\infty} f(x, p, t_1)\, \varphi\big(p - a + \beta t_2 \psi(p), t_2\big)\, dp = f(x, a, t_1 + t_2). \qquad (7.4''')$$

To solve this equation, Burger replaced p with $\xi + a$ (ξ was small because of the short interval t_2) and expanded the two factors of the integrand on the left side into Taylor series of ξ and t_2:

$$f(x, a + \xi, t_1) = f(x, a, t_1) + \xi \frac{\partial f}{\partial a} + \frac{1}{2} \xi^2 \frac{\partial^2 f}{\partial a^2} + \ldots$$

$$\varphi\big(\xi + \beta t_2 \psi(a + \xi), t_2\big) = \varphi(\xi, t_2) + \left[\beta\psi(a) + \beta\xi \frac{\partial \psi(a)}{\partial a}\right] \frac{\partial \varphi(\xi, t_2)}{\partial \xi} t_2 + \ldots$$

He expanded the term on the right side of (7.4"') with a Taylor series of t_2:

$$f(x, a, t_1 + t_2) = f(x, a, t_1) + t_2 \frac{\partial f(x, a, t_1)}{\partial t_1} + \ldots$$

Plugging the Taylor series into (7.4"'), keeping only the leading terms, performing the integral over ξ, and truncating the second- and higher-order terms, Burger obtained

[24] Ibid., 647–648.

$$\frac{\partial f(x, a, t_1)}{\partial t_1} = D\frac{\partial^2 f}{\partial a^2} - \beta\psi(a)\frac{\partial f}{\partial a} - \beta f\frac{\partial \psi}{\partial a}.$$

This was equivalent to

$$\frac{\partial f}{\partial t} = D\frac{\partial^2 f}{\partial a^2} - \beta\frac{\partial}{\partial a}(\psi f). \tag{7.7}$$

This equation was identical to (7.6), Smoluchowski's formulation of the Fokker–Planck equation. Burger demonstrated an alternative approach to produce this governing equation for Brownian motion with an external force.[25]

Ornstein's Employment of the Langevin Equation

Burger's paper at the KNAW meeting in April 1917 raised his supervisor's interest in Brownian motion. At another KNAW meeting in December, Ornstein presented his own study of the subject. The Utrecht professor's paper began with a note that Smoluchowski had deduced a differential equation for the probability function of Brownian particles with an external force "by a phenomenological method." Shortly afterward, his student Burger deduced the same equation "following a method, which takes the essence of the function of probability more into consideration." Ornstein remarked, however, "both deductions do not stand in direct connection with the mechanism of the Brownian motion." A method in direct connection with such a mechanism, in his view, was: "starting from a relation which Mrs. De Haas-Lorentz has used in her dissertation, to determine the average square of the distance accomplished, one is able to determine the function of probability of the Brownian motion."[26] Such a relation was what scientists by this time called the "Langevin equation" (5.12'). As we have seen in chapters 5 and 6, the Langevin equation coped directly with the law of motion for a Brownian particle by treating thermal agitation as a random force, and de Haas-Lorentz devised a procedure to compute the statistical properties of Brownian particles from the equation. Ornstein's objective was to advance de Haas-Lorentz's grappling with the Langevin equation for rederiving the Fokker–Planck equation.

[25] Ibid., 648–649.

[26] Leonard Ornstein, "On the Brownian motion," *Proceedings of the Huygens Institute—Royal Netherlands Academy of Arts and Sciences (KNAW)*, 21:1 (1919), 96.

Ornstein's starting point was the Langevin equation, albeit in terms of a Brownian particle's speed, not its position as in the equation's original form (5.12'):

$$m\frac{du}{dt} = -wu + mF, \tag{7.8}$$

where u denoted the particle's velocity, m denoted its mass, w represented the effect of the fluid's friction on the particle, and F represented the force due to the particle's random collisions with fluid molecules. In this expression, w followed Stokes's formula of friction $w = 6\pi\mu a$ when the particle was a sphere of radius a (μ was the fluid's coefficient of friction); F was normalized with m. At first, Ornstein solved the ordinary differential equation (7.8) and examined various statistical averages from the solution: the particle's mean velocity, mean square velocity, mean displacement, mean square displacement, etc. He also discussed the question raised by Johannes van der Waal and A. Snethlage in 1916 that Einstein's theory of Brownian motion generated a mean square velocity increasing indefinitely with time, whereas this quantity should approach kT/m at thermal equilibrium according to statistical mechanics. Ornstein demonstrated that the outcome from statistical mechanics could be obtained if one took the statistical averages in a careful way.[27]

After this exercise, Ornstein sought to deduce the governing equation for the probability density of Brownian particles. Whereas Smoluchowski and Burger focused on the probability function at a given particle *displacement* (or *position*) and time $f(x, t)$, Ornstein examined the probability function at a given particle *velocity* and time $f(u, t)$. To do so, he considered a short time interval μ at the beginning of which the particle had an initial speed u_0. Taking a time integral of (7.8) over this interval, he obtained $u - u_0 = -\alpha u_0\tau + X$, or $u = u_0(1 - \alpha\tau) + X$, where $\alpha = w/m$ and $X \equiv \int_0^\tau F(t)\, dt$. Ornstein claimed that given the homogeneous nature of thermal agitation, the accumulated force X should have a probability distribution $\varphi(X)$ following the aforementioned relations: $\int_{-\infty}^{\infty}\varphi(X)\, dX = 1$, $\int_{-\infty}^{\infty}X\varphi(X)\, dX = 0$, and $\int_{-\infty}^{\infty}X^2\varphi(X)\, dX = \vartheta\tau$. Ornstein considered the probability density function for particles with a velocity around u' at time t and that for particles with a velocity around u at time $t + \tau$. In the deterministic case, u' and u followed $u'(1 - \alpha\tau) = u - X$, from Ornstein's solution to the Langevin equation. In the stochastic case, the particles spread within a velocity window du' at time t and another velocity window du at time $t + \tau$. And its deterministic counterpart implied that $du'(1 - \alpha\tau) = du$, or

[27] Ibid., 96–103.

$du' = (1 + \alpha\tau)\, du$ for small τ. The probability density at time $t + \tau$, $f(u,t+\tau)du$, Ornstein argued, equaled the probability density at the earlier time t, $f(u',t)du'$, multiplied by the probability density associated with the uncertainty that the random force introduced onto the particles. In other words,

$$f(u, t + \tau)\, du = du' \int f(u', t)\, \varphi(X)\, dX = (1 + \alpha\tau)\, du \int f(u', t)\, \varphi(X)\, dX.$$

To produce a governing equation for f, Ornstein employed $u' = (u - X)/(1 - \alpha\tau) \cong (1 + \alpha\tau)\, u - X$, expanded the terms with a Taylor series of τ (on the left side of the equation) and of X (on the right side), retained only the leading orders, and integrated over X using the familiar relations about $\varphi(X)$. Once again, he obtained a governing equation in a similar form as Fokker's, Planck's, Smoluchowski's, and Burger's:[28]

$$\frac{\partial f(u, t)}{\partial t} = \frac{\vartheta}{2} \frac{\partial^2 f(u, t)}{\partial u^2} + \alpha \frac{\partial}{\partial u} \left[u f(u, t) \right]. \tag{7.9}$$

The previous versions of the Fokker–Planck equation had a more general form for the external force g in (7.6) or y in (7.7). In Ornstein's formulation, the external force was the Stokes resistance wu that was proportional to the particle speed.

Between 1914 and 1917, physicists tackled the subject of the Fokker–Planck equation for Brownian motion again and again. Despite the existence of its primitive form in Einstein's work in 1905–06, Fokker, Planck, Smoluchowski, Burger, and Ornstein devoted efforts to derive the equation, using one approach after another—at least five distinct versions were published on this topic. Why did they bother to do such seemingly repetitive work? What was the point of revisiting the equation so many times? In comparison with Einstein's work, what novelties did they bring to understanding random thermal fluctuations?

In 1957, Thomas Kuhn wrote an essay titled "Energy conservation as an example of simultaneous discovery." He examined why within a period of less than twenty years in the first half of the nineteenth century, more than twelve individuals proposed the hypothesis of energy conservation and mutual convertibility between different forms of a single "force." Instead of construing this phenomenon as a simultaneous discovery of the same effect, Kuhn contended that different scientists' findings and assumptions (conservation of *vis viva*,

[28] Ibid., 103–104.

mechanical equivalent of heat, animal heat, mutual transformation between electricity and magnetism, thermodynamics) constituted different elements of what would later become a general principle.[29]

The case of the Fokker–Planck equation bears similarities to the history of energy conservation under Kuhn's scrutiny. Although all five scientists derived the governing equation for the probability distribution of Brownian motion as Einstein had attempted to do, they extended or circumvented Einstein's goals in different ways. Fokker and Planck tried to employ Einstein's theoretical structure for the problem of irradiating molecular dipoles. Smoluchowski directed focus on the important situations that Einstein had not considered: when the particles were subjected to an external force. Burger reworked Smoluchowski's outcomes through a mathematical procedure that he considered more rigorous, by utilizing a Brownian motion's mutual independence at distinct times, homogeneity of thermal agitations, and superposition of the force-driven and fluctuating components of a Brownian motion. Ornstein tackled the problem via the Langevin equation, an angle that he believed to be closer to the particles' dynamics. Their works would surely constitute different elements of the Brownian-motion theory.

An obvious value of the detailed inspection of the Fokker–Planck equation was its implications for the interpretation of new experimental data. Since Perrin's crucial experiment in 1908–09 that confirmed Einstein's theory of Brownian motion, experimenters working on colloidal solutions had known about the important effect of gravity as an additional force on the suspended particles. The call for engagement with updated experimental results motivated Smoluchowski, Burger, and Ornstein to look closely into the Fokker–Planck equation. For example, after producing theoretical results, Burger immediately moved on to compare his prediction with a few sets of published experimental data on the number density of suspended particles in a colloidal solution at different heights and times.[30] As a partial differential equation that governed the particles' probability distribution, moreover, the Fokker–Planck equation was a better fit to deal with certain kinds of problem situations than the alternative Langevin equation, especially for those that posed a boundary condition. In Fürth's work in 1917, for example, he used the solutions to the Fokker–Planck

[29] Thomas Kuhn, "Energy conservation as an example of simultaneous discovery," in Marshall Clagett (ed.), *Critical Problems in the History of Science* (Madison: University of Wisconsin Press, 1969), 321–356.

[30] Burger, "On the theory of the Brownian movement and the experiments of Brillouin" (1917), 654–658. Here Burger referred to the French physicist Léon Brillouin's experiment counting particles of gamboge in a mixture of glycerin and water.

equation to explain the experimental data when a "wall" was introduced into a colloidal solution to force a zero-density condition in the middle of the volume.[31]

Moreover, the multiple revisits of the Fokker–Planck equation and the repetitive attempts to deduce it from different presuppositions as such testified to a process of theory building. Although the scientists reached similar conclusions, their routes toward them showcased various ways of modeling the random motions of thermally agitated particles, the differential alterations of such motions due to an external force, the connection of this dynamics with the particles' density flow, and distinct ways of conducting perturbation analysis. By the late 1910s, these aspects became notable elements of the theoretical framework for random fluctuations due to thermal agitations.

George Uhlenbeck and Physics of Random Fluctuations

Treating random fluctuations as problems of dynamics and statistical mechanics was an encompassing research program in the 1910s–30s. It included not only deriving Brownian particles' probability distribution, but also solving the Fokker–Planck equation and the Langevin equation under various boundary conditions or initial conditions, examining the time-evolution of the solutions, conducting statistical averages on the solutions, and applying the outcomes to the studies of measuring instruments' precision and comparing theoretical predictions with experimental data. As a scientist active in all these aspects, the Dutch-American physicist George Uhlenbeck was a representative figure in physics of random fluctuations in the 1920s–30s.

Uhlenbeck the Statistical Physicist

George Eugene Uhlenbeck was born in 1900 at Batavia. Son of an infantry officer in the Royal Netherlands East Indies Army, George spent his first years in the Southeastern Asian colony. In 1906, his father was retired from the military post, and the whole family moved back to the Netherlands. They settled at The Hague, where George attended school. After high school, Uhlenbeck was matriculated at the Technical University of Delft in 1918 and majored in chemistry. He was not pleased with this choice. Clumsy with lab work, he preferred

[31] Reinhold Fürth, "Einige Untersuchungen über Brownsche Bewegung an einem Einzelteilchen," *Annalen der Physik*, 53 (1917), 177–213.

theory and thinking, and wanted to study at an "academic" university. In 1919, he transferred to the University of Leyden and switched major to physics.[32]

Uhlenbeck found Leyden "a kind of paradise." The course load was lighter, and physics was his favorite subject. Since high school, he had been interested in the kinetic theory of gases, because he thought it really explained phenomena. He read Boltzmann's *Vorlesungen über Gastheorie* and Gibbs's *Statistical Mechanics*. By this time, Lorentz had been retired from the university, moved to Haarlem, and dedicated himself to museum work and civic service. Although he gave a lecture at Leyden every week on the recent development of physics, the focus of research and teaching in theoretical physics in town had been shifted to Paul Ehrenfest, an Austrian Jew who was Boltzmann's former student in Vienna and succeeded Lorentz's position. Uhlenbeck began to work with Ehrenfest for a graduate study in 1920.[33]

Ehrenfest exerted a lifelong influence on Uhlenbeck. Ehrenfest's inheritance of Boltzmann's intellectual tradition (the "Vienna School" discussed in chapter 5) of statistical physics passed on to Uhlenbeck. As witnesses of the quantum revolution, they both explored the integration of quantum mechanics into statistical mechanics. Ehrenfest's style of research also impressed Uhlenbeck. The latter recalled his inspiring experience attending Ehrenfest's lectures and famous Wednesday colloquia, at which the Leyden professor often bypassed mathematical details and asked his students or guest speakers, "*Was ist der Witz*?" ("What is the wit?") Ehrenfest's direct grappling with concepts and intuiting of physical processes without overmediation of mathematics might not be Uhlenbeck's own style. Uhlenbeck once commented (which sounded more like a complaint) in an interview that Ehrenfest:[34]

> had, somehow, no technique. Nothing was in his fingers. He always had to think it out completely from the beginning. Although he knew mathematics it was not simple for him. He was not a computer. He could not compute.

[32] Max Dresden, "Obituaries: George E. Uhlenbeck," *Physics Today*, 42:12 (1989), 91–92; K. van Berkel, "Uhlenbeck, George Eugene (1900-1988)," *Biografisch Woordenboek van Netherland*, 6 (2013), available at http://resources.huygens.knaw.nl/bwn1880-2000/lemmata/bwn6/uhlenbeck (accessed April 8, 2022); Thomas Kuhn, "Oral history transcript of Dr. George Uhlenbeck, Session V," interview on December 9, 1963, Niels Bohr Library and Archives, American Institute of Physics, Washington, D.C., available at https://www.aip.org/history-programs/niels-bohr-library/oral-histories/4922-5 (accessed April 8, 2022).

[33] George Uhlenbeck, "Fifty years of spin: personal reminiscences," *Physics Today*, 29:6 (1976), 44.

[34] Thomas Kuhn, "Oral history transcript of Dr. George Uhlenbeck, Session I," interview on March 30, 1962, Niels Bohr Library and Archives, American Institute of Physics, Washington, D.C., available at https://www.aip.org/history-programs/niels-bohr-library/oral-histories/4922-1 (accessed April 8, 2022).

> That's the one thing I never learned from him. I had to learn it all by myself later on.

Uhlenbeck contrasted Ehrenfest's lack of computing skills, in sorrow, with the *tour de force* in mathematics that Max Born, Arnold Sommerfeld, and H.A. Lorentz had exhibited. On another occasion, however, he praised Ehrenfest's intuitive approach and remembered from his student years that:[35]

> although perhaps we did not learn how to compute, we certainly learned what the real problems were. It is difficult to say how this was accomplished. One reason was just the absence of technicalities. Only the fundamentals were carefully developed and drilled into one's mind; the rest of the time Ehrenfest would give wonderfully short bird's-eye views of various topics with a few characteristic results and with references, to whet the appetite of the student. In my opinion it is the best way to treat a subject, much better than the rigorous, complete, and systematic way which is in vogue in our universities.

The paradise in Leyden did not last for long. In 1922–25, Uhlenbeck experienced a period of wanderlust mixed with financial stress and an identity crisis. Soon after entering the doctoral program, he began to teach mathematics at a local gymnasium to make ends meet. This took up much of his available time and delayed his research progress. Then, he accepted an offer to serve as private tutor for the son of the Dutch Ambassador in Italy. In Rome, Uhlenbeck took lessons in Italian, became highly interested in Ancient and Medieval arts, and seriously considered switching career to art history. He eventually gave up art history for his lack of patience with Latin and decided to return to Leyden in 1925 to finish up his Ph.D. degree.[36]

Upon Uhlenbeck's return, Ehrenfest introduced him to another doctoral student, Samuel Goudsmit. Uhlenbeck and Goudsmit quickly became close friends and began to cooperate on research. In summer 1925, they were well into the quantum theory for atomic spectra. They were both aware of the riddle of the gyromagnetic ratio, the unexplained deviation of measured gaseous spectral lines under a magnetic field from Sommerfeld's and Schrödinger's quantum theories. To explain this phenomenon, Wolfgang Pauli, Werner Heisenberg, and Alfred Landé introduced a fourth atomic quantum number. In a casual conversation, Uhlenbeck and Goudsmit came up with the idea

[35] George Uhlenbeck, "Reminiscences of Professor Paul Ehrenfest," *American Journal of Physics*, 24 (1956), 431–432.

[36] Dresden, "Uhlenbeck" (1989), 92; Uhlenbeck, "Fifty years of spin" (1976), 44–45.

that this quantum number could correspond to a rotating electron's angular momentum.

This idea prompted an intense collaboration. At this time, Goudsmit also served as an in-house theoretician at Pieter Zeeman's laboratory in Amsterdam. Every week, he spent a day or two in Leyden, worked with Uhlenbeck at length, and then rushed back to Amsterdam to work at Zeeman's lab for the rest of the week. Uhlenbeck and Goudsmit deduced the formal expression of an electron's innate angular momentum and coined the name "spin" for it. The spectral anomaly was a consequence of the coupling between electrons' spins and the external magnetic field. Neat and explanatory as it was, this model still had significant problems—an electron's rotating speed ought to exceed the velocity of light if its angular momentum could reach the theoretical level for spin. Yet, Ehrenfest encouraged them to publish the results, regardless of potential mistakes. Uhlenbeck and Goudsmit's paper on the electronic spin quickly became a classic in quantum physics. The two young researchers made their international fame.[37]

While Uhlenbeck was working on the theory of electronic spin, he also undertook two other projects. In one, he was instructed by Ehrenfest to examine the inverse-scattering for wave equations of higher dimensions than three. In the other, he looked into the statistical structure of a strange "condensate"—a transition from an ordinary molecular state of matter to a highly coherent and uniform quantum state—that Santyendra Nath Bose and Einstein had proposed recently. Uhlenbeck argued that a mathematical mistake in Bose and Einstein's original theory—one that conflated a discrete sum with a continuous integral—led to the incorrect prediction of the condensate. According to a commentary later, Uhlenbeck's critique "delayed the general acceptance of the Bose-Einstein condensate for 10 years."[38]

Uhlenbeck and Goudsmit's paths overlapped in the following years. In 1925, Uhlenbeck turned his research on the Bose–Einstein condensate into a doctoral thesis and spent one more year in Leyden finishing it up while serving as Ehrenfest's research assistant. Goudsmit still concentrated on atomic spectroscopy but moved from Zeeman's laboratory to Walther Gerlach's in Tübingen. In early 1927, both Uhlenbeck and Goudsmit spent months at Niels Bohr's institute in Copenhagen, writing their dissertations. They received their Ph.D. from the University of Leyden on the same day in July, secured teaching

[37] Uhlenbeck, "Fifty years of spin" (1976), 46–47.

[38] Ezechiel Godert David Cohen, "George E. Uhlenbeck and statistical mechanics," *American Journal of Physics*, 58 (1990), 619.

positions at the Department of Physics, University of Michigan, Ann Arbor, and boarded the same ship to the US in the fall.[39]

Uhlenbeck's career took off at the University of Michigan. Like Goudsmit, he started as an instructor in 1927, and was soon promoted to professor. Except for 1935–38, when he served as a professor at the University of Utrecht, he stayed in Ann Arbor until 1960. In Michigan, Uhlenbeck's research focused on statistical mechanics and quantum physics: the mathematical foundation of phase transition; new formulations, approximations, and expansions of statistical mechanics in light of quantum physics; positrons; nuclear physics; Boltzmann's transport theory; and inverse scattering. These topics bore a stark character of statistical methodology.[40]

Uhlenbeck's Metrological Angle to Brownian Motion

Soon after arriving in Ann Arbor, Uhlenbeck began to work on Brownian motion. Compared with the aforementioned subjects, the studies of Brownian motion did not strike many contemporary physicists as particularly "frontier" research. When he told Wolfgang Pauli that he was working on Brownian motion, Pauli immediately replied with disdain: "*Brownian motion—Desperazions Physik*!" ("Brownian motion—desperate physics!")[41] Then, why did Uhlenbeck pursue the studies of Brownian motions? His student E.G.D. Cohen believed that his interest in the topic came, at least in part, through Leonard Ornstein.[42] Uhlenbeck himself offered a somehow different account. Years later, when he received an interview from Kuhn as part of the well-known project of the oral history for quantum physics, Uhlenbeck was asked why he studied Brownian motion. He gave the following answer:[43]

> There were definite questions, there were various questions, you see, which were not answered, and it came also partially through the work in Michigan on the noise questions, Johnson noise. They made very fine experiments there and I was then clearly the man, since it was statistics and I was supposed to be the theoretical adviser about it. So I studied all that very carefully and that brought me to the Brownian motion. I got a dissertation out of it for someone

[39] Uhlenbeck, "Fifty years of spin" (1976), 48.

[40] George Ford, "George Eugene Uhlenbeck, 1900–1988," *Biographical Memoirs of the National Academy of Sciences*, 91 (2009), 1–22.

[41] Kuhn, "Uhlenbeck V" (1963).

[42] Cohen, "Uhlenbeck" (1990), 621.

[43] Kuhn, "Uhlenbeck V" (1963).

> and then when I went back to Holland—the second time, I think, or the first time—I worked with Ornstein on it where I did most of the work, but it was the methods which he had invented, all right. At that time, of course, nobody looked at that paper really; it only became known in the war when all these noise questions came up again.

Under this context, Uhlenbeck's treatments of the Brownian motions had two distinct features. First, he almost always collaborated with other researchers when he tackled this subject—with his peers such as Goudsmit and Ornstein in the early years and his students such as G.A. van Lear and Ming Chen Wang in the later years. Second, his studies in this area were preoccupied with finding the refined time-evolution of physical systems when thermal fluctuations influenced its dynamics. This epitomized precisely the "time-domain" approach. In Cohen's review of his mentor's works, he identified two intellectual traditions in statistical physics since the nineteenth century: the "Gibbs tradition" that focused on the microscopic interpretation of equilibrium, and the "Boltzmann tradition" that focused on the *approach* to equilibrium rather than equilibrium as such. To Cohen, Uhlenbeck's work fit squarely within the Boltzmann tradition.[44] Although Cohen primarily referred to Uhlenbeck's examination of the transport phenomena and the dynamical formulation in statistical mechanics, Uhlenbeck's studies of Brownian motions carried the same mark of the Boltzmann tradition in its emphasis on time-evolution.

Uhlenbeck's entry point to the studies of Brownian motion was mechanical experiments. Through Goudsmit's introduction, he noted certain trials conducted at Zeeman's laboratory in Amsterdam and Gerlach's in Tübingen in the 1920s, where Goudsmit used to work. Zeeman and his assistant A. Houdijk (or Houdyk) registered the motion of a suspended wire's loose end, while Gerlach and his assistant E. Lehrer photographed the motion of a little mirror fixed on a very fine wire. In both cases, the experimenters observed random fluctuations in their registered motions, despite their efforts to stabilize the system such as rarifying the air and controlling the wobbling structure's fixed end (Figure 7.1). Such fluctuations were attributed to the wire or mirror's Brownian motion in response to air molecules' random collisions. Uhlenbeck and Goudsmit were not the first researchers to look into these phenomena. The

[44] Cohen, "Uhlenbeck" (1990), 622. Jürgen Renn also observed a strong tendency in Boltzmann's and Einstein's works to reduce the problems in statistical mechanics to those in dynamics, as both held that the statistical techniques were the necessary simplifications for grappling with the whole of all particles' dynamic trajectories in a physical system. See Renn, "Einstein's controversy with Drude and the origin of statistical mechanics" (1997).

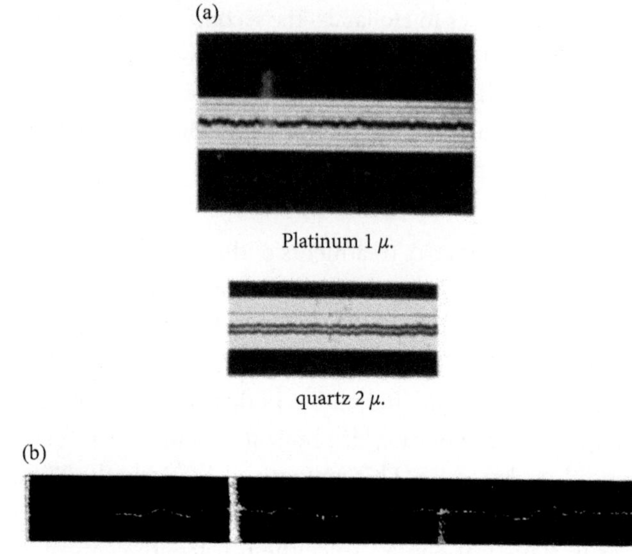

Figure 7.1 The wobbling motion of the object registered at Zeeman and Houdijk's experiment (a) and Gerlach and Lehrer's experiment (b). A. Houdijk and Pieter Zeeman, "The Brownian motion of a thread," *Proceedings of the Huygens Institute—Royal Netherlands Academy of Arts and Sciences (KNAW)*, 28:1 (1925), 54, permission from the Royal Netherlands Academy of Arts and Sciences; W. Gerlach and E. Lehrer, "Über die Messung der rotatorischen Brownschen Bewegung mit Hilfe einer Drehwage," *Naturwissenschaften*, 15:1 (1927), 15, permission from Springer Nature.

experimental configurations in Amsterdam and Tübingen bore similarity to the mechanism of a pointer or indicator in a measuring instrument like a galvanometer. Thus, the inquiries into these random fluctuations had practical implications for understanding the sensitivity of measuring instruments.[45]

In chapter 6, we have seen Einstein's and de Haas-Lorentz's speculations in the early 1910s concerning the effect of electrons' Brownian motions on the precision of electrometers. Historian Martin Niss further examined certain theoretical discussions on the same issue in the 1910s–30s. Niss's study showed that the Swedish physicist Gustaf Ising produced a doctoral thesis in 1917–19 at the University of Uppsala on the limit of galvanometers' and electrometers' performances due to random microscopic actions. Ising employed

[45] A. Houdijk and Pieter Zeeman, "The Brownian motion of a thread," *Proceedings of the Huygens Institute—Royal Netherlands Academy of Arts and Sciences (KNAW)*, 28:1 (1925), 52–54; Walther Gerlach and E. Lehrer, "Über die Messung der rotatorischen Brownschen Bewegung mit Hilfe einer Drehwage," *Naturwissenschaften*, 15:1 (1927), 15.

Smoluchowski's work to estimate the mean square deviation of the pointer's position as an outcome of equipartition. Later, Frits Zernike at the University of Groningen in the Netherlands debated with Ising about whether the specific source of such deviations came from the fluctuating electric current due to free electrons' Brownian motion or random collisions with air molecules on the pointer wire. But no matter whether the cause of fluctuations was electrical or mechanical, physicists by the early 1930s generally accepted the view that thermal agitations were a central factor that limited the sensitivity of measuring instruments. This was also around the same time when Heisenberg's uncertainty principle was widely discussed among physicists, who seriously considered quantum fluctuations as another source to limit the precision of measurement.[46]

Against this background, Uhlenbeck and Goudsmit published a paper on the Brownian motion of a suspended solid in 1929.[47] They were interested in the rotation of a small mirror suspended on a fine wire, which was the setting for Gerlach's measurement in Tübingen. The mirror's irregular rotation was understood as a Brownian motion. There had been a standard approach to characterize this motion: the Einstein–Smoluchowski theory predicted via equipartition that the mean square deflection of the wobbling mirror from its equilibrium position was a constant proportional to the temperature of its surrounding gas, but independent of other physical factors such as the gas molecular weight and pressure. Uhlenbeck and Goudsmit were not satisfied with the Einstein–Smoluchowski theory, for Gerlach's experimental data showed more than a constant average. As they stated, Gerlach's experiments "give more than merely the average square deviation; the registered curves show to some extend [*sic*] at least the time-dependence of the irregular motion." In fact, Gerlach indicated that "the general appearance of these curves is quite different at different pressures of the surrounding gas, though the average square deviation remains the same for any given temperature." To the two young physicists, the problem was therefore "to give a more detailed theory of these curves."[48]

An inspiration to solving this problem came from Schottky's treatment of shot noise in 1917, as we have seen in chapter 6. Uhlenbeck and Goudsmit noted an interesting analogy between Gerlach's experimental configuration

[46] Martin Niss, "Brownian motion as a limit to physical measuring processes: A chapter in the history of noise from the physicists' point of view," *Perspectives on Science*, 24:1 (2016), 29–44.

[47] George Uhlenbeck and Samuel Goudsmit, "A problem in Brownian motion," *Physical Review*, 34 (1929), 145–151.

[48] Ibid., 145–146.

and Schottky's problem situation, which was materialized in Hartmann's and Hull and Williams's experiments. In Schottky's case, the fluctuating electrical current due to random movements of electrons in a thermionic tube was coupled inductively with an amplifier, which contained a frequency-selective circuit (as a bandpass filter) with a single characteristic frequency. In Gerlach's case, thermally agitated gas molecules collided randomly with the mirror and produced its fluctuating angular momentum. But the mirror itself (combined with the wire to which it was fixed) was a mechanical system that collected and magnified the small rotational fluctuations with a frequency response similar to a bandpass filter with a single characteristic frequency. The object functioned like an amplifier—albeit of angular momentum instead of electrical current. Because of this similarity, Uhlenbeck and Goudsmit believed that the mathematical structure in Schottky's theory of shot noise could be employed directly in tackling their problem for the rotational Brownian motion. The sharp frequency response of Gerlach's mirror system rendered his experimental results contingent upon the rotational Brownian motion within a narrow bandwidth centered on the system's resonating frequency. Although Uhlenbeck and Goudsmit aimed to find "the time dependence of the irregular motion" (a *raison d'être* for the time-domain approach), Schottky's Fourier analysis of Brownian motion (which characterized the frequency-domain approach) fit their goal.[49]

Uhlenbeck and Goudsmit embarked on their theoretical pursuit with a formulation for the Gerlach system's equation of motion:

$$I\frac{d^2\phi}{dt^2} + f\frac{d\phi}{dt} + D\phi = M(t). \tag{7.10}$$

In this equation, ϕ represented the mirror's angle of deflection, and I represented its angular momentum. In addition to the mirror's inertia $I\frac{d^2\phi}{dt^2}$, three torques were exerted: the friction $f\frac{d\phi}{dt}$, the elastic torsion $D\phi$, and the fluctuating torque $M(t)$ caused by gas molecules' random collisions. With a proper normalization, (7.10) became

$$\frac{d^2\phi}{dt^2} + r\frac{d\phi}{dt} + \omega^2\phi = T(t), \tag{7.10'}$$

where $r = f/I$, $\omega = \sqrt{D/I}$, and $T(t) = M(t)/I$.[50]

[49] Ibid., 146.
[50] Ibid., 146–147.

Equation (7.10) was in the form of the Langevin equation. Since the elastic torsion bestowed a characteristic frequency $\omega = \sqrt{D/I}$ on the mirror system and thus a bandpass effect on ϕ, Uhlenbeck and Gourdsmit followed Schottky's method to solve (7.10') through Fourier analysis. Key to this scheme was the spectral expansion of the random force $T(t)$. The two Dutch scientists assumed that the observation was conducted within a period τ much longer than the system's characteristic period $2\pi/\omega$. Then $T(t)$ could be expressed with a Fourier series:

$$T(t) = \sum_{n=0}^{\infty} [A_n \cos(\omega_n t) + B_n \sin(\omega_n t)],$$

where $\omega_n = 2\pi n/\tau$, $A_n = (2/\tau)\int_0^{\tau} T(t)\cos(\omega_n t)\,dt$, and $B_n = (2/\tau)\int_0^{\tau} T(t)$ $\sin(\omega_n t)\,dt$. Plugging this expression into (7.10'), decomposing ϕ into another Fourier series, and comparing each frequency component on both sides of the equation, Uhlenbeck and Goudsmit obtained the solution of (7.10'):[51]

$$\phi(t) = \sum_{n=0}^{\infty} \phi_n(t) = \sum_{n=0}^{\infty} \frac{\left\{\left[A_n\left(\omega^2 - {\omega_n}^2\right) - B_n r\omega_n\right]\cos(\omega_n t) + \left[A_n r\omega_n + B_n\left(\omega^2 - {\omega_n}^2\right)\right]\sin(\omega_n t)\right\}}{\left(\omega^2 - {\omega_n}^2\right)^2 + r^2{\omega_n}^2}. \tag{7.11}$$

Since the source term $T(t)$ was random, its Fourier coefficients $\{A_n, B_n\}_{n=0}^{\infty}$ were random, too. This rendered (7.11) useless for pinning down any observable quantity. What was measured in an experiment had to be the statistical averages of $\phi(t)$, its mean square, or the average or mean square of any of its frequency components. Uhlenbeck and Goudsmit believed that the most useful quantity to compare with experimental data was the mean square of $\phi(t)$'s frequency components $\phi_n(t)$. This "average" ought to be taken both over time and over ensemble. The first operation was straightforward. The time average of $\phi_n(t)^2$ was

$$\overline{{\phi_n}^2} = \frac{1}{2}\frac{{A_n}^2 + {B_n}^2}{\left(\omega^2 - {\omega_n}^2\right)^2 + r^2{\omega_n}^2}.$$

The more challenging task was to take the ensemble average of $\overline{{\phi_n}^2}$, which amounted to taking the ensemble average of ${A_n}^2 + {B_n}^2$. This average required a closer inspection of the statistical properties of $T(t)$. Here Uhlenbeck and

[51] Ibid., 148.

Goudsmit followed Schottky's scheme. In reality, they argued, the fluctuating source $T(t)$ comprised a huge number of randomly arriving sharp peaks (like shot current) corresponding to the impulse moments transferred to the mirror by stochastic molecular collisions. Like Schottky, Uhlenbeck and Goudsmit introduced a "physically infinitely small" time-element Δt that was much shorter than the system's characteristic period $2\pi/\omega$, but was long enough so that many collisions occurred within it. Under the premise that $\Delta t \ll 2\pi/\omega$, they felt comfortable replacing the continuous integrals $A_n = (2/\tau)\int_0^\tau T(t)\cos(\omega_n t)\,dt$ and $B_n = (2/\tau)\int_0^\tau T(t)\sin(\omega_n t)\,dt$ with the discrete sums $A_n = (2/\tau)\sum_{i=1}^{z} T(t_i)\cos(\omega_n t_i)\,\Delta t$ and $B_n = (2/\tau)\sum_{i=1}^{z} T(t_i)\sin(\omega_n t_i)\,\Delta t$, where $t_i = i\Delta t$ was the ith time element. In addition, they asserted three statistical properties of $T(t_i)$:

- $T(t_i)$ had an equal chance being positive or negative, so that its ensemble average $\langle T(t_i)\rangle = 0$.
- The random forces due to molecular collisions were statistically independent of each other at different time elements: $\langle T(t_i)\,T(t_j)\rangle = 0$.
- From the Maxwell distribution for gas molecules, Uhlenbeck and Goudsmit managed to calculate the theoretical value for the mean square of the molecular force at any time. They obtained that $\langle T(t_i)^2\rangle = (4m\bar{v}p/I\rho)\,\Delta t$, where m was a gas molecule's mass, $\bar{v}$ was its mean speed, p was gas pressure, and ρ was gas density. Similarly, they obtained the theoretical value of the friction coefficient $r = (2m\bar{v}p\rho/IkT)$.[52]

Utilizing the three properties and the well-known relation that $\bar{v} = \sqrt{(8kT/\pi m)}$, Uhlenbeck and Goudsmit obtained what they called the "final formula":

$$\langle \phi_n{}^2\rangle = \frac{\sqrt{\pi m}\,(8kT)^{2/3}p}{\rho I\tau}\,\frac{1}{\pi kT(\omega^2-\omega_n{}^2)^2 + 32p^2\rho^2\omega_n{}^2}. \tag{7.12}$$

Uhlenbeck and Goudsmit were convinced that this "final formula" confirmed Gerlach's observation that the extent of the mirror's fluctuating deflection depended on the pressure of its surrounding gas. To them, the mean square amplitude at a certain frequency, especially the one close to the mirror system's characteristic frequency, best represented what was measured in Gerlach's experiment, since the system had a clear frequency selection.[53]

[52] Ibid., 147–148.
[53] Ibid., 149.

A General Theory of Random Fluctuations

While Uhlenbeck focused on a particular physical scenario—a suspended mirror—in his collaboration with Goudsmit in 1929, his attention was soon directed to a more general theory of random fluctuations. Schottky's spectral analysis did not seem a proper method for general cases. Around this time, Leonard Ornstein introduced to Uhlenbeck the method of solving the Langevin equation and deriving the Fokker–Planck equation that he and his student Herman Burger had developed in 1917–19 and elaborated in the 1920s. This contact led to a collaboration between Uhlenbeck and Ornstein. In 1930, they published an article titled "On the theory of the Brownian motion" in *Physical Review*. Aiming to present a general theory of thermally agitated random fluctuations, this paper would become a classic in the studies of Brownian motions.[54]

Uhlenbeck and Ornstein's baseline for a general theory of Brownian motion was the Einstein–Smoluchowski doctrine: A free Brownian particle suspended in a fluid had a mean square displacement $\langle s^2 \rangle = \left(2kT/f\right) t$, where f was the friction coefficient, t was the elapsed time, T was the fluid's temperature, and k was the Boltzmann constant (see chapter 5). Accordingly, the Brownian particles spread out with time like diffusion. Uhlenbeck and Ornstein argued that the Einstein–Smoluchowski theory was only valid when the elapsed time was sufficiently long for the physical system to reach equipartition and the mean energy of each degree of freedom to be kT. *Before* equilibrium, however, the quantities associated with the Brownian particles varied with time. Finding the transient properties of thermal agitations before equilibrium was Uhlenbeck and Ornstein's objective. Mathematically, their work involved solving the Langevin equation and Fokker–Planck equation and observing the solution's statistical features.[55]

To obtain a more accurate temporal evolution of Brownian motion than the Einstein–Smoluchowski theory, Uhlenbeck and Ornstein considered three representative cases. First, they examined the probability distribution for the velocities of Brownian particles freely suspended in a fluid. The equation of motion for a particle in this case was

$$\frac{du}{dt} + \beta u = A(t), \tag{7.13}$$

[54] George Uhlenbeck and Leonard Ornstein, "On the theory of the Brownian motion," *Physical Review*, 36 (1930), 823–841.

[55] Ibid., 823–824.

where $\beta = f/m$, and $A = F/m$ was the mass-normalized stochastic force due to molecular collisions. Ornstein had obtained a standard solution for this Langevin equation in 1917:

$$u = u_0 e^{-\beta t} + e^{-\beta t} \int_0^t e^{\beta \xi} A(\xi)\, d\xi, \qquad (7.14)$$

where u_0 was the particle's initial velocity $u_0 = u(t = 0)$. To obtain the probability distribution of u, Uhlenbeck and Ornstein employed a "method of moments"—a well-known theorem in probability theory since the late nineteenth century: one could construct the probability distribution of a random variable u from the mean values of its first, second, ... orders $\langle u \rangle$, $\langle u^2 \rangle$, ... via the characteristic function (see chapter 5). To proceed in steps, they first conducted a statistical average over the ensemble of A while keeping u_0 a constant. From (7.14), the first-order mean was $\langle u \rangle^{u_0} = u_0 e^{-\beta t}$, assuming the statistical average of the random force $\langle A(\xi) \rangle = 0$ because the molecular collisions had an equal chance to incur a positive or negative momentum transfer. In this notation, $\langle u \rangle^{u_0}$ means taking the statistical average of u while keeping u_0 a constant.[56]

The second-order statistical average of u was more complicated:

$$\langle u^2 \rangle^{u_0} = {u_0}^2 e^{-2\beta t} + e^{-2\beta t} \int_0^t \int_0^t e^{\beta(\xi+\eta)} \langle A(\xi) A(\eta) \rangle d\xi d\eta.$$

This was determined by the statistical average $\langle A(\xi) A(\eta) \rangle$. Uhlenbeck and Ornstein did not know the exact form of this average. But they believed that the random force at different instants had very low correlation with each other, because of the "memoryless" feature of thermal agitations. Thus, $\langle A(\xi) A(\eta) \rangle$ would be a function of only $\xi - \eta$ (denoted by $\phi_1(\xi - \eta)$), which was peaked at $\xi - \eta = 0$ and small otherwise. Without specifying the form of $\phi_1(\xi - \eta)$, Uhlenbeck and Ornstein defined $\tau_1 \equiv \int_{-\infty}^{\infty} \phi_1(w)\, dw$. Then the mean square of u became

$$\langle u^2 \rangle^{u_0} = {u_0}^2 e^{-2\beta t} + \frac{\tau_1}{2\beta}(1 - e^{-2\beta t}). \qquad (7.15)$$

[56] Ibid., 827.

To determine the value of τ_1, they resorted to equipartition: For $t \to \infty$, the mean square of u corresponded to the equipartition energy kT:

$$\lim_{t\to\infty}\langle u^2\rangle^{u_0} = \frac{kT}{m} = \frac{\tau_1}{2\beta}.$$

This gave $\tau_1 = 2\beta kT/m$. Plugging this result into the mean square led to[57]

$$\langle u^2\rangle^{u_0} = \frac{kT}{m} + \left(u_0{}^2 - \frac{kT}{m}\right)e^{-2\beta t}. \tag{7.16}$$

From u's first- and second-order moments, Uhlenbeck and Ornstein were able to derive u's moments $\langle u^n\rangle^{u_0}$ for $n > 2$, from which they could construct u's probability distribution. After this exercise, they found that $u - u_0 e^{-\beta t}$ followed a zero-mean Gaussian distribution with variance $\frac{kT}{m}(1 - e^{-2\beta t})$.[58]

Solving the Langevin equation was not the only approach. Uhlenbeck and Ornstein demonstrated that the same problem could be solved via the Fokker–Planck equation. They constructed the Fokker–Planck equation via the same technique that Burger and Ornstein had used in 1917–19. The outcome was a diffusion equation for u's probability density G:

$$\frac{\partial G}{\partial t} = \beta\frac{\partial}{\partial u}(uG) + \frac{\tau_1}{2}\frac{\partial^2 G}{\partial u^2}. \tag{7.17}$$

The solution of (7.17) required information about the distribution of G at $t = 0$. Employing a standard technique for solving linear partial differential equations, Uhlenbeck and Ornstein maintained that the solution of (7.17) was in the form of

$$G(u,t) = \int_{-\infty}^{\infty} g(u,v,t)f(v)\,dv,$$

where $f(u) = G(u, t{=}0)$ represented G's initial condition and g a certain "Green's function" as the particular solution of the equation when $f(u)$ was an impulse. After intricate computations, Uhlenbeck and Ornstein demonstrated that the solution $G(u, t)$ had the same form of Gaussian distribution as they obtained from solving the Langevin equation.[59]

57 Ibid., 828.
58 Ibid., 828.
59 Ibid., 830.

In the second case, Uhlenbeck and Ornstein considered the same physical scenario whose dynamics was governed by the equation of motion (7.13). But instead of computing the probability distribution of a Brownian particle's velocity, they calculated the probability distribution of the particle's displacement $s = x - x_0$ (x_0 represented the particle's initial position and x represented its position at time t). The particle position x could be expressed as an integral of the velocity u in (7.14) over time. Following the same procedure as that in the first case, they obtained that $s - (u_0/\beta)(1 - e^{-\beta t})$ followed a zero-mean Gaussian distribution with a variance $(kT/m\beta^2)(2\beta t - 3 + 4e^{-\beta t} - e^{-2\beta t})$.[60]

In the third case, Uhlenbeck and Ornstein considered the physical scenario when Brownian particles were subject to a position-dependent external force $K(x)$:

$$\frac{du}{dt} + \beta u = A(t) + \frac{1}{m}K(x). \tag{7.18}$$

The most typical external force was an elastic restoration: $\frac{K(x)}{m} = -\omega^2 x$. Smoluchowski in 1915 and Ornstein in 1919 had dealt with this case when the system had reached or was close to equipartition so that the initial condition played no effect. Here Uhlenbeck and Ornstein wanted a more exact solution. They followed the same procedure for solving the Langevin equation as in the previous two cases. The results were more complicated, for the mean and mean square of u involved exponential decays $e^{-\beta t}$ and $e^{-2\beta t}$ and harmonic oscillations $\cos(\omega t)$ and $\sin(\omega t)$. This complexity prevented Uhlenbeck and Ornstein from computing u's higher-order statistics and constructing its probability density function. Yet, by tracing the temporal variation of $\langle u \rangle$ and $\langle u^2 \rangle$ they were able to observe how the system approached equipartition. Uhlenbeck and Ornstein applied a language of statistical mechanics when they performed the average $\langle u \rangle$ and $\langle u^2 \rangle$. Such averages were undertaken over different realizations of the random force A, and distinct values of the initial position x_0 and velocity u_0. To them, the harmonic oscillators with similar dynamical condition but different values of A, x_0, and u_0 formed a "canonical ensemble." Its subset with the same value of x_0 or u_0 formed a "sub-ensemble."[61]

Uhlenbeck and Ornstein's development of a general theory of thermal fluctuations in idealized problem situations—free particles in a fluid or those subject to an elastic force—prompted further inquiries into Brownian motions

[60] Ibid., 832.
[61] Ibid., 834–837.

in more complex scenarios. Starting in 1929, Uhlenbeck worked with a Ph.D. student George van Lear at the University of Michigan on the same topic. In a reference letter for van Lear's fellowship application, Uhlenbeck wrote:[62]

> Mr. van Lear worked the last year under my direction on the theory of the Brown Motion of strings and in general of continuous media. He has done this very well. At the same time he was much interested in the experimental application of the problem, especially with regard to the electrical Brownian motion and the limit of sensitivities of galvanometers. Together with Dr. Hardy, who worked experimentally on the last problem, he has done much to clean up the situation.

Uhlenbeck asked van Lear to examine the effect of molecular thermal agitations for a suspended string or rod of a notable volume and weight that could not be treated as a point mass or a spinning thread around a single axis. Compared with Uhlenbeck and Ornstein's treatment in 1929, Uhlenbeck and van Lear's new problem had infinite degrees of freedom. The dynamics for a wobbling string or rod was a second-order partial differential equation with both time t and position x along the elastic body as variables. The fluctuating force $F(x, t)$ caused by gas molecules' random collisions was a stochastic function of both x and t. This equation was more complicated than the Langevin equation (7.13) that depended only on t. Its solution was more involved; but it still invoked the core of Uhlenbeck and Ornstein's approach.

Van Lear and Uhlenbeck expanded the analytical solution of the two-dimensional equation of motion with a series in terms of its spatial part's eigenfunctions. Each eigenfunction in the expression corresponded to a mode of oscillation for the string or rod. Then following Uhlenbeck and Ornstein, they performed different orders of statistical average over this "canonical ensemble" of oscillation modes, utilizing the memoryless property of the Brownian force. Plugging in the result from equipartition that each mode of oscillation had an energy kT at equilibrium, they obtained the mean square amplitude of each oscillating mode in the expression for the displacement distribution along the wobbling solid. Accordingly, the mean, mean square, and higher-order statistics of this displacement varied with time (and space). Its temporal evolution (with both exponential decays and harmonic oscillations) once again testified to the intricate dynamics of the fluctuating system before it

[62] Letter, Uhlenbeck to Gus, 1930,Box 1, Folder "Correspondence 1930," George Uhlenbeck Papers, Manuscript Collection, Library of the University of Michigan, Ann Arbor, US.

settled down at equilibrium. Van Lear and Uhlenbeck published their results in 1931.[63]

Van Lear and Uhlenbeck's theoretical examination of Brownian motion engaged experimental work. In the 1931 paper, they pointed out that their theoretical prediction for the mean square Brownian displacement of a suspended wire fit well with Houdijk's experimental data in Amsterdam.[64] This was no happy coincidence. Van Lear had been in close contact with James Hardy at the University of Michigan, who had been performing experiments on Brownian motions of solid bodies. Before submitting the 1931 paper in August, moreover, Uhlenbeck wrote to Houdijk and asked him to provide detailed information about his experiment: the exact wire length, the radius of the wire's cross section, Young's modulus for the wire material, the value of the Boltzmann constant that Houdijk used in his experimental design, ambient temperature, and air's damping constant.[65]

Van Lear and Uhlenbeck's paper in 1931 caught Ornstein's attention. Two months after its publication, Ornstein sent a letter to Uhlenbeck, indicating that he read this work with "great interest." Ornstein suggested a way for the theory of Brownian motion to engage quantum physics. In van Lear and Uhlenbeck's paper, the average energy of each oscillation mode at equilibrium equaled kT due to equipartition. To Ornstein, these modes were sufficiently microscopic to exhibit quantum behavior like Planck's black-body radiators. Thus, kT should be replaced by the "Debye function" $\omega_n/\left(e^{\omega_n/kT}-1\right)$, where ω_n was the frequency of the quantized oscillator.[66]

In addition to the Brownian motion of a suspended solid body in air, van Lear was concerned with a practical problem. The Ph.D. student was writing a dissertation on the effect of Brownian motion on the sensitivities of electrical gauging instruments, including a galvanometer to measure weak electric current, and a radiometer to measure radiation. Van Lear considered random fluctuations connected *both* to collisions of air molecules *and* to thermal noise. The idea of random fluctuations obstructing the precision of electrical measurement was not new. Einstein's note in 1907 and de Haas-Lorentz's thesis in 1912 entertained the concept and estimated the limit of instrumental sensitivity. Ising's and Zernike's works in the 1910s–20s pointed out further that the noise had two origins, one from thermal agitations of gas molecules

[63] George van Lear and George Uhlenbeck, "The Brownian motion of strings and elastic rods," *Physical Review*, 38 (1931), 1583–1598.

[64] Ibid., 1593–1594.

[65] Letter, Uhlenbeck to Houdyk (Houdijk), July 7, 1931, Box 1, Folder "Correspondence 1931," Uhlenbeck Papers.

[66] Letter, Ornstein to Uhlenbeck, December 26, 1931, Box 1, Folder "Correspondence 1931," Uhlenbeck Papers.

surrounding the instrument, and the other from thermal agitations of electrons in its electrical circuits.

When van Lear submitted his thesis in 1931, Uhlenbeck passed it to Ornstein, who read it and expressed in a letter his agreement with van Lear's theoretical premise. The electromotive force driving the pointer of a galvanometer, Ornstein declared, was a superposition of the regular electromotive force—which often had a harmonic waveform—and a fluctuating Brownian motion (Figure 7.2).[67] While the idea appeared to be straightforward, its mathematical representation was challenging. Van Lear continued to work on it with Uhlenbeck even after he obtained his Ph.D. and landed a teaching position in the Physics Department at the University of Oklahoma. Van Lear published an article in 1933 on Brownian motion in a resonance radiometer, an instrument comprising a thermocouple connected optically and mechanically with a galvanometer for measuring radiation. He gave a formulation of the instrument's equation of motion, in which the random electromotive force was the sum of both Brownian motions due to molecular collisions and thermal noise in the electric current. And he employed Uhlenbeck and Ornstein's mathematical

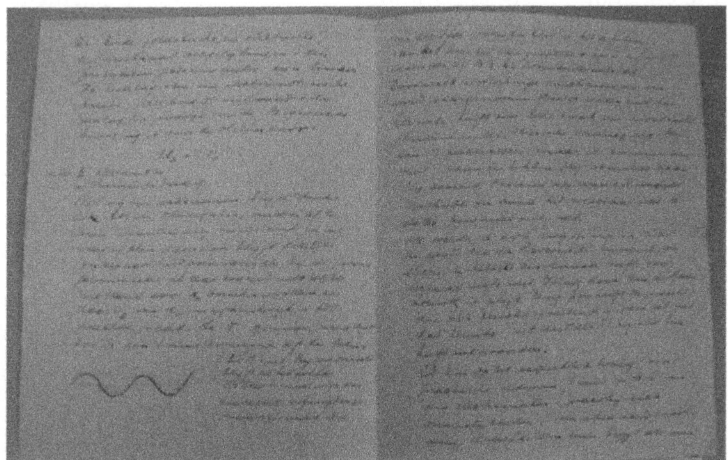

Figure 7.2 Leonard Ornstein's claim that the electromotive force driving a galvanometer was a superposition of a regular harmonic waveform and a fluctuating Brownian motion. Letter, Ornstein to Uhlenbeck, July 16, 1931, Box 1, Folder "Correspondence 1931," George Uhlenbeck Papers, Manuscript Collection, Library of the University of Michigan, Ann Arbor, Michigan, US.

[67] Letter, Ornstein to Uhlenbeck, July 16, 1931, Box 1, Folder "Correspondence 1931," Uhlenbeck Papers.

procedure to obtain the statistical attributes of the instrument's indicator.[68] Yet, van Lear was not satisfied with this paper. In his letters to Uhlenbeck in 1931–32, he repetitively mentioned that there was a discrepancy of up to 50% between his theoretical predictions and James Hardy's experimental data. Finding no way to reduce the discrepancy, van Lear eventually had to give up his original plan for a comparative study between theory and experiment, but rather made the paper a purely theoretical discussion.[69]

George Uhlenbeck's work on Brownian motion in the 1920s–30s epitomized the dynamical approach to random fluctuations. Together with his collaborator Samuel Goudsmit, he applied Schottky's spectral analysis of electronic noise to the studies of a wobbling mirror. He extended Leonard Ornstein's scheme of tackling the Langevin equation and the Fokker–Planck equation into a general theory of Brownian motion. And he worked with his student George van Lear to employ this general theory in experimental problem situations ranging from thermal agitations on a suspended solid body to electrical measuring instruments' limit of sensitivity. In all of these endeavors, Uhlenbeck—and those Dutch physicists surrounding him in Ann Arbor, Leyden, Utrecht, and Amsterdam—showed a consistent focus: formulating the dynamical equation for an object under random fluctuations, solving it, and examining the detailed temporal variations of the object's statistical attributes before it reached equilibrium. Inheriting Einstein's, Smoluchowski's, and Langevin's agendas, Uhlenbeck's research program was a canonical approach to random noise from the perspective of statistical physics.

Around the same time, telephone and radio engineers also worked on theoretical problems of random noise. Yet, their approach differed fundamentally from physicists' method. The electrical engineers paid attention neither to the dynamics of the physical processes that caused noise nor to the temporal variations of the random fluctuations' statistical attributes. Rather, they were concerned with the energy distribution of random noise at different frequencies. This "frequency response" of noise was important because it could help determine the performance of telecommunication system random fluctuations—e.g. the signal-to-noise ratio—when different electrical circuits, modulation schemes, or filter designs were employed. The mathematical theory of noise for electrical engineers in the 1920s–30s thus exhibited very different characteristics.

[68] George van Lear, "The Brownian motion of the resonance radiometer," *Review of Scientific Instruments*, 4 (1933), 21–27.

[69] Letters, van Lear to Uhlenbeck, November 24, 1931, Box 1, Folder "Correspondence 1931"; January 28, 1932 and March 5, 1932, both in Box 1, Folder "Correspondence 1932-33," Uhlenbeck Papers.

8

Spectrum of Random Noise

Theories of Noise and Telecommunications Technology

As Fokker, Planck, Uhlenbeck, Ornstein, Goudsmit, Burger, van Lear, and other physicists in the Netherlands, US, and Germany examined Brownian motion during the 1910s–30s, another expert community also showed a significant interest in the theories of random fluctuations. They were engineers working on telephony, wireless telegraphy, and radio. The types of random fluctuations that interested them were not thermal agitations of drifting particles in a fluid, suspended objects, or mechanical and electrical measuring instruments, but disturbing noise that degraded the technological reproduction of sound and voice. Among the kinds of noise that plagued electrical telecommunications, atmospherics and electronic noise were the core subjects of their inquiries.

Random noise in sound reproduction was not new to engineers by this time. Chapter 4 unveils the metrological aspect of this development in which American naval researchers began systematic observations of statics in the 1910s, while staff at AT&T and the British Radio Research Board in the 1920s–30s introduced new devices and methods using ear balancing and self-registration to measure the quantitative properties of environmental din and noise in telecommunication. In chapter 6, we have seen researchers' work at Siemens, AT&T, and General Electric that identified shot noise and thermal noise in radio and telephony, developed accounts of them based on statistical physics and the theory of Brownian motion, and devised experiments to verify these accounts.

By the 1920s, such explorations of random noise entered the practice of telecommunication engineering and became a central consideration in the design of radio, telephony, and telegraphy. A sizable theoretical literature concerning random noise emerged in the engineering field. Unlike physicists' publications on Brownian motion, these technical documents and discussions rarely examined the causes and mechanisms of random noise, its "equations of motion," its dynamic models, and its evolution over time—the papers by Schottky, Johnson, Nyquist, Hartmann, Hull, Williams, Ballantine, and others on electronic noise were widely believed to have already addressed such issues. Rather, the engineering researchers' theoretical deliberations concentrated on the impacts of random noise on the performance of telecommunication systems, and plausible ways—or the lack of which—to reduce such impacts. To meet this goal, engineers treated noise as a bundle of distinct single-frequency components, each of which had an undetermined amplitude and phase. Sometimes an even simpler assumption of noise as a single sinusoid was staged. This "frequency-domain" approach to noise focusing on spectral analysis was related to the core technological architecture of telecommunication systems in the early twentieth century: a signal converted from sound comprised distinct harmonic components. It was placed at a specific position in the spectrum; entered a receiving system along with noise; went through various stages of amplification, frequency selection, and frequency down-conversion to ensure that the output was close to the original signal; and was transduced back to sound. Under this technological framework, engineers' major concerns about noise were questions such as: how noise behaved when it passed through an amplifier, a filter, a modulator, a demodulator, or a detector, which were all frequency-selective components. Knowing the spectral characters of random noise thus became a pragmatic priority, while its dynamics and temporal evolution offered little help for engineering purposes.

This theoretical work on random noise in telecommunication systems was developed in the context of the changing sensibilities with respect to the technologized soundscape during the interwar years. The introduction of commercial radio broadcasting and wireless telephony after World War I opened a new page for mechanical reproduction of sound. As music, news, speeches, dramas, and quotidian conversations became consumer products of telecommunicated sounds, the users and developers of radio and telephony followed the phonograph-gramophone enthusiasts to instigate the notion and caliber of "high fidelity" for the technology they used. The emerging sensibility of fidelity in phonography was closely connected to Edison Phonograph Company's systematic probing of surface noise's effects on music records in the 1910s (chapter 3). Similarly, the theoretical studies of fluctuating noise's effects on radio and telephony had a strong underpinning in the demand for quality of reproduced sounds in telecommunication systems. For this reason, as Mischa Schwartz observed, engineers' theoretical work on noise during this period was embedded in the design, assessment, and revision of novel schemes aiming to enhance the performance of telecommunication systems, such as single-sideband amplitude modulation (AM) and frequency modulation (FM).[1]

John Carson's Research on Noise at AT&T

The research at AT&T in the 1920s–30s epitomized the frequency-domain approach to random noise in telecommunication systems. By the dawn of World War I, AT&T and Western Electric had monopolized the American telephone business and enacted one of the country's first corporate research establishments. Toward the end of the war, the telephone cartel experimented with ship-to-shore and transoceanic wireless telephony and ventured into the production of equipment and receiving sets for radio broadcasting. As the quality of voice communications was an increasingly pressing issue, the problems of noise became more acute.

A representative figure for AT&T's theoretical investigations of noise in telecommunication was John Carson, whose career embodied the company's strategy to fuel technological and business successes with scientific research. Born in 1886 at Pittsburgh, John Renshaw Carson attended Princeton University, where he received a B.S. (1907), a degree in electrical engineering (1909), and an M.S. (1912). From 1912 to 1914, he taught physics and electrical engineering at his alma mater. He entered American Telephone and Telegraph in 1915 and stayed at the firm until retirement: first as a research staff member of AT&T's Department of Development and Research until 1925, and then as a scientist at Bell Telephone Laboratories. Carson was a rare breed among his corporate engineering colleagues: he was a theorist, perhaps "in part because of impaired hearing and generally poor health," according to a historian.[2] Throughout his career, Carson worked on subjects touching upon various aspects of telecommunication engineering: he developed a mathematical tool known as operational calculus which originated from the British Maxwellian researcher Oliver Heaviside and applied it to the analysis of electric circuits in radio and telephony. He studied propagation of radio waves between a

[1] Mischa Schwartz, "Improving the noise performance of communication systems: 1920s to early 1930s," *IEEE Communications Magazine*, 47:12 (December 2009), 16–20; "Armstrong's invention of noise-suppressing FM," *IEEE Communications Magazine*, 47:4 (April 2009), 20–23.

[2] James Brittain, "John R. Carson and conservation of radio spectrum," *Proceedings of the IEEE*, 84:6 (1996), 909.

conducting wire and the ground, along transmission lines, and in other configurations. He introduced a theory for electrical filters. He participated in the company's waveguide design.[3]

Carson's interest in noise was embedded in his preoccupation with improving the efficacy of voice wireless communications through theoretically informed schemes of what were then called the "signaling" systems. When radio waves were first used in telecommunication in the 1890s, the form of application was wireless telegraphy, in which transmission of signals was accomplished via turning on and off telegraph keys to produce dots and dashes. The technical design was more complex when inventors and engineers at the beginning of the twentieth century attempted wireless transmission of voice. How to send an audio-frequency sound via a higher radio-frequency "carrier" wave became a central question of "signaling." Wireless telephony was facilitated by the inventions that generated continuous radio-frequency carriers in the 1900s–10s, including the high-frequency alternating-current dynamos, electric arc oscillators, and thermionic-tube oscillators. At the time, the dominant method for placing a transmitted signal onto the carrier was so-called amplitude modulation (AM): using a low-frequency signal to modify a carrier's amplitude so that the transmitted wave looked like a continuous series of high-frequency oscillations enveloped by the slowly varying signal. To express AM mathematically, for radio-frequency carrier $A_c \cos\left(2\pi f_c t + \theta_c\right)$ and voice signal $s(t)$, the amplitude-modulated waveform was $s(t) \cdot A_c \cos\left(2\pi f_c t + \theta_c\right)$.[4]

When Carson joined AT&T's Department of Development and Research in New York City in 1915, he came up with an invention to improve the efficiency of AM. The Princeton-trained junior researcher noticed that conventional AM not only shifted the signal spectrum to somewhere around the carrier frequency f_c but also *duplicated* that spectrum. Take the example when the voice signal was a monotone $s(t) = A_s \cos\left(2\pi f_s t\right)$. The amplitude-modulated waveform $A_s \cdot A_c \cos\left(2\pi f_c t + \theta_c\right) \cos\left(2\pi f_s t\right)$ comprised two frequency components—one at f_c+f_s and the other at f_c-f_s. This was a waste of spectrum resource, for the same signal $s(t)$ could be recovered using only one of the two sidebands at f_c+f_s or f_c-f_s. The gist of Carson's invention was to produce a single-sideband waveform and to recover from it the original signal. The patent Carson filed in 1915

[3] Ibid., 909–910.

[4] For the early history of AM, see Aitken, *The Continuous Wave* (1985), 28–249.

became the technical backbone of what was later known as "single-sideband amplitude modulation."[5]

Single-sideband amplitude modulation saved half of conventional AM's spectrum resource and attained considerably higher efficiency. Soon after its invention, single-sideband amplitude modulation was employed in AT&T's working wireless communication systems. But what was the performance of this invention in terms of its withstanding noise? In 1923, Carson published an estimated comparison of the noise-withstanding performance between single-sideband AM and conventional, double-sideband AM in radio telephony. Like Arnold and Espenschied's experiment with AT&T's transatlantic wireless telephony around the same time, Carson utilized the signal-to-noise ratio—or, in his own word, the "signal-to-static-interference ratio," for he considered only the dominant type of random noise (static) in long-range wireless telephony—as a measure of a communication system's performance against noise. Carson assumed that static as random noise permeated uniformly across the entire working bandwidth of a wireless receiver. Following a wisdom of the trade, therefore, he claimed that the energy of the received static noise was proportional to the receiver's bandwidth. The broader the bandwidth, the higher noise energy a wireless receiver picked up. As a result, a double-sideband AM had twice the in-take noise energy as that of a single-sideband AM, when all other factors remained equal. Whether a single-sideband system could improve its signal-to-noise ratio over a double-sideband system thus depended on whether an ingenious way could be devised to enhance the signal energy.

In ordinary AM in which a signal directly rode on the carrier's amplitude $s(t) \cdot A_c \cos(2\pi f_c t + \theta_c)$, the carrier frequency f_c did not appear in the transmitted spectrum and the whole scheme was called "carrier suppression." In that case, the signal energy for single-sideband AM was simply half of that for double-sideband AM, and the signal-to-noise ratios in both systems were equal. The single-sideband system did not advance an improvement over the noise-withstanding performance. Nevertheless, when the AM scheme was slightly modified so that the amplitude of the transmitted wave contained not only the signal waveform but also a constant carrier, $[B + s(t)] \cdot A_c \cos(2\pi f_c t + \theta_c)$, the single-sideband system's signal energy could be enhanced compared with its counterpart in the double-sideband system.

[5] John Carson, "Method and means for signaling with high-frequency waves," US Patent 1,449,382 (March 27, 1923).

In this scenario, Carson asserted that the signal-to-noise ratio for a single-sideband system was $(1 + c)$ times of the signal-to-noise ratio for the corresponding double-sideband system, where c was the ratio of the unmodulated carrier energy to the sideband energy in AM.[6]

This deliberation on wireless telephony's signal-to-noise ratio was connected to a larger project of Carson's. In 1923, Carson and his AT&T colleague Wisconsin-trained physicist Otto Zobel investigated the transient behavior of electrical filters that made up a wireless receiver. Part of the problem that concerned them was what happened when transient disturbances like "static in radio transmission and noise in wire transmission" passed through a filter. From Carson and Zobel's understanding, disturbances of these kinds were random. In their own words:[7]

> By this it is meant that the interfering disturbance, which may be supposed to originate in a large number of unrelated sources, varies in an irregular, uncontrollable manner, and is characterized statistically by no predominant frequency. Consequently the wave form of the applied force at any particular instant is entirely indeterminate. This fact makes it necessary to treat the problem as a statistical one, and deal with mean values.

This characterization of random noise shaped Carson and Zobel's treatment.

In Carson and Zobel's picture, random noise—no matter whether it was static in radio communication or electronic noise in wired telephony—within a long period of time from $t = 0$ to $t = T$ comprised a sequence of pulses that were mutually uncorrelated with one another and had random arriving times:

$$\sum(t) = f_1(t) + f_2(t) + \ldots + f_n(t), \tag{8.1}$$

where the shapes of the waveforms $f_1, f_2, \ldots, f_n$ were indeterminate but followed the same statistical distribution. To tackle this noise in the frequency domain, Carson and Zobel expressed (8.1) in terms of its Fourier integral

$$\sum(t) = \frac{1}{\pi}\int_0^{\infty} |F(\omega)| \cos[\omega t + \theta(\omega)]\, d\omega. \tag{8.2}$$

[6] John Carson, "Signal-to-static-interference ratio in radio telephony," *Proceedings of the IRE*, 11 (June 1923), 271–274.

[7] John Carson and Otto Zobel, "Transient oscillations in electric wave-filters," *The Bell System Technical Journal*, 3 (July 1923), 23.

Then they defined the noise's energy spectrum as

$$R(\omega) = \frac{1}{T}|F(\omega)|^2. \tag{8.3}$$

According to Carson and Zobel's prior investigation, when a deterministic signal (e.g. a sinusoidal wave) with an energy spectrum $R_s(\omega)$ passed through a filter (or a linear circuit in general), the signal energy per unit time (i.e. the signal power) at the output of the filter would be

$$E_s = \frac{1}{\pi}\int_0^\infty \frac{R_s(\omega)}{|Z(i\omega)|^2}d\omega. \tag{8.4}$$

where $Z(i\omega)$ was the filter's transfer impedance as the ratio of its output voltage over its input current at angular frequency ω. Taking for granted that random noise passing through the filter would exhibit the same form in its mean output power as (8.4):

$$E_n = \frac{1}{\pi}\int_0^\infty \frac{R(\omega)}{|Z(i\omega)|^2}d\omega = \frac{1}{\pi T}\int_0^\infty \frac{[F(\omega)]^2}{|Z(i\omega)|^2}d\omega. \tag{8.4'}$$

From (8.4) and (8.4'), the signal-to-noise ratio (in terms of power) at the output of the filter could be calculated.[8]

Key to determining noise power at the filter output was the incoming noise's energy spectrum $R(\omega)$. Carson and Zobel admitted that they did not know the exact form of $R(\omega)$, which could only be determined through empirical measurements. Yet, they believed that the general properties of random noise could help lay out certain quantitative features of the noise power spectrum $R(\omega)$. Just like the cases of shot and thermal noise, random fluctuations as a collection of many pulses with random arriving times and undetermined shapes should not have a "preference" over a specific frequency. Thus, $R(\omega)$ should be flat and smooth over frequencies. To be consistent with physical intuition and energy conservation, the noise's energy spectrum should disappear at high frequencies. Carson and Zobel stipulated that "$R(\omega)$ is a continuous, finite function of ω which converges to zero at infinity and is everywhere positive. It possesses no maxima or minima, and its variation with respect to ω, where it exists, is small." They contended that "these properties of $R(\omega)$ are believed to be evident from physical considerations."[9]

[8] Ibid., 23–24.
[9] Ibid., 25.

Since the noise energy spectrum $R(\omega)$ was much flatter than the filter's frequency response $Z(i\omega)$, $R(\omega)$ in (8.4') could be replaced with its value at the filter bandwidth's central angular frequency ω_m. Because this $R(\omega_m)$ was independent of ω, it could be taken out of the integral to produce a simpler expression for the output noise power:

$$E_n \cong \frac{R(\omega_m)}{\pi} \int_0^{\infty} \frac{1}{|Z(i\omega)|^2} d\omega. \tag{8.5}$$

In this expression, the effect of input noise and that of the filter could be separated as distinct multiplicative factors. If the electrical circuit was an ideal bandpass filter with a flat response between angular frequencies ω_1 and ω_2 and zero at other frequencies, then the filter factor as the integral in (8.5) was proportional to the bandwidth $\omega_2-\omega_1$:[10]

$$E_n \cong \frac{R(\omega_m)}{\pi |Z(i\omega_m)|^2} (\omega_2 - \omega_1). \tag{8.6}$$

That means, the wider the circuit's bandwidth, the higher the noise level at the output.

After Carson and Zobel's paper in 1923, Carson quickly undertook a follow-up study of frequency-selective circuits and static interference. In June 1924, he presented the results of his theoretical work at the Annual Convention of the American Institute of Electrical Engineers (AIEE) in Chicago. Here he reaffirmed the results about filtering noise he and Zobel had obtained in 1923 through a more formal manner. Moreover, he attempted to deduce the character of the noise energy spectrum $R(\omega)$ with the mathematical properties of randomly arriving pulses. Reformulating his and Zobel's previous idea that the noise was a series of randomly arriving pulses, he had

$$\Phi(t) = \sum_{r=1}^{N} \phi_r(t - t_r), \tag{8.7}$$

where ϕ_r was the r-th pulse and t_r was its arriving time between 0 and T. The cosine and sine components of the r-th pulse ϕ_r's Fourier integral were

$$C_r(\omega) = \int_0^{\infty} \phi_r(t) \cos(\omega t)\, dt,\ S_r(\omega) = \int_0^{\infty} \phi_r(t) \sin(\omega t)\, dt.$$

[10] Ibid., 25–27.

Then the energy spectrum of $\Phi(t)$ was the magnitude square of its Fourier integral:[11]

$$|F(\omega)|^2 = \sum_{r=1}^{N} \left[C_r(\omega)^2 + S_r(\omega)^2\right] + \sum_{r=1}^{N} \sum_{s=1 s\neq t}^{N} \cos\left[\omega\left(t_r - t_s\right)\right]\left[C_r(\omega)\, C_s(\omega) + S_r(\omega)\, S_s(\omega)\right]. \qquad (8.8)$$

The single-sum term in (8.8) corresponded to the sum of individual pulses' energy spectrum $|f_r(\omega)|^2$, where $f_r(\omega)$ was the Fourier integral of $\phi_r(t)$. To compute the double-sum in (8.8), Carson first averaged over random and uncorrelated arriving times t_r and t_s. Doing so gave a highly selective function with a large value at $\omega = 0$ but which vanished at other frequencies when the total time T was long. Thus, the computation of the double-sum amounted to the calculation of $[C_r(0)\, C_s(0) + S_r(0)\, S_s(0)]$, which corresponded to the direct-current components of random noise. Carson further assumed that an electric filter or another frequency-selective circuit at the front end of a telecommunication system usually operated at much higher radio frequencies than direct current. Thus, he discarded the double-sum in (8.8) and kept only the single-sum. Moreover, although individual pulses differed substantially from one another, over a large number the individual differences should be smoothed out. A mean energy spectrum $|f(\omega)|^2$ for each pulse should stand out through an averaging process. The statistical average of the noise energy spectrum in (8.8) became:[12]

$$|F(\omega)|^2 = N|f(\omega)|^2. \qquad (8.9)$$

The question about the shape of the noise energy spectrum therefore amounted to the question about the statistical average of an individual noise pulse's energy spectrum. Once again, Carson did not know the answer to this question. But he resorted to the same reasoning that the random character of these individual pulses in radio static or electronic noise should prevent its mean energy spectrum from taking preference at any frequencies. As he stated, "the fact that static is encountered at all frequencies without any sharp changes in its intensity as the frequency is varied, and that the assumption of a systematic wave form for the elementary disturbances would be physically unreasonable, constituted strong inferential support of the hypothesis

[11] John Carson, "Selective circuits and static interference," *Transactions of the AIEE*, 43 (June 1924), 792.

[12] Ibid., 793.

underlying equation (27)"[13] (our equation (8.5)). In this statement, Carson's identification of random disturbances with waveforms devoid of systematic features and spreading equally across all frequencies reminds us of Helmholtz's definition of noise half a century before (chapter 2). To provide empirical support for his assertion, Carson cited Watson-Watt and Appleton's experimental data with atmospherics waveforms in 1923 using oscilloscopic display (chapter 4) that indicated individual static pulses of "rather widely variable amplitudes and durations."[14]

Carson reached a dire conclusion from his analysis that:[15]

- Even with ideal selective circuits, an irreducible residual of interference was absorbed, and this residual increased linearly with the frequency range needed for signaling.
- For statics or random interference, it was useless to employ extremely high selectivity. The gain, as compared with circuits of only moderate selectivity, was small and inevitably accompanied by disadvantages such as sluggishness of response with consequent slowing down of possible speed of signaling.

When asked by an L.W.W. Morrow "if we are faced with the proposition of always suffering from static interference and is there no remedy?" during the discussion after the AIEE presentation, Carson's answer was even more pessimistic:[16]

> The question of static interference is the same problem we come across in telephone transmission, that is, when the signal comes so weak that the interference energy level becomes comparable with the signal, we have got to either face interference or we have got to raise our power. To my mind, the attempt to actually eliminate static by any form of selective circuit would be comparable with the attempt to find a perpetual motion machine. It is there and we have certainly got to receive a frequency range for signaling and from the very nature of the thing, must receive the interference in that range.

Carson noted that perhaps we could reduce considerably particular kinds of static, such as the atmospheric interference coming from a specific geographical region, by devising a dedicated physical arrangement like a directional

[13] Ibid., 794.
[14] Ibid., 794.
[15] Ibid., 789.
[16] Ibid., 796.

antenna to steer the receiving beam pattern away from the region. But a general treatment of noise from the perspective of signals and systems had its constraint and limitation, because noise was always there. Here Carson's thinking of noise entered an ontological level. His idea about the intrinsic performance limit imposed by noise in radio and telephony was close to the thought of de Haas-Lorentz, Ising, Zernike, Uhlenbeck, and van Lear concerning the fundamental limit of precision associated with measuring instruments due to molecular Brownian motions.[17]

Carson's study of noise was symptomatic of an increasing anxiety at the time. In the 1920s–30s, long-distance wireless telephony had evolved from a prototype into a regular operation. The first transatlantic radio telephony was enacted by AT&T and the British General Post Office between New York and London. Radio broadcasting had been transformed from an undertaking by amateurs to a major commercial presence. Stations and studios were established in big cities around the world to broadcast music, news, speeches, and dramas.[18] The demand for the quality of reproduced sound rose; and the preoccupation with high fidelity emerged. Cracking, howling, hissing, and humming—the normal state of affairs for business telegraph operators, soldiers, sailors, and radio hams—became less tolerable for interwar people who used telephone and radio not only for exchanging information but also for sociability and entertainment. Carson's work provided a technical framework to analyze the effects of noise in telecommunications. Other contemporary engineering scientists' research—e.g. Schottky's and Nyquist's—uncovered noise's physical causes and character. Yet, while some outcomes of these studies suggested mild mitigations to the problems of noise (e.g. leveling up the signal power, narrowing down the transmitted bandwidth), most of them pointed to an inevitable presence of disturbances to sound reproduction: random fluctuations posed a fundamental limit to the efficacy of telecommunication. Noise was always there. No systematic means were available to overcome its debacle. The users of the radio and telephone simply had to bear with it.

[17] In addition to publishing his 1924 talk in the *Transactions of the AIEE*, Carson also published the same paper in *The Bell System Technical Journal*. See John Carson, "Selective circuits and static interference," *The Bell System Technical Journal*, 4 (April 1925), 265–279.

[18] For AT&T's wireless telephony in the 1920s–30s, see Fagen, *A History of Engineering and Science in the Bell System: The Early Years (1875–1925)* (1975), 391–412. For the early days of radio broadcasting, see Sterne, *Audible Past* (2003), 215–286.

FM Radio and Noise

Carson's pessimism was proven wrong, fortunately. In the 1920s–30s, a new invention moved telecommunication technology out of the debacle and offered a significant improvement in the quality of sound reproduction and the reduction of noise. The invention was frequency modulation (FM). One of its inventors was Edwin Armstrong.

Edwin Howard Armstrong was in the generation of American inventor-entrepreneurs representing the United States' rapid rise of industry and technology at the turn of the twentieth century. A native of New York, Armstrong attended Columbia University, majored in electrical engineering, and graduated in 1913. Unlike many inventors, Armstrong did not work for any corporations or create his own. After graduation, he worked at his alma mater throughout his career: first as a laboratory assistant for Professor Michael Pupin (Helmholtz's student in Berlin known for the invention of a loading coil to reduce telephone signal distortion), then as an independent researcher running his own lab, and finally as a professor of electrical engineering (without teaching obligations). By the early 1920s, Armstrong was renowned among electrical engineers and technologies for two inventions in radio communication: the "regenerative" circuit using a positive-feedback thermionic-tube amplifier to produce monotonic oscillations at radio frequencies, and the "superheterodyne" receiver mixing incoming radio waves with a harmonic wave at a higher frequency to convert the former down to the intermediate-frequency range. While these two inventions brought fame to Armstrong, they also drew him into patent disputes with competitors. Planning to remove himself out of the fiascos of patent controversies, he wanted to tackle a new subject, and found FM.[19]

The idea of frequency modulation traced back to the early days of wireless communications. When arc oscillators were introduced to replace spark-gap dischargers for wireless telegraphy in the early 1900s, engineers explored the arrangement of sending different telegraph keys (dot or dash on the one hand and space on the other) with different radio frequencies. This led to the thought of sending a voice via modulating a radio wave's frequency by attaching a microphone on a radio transmitter's oscillating circuit. Translating time-evolving sound signals into variations on carrier frequency became feasible as thermionic-tube oscillators were available in the early 1910s. But

[19] For Armstrong's early career, see Lawrence Lessing, *Man of High Fidelity: Edwin Howard Armstrong* (New York: Bantam, 1969), 1–156.

frequency modulation did not seem promising as a signaling scheme for sound reproduction at the time. Leading electrical engineers argued that FM was inferior to AM.[20]

In 1922, John Carson produced a theoretical comparison between FM and AM. He started with the claim that FM could be accomplished when the signal to be transmitted changed a variable capacitor's capacitance in the radio-frequency oscillator. For a sinusoidal signal $h \cdot \sin(pt)$, the transmitted radio frequency varied harmonically as $\omega_0 [1 + h \cdot \sin(pt)]$. For a small signal magnitude h, this corresponded to a more complicated differential equation $(d^2I/dt^2) + \omega_0{}^2 [1 + 2h \cdot \sin(pt)] I = 0$ than the governing equation for a simple harmonic oscillator. Carson noted that this expression was a Mathieu equation, a special type of differential equation.

The solution of this Mathieu equation could be represented as an infinite series containing sinusoidal components at angular frequencies ω_0, $\omega_0 \pm p$, $\omega_0 \pm 2p$, $\omega_0 \pm 3p$, ..., etc. When the signal frequency p was low, the coefficients of the terms in this infinite series could be approximated as Bessel functions, and their sum amounted to a formula $I = A \cos[\omega_0 t - (h\omega_0/p)\cos(pt)]$. But Carson demonstrated that for the low-intensity signal $h \cdot \sin(pt)$, the transmitted waveform I had a frequency range centering at ω_0 and spanning across $2p$. The required bandwidth for FM was thus twice that for the single-sideband AM he had introduced. This made FM a signaling scheme that consumed more spectrum resource than single-sideband AM. According to his analysis, moreover, the signal transmitted through FM was more prone to distortion than that through AM. Carson contended that the two shortcomings of FM—wider bandwidth requirement and more distortion—applied not only to sinusoidal signals, but also to arbitrary signals with small magnitudes and audio bandwidth. Thus, FM was "inferior to" AM.[21]

Carson's theoretical assessment, along with a few other mathematical analyses, set an unfavorable tone against the further development of FM.[22] Yet, Armstrong was not deterred by the theorists' warnings. Between 1928 and 1936, he devoted himself to the design, building, testing, and promotion of an FM radio communication system.[23] Armstrong's primary motivation to

[20] For the early development of FM and its difficulties, see Edwin Armstrong, "A method of reducing disturbances in radio signaling by a system of frequency modulation," *Proceedings of the IRE*, 24:5 (May 1936), 689–691.

[21] John Carson, "Notes on the theory of modulation," *Proceedings of the IRE*, 10:2 (February 1922), 57–64.

[22] Armstrong, "Frequency modulation" (1936), 690–691.

[23] For Armstrong's own account of this process, see ibid. A historical/journalistic account of this process is available in Lessing, *Man of High Fidelity* (1969), 103–132.

venture into FM was to tackle the problems of random noise in wireless communication. He was introduced to the problems of static noise in the 1910s when he worked with his adviser Pupin, who was interested in reducing atmospheric interferences in wireless telegraphy.[24] In 1928, Armstrong explored a new method of doing so. He noted that the energy of atmospheric interferences—a "crash" or bursts of statics—distributed uniformly over a range of frequencies. Based on this observation, he entertained the idea of frequency shift keying, in which a dot or dash in a telegraphic message ("key down") was transmitted via a radio frequency at, say, 20,000 Hz, while a space ("key up") was transmitted via a frequency at, say, 20,060 Hz.

At the receiver, this signaling scheme amounted to two distinct signal paths, one of which demodulated the parts of the binary waveform corresponding to "key down" (i.e. dots or dashes), whereas the other demodulated the parts of the waveform corresponding to "key up" (i.e. spaces). Then Armstrong introduced a balancing device that subtracted the signal path of "key down" from that of "key up." Subtracting the two received signals would not change the telegraphic waveform, since the dots and dashes were on the positive side and spaces were on the negative side. But the amounts of atmospheric disturbances that affected the two signals were about the same due to random noise's uniform energy distribution across frequencies. Thus, subtraction in the balancing device should more or less *cancel* the noise plaguing the two telegraphic signals. Hence the output should be much cleaner. Armstrong implemented a prototype balanced frequency-shift keying radio telegraphic system in his Hartley Research Laboratory at Columbia University, and indeed observed an improvement with respect to noise reduction (Figure 8.1).[25]

Soon after Armstrong published his finding, Carson openly challenged his claim of improving the noise situation with balanced frequency-shift keying. Once again, Carson questioned the validity of reducing random noise with a novel scheme of modulation. In 1928, he wrote an article to analyze the effectiveness of Armstrong's design for noise reduction. Not surprisingly, Carson's conclusion was entirely negative. Without losing generality, he tackled the situation in which dots or dashes were sent. In this situation, the electrical current on the key-down channel of the receiver comprised both the telegraphic signal component and a noise component, whereas the key-up channel had only a noise component. Carson modeled the signal component as a sinusoidal carrier amplitude-modulated by the sequence of telegraph keys. The noise

[24] Lessing, *Man of High Fidelity* (1969), 101–102.

[25] Edwin Armstrong, "Methods of reducing the effect of atmospheric disturbances," *Proceedings of the IRE*, 16:1 (January 1928), 15–26.

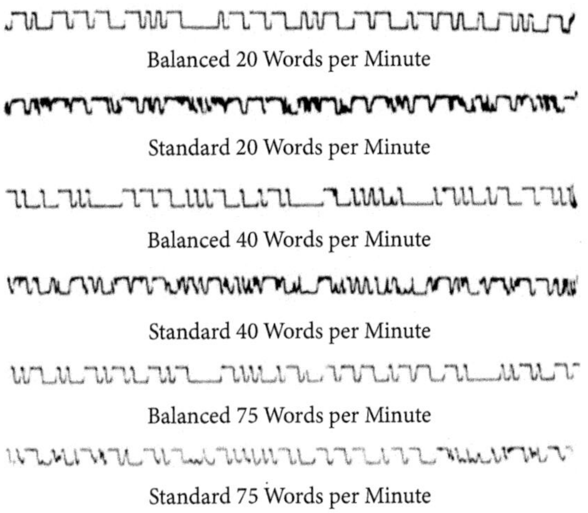

Balanced 20 Words per Minute

Standard 20 Words per Minute

Balanced 40 Words per Minute

Standard 40 Words per Minute

Balanced 75 Words per Minute

Standard 75 Words per Minute

Figure 8.1 Results from Armstrong's 1928 experiment with noise reduction using frequency-shift keying. The telegraphic waveforms under the "standard" condition (without the balancing subtraction) look noisier than those under the "balanced" condition (with the balancing subtraction). Edwin Armstrong, "Methods of reducing the effect of atmospheric disturbances," *Proceedings of the IRE*, 16:1 (January 1928), 20, Figure 5.
Permission from IEEE.

component he modeled was also a harmonic carrier amplitude-modulated by a randomly varying fluctuation. In other words, the current at the key-down channel was

$$W(t) = S(t) \sin(\omega t) + J(t) \sin[\omega t + \phi(t)],$$

where $S(t)$ was the sequence of telegraph keys, $J(t)$ was noise's amplitude, $\phi(t)$ was noise's phase, and ω was the carrier frequency at the channel. The current at the key-up channel was

$$W_1(t) = J_1(t) \sin[\omega_1 t + \phi_1(t)],$$

where $J_1(t)$ was noise's amplitude, $\phi_1(t)$ was noise's phase, and ω_1 was the carrier frequency at the channel.[26]

Carson did not employ the more sophisticated model of noise as a composite of frequency components with a statistically uniform distribution—the

[26] John Carson, "The reduction of atmospheric disturbances," *Proceedings of the IRE*, 16:7 (July 1928), 972.

one he had developed with Zobel in 1923–24 for inspecting the effect of filtering on random disturbances ((8.2)–(8.6)). His model for noise here—a fast-oscillating carrier wave with fluctuating amplitude—was much simpler. This assumption was consistent with the receiver architecture that selected only noise's spectral components near the carrier frequency ω or ω_1. In any case, this assumption brought convenience and made the form of noise susceptible to straightforward mathematical manipulations. Passing the currents at both channels through a homodyne detector for frequency down-conversion and a balancing device that subtracted one output from the other, Carson obtained

$$\frac{1}{2}S(t) + \frac{1}{2}J(t)\cos\phi - \frac{1}{2}J_1(t)\cos\phi_1.$$

Here the first term represented the telegraph signal, whereas the second and third terms represented static noise at the two channels. Armstrong's design implicated that the noise components at the two channels $\frac{1}{2}J(t)\cos\phi$ and $\frac{1}{2}J_1(t)\cos\phi_1$ could cancel each other due to the balancing operation. Yet, Carson argued that since noise's phases at the two channels ϕ and ϕ_1 corresponded to its properties around two frequencies ω and ω_1, they should be random and statistically independent of each other. This implied that "the two interference components are equally likely to add or subtract so that no gain by balancing results on the average." Through a more careful calculation, Carson demonstrated that the balanced current's "ratio of interference to signal" (the reciprocal of the signal-to-noise ratio) was twice the standard receiver's ratio. In terms of signal-to-noise ratio, Armstrong's balanced receiver was twice worse than a standard receiver.[27]

While Armstrong did not openly refute Carson's challenge, he nonetheless continued to work on noise reduction via signaling. Carson might have the sophisticated mathematical tools of Bessel functions and the Mathieu equation to demonstrate the futility of engineering improvement. But Armstrong had experimental data to show otherwise. In the same year when he published his work on frequency-shift keying for telegraph noise reduction, he launched a full-fledged trial on frequency modulation of voice and sound signals. Between 1928 and 1933, Armstrong devoted his full time to the development of an FM system in his laboratory at Columbia University.

As a prototype transmitter and receiver with a 44-MHz carrier and a frequency swing of 150 kHz became available, he invited David Sarnoff, President

[27] Ibid., 973–974.

of the Radio Corporation of America (RCA), to see a demonstration. Armstrong's demo of FM on Christmas Day of 1933 at Columbia University impressed Sarnoff. Despite the lawsuit between Armstrong and RCA concerning the patent of regenerative circuits, Sarnoff still agreed to help with the further testing of FM. Through this assistance, Armstrong ended up using the National Broadcasting Company's experimental station on top of New York's Empire State Building to set up his transmitter. Two sites for the receiver were selected: one in Westhampton, Long Island, and the other in Haddonfield, New Jersey. In 1934–35, Armstrong and his team conducted hundreds of transmission trials at these sites. The data from the experiments gave him confidence to publicize his invention.[28]

Armstrong announced his invention at the Institute of Radio Engineers' (IRE's) annual meeting in the auditorium of its Engineers Building on 39th Street in New York on the evening of November 5, 1935, two years after his four patents on frequency modulation were issued. At the meeting, he gave a paper to report the design of his FM system and the results from his experiments in 1933–34. Yet, Armstrong arranged clandestinely a surprise at his presentation. The night before the meeting, he sneaked an FM receiver into the auditorium, and tuned it to an FM transmitter at 110 MHz that he and his collaborators had installed at a ham radio station in Yonkers, miles north of the city. When Armstrong got a silent notice from his partners in the room during his talk, he declared to the audience that he was going to show what his FM system was like; and suddenly, sounds came out of the receiver. Armstrong's demonstration, as the science writer Lawrence Lessing described, showcased a hi-fi regime of sound reproduction that connoted a different experience with noise. Lessing's picturesque narrative is worth a full quote:[29]

> The demonstration that ensued became a part of the Major's standard repertoire in showing off the remarkable properties of his new broadcasting system. A glass of water was poured before the microphone in Yonkers; it sounded like a glass of water being poured and not, as in the "sound effects" on ordinary radio, like a waterfall. A paper was crumpled and torn; it sounded like paper and not like a crackling forest fire. An oriental gong was softly struck and its overtones hung shimmering in the meeting hall's arrested air. Sousa marches were played from records and a piano solo and guitar number were performed by local talent in the Runyon livingroom. The music was

[28] Lessing, *Man of High Fidelity* (1969), 157–165; Armstrong, "Frequency modulation" (1936), 717.
[29] Lessing, *Man of High Fidelity* (1969), 170.

> projected with a "liveness" rarely if ever heard before from a radio "music box." The absence of background noise and the lack of distortion in FM circuits made music stand out against the velvety silence with a presence that was something new in auditory experiences.

The FM system could transmit sounds with considerably lower noise levels than what most users of sound reproduction technologies were familiar with.

Armstrong's technology to achieve this degree of high fidelity comprised an FM transmitter and receiver. These apparatuses were state-of-the-art electronic circuitry. The transmitter included a radio-frequency vacuum-tube oscillator, whose output phase—or more precisely, its output frequency as the time derivative of its phase—was modified by the input audio signal via a balanced modulator. This frequency-modulated waveform was then amplified and sent out by an antenna. At the receiver, the radio-frequency FM signal entered an antenna, was heterodyned down to an intermediate frequency, and amplified. This intermediate-frequency wave passed through a vacuum-tube *limiter* that truncated its maximum amplitude at a fixed level to prevent unwanted amplitude modulation. The output from the limiter passed through another electronic circuit known as a *discriminator*, which converted the signal's frequency variation into an amplitude variation through a (time) differential operation. The output from the discriminator looked like an amplitude-modulated signal. The remaining blocks of Armstrong's FM receiver were similar to the baseband part of an AM receiver.[30]

In an ideal situation when only the voice signal was present, Armstrong's transmitter could produce a waveform frequency-modulated by the signal, and his receiver could demodulate the waveform to reproduce the signal. But why could FM achieve apparently much better noise reduction than AM? What aspects of FM helped with combating noise? While the general public would not have a chance to experience FM until its commercialization after World War II, its outstanding performance nonetheless impressed the engineering community by the mid-1930s. After Armstrong's demonstration in 1936, engineering researchers made various attempts to understand why FM provided a much more effective means of noise reduction than AM. These discussions were tied to the frequency-domain analysis of noise.

Armstrong himself entertained an intuitive answer to this question. When he began to develop FM in the mid-1920s, he thought that the frequency-modulated signals could overcome the effects of static disturbances because

[30] Ibid., 168; Armstrong, "Frequency modulation" (1936), 692–700.

signal waves were fundamentally different in character from those of statics.[31] In his report at the IRE meeting in 1936, he did not develop a theory of noise reduction for his FM system. Yet, he did produce many measurements and obtained useful empirical rules about the noise properties of FM compared with AM. The most important empirical rule he found was that the signal-to-noise ratio at the output of the receiver was proportional to the square of the range of the sweeping frequencies in the FM signal—i.e. the square of the radio-frequency signal's operating bandwidth. In contrast, the signal-to-noise ratio of an AM system did not have an obvious dependence on the signal's operating bandwidth. Thus, a wideband FM system had an advantage over an AM system in terms of noise suppression. This observation constituted Armstrong's primary discovery: the FM system advanced a substantial improvement in resisting the effects of noise by spreading its signal over a broader spectrum.[32]

Yet, this feature was not the whole story about the noise-related performance of FM. From his experimental data, Armstrong also noted a peculiar relationship between the signal-to-noise ratio at the input of the FM receiver and that at the output of the receiver. The two quantities did not change in proportion to each other, but co-varied nonlinearly. When the input signal-to-noise ratio exceeded a threshold, the FM receiver's output signal-to-noise ratio showed a significant improvement and was substantially better compared to the much more uniform signal-to-noise ratio for the AM receiver. But when the input signal-to-noise ratio was below this threshold, the output signal-to-noise ratio did not have an obvious improvement at all, and FM's noise-suppressing performance at this condition was inferior to that of AM (Figure 8.2).[33]

Shortly after Armstrong's invention became popular in the technological community, engineering researchers undertook mathematical analyses of FM and how random noise affected its performance. Carson and Thornton Fry, the mathematical figureheads of Bell Labs, led this line of investigation. Having heard of Armstrong's experimentation before his renowned IRE talk in November 1936, Fry produced an AT&T internal research report on the computation of audible noise in frequency-modulated systems with and without amplitude limitation in November 1935. A few months later, Carson wrote another internal report on the comparison of FM and AM with respect to their performances against atmospherics. In 1937, Carson and Fry combined their

[31] Lessing, *Man of High Fidelity* (1969), 157.
[32] Armstrong, "Frequency modulation" (1936), 709.
[33] Ibid., 711.

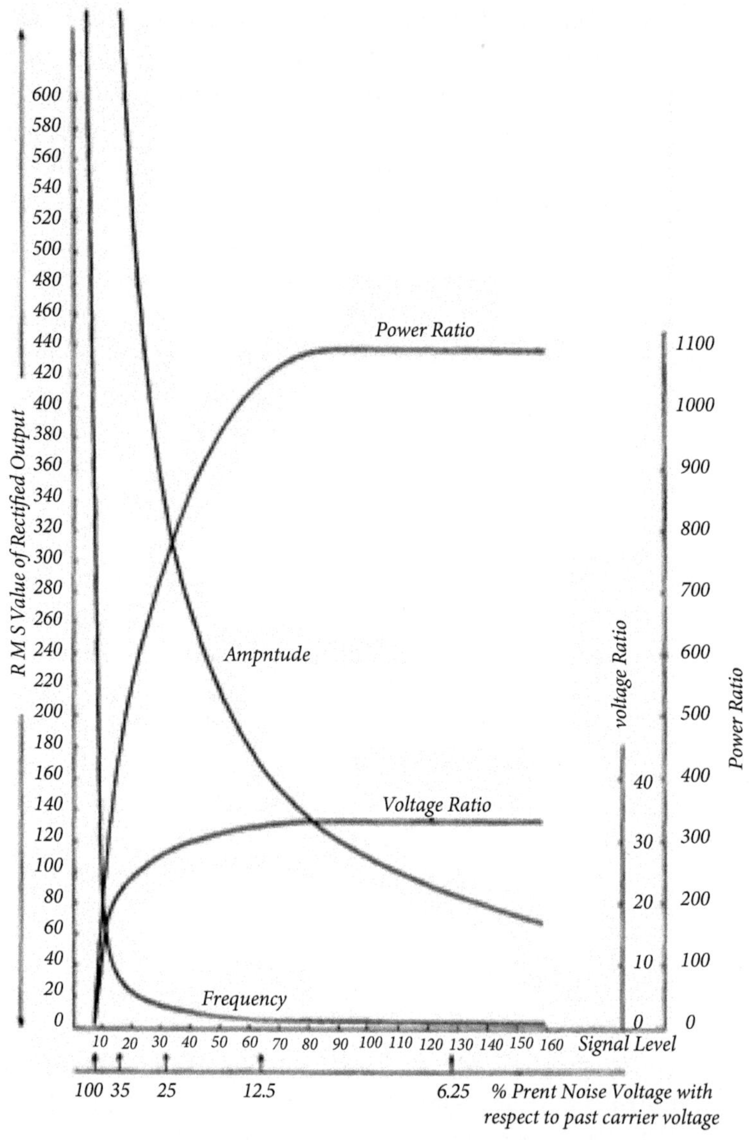

Figure 8.2 Signal-to-noise ratios at the FM receiver output in Armstrong's experiment during 1933–34. The curves marked by "Power Ratio" and "Voltage Ratio" correspond to the signal-to-noise power and amplitude ratios at the FM output. The curves marked by "Frequency" and "Amplitude" represent the noise amplitude levels at the output of an FM and a AM receiver, given a fixed signal amplitude. Edwin Armstrong, "A method of reducing disturbances in radio signaling by a system of frequency modulation," *Proceedings of the IRE*, 24:5 (May 1936), 711, Figure 18.
Permission from IEEE.

respective work and co-wrote an article published in *The Bell System Technical Journal*.[34]

In Carson and Fry's model, the wave frequency-modulated by the audio signal $s(t)$ was

$$W = \exp\left\{i\left[\omega_c t + \lambda \int_0^t s(t')\,dt'\right]\right\}. \tag{8.10}$$

The low-frequency signal $s(t)$ in this waveform was "implicit," for the signal did not show up in the waveform's amplitude variation. Armstrong's discriminator rendered the signal "explicit." To Carson and Fry, the discriminator took a time derivative of W in (8.10):

$$\frac{dW}{dt} = [\omega_c + \lambda s(t)] \exp\left\{i\left[\omega_c t + \lambda \int_0^t s(t')\,dt'\right]\right\}, \tag{8.11}$$

where ω_c was the carrier (angular) frequency and λ was the modulation index. While the first term in (8.11) still had the same form as the frequency-modulated wave, the second term had an amplitude variation proportional to the signal $s(t)$. Passing this output from the discriminator to an amplitude-detecting device and filters retrieved the signal.[35]

To analyze the effect of atmospherics, Carson and Fry assumed that the relevant noise present at the front end of the receiver was in the form of

$$W_n = A_n \exp\left[i(\omega_c t + \omega_n t + \theta_n)\right], \tag{8.12}$$

where A_n, ω_n, and θ_n were the amplitude, frequency deviation (from the carrier), and phase of noise, respectively. The phase of noise was random and uniformly distributed at all angles. Carson and Fry added noise of the form in (8.12) to the signal in (8.11), and delivered the sum into the aforementioned discriminator, amplitude detector, and filter to see the corresponding power levels of signal and noise at the output. After mathematical analysis, the AT&T researchers concluded that the amplitude limitation of the

[34] Thornton Fry, "Alternative method of computing audible noise in frequency modulated systems with and without amplitude limitation," Technical Report (November 27, 1935), in Case No. 20,878, volume A; John Carson, "Amplitude vs. frequency modulation: Relative effectiveness of limiters against disturbances of large amplitudes and short durations," Technical Report (August 29, 1936), in Case No. 36,760, volume A, both in AT&T Archives. For their publication, see John Carson and Thornton Fry, "Variable frequency electric circuit theory with application to the theory of frequency modulation," *The Bell System Technical Journal*, 16 (October 1937), 513–540.

[35] Carson and Fry, "Frequency modulation" (1937), 524–525.

frequency-modulated waveform was a determining factor for the FM system's high performance in terms of signal-to-noise ratio.[36]

The rave to seek the reason for FM's noise-suppressing advantage affected other Bell researchers. In October 1939, Ralph Potter provided his account in another internal technical report. Once again, he assumed the noise to be in the form of a single sinusoid like (8.12). He used a vector diagram to explain what happened to the relationship between signal and noise at distinct stages of the FM receiver: the limiter functioned effectively like a negative amplitude modulator that decomposed noise into two components with different amplitudes. This amplitude difference was further magnified by the discriminator. But after amplitude detection and filtering, only the difference between the two components was relevant to the audio output, and that difference was considerably smaller than the original noise level. This demonstrated

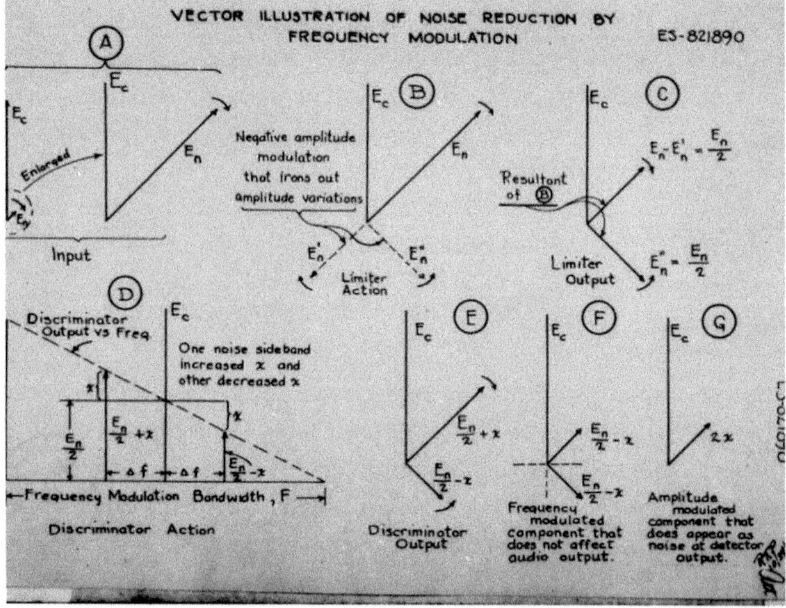

Figure 8.3 Ralph Potter's vector-diagram illustration of why FM could suppress noise. Ralph Potter, "Noise reduction advantage of frequency modulation," Technical Report (October 27, 1939), 7, in Case No. 36760, volume A, AT&T Archives. Courtesy of AT&T Archives and History Center.

[36] Ibid., 529–534.

why the presence of noise was much weaker at the output than at the input (Figure 8.3).[37]

Harry Nyquist provided his account of FM's noise suppression, too. In the technical report he submitted to Bell Labs in December 1935, he argued that when atmospherics was weaker than the FM signal, the latter simply shifted the phase of the former and accordingly displaced the zeros of the signal. Since the frequency of the signal wave was determined by the separations between zeros, such a displacement did not really degrade the quality of the transmitted message or voice. When noise was stronger than the signal, on the other hand, the former created or annihilated zeros of the signal. This modified more notably the voice or message to be transmitted. To Nyquist, such a disparity explained why FM had a superior performance in noise suppression when the incoming signal-to-noise ratio exceeded a threshold but became much worse at a lower input signal-to-noise ratio. To evaluate such phase shifts, he conducted a further mathematical analysis, assuming, once again, that noise was a harmonic wave with a single yet undetermined frequency, amplitude, and phase.[38]

Outside AT&T, Murray Crosby at RCA also attempted to develop a communication theory for an FM system under the influence of noise. To assess the signal-to-noise ratio at the output of the FM receiver, he supposed, like the Bell researchers, that the atmospherics was a harmonic wave in the form of (8.12). Adding this noise onto the FM signal, he then passed the composite waveform through the discriminator and frequency-detecting block, identified the strength of the signal and noise components at the output, and calculated the corresponding signal-to-noise ratio. Crosby's theoretical results also reached the same conclusion as Armstrong's experimental findings had indicated.[39]

During the 1930s, theoretical analyses of FM—a significant technological breakthrough to reduce noise in sound reproduction—unanimously treated random noise as a single harmonic oscillation. This was the simplest model for noise that the frequency-domain approach could get. Of course, the dynamics and temporal evolution of noise that concerned statistical physicists were beyond the scope of electrical engineers' consideration. Even Carson's and Nyquist's more sophisticated frequency-domain model of atmospheric and electronic noise with a uniform, "white" spectrum did not appear in FM

[37] Ralph Potter, "Noise reduction advantage of frequency modulation," Technical Report (October 27, 1939), in Case No. 36,760, volume A, both in AT&T Archives.

[38] Harry Nyquist, "Frequency modulation – noise effect," Technical Report (December 2, 1935), in Case No. 36,760, volume A, both in AT&T Archives.

[39] Murray Crosby, "Frequency modulation noise characteristics," *Proceedings of the IRE*, 25:4 (April 1937), 472–514.

theorists' discussions. This focus on the simplest form and absence of elaborate modeling was a strong propensity among electrical engineers in the treatment of noise. It had a sharp contrast not only with the physicists' time-domain approach, but also with the approach to random fluctuations taken by mathematicians around the same period.

9
A Mathematical Foundation of Fluctuations

Stochastic Processes, Measure Theory, and Mathematical Rigor

The theories of noise developed by physicists and engineers in the 1910s–30s focused on two aspects of random fluctuations. To statistical physicists following the research tradition of Maxwell, Gibbs, and Boltzmann, the priority in theory construction was to find the governing dynamical equations for Brownian-like motion, to obtain their temporal evolution from solving the equations, and to inspect the transient and asymptotic behavior of such motions in a statistical sense. To electrical engineers preoccupied with technical problems of interference reduction in telephone and radio technologies, the urgent tasks were to determine the power density of random noise at different frequencies and to assess the effects of noise when it passed through amplifiers, filters, modulators, demodulators, or other signal-processing devices. These "time-domain" and "frequency-domain" approaches connected to the intellectual legacy or pragmatic need of physicists and engineers configured the types of questions that theorists asked about noise.

Yet, a third expert group around the same time also showed a significant interest in the theoretical issues related to random fluctuations, albeit from a different angle. This group comprised mathematicians, and the third view may be labeled as mathematical. Although physicists and engineers used various mathematical tools in their theorization of noise, the kinds of theoretical issues about random fluctuations that caught mathematicians' attention were subtler, more formal, and more abstract. Influenced by the intellectual movements—real analysis, complex analysis, theory of functions, set theory, axiomatization, mathematical logic—since the enactment of "pure mathematics" as a disciplinary identity in the mid-nineteenth century,[1] the mathematicians working

[1] For mathematicians' intellectual traditions formed since the mid-nineteenth century, see Klein, *Mathematical Thought from Ancient to Modern Times* (1972), 947–1211.

on noise were much more concerned with the formal structures and peculiar attributes of random fluctuations as mathematical entities than with computation of specific trajectories or spectra. They asked, for example: What kind of function is the path of a Brownian particle or the waveform of noisy current? How do we specify the probability associated with a particular path or waveform of this sort? What is the abstract space that best represents all these paths? What are the structural features of such a space? Can we employ the same mathematical methods—such as spectral analysis and Taylor series expansion—as for the analysis of regular functions for the analysis and representation of random fluctuations? If not, how should the methods be modified? Answering these questions might not provide immediate help for physicists and engineers in predicting experimental effects or designing technological systems. To mathematicians, however, these questions revealed foundational issues underlying noise, and resolving them would pave a way toward a more rigorous theory of random fluctuations.

The development of the mathematical theory of random fluctuations is an encompassing, extensive, multi-sited, and multi-routed historical process. If we emphasize early explorations of measure theory and its connection to a formal representation of probability, then we should inspect the intellectual genealogy of the Paris school of functional analysis from the 1890s to the 1920s: Émile Borel, Henri Lesbesgue, and Louis Bachelier's nemesis Paul Lévy.[2] If we want to follow the beginning of the set-theoretic axiomatization for probability and stochastic processes and their systematic treatments built on conditional probabilities, Markov chain, and forward and backward propagation, then we will encounter the Soviet math giants Andrey Kolmogorov and Aleksandr Khinchin in Moscow during the 1920s–40s.[3] If we are interested in the mathematization of the ergodic hypothesis that Boltzmann and Gibbs entertained as a premise of statistical mechanics, then George Birkoff's and John von Neumann works in Boston and Princeton during the 1930s were central.[4] If our curiosity lies in the development of differential and integral calculus for stochastic processes that later constituted a backbone of mathematical finance involving time-series data processing, then the contributions

[2] For the early history of the theory of measure, see Thomas Hawkins, *Lesbesgue Theory of Integration: Its Origin and Development* (New York: Chelsea Publishing Company, 1979).

[3] Kendall, "Kolmogorov" (1990).

[4] For a brief review of Birkoff's ergodic theorem, see Norbert Wiener, Armand Siegel, Bayard Rankin, and William Ted Martin, *Differential Space, Quantum Systems, and Prediction* (Cambridge, MA: MIT Press, 1966), 3–8.

from Joseph Doob in Urbana, Illinois, and Kiyosi Itô(伊藤清). in Tokyo, among others during the 1930s–50s, were pivotal.[5]

Delving into such a complex history is beyond the scope of this book. In this chapter, we focus on the work of an individual on the mathematical theory of noise during the interwar period: the American mathematician Norbert Wiener. Concentrating on Wiener is relevant for our purpose: He was among the first researchers to construct a theory of random fluctuations as mathematical entities with abstract and formal structures out of concrete physical and engineering knowledge on Brownian motion and electronic noise. Also, his theory of fluctuations would eventually become a powerful tool for filtering and reduction of noise in gunfire control, radar, and wireless telephony during World War II and its further generalization into an informational worldview (part IV). Arguably, his work epitomized the mathematical approach to noise in the 1920s–30s.

Norbert Wiener's Route toward Random Noise

Norbert Wiener is one of the most visible figures in the history of information science and technology. He is best known as the founder of cybernetics, a doctrine of modeling biological, mental, social, and cultural phenomena with feedback-control systems via communication of information. Given the strong influence of cybernetics in realms from academic research and policy making to social imaginaries and popular cultures during the postwar and Cold War years, most historical accounts of Wiener's work and life have directed attention to his role in the origin of this informational worldview. Peter Galison examined Wiener's development of an antiaircraft gun director during World War II as an ontological prelude to his cybernetics. Steve Heims and Ronald Kline followed Wiener's involvement in the formation of the first postwar academic community—the group surrounding the Macy Conferences—dedicated to research and promotion of cybernetics. In his biography by journalists Flo Conway and Jim Siegelman, Wiener is referred to as a "dark hero of the information age."[6]

Casting a spotlight on Wiener's work before the Great War shows a different picture, though. There, he was neither a mastermind contemplating a theory

[5] For a historical review of the theory of stochastic calculus, see Davis and Etheridge, *Louis Bachelier's* Theory of Speculation (2006), 80–115; Jarrow and Protter, "Stochastic integration" (2004).

[6] Galison, "Ontology of the enemy" (1994), 228–266; Kline, *Cybernetic Moment* (2015), 9–67; Heims, *Cybernetic Group* (1991), chapter 1; Flo Conway and Jim Siegelman, *Dark Hero of the Information Age: In Search of Norbert Wiener, the Father of Cybernetics* (New York: Basic Books, 2006).

of the world nor a savant building an academic discipline, but a junior mathematician struggling his way through a mist of problems in probability theory, stochastic processes, and random fluctuations. By inspecting his theorization of noise, we can see a clear intellectual trajectory from statistical characterization, representation, and analysis of noise, via its engineering manipulations through filtering and prediction, and eventually to the modeling of feedback control processes in which both the controllable and the disturbances were conceived as innately random time series.[7]

Norbert Wiener was born into a family of Eastern-European Jewish immigrants in 1894 at Columbia, Missouri. His father Leo was an academic. When Norbert was little, the family moved to Cambridge, Massachusetts, where Leo taught Slavic languages at Harvard University.[8] The young Norbert was a child prodigy. His father found his giftedness and homeschooled him. Leo's curriculum was overwhelming. By his early teens, Norbert had learned several European languages and advanced mathematics, and read philosophical classics. At twelve, he was enrolled at Tufts College to study biology and mathematics. He obtained his B.A. in three years and entered the graduate school at Harvard planning to concentrate on zoology. After discovering that laboratory work did not suit him, he transferred to Cornell University and switched to philosophy, but moved back to Harvard a year later. He obtained his M.A. in 1912 and Ph.D. in 1913 in philosophy.[9]

Wiener's doctoral dissertation concerned mathematical logic. Upon graduation, he obtained a postdoctoral fellowship to study at Cambridge University, intending to research mathematical logic with Bertrand Russell, who was world-famous for his philosophical program of analyzing and constructing all expressions, concepts, and statements in everyday and formal languages with logical terms. Wiener's visit to Cambridge in 1913–14 was formative for his intellectual development. His plan of pursuing mathematical logic with Russell did not proceed well. The nineteen-year-old "Wunderkind" got lost in the subject's tedious intricacies. The interpersonal atmosphere at Cambridge did not help, either. As he complained to his father, "Russell is an iceberg. His mind impresses one as a keen, cold, narrow logical machine that cuts

[7] In his edited volume of Wiener's collected papers, his student Pesi Masani provided a technical overview and comments on his mathematical work. See Pesi Masani (ed.), *Norbert Wiener: Collected Works with Commentaries*, Volumes 1 and 2 (Cambridge, MA: MIT Press, 1976 and 1979), "Introduction and acknowledgement."

[8] Irving Ezra Segal, "Norbert Wiener," *Biographical Memoirs of the National Academy of Sciences of the United States of America*, 61 (1992), 390–391.

[9] Ibid., 392–393. Wiener himself provided an intimate description of his childhood and college and graduate studies in his autobiography, Norbert Wiener, *Ex-Prodigy: My Childhood and Youth* (Cambridge, MA: MIT Press, 1964).

the universe into little packets"; and the students "are as close as oysters." Fortunately, he got to know Godfrey Harold Hardy, a leading English mathematician. Hardy's courses in mathematical analysis inspired him. Hardy encouraged and praised the smart but hesitating young Wiener, as he did for the Indian genius Srinivasa Ramanujan. Wiener decided to be a career mathematician.[10]

Wiener's visit to Cambridge came to an end as World War I broke out. Between 1914 and 1919, he wandered around New York, Massachusetts, and Maryland taking several jobs. In 1919, he accepted an offer from Massachusetts Institute of Technology (MIT). He would spend his entire career there.[11] When Wiener joined MIT, it was not widely considered as a center of American mathematical studies. While the departments of mathematics at Harvard and Princeton had nurtured cutting-edge research comparable to their European counterparts in analysis and topology, the department of mathematics at MIT still functioned largely as a provider of teaching, service, and consultation for the Institute's projects in engineering. This limitation discomforted Wiener in his first few years at MIT. Yet, the necessity to work closely with engineers and experimentalists at MIT exerted a lifelong influence on Wiener's mathematical approach, which often started with considerations of concrete physical or technical problems and continued to maintain connections to such problems despite highly abstract derivations.[12]

Wiener began mathematical research as soon as he settled down at MIT. His first topic had to do with the probabilistic structure of functions with apparently random variations. Several threads of thoughts led him to this subject. The most obvious one was his exposure to the problems of Brownian motion. Wiener came to know about Einstein's work on Brownian motion at Cambridge in 1913 when Russell assigned him to read papers important to the foundation of modern science.[13] Meanwhile, he learned from Hardy about a novel approach to integration named Lebesgue's theory of measure.

Invented by Borel and Lebesgue in the 1890s, the theory extended the conventional scheme of the Riemann sum to a definition of a function's integral. When a non-negative real function $f(x)$ is integrated over a fixed domain,

[10] Segal, "Wiener" (1992), 393–394; Norbert Wiener, *A Life in Cybernetics—Ex-Prodigy: My Childhood and Youth, and I am a Mathematician: The Later Life of a Prodigy* (Cambridge, MA: MIT Press, 2018), 141–157. For the citations of Wiener's writing to his father, see Letters, Norbert Wiener to Leo Wiener, October 15, 1913 and October 25, 1913, Box 1, Folder 5, "Correspondence 1913," Norbert Wiener Collection (MC22), MIT Archives, Cambridge, MA, US.

[11] Segal, "Wiener" (1992), 395–396; Wiener, *Life in Cybernetics* (2018), 159–217.

[12] Segal, "Wiener" (1992), 389–390; Wiener, *Life in Cybernetics* (2018), 242–245.

[13] Wiener, *Life in Cybernetics* (2018), 155.

e.g. the interval $[a, b]$, the Riemann sum divides the domain into many small pieces marked by points $a < x_1 < x_2 < \ldots < x_N < b$, and the integral converges asymptotically to a sum when the small pieces become infinitesimal:

$$\int_a^b f(x)\,dx = \lim_{N\to\infty} \sum_{n=1}^{N} f(x_n)\,(x_n - x_{n-1})\,.$$

In contrast, the Lebesgue theory does not divide the domain into adjacent intervals $a < x_1 < x_2 < \ldots < x_N < b$. Rather, it partitions $[a, b]$ into small segments, each of which is a set $\{x\}$ that satisfies a common property in relation to their functional mapping; for example, $\{x: f(x) > y_n\}$. The size of this set (or its length, when it is an interval on the real line) can be defined as a *measure* of the set—e.g. $\mu(\{x: f(x) > y_n\})$. When the division on the function's range y becomes infinite, we can define another integral in terms of the measure:

$$\int f d\mu = \lim_{N\to\infty} \sum_{n=1}^{N} (y_{n+1} - y_n)\,\mu\left(\{x : f(x) > y_n\}\right).$$

The Lebesgue integral and the Riemann integral should be identical.[14] While this definition of measure and Lebesgue integral is on the real line, it can be extended to a more general space.

Although the Lebesgue measure theory was originally developed for mathematical analysis of integration, researchers at the beginning of the twentieth century started to apply it to the formulation of probability theory and mathematical statistics. The probability of an event, they noted, could be conceived as a Lebesgue measure of a set of points in an abstract "sample space" or "measure space," where all the points in the set had a one-on-one correspondence with a particular and distinct condition of the event (as long as the measure of any set in such a space was between 0 and 1, the measure of the entire space was 1, and the measure of the null set was 0). In this regard, the measure theory of integration was a natural platform to formulate the statistical average over curves of complicated and irregular shapes.

Wiener was notified about recent works on the Lebesgue theory of measure by a fellow mathematician I. Barnett in Cincinnati. From Barnett, Wiener learned that the theory of measure enabled a representation of the statistical average of not only points but also curves.[15]

[14] Hawkins, *Lebesgue's Theory of Integration* (1979), ix–xv.
[15] Wiener, *Life in Cybernetics* (2018), 245–246.

The last piece of the puzzle came from the Cambridge mathematician Geoffrey Ingram Taylor. According to Wiener's own reminiscence, around 1920:[16]

> I was an avid reader of the journals, and in particular of the *Proceedings of the London Mathematical Society*. There I saw a paper by G. I. Taylor, later to become Sir Geoffrey Taylor, concerning the theory of turbulence. This is a field of essential importance for aerodynamics and aviation, and Sir Geoffrey has for many years been a mainstay of British work in these subjects. The paper was allied to my own interests, inasmuch as the paths of air particles in turbulence are curves and the physical results of Taylor's papers involve averaging or integration over families of such curves.

Furthermore, he continued:[17]

> With Taylor's paper behind me, I came to think more and more of the physical possibilities of a theory for averages over curves. The problem of turbulence was too complicated for immediate attack, but there was a related problem which I found to be just right for the theoretical considerations of the field which I had chosen for myself. This was the problem of the Brownian motion, and it was to provide the subject of my first major mathematical work.

Russell's introduction to Einstein's work on Brownian motion, Hardy's lectures on the Lebesgue theory, Barnett's exposé on the recent development of averaging curves via measure, and Taylor's paper on turbulence converged to Wiener's first mathematical research on random fluctuations.

Brownian Motion and Differential Space

Wiener began to work on the problem of finding a functional's average circa 1919. A notion dating back to the eighteenth century, a functional refers to a mathematical relationship that takes a function (or functions in the case of multiple arguments) rather than a number as an argument and maps it onto a range of numbers. For instance, if a one-dimensional rod has a mass-density profile $m(x)$ over position x, then the location of its center of mass is determined by the entire density profile. That is, the location of the rod's center of

[16] Ibid., 247.
[17] Ibid., 247.

mass $x_c = M[m(x)]$ is a functional that takes the mass-density profile $m(x)$ as its argument.

For an ordinary function $f(u)$, the calculation of its average is straightforward, as long as we have a "weight" $w(u)$ corresponding to each value of u: the mean is $\langle f(u) \rangle = \sum_i f(u_i)\, w(u_i)$ for discrete u and $\langle f(u) \rangle = \int f(u)\, w(u)\, du$ for continuous u. For a functional $F[f(u)]$, the calculation of its average requires a weight corresponding to each function $f(u)$, which is far from clear when u takes continuous values.

By the time when Wiener began to inquire into this subject, an English mathematician P.J. Daniell teaching at Rice Institute of Technology in Houston had developed a theory to deal with the average of a functional $F[f(t)]$ for continuous t. Daniell's approach was to sample the function $f(t)$ at $t = t_1, t_2, \ldots t_N$. The functional $F[f(t)]$ was approximated as a function with discrete arguments $F\left[f(t)\right] \cong \Phi\left[f(t_1), f(t_2), \ldots, f(t_N)\right]$. The average of this function was easy to obtain, since it was straightforward to specify the weight corresponding to each individual $f(t_i)$:

$$\langle \Phi\left[f(t_1), f(t_2), \ldots, f(t_N)\right] \rangle = \int d\left[f(t_1)\right] w\left[f(t_1)\right] \ldots \int d\left[f(t_N)\right] w\left[f(t_N)\right] \times \Phi\left[f(t_1), f(t_2), \ldots, f(t_N)\right]. \tag{9.1}$$

Daniell's theory asserted that when sampling was infinitely dense in t so that N was infinite, this discretized average would converge to the average of the functional $\langle F\left[f(t)\right] \rangle$. Through Barnett's introduction, Wiener's initial work on the subject followed closely this theory. Published in 1920, his first paper on the average of a functional provided a minor revision of Daniell's approach.[18]

Yet, Wiener swiftly changed his direction of thought, as he steered his focus onto the connection between this abstract mathematical problem and a concrete physical scenario. In Daniell's theory, the function $f(t)$ could take any form. By 1921, Wiener narrowed down to a particular class of functions: Brownian motion. Suppose a body of particles undertook one-dimensional Brownian motion. The displacement of each particle over time t was represented by a function $x = f(t)$. The family of all functions corresponding to the Brownian particles' trajectories under the same physical state (temperature, viscosity, etc.) constituted a domain in which a functional could be defined and an average could be performed. When an average was calculated in this domain, there was a natural definition of the "weight" associated with each curve: the weight could be construed as the curve's probability. To advance

[18] Norbert Wiener, "The mean of a functional of arbitrary elements," *Annals of Mathematics*, 2:22 (1920), 66–72.

this mathematical framework, Wiener invoked Einstein's theory of Brownian motion. He reiterated the longstanding wisdom that the probability for the displacement of a Brownian particle $x = f(t)$ to fall on a value must be independent of:[19]

- the position from which the particle starts to wander,
- the time when the particle starts to wander, and
- the direction in which the particle wanders.

As Einstein, Smoluchowski, and Langevin had found, these "memoryless" properties of Brownian motion led to a theorem that the probability for the displacement $x = f(t)$ at time t to be between x_0 and x_1 was an integral over a Gaussian distribution with a variance proportional to the elapsed time: $\left(1/\sqrt{\pi ct}\right)\int_{x0}^{x_1} e^{-(x^2/ct)}dx$, where c was a constant coefficient. They had also shown that the probability of the displacement at time $t_1 + t_2$ being between x_0 and x_1 was an integral of the same form, albeit with a variance proportional to $t_1 + t_2$.[20]

With these results, Wiener considered the assemblage of all Brownian curves $\{f(t)\}$ between $t = 0$ and $t = 1$ that began with $f(0) = 0$. And he considered a functional $F[f]$ that could take any curve from this assemblage. His aim was to compute the average of this functional F over the assemblage. According to Daniell's theory, Wiener could have selected a discrete set of sampling instants $t_1, t_2, \ldots t_N$ between $t = 0$ and $t = 1$, taken an average of F over $f(t_1), f(t_2), \ldots, f(t_N)$, and obtained the functional's average as a limit of the discrete case when N became infinite. Because of a special property of Brownian motion, however, Wiener pursued an alternative scheme: He noted that for Brownian motion, $f(t_1), f(t_2), \ldots, f(t_N)$ were not statistically independent of one another, but $f(t_1), f(t_2) - f(t_1), \ldots, f(t_N) - f(t_{N-1})$ were. The "weights" associated with the Brownian trajectories satisfying $x_1 = f(t_1), x_2 = f(t_2), \ldots, x_N = f(t_N)$ equaled the probability density for these curves. This probability density was separable into factors corresponding to the Gaussian distributions for $x_1, x_2 - x_1, \ldots, x_N - x_{N-1}$:

$$\left(1/\sqrt{\pi^N c^N t_1 (t_2 - t_1)\ldots(t_N - t_{N-1})}\right) e^{-(x_1{}^2/ct_1)} e^{-[(x_2-x_1)^2/c(t_2-t_1)]} \ldots e^{-[(x_N-x_{N-1})^2/c(t_N-t_{N-1})]}.$$

[19] Norbert Wiener, "The average of an analytical functional," *Proceedings of the National Academy of Science*, 7:9 (1921), 253.
[20] Ibid., 253.

To start with a simple case, Wiener first supposed that the functional depended only on $f(t)$ at $t_1, t_2, \ldots, t_N$, so that $F[f(t)] = \Phi[f(t_1), f(t_2), \ldots, f(t_N)] = \Phi(x_1, x_2, \ldots, x_N)$. Then Wiener replaced (9.1) from Daniell's theory with another integral to express the functional's average:

$$\begin{aligned} A\{F\} &= \frac{1}{\sqrt{\pi^N c^N t_1 (t_2 - t_1) \ldots (t_N - t_{N-1})}} \int_{-\infty}^{\infty} dx_1 \ldots \int_{-\infty}^{\infty} dx_N e^{-\frac{x_1^2}{ct_1}} e^{-\frac{(x_2-x_1)^2}{c(t_1-t_2)}} \\ &\ldots e^{-\frac{(x_N-x_{N-1})^2}{c(t_N-t_{N-1})}} \times \Phi(x_1, x_2, \ldots, x_N) \\ &= \frac{1}{\sqrt{\pi^N c^N t_1 (t_2 - t_1) \ldots (t_N - t_{N-1})}} \int_{-\infty}^{\infty} dy_1 \ldots \int_{-\infty}^{\infty} dy_N e^{-\frac{y_1^2}{ct_1}} e^{-\frac{y_2^2}{c(t_1-t_2)}} \\ &\ldots e^{-\frac{y_N^2}{c(t_N-t_{N-1})}} \times \Phi(y_1, y_1 + y_2, \ldots, y_1 + y_2 + \ldots + y_N) \end{aligned} \tag{9.2}$$

where $y_1 = x_1, y_2 = x_2 - x_1, \ldots, y_N = x_N - x_{N-1}$.[21]

In general, the functional $F[f(t)]$ did not necessarily depend only on f's discrete samples $f(t_1), f(t_2), \ldots, f(t_N)$. Wiener was more interested in the broader condition in which $F[f(t)]$ was an analytical functional and expressible with a generalized Taylor series:

$$F[f] = a_0 + \int_0^1 f(t_1)\varphi_1(t_1)\,dt_1 + \int_0^1 \int_0^1 f(t_1)f(t_2)\varphi_1(t_1, t_2)\,dt_1 dt_2 + \ldots$$

The average of the functional thus became

$$\begin{aligned} A\{F[f]\} &= a_0 + \int_0^1 A\{f(t_1)\}\varphi_1(t_1)\,dt_1 \\ &+ \int_0^1 \int_0^1 A\{f(t_1)f(t_2)\}\varphi_1(t_1, t_2)\,dt_1 dt_2 + \ldots \end{aligned} \tag{9.3}$$

And the average of the random variables $f(t_1), f(t_1)f(t_2), \ldots$ could be obtained from (9.1).[22]

To Wiener, the average of the analytical Brownian functionals, as described in (9.3), constituted an algebraic operation with several properties, includ-

[21] Ibid., 253–254.
[22] Ibid., 255–256.

ing: $A\{F_1\} + A\{F_2\} = A\{F_1 + F_2\}$, $A\{cF\} = cA\{F\}$, $A\{\sum_{n=1}^{\infty} F_n\} = \sum_{n=1}^{\infty} A\{F_n\}$, $A\{\int_a^b F_u du\} = \int_a^b A\{F_u\} du$, and that the functional average for discrete samples in (9.2) converged to the average of a continuous functional $A\{F\}$ when N became infinite. To attain mathematical rigor, Wiener devoted effort to proving these properties of the functional average defined over Brownian curves.[23]

Wiener developed this scheme in 1921 to tackle Brownian motion. He immediately applied the method to compute Brownian particles' average displacements as a function of time.[24] Yet, his real heart was not in the question of trajectory tracing that concerned Uhlenbeck and other physicists. Rather, Wiener was more interested in using his newly found method to explore the abstract representational structure of Brownian motion. In 1923, he published an article titled "Differential-space" to address this problem.

Wiener began to conceive the notion of differential space in 1920, when he went to study briefly with the French mathematician Maurice Fréchet before the International Mathematical Congress in Strasbourg. Fréchet wanted Wiener to work on much more generic Banach space that aligned more closely with "pure" mathematics. Wiener did not believe that "Fréchet appreciated the importance of differential space when I first mentioned the theory to him." But the senior mathematician introduced Wiener to Paul Lévy, who was teaching at l'École Polytechnique in Paris and a leading expert on measure theory of probability. Unlike Fréchet, Lévy quickly recognized the significance of Wiener's notion of differential space, and would become one of his closest friends and supporters.[25]

Wiener's concept of differential space was a point of contact between his work on Brownian motion and Daniell's and Lévy's theories of functional analysis. According to them, the mean value of a functional $U[x(t)]$ over continuous functions $x(t)$ between $t = 0$ and $t = 1$ was the result of the limit of a similar average over discrete samples $x(t_1)$, $x(t_2)$, ..., $x(t_N)$ for $0 < t_1 < t_2 < ... < t_N$. Following this logic, Wiener considered again the assemblage of Brownian motions $\{f(t)\}$ for t between 0 and 1 and $f(0) = 0$. He divided the interval [0,1] into N equal parts and let $t_n = n/N$. A Brownian trajectory $f(t)$ could be approximately represented with a denumerable set $\{f(0), f(1/N), f(2/N), ..., f((N-1)/N), f(1)\}$. Yet, this representation was not convenient, since the members of the set were statistically dependent upon one another

[23] Ibid., 256–260.

[24] Norbert Wiener, "The average of an analytic function and the Brownian movement," *Proceedings of the National Academy of Science*, 7:10 (1921), 294–298.

[25] Wiener, *Life in Cybernetics* (2018), 267.

in general. Instead, Wiener claimed, a more appropriate representation comprised another set with statistically independent terms:

$$\begin{aligned} y_1 &= f\left(\frac{1}{N}\right) - f(0) \\ y_2 &= f\left(\frac{2}{N}\right) - f\left(\frac{1}{N}\right) \\ y_N &= f(1) - f\left(\frac{N-1}{N}\right) \end{aligned} \tag{9.4}$$

Here, each element was the *difference* between the curve's values at two adjacent sampling instants. In other words, a Brownian curve $f(t)$ was approximately represented by $[f(1/N)-f(0), f(2/N)-f(1/N), \ldots, f(1)-f((N-1)/N)]$ $(f(0)=0)$. This N-tuple could be understood as a point in the N-dimensional space spanned by $\{y_1, y_2, \ldots, y_N\}$. Taking N to infinity, the N-tuple representation became exact, and a particular trajectory of a Brownian particle could be expressed as a point in this infinite-dimensional space. The whole assemblage of Brownian curves mapped onto it. Wiener coined the name "differential space" for the space specified by $f(1/N) - f(0)$, $f(2/N) - f(1/N), \ldots, f(1) - f((N-1)/N)$ when N became infinity.[26]

Wiener's crucial insight on differential space was that for Brownian curves $f(t)$, it was not the values of $f(t)$ at distinct sampling times that were uniformly and independently distributed, but their differences. These differences acted as dimensions for the space that represented the Brownian curves.[27]

Moreover, the formulation of differential space allowed Wiener to pose a definition of probability compatible with the definition of measure in Lebesgue's theory. Since the differential space spanned by $\{y_1, y_2, \ldots, y_N\}$ was uniform, homogeneous, and independent along any dimension, the probability for a set of Brownian curves should equal the set's hyper-volume in the space. This hyper-volume was a Lebesgue measure.[28]

This probability measure became a powerful tool for Wiener to unveil mathematical properties of Brownian motion. Following Lévy, Wiener equated the probability of the quantity $f(a) - f(0)$ falling between α and β to the Lebesgue measure of the region corresponding to $\alpha \leq f(a) - f(0) \leq \beta$ on the hyper-sphere $\sum_{n=1}^{N} {y_n}^2 = Nr^2$ when N became infinite. He noted that $f(a) - f(0) = \sum_{n=1}^{N_a} y_n$. With a change of coordinates, he made $f(a) - f(0)$ one of the dimensions. And

[26] Norbert Wiener, "Differential-space," *Journal of Mathematics and Physics*, 2 (1923), 131–136.
[27] Ibid., 137.
[28] Ibid., 135–136.

the measure of the region of interest under the new coordinate system became a simpler hyper-volume integral. Carrying out this integral, Wiener obtained the result that the measure (probability) for $\alpha \leq f(a) - f(0) \leq \beta$ was

$$\frac{1}{r\sqrt{2\pi a}} \int_{\alpha}^{\beta} e^{-\frac{u^2}{2ar^2}} du.$$

This matched the central outcome of the Einstein–Smoluchowski theory of Brownian motion.[29]

After establishing the normal distribution, Wiener turned to the probability density of the hyper-sphere's radius square $q \equiv \sum_{k=1}^{N} {y_k}^2 = \sum_{k=1}^{N} \{f(k/N) - f((k-1)/N)\}^2$. Again, the probability that q fell between α and β was the Lebesgue measure of the corresponding region in the differential space. Following a similar procedure of coordinate change, he obtained a more complicated form of integral. He carried out the integral, and found that for a very large N, q was almost certainly Nr^2, the number of dimensions times a constant. The probability for q to be otherwise converged to 0 as N became infinite. This discovery legitimized his earlier choice to treat the hyper-sphere $\sum_{n=1}^{N} {y_n}^2 = Nr^2$ as the sampling space: all Brownian curves in the differential space gravitated toward the surface of the hyper-sphere as the dimensions became infinite. This assertion was not a surprise, if we consider Brownian trajectories as random walks and employ the law of large numbers.[30]

So far, Wiener's mathematical tool of differential space had been used to represent Brownian motions in an alternative framework and to reconstruct their known properties. But these elements did not exhaust the significance of his theory. The most famous product of Wiener's differential space was his discovery of a *new* property of Brownian motion. His intuitive idea about this property came from an experimental observation. Quoting Perrin's claim about the irregularity of Brownian motion, Wiener stated that:[31]

> One realizes from such examples how near the mathematicians are to the truth in refusing, by a logical instinct, to admit the pretended geometrical demonstrations, which are regarded as experimental evidence for the existence of a tangent at each point of a curve.

[29] Ibid., 136–137.
[30] Ibid., 138–143.
[31] Ibid., 133.

According to Wiener's interpretation of Perrin's claim, Brownian motions might provide a case for the kind of curves (functions) that were continuous but did not have a well-defined tangent (derivative). Mathematicians might have contemplated about such pathological functions. But Brownian motion could be a physical realization of them.

The non-differentiability of Brownian motions was deceptively simple to demonstrate under Wiener's differential space. Again, he considered $q = \sum_{k=1}^{N} y_k{}^2 = \sum_{k=1}^{N} \{f(k/N) - f((k-1)/N)\}^2$. If $f(t)$ had a first-order derivative everywhere between $t = 0$ and $t = 1$, then it must be smooth enough so that the maximum of $|f(k/N) - f((k-1)/N)|$ over all k's existed and was finite. Thus, $\sum_{k=1}^{N} \{f(k/N) - f((k-1)/N)\}^2 \leq \max_k |f(k/N) - f((k-1)/N)| \sum_{k=1}^{N} |f(k/N) - f((k-1)/N)|$. Moreover, the finite derivative between $t=0$ and $t=1$ implied that $|f(k/N) - f((k-1)/N)| \leq T/N$, where T was the maximum derivative of f between $t=0$ and $t=1$. This led to $\sum_{k=1}^{N} \{f(k/N) - f((k-1)/N)\}^2 \leq \max_k |f(k/N) - f((k-1)/N)|\, T$. When N approached infinity, $\max_k |f(k/N) - f((k-1)/N)|$ became 0. This meant that $\lim_{N\to\infty} \sum_{k=1}^{N} \{f(k/N) - f((k-1)/N)\}^2 = 0$. Yet, he had just established that $\lim_{N\to\infty} (1/N) \sum_{k=1}^{N} \{f(k/N) - f((k-1)/N)\}^2 = r^2$. To avoid the contradiction, Wiener concluded, the assumption that $f(t)$ was differentiable everywhere between $t = 0$ and $t = 1$ must not hold. The same reasoning could be used to prove that the condition of differentiability was invalid for *any* interval between $t = 0$ and $t = 1$. Thus, one could not find any differentiable segment in a Brownian curve. Brownian motion was continuous but non-differentiable. Perrin was right.[32]

After his 1922 paper on differential space, Wiener furthered the research on the subject and introduced a scheme of dividing the range of $x = f(t)$ into a nested $\{I_n\}$, where the sets $I_1, I_2, \ldots, I_N$ partitioned the range $[x = -\infty,\ x = \infty]$, and I_{n+1} was embedded within I_n. Using this division, he showed that the average of the discretized functional would converge to the average of the functional when N became infinite.[33]

Wiener's proof of Brownian motion's non-differentiability exemplified the relevance of constructing the meta-structure of random fluctuations to further understanding of their properties. This was a finding about Brownian motion that could not be made obvious through physicists' framework of the

[32] Ibid., 143.

[33] Norbert Wiener, "The average value of a functional," *Proceedings of the London Mathematical Society*, 22 (1924), 454–467.

Fokker–Planck equation, Langevin equation, and Markov process. The Lebesgue theory of measure and the average of a functional in "pure mathematics" played a significant part in this discovery.

Shortly after Wiener developed his theory of Brownian motion, he shared the results with a few colleagues and discussed with them plausible directions to proceed. As soon as he published his 1922 paper on differential space, he sent it to Lévy. The French mathematician replied in November with an enthusiastic tone:[34]

> It's my greatest pleasure to learn about the result of your research. It verifies once again that it is by physics that mathematics is destined to innovate. [Il se vérifie une fois de plus que c'est par physique que les mathématiques sont destinées à se renouveller.] The studies of the Brownian motion you have brought pose a problem that is analogous to the one I have treated, but also with important differences.

In a follow-up letter in July 1923, Lévy further inquired whether Wiener had considered constructing a theory based on the difference not between a Brownian particle's positions at adjacent instants $x(t_{n+1}) - x(t_n)$, but between its velocities at adjacent instants $v(t_{n+1}) - v(t_n)$.[35] Wiener also sent his 1924 paper "The average value of a functional" to G.I. Taylor. Admitting that he had not known about Einstein's work on Brownian motion, Taylor wrote back to Wiener to discuss the possibility of extending the latter's method. In particular, the English mathematician did not understand what Wiener meant when he posed the condition that the probability for a Brownian particle to wander a given distance in a given time was independent of the *direction* in which it wandered.[36] Taylor also wanted to know whether Wiener was considering another type of motion in which the segments between successive flights were correlated to each other.[37]

Responses of this kind were rare. As a biographer of Wiener indicated, "the novelty of his Brownian motion theory was such that it was not widely appreciated at the time, and the few who did, such as H. Cramer in Sweden and P. Lévy in France, were outside the United States."[38] When Wiener built up

[34] Letter, Paul Lévy to Wiener, November 21, 1922, Box 2, Folder 23, "Correspondence 1922," Wiener Collection; original in French, my translation.

[35] Letter, Lévy to Wiener, July 29, 1923, Box 2, Folder 24, "Correspondence 1923," Wiener Collection.

[36] Letter, G.I. Taylor to Wiener, August 4, 1924, Box 2, Folder 25, "Correspondence 1924," Wiener Collection.

[37] Letter, Taylor to Wiener, October 6, 1924, Box 2, Folder 25, "Correspondence 1924," Wiener Collection.

[38] Segal, "Wiener," 598.

his reputation as a mathematician in the 1920s, he was much better known by his peers from his work on more classical topics such as potential theory. The situation started to change toward the end of this decade, as more mathematicians looked into the formulation of probability and random fluctuations through more rigorous lenses of set theory, measure theory, and functional analysis. Andrey Kolmogorov's formal representation of stochastic processes in 1931[39] and his canonical definition of probability in terms of set-theoretical axioms on a Lebesgue measure in a sample space in 1933[40] constituted a major milestone in this development. By the 1930s, Wiener's work on Brownian motion gained a wider acknowledgment, thanks to this trend in mathematical research.

Retrospectively, Wiener's work would be viewed as one of the first to give a formal specification of a stochastic process; and Brownian motion, later known as the "Wiener process," was one of the first explicitly defined stochastic processes. Yet, Wiener's angle was not a broad conceptualization of all random processes—a project more suitable to individuals like Kolmogorov. Rather, the MIT mathematician's agenda was to scrutinize a particular kind of random fluctuations—Brownian motion—and characterize it with the daunting mathematical tools he developed. No matter how abstract Wiener's work became, the source of inspiration and, as Lévy remarked, the destiny of mathematical innovation were considerations of physical scenarios.

Years later, Wiener reflected on the nature of his work on Brownian motion and differential space in relation to a broader intellectual movement. He maintained that since the late nineteenth century, some mathematicians attempting to unveil the foundation of their undertakings had held a "postulationist" conviction that the whole system of mathematical knowledge could be deduced from a set of explicitly stated general principles. On the other hand, some others had held a "constructionist" conviction and believed that abstract mathematical knowledge was constructed piece by piece from studies of more concrete situations. Wiener admitted that he was highly influenced by the postulationist tradition of Whitehead, Russell, and Hilbert in his training as a mathematician. But he often made contributions to mathematics through the constructionist route, as he tended to work on specific scenarios—some of which were closely connected to physics and engineering—and extended the outcomes to more abstract forms. His theory of differential space for

[39] Andrei Kolmogorov, "Über die analytischen Methoden in der Wahrscheinlichkeitsrechnung," *Mathematische Annalen*, 104 (1931), 415–458.

[40] Andrei Kolmogorov, *Foundation of the Theory of Probability*, translated by Nathan Morrison (New York: Chelsea, 1956).

Brownian motion exemplified this approach: starting with a consideration of suspended particles in a fluid, he arrived at a conclusion concerning the general formulation of stochastic processes in an infinite-dimensional functional space.[41]

Generalized Harmonic Analysis

When Wiener was working on a theory of differential space for Brownian motion in the first half of the 1920s, the problem of spectral analysis for irregular waveforms also came to his attention. In his 1923 paper on differential space, he devoted a section to the functional average over the trajectories' Fourier series.[42] The step from inspecting the meta-structure of the functional space spanned by Brownian motion to studying the frequency components of random time series looked straightforward. Yet, Wiener's move toward spectral analysis was shaped by multiple factors. Above all, the engineering environment at MIT put demands on him to work on problems in applied mathematics. His examination of the mathematical basis of Heaviside's operational calculus—a system of notations to express derivatives and integrals in terms of operators similar to algebraic symbols—was a response to requests from colleagues in electrical engineering. Although his research on potential theory drew on a longstanding research tradition in mathematical analysis, its practical implications for electrical technology were clear.

As a math professor at MIT, Wiener was expected to assist his colleagues tackling theoretical issues in radio and telephony. The head of the Department of Electrical Engineering Dugald Jackson asked for Wiener's help with sorting out a proper foundation for communication theory. Doing so required a detailed analysis of signals and interferences in the frequency domain. Jackson's request prompted Wiener to think carefully about random noise's spectra, and to connect with the issues for engineering researchers undertaking the frequency-domain approach to noise.[43]

The spectral analysis of random noise was familiar to electrical engineers. As we have seen in chapter 8, their approach was either to suppose that the noise had the same power or energy at all frequencies (a flat spectrum like shot noise or thermal noise), or to assume that noise was a sinusoid with a random intensity, frequency, and phase (as in the estimate of FM's

[41] Wiener, *Life in Cybernetics* (2018), 260; Wiener et al., *Differential Space, Quantum Systems, and Prediction* (1966), 11–14.

[42] Wiener, "Differential-space" (1923), 170–174.

[43] Wiener, *Life in Cybernetics* (2018), 277–278.

signal-to-noise ratio). To Wiener, taking any of these assumptions would overlook an underlying mathematical problem. By the early twentieth century, mathematicians knew of two types of functions susceptible to spectral analysis. The first type was a periodic function $f(t)$ satisfying $f(t+T) = f(t)$ for any t, with period T. This function could be expressed as a discrete sum of harmonic components, each of which had a frequency equaling an integer multiple of a fundamental frequency $(1/T)$: $f(t) = \sum_{n=-\infty}^{\infty} a_n e^{in\omega t}$, where $\omega = (2\pi/T)$, and $a_n = (1/T)\int_{-T/2}^{T/2} f(t)\, e^{-in\omega t} dt$. The denumerable set $\{a_n\}$ constituted $f(t)$'s *Fourier series*. They represented the amplitudes and phases of the function's frequency components.

The second type was a non-periodic function $g(t)$ with finite "energy." That is, the function's square integral over all t's, $\int_{-\infty}^{\infty} [g(t)]^2 dt$, was finite. To have finite energy, this function $g(t)$ must have non-zero values only within a finite duration of t. The spectrum of this kind of functions was expressed in terms of an integral $G(u) = \lim_{A\to\infty} \left(1/\sqrt{2\pi}\right) \int_{-A}^{A} g(t)\, e^{-iut} dt$, which was called a *Fourier integral* or *Fourier transform*. In the 1910s, the Swiss mathematician Michel Plancherel proved that a finite-duration function $g(t)$ could be almost certainly reconstructed from its Fourier integral: $g(t) = \underset{A\to\infty}{\text{l.i.m.}} \left(1/\sqrt{2\pi}\right) \int_{-A}^{A} g(t)\, e^{-iut} dt$, where l.i.m. stood for "limit in mean" (mean over t of the difference between the two functions on both sides of the equation would converge to 0 as A reached infinity).[44]

By the early twentieth century, scientists employed two types of spectral analysis. When they encountered a regular, periodic function, they decomposed it into a *Fourier series* that represented a spectrum of discrete lines. When they dealt with an irregular function of a finite duration, they computed its continuous spectrum with a *Fourier integral.*

Random fluctuations were different from both, though. Radio atmospherics, electronic noise, and Brownian motion were not periodic. So the harmonic analysis in terms of Fourier series could not be applied. Yet, they were not duration-limited, either. Any of these fluctuations could last indefinitely no matter how long the period of observation was. Since they did not have finite "energy," harmonic analysis in terms of a Fourier integral was unavailable, either. Many irregular variations had this property: incoherent white light, chronicle changes of meteorological data, turbulences in a fluid, etc.

[44] Norbert Wiener, "Generalized harmonic analysis," *Acta Mathematica*, 55 (1930), 120–126.

Wiener became aware of the complex issues with harmonic analysis circa 1925. He looked into the relevant studies of shot noise, white light, meteorological data, and turbulence, and found the state of the scholarship confusing. As early as the 1880s, Louis Georges Gouy and British physicist Arthur Schuster assumed that white light—light with a uniform power distribution over frequencies—might be viewed as a long train of irregular pulses. To match experimental data, Gouy represented white light with a Fourier series, while Rayleigh represented it with a Fourier integral. In contrast, Schuster believed that white light's monochromatic components from a grating structure were created, not analyzed, by the device. Schottky asserted that shot noise's power spectrum was a flat continuous curve, while Fry denied the plausibility of spectral analysis for electronic noise.[45] In a letter to his sister Bertha on July 21, 1925, Wiener wrote that "the theory of white light is an awful mess. Nobody rightly knows what spectrum analyzer means, mathematically speaking." Although Poincaré, Rayleigh, Gouy, and Schuster attempted to resolve the issues, Wiener continued, "they themselves didn't have the mathematical tools to do a clean up." He reported to his sister, with pride, that he was looking closely into "Woods' *Physical Optics*, chapter on white light. I have an absolute clean up on that. So your little brother is a physicist."[46]

Wiener's "absolute clean up" was a scheme of harmonic analysis that differed both from Fourier series and from Fourier integral. He published his first paper on the subject in 1925.[47] His point of departure was the work of the Danish mathematician Harald Bohr (brother of the leading quantum physicist Niels Bohr) on "almost periodic" functions. There were numerous phenomena in dynamics (e.g. a planet's orbit) in which variations of relevant quantities seemed to follow regular patterns but did not come back exactly to the same state in any periodic fashion. Aiming to unveil the spectral characteristics of such patterns, Bohr attempted to represent them with integrals of trigonometric functions. This was also Wiener's approach to harmonic analysis of irregular motions in the mid-1920s. The functions $f(t)$ he considered were neither periodic nor duration-limited but permitted the existence of a mean square $\lim_{A\to\infty} (1/2A) \int_{-A}^{A} [f(t)]^2 dt = M\{[f(t)]^2\}$. This kind of functions could not be represented by a Fourier-series expansion in general; and their Fourier

[45] Ibid., 127; Norbert Wiener, "The harmonic analysis of irregular motion (second paper)," *Journal of Mathematics and Physics*, 5 (1926), 182–183.

[46] Letter, Norbert Wiener to Bertha Wiener, July 21, 1925, Box 2, Folder 26, "Correspondence January-July 1925," Wiener Collection.

[47] Norbert Wiener, "On the representation of functions by trigonometrical integrals," *Mathematical Zeitschrift*, 24 (1925), 575–617.

integrals might not exist. Yet, Wiener found two well-defined functions that allowed an expression of $f(t)$ in terms of trigonometric integrals:

$$\Gamma(u) = \frac{1}{\pi}\int_{-\infty}^{\infty} f(t)\,\frac{1-\cos(ut)}{t^2}dt,\ \Delta(u) = \frac{1}{\pi}\int_{-\infty}^{\infty} f(t)\,\frac{ute^{-t^2}-\sin(ut)}{t^2}dt. \tag{9.5}$$

He proved that $f(t)$ converged in the mean to the sum of their trigonometric integrals:

$$f(t) = \underset{A\to\infty}{\text{l.i.m.}} \left\{\int_{-A}^{A} \cos(ut)\, d\Gamma'(u) + \int_{-A}^{A} \sin(ut)\, d\Delta'(u)\right\}, \tag{9.6}$$

where $\Gamma'(u)$ and $\Delta'(u)$ were the derivatives of $\Gamma(u)$ and $\Delta(u)$ respectively; $d\Gamma'(u)$ and $d\Delta'(u)$ were the differentials in the Stieltjes integrals. To him, $\Gamma''(u)$ and $\Delta''(u)$ represented $f(t)$'s power spectrum, since intuitively,

$$f(t) = \underset{A\to\infty}{\text{l.i.m.}} \left\{\int_{-A}^{A} \cos(ut)\,\frac{d^2\Gamma(u)}{du^2}du + \int_{-A}^{A} \sin(ut)\,\frac{d^2\Delta(u)}{du^2}du\right\}.$$

To corroborate this theory, he considered the case when $f(t)$ was a finite sum of trigonometric functions. Computing $\Gamma(u)$ and $\Delta(u)$ in this case indeed led to a discrete power spectrum usually attributed to periodic functions.[48]

Next, Wiener examined a linear transformation of an irregular function $f(t)$, $h(t) = \int_{-\infty}^{\infty}\Phi(t-\tau)f(\tau)\,d\tau$. Whether the spectrum of $h(t)$ existed and its form (if it existed) depended on the irregular motion $f(t)$ and the impulse response of the linear transformation $\Phi(t)$. To Wiener, the most meaningful way to proceed was to assume the form $f(t)$ that properly represented irregular motions in natural circumstances. Here his prior work on differential space for Brownian motion became relevant. In his 1923 theory, within the interval $t \in [0,1]$ a Brownian motion $f(t)$ starting from t_1 and ending at t_2 had a Gaussian distribution depending only on $t_2 - t_1$. Generalized harmonic analysis nonetheless required consideration of functions defined on the whole real axis, $t \in [-\infty,\infty]$. Based on $f(t)$, Wiener constructed a function $g(t)$ that satisfied his previous conditions for Brownian motion but was defined for $t \in [-\infty,\infty]$. With this fluctuation, he went through involved mathematical manipulations to prove that the corresponding linear transformation

[48] Norbert Wiener, "The harmonic analysis of irregular motion," *Journal of Mathematics and Physics*, 5 (1926), 99–121.

$h(t) = \int_{-\infty}^{\infty} \Phi(t-\tau) g(\tau)\, d\tau$ had well-defined spectral functions as described in (9.5) and (9.6), which took the form when combining together:

$$R\{h; u\} = \frac{r^2}{\pi} \int_{-u}^{u} \frac{[\psi(\nu)]^2}{\nu^2} d\nu, \tag{9.7}$$

where $\psi(\nu) = \int_{-\infty}^{\infty} \Phi(\tau) e^{-i\nu\tau} d\tau$ was the Fourier integral of the linear transform's impulse response, and r^2 marked the magnitude of g's mean square.[49]

Wiener believed that these results were particularly useful in practical applications. In his own words:[50]

> A linear system such as a resonator may be regarded as generating a linear transformation of one function, representing the impressed displacement, into another function, representing the emergent displacement. For example, the horn of the phonograph generates a transformation of the displacement of the diaphragm into the displacement of the air at a specified place. Precisely similar cases occur in electrical systems.

The linear system characterized by $\Phi(t)$ could be an acoustic resonator, an electronic filter or amplifier, or an optical grating. Taking random fluctuations $g(t)$ like cacophony, shot noise, or white light as the input, the linear system would produce an output whose spectral properties could be determined by (9.7). As Wiener pointed out, a particular application of this theory was to consider "the response of an acoustic resonator to a noise consisting of a series of sharp impulses to the air, such as those produced by the grains of sand in a sand-blast." Another was to examine shot noise—also a train of random pulses in electric current—from a vacuum tube passing through "a linear system of inductances, capacities, and resistances" that formed the frequency-selective part of an electronic amplifier circuit. In both cases, he claimed, the input noise resembled Brownian motion. And (9.7) led to

$$R\{h; u\} = \frac{r^2}{\pi} \int_{-u}^{u} \left[\int_{-\infty}^{\infty} \Phi(t)\, e^{i\omega t} \right]^2 d\omega. \tag{9.8}$$

Equation (9.8) showed that when shot noise passed through a filter or when some grinding noise passed through an acoustic resonator, the spectrum of its output was basically the spectrum of the filter's or resonator's impulse

[49] Wiener, "Harmonic analysis of irregular motion (second paper)," 159–180.
[50] Ibid., 180.

response, $[\int_{-\infty}^{\infty} \Phi(t)\, e^{i\omega t}]^2$. This conclusion was consistent with engineering researchers' common treatment of random noise as irregular motions with a "flat" spectrum, so that the spectrum of its output from a frequency-selective device was identical to the device's own frequency response. Wiener provided a mathematical ground for this conviction.[51]

Moreover, Wiener contended that his theory could also be used to understand the nature of white light's incoherence. Physicists had found that unlike monochromatic light, white natural light was not coherent, and could not add together directly. If two light beams were monochromatic and coherent to each other in the form of $h_1(t) = A_1 \cos(\omega_1 t + \phi_1)$ and $h_2(t) = A_2 \cos(\omega_2 t + \phi_2)$, then the intensity of their superposition would be the magnitude of the direct sum $h_1 + h_2$. If the two beams were white light, then h_1 and h_2 were incoherent to each other, meaning that they did not have a fixed phase relationship. Instead of $|h_1 + h_2|$, the intensity of the two superposed beams should be $\sqrt{|h_1|^2 + |h_2|^2}$. Wiener was convinced that his theory could offer an account of this longstanding observation. To him, white natural light resembled a Brownian motion or shot noise, and the optical device for intensity measurement was a linear system. For two beams of white light, the output from the device was $\int_{-\infty}^{\infty} \Phi(t-\tau)\, g_1(\tau)\, d\tau + \int_{-\infty}^{\infty} \Phi(t-\tau)\, g_2(\tau)\, d\tau$. Its square intensity was $\{\int_{-\infty}^{\infty} \Phi(t-\tau)\, [g_1(\tau) + g_2(\tau)]\, d\tau\}^2$, equaling $\{\int_{-\infty}^{\infty} \Phi(t-\tau)\, g_1(\tau)\, d\tau\}^2 + \{\int_{-\infty}^{\infty} \Phi(t-\tau)\, g_2(\tau)\, d\tau\}^2 + 2\int_{-\infty}^{\infty} \Phi(t-\tau)\, g_1(\tau)\, d\tau \int_{-\infty}^{\infty} \Phi(t-\tau')\, g_2(\tau')\, d\tau'$.

Because of the properties of g_1, g_2, and Φ, when taking the mean $\lim_{A\to\infty} (1/2A) \int_{-A}^{A} d\tau$ over the light square intensity, the third term involving g_1 and g_2 would disappear asymptotically. This explained why for incoherent white light, the formula was $\sqrt{|h_1|^2 + |h_2|^2}$, not $|h_1 + h_2|$.[52]

Wiener's theory of generalized harmonic analysis opened a new direction to explore the frequency responses of irregular functions and their plausible applications to material scenarios involving random fluctuations. But his spectral formulation in (9.5) was complicated. To derive useful results from this expression required daunting mathematical efforts. Within a year, he embarked upon a revision of this theory. The drive for change came from new stimulations. In 1925, Wiener obtained a Guggenheim Fellowship, which gave him a chance to visit the University of Göttingen for one year. There, he had an opportunity to work with some leading German mathematicians: David Hilbert, Edmund Landau, and Richard Courant. Through correspondence

[51] Ibid., 180–182.
[52] Ibid., 185.

and short-term visits, he also resumed closer contacts with his previous collaborators in England and France, including Hardy, Taylor, and Lévy. Wiener talked formally and informally with these mathematicians about his work on generalized harmonic analysis and obtained feedback.[53]

Meanwhile, Wiener was exposed to a few works on mathematics of random fluctuations; all of which pointed to a scheme to identify irregular signals' periodicities. He found from Taylor's 1920 article on turbulence the notion of "correlation between a function and itself under displacement" which could be used to study irregular motions in a fluid.[54] Likely through the Göttingen professors' introduction, Wiener came across the publications by Hermann Weyl—a graduate of Göttingen under Hilbert's supervision who was then teaching at Zurich—that used the same correlation to study almost periodic functions.[55] Perhaps the strongest influence came from Wiener's awareness of a statistical tool developed by Schuster.[56] Dubbed "periodogram," the tool was used by Schuster to find hidden periodicities of meteorological data. For time-series data represented by a function $f(t)$, the periodogram was a plot of $f(t)$'s correlation function:

$$\varphi(t) \equiv \lim_{A\to\infty}\frac{1}{2A}\int_{-A}^{A} f(t+\tau)f^{*}(\tau)\,d\tau, \tag{9.9}$$

where f^{*} was f's complex conjugate.

With respect to the correlation of an irregular function, Wiener defined[57]

$$S(\omega) \equiv \frac{1}{2\pi}\int_{-\infty}^{\infty} \varphi(t)\,\frac{e^{i\omega t}}{it}dt. \tag{9.10}$$

What followed from this definition in 1925 was a *tour de force* for the proofs of existence and convergence utilizing the intricate tools of Tauberian theorems in mathematical analysis. His central task was to replace (9.5) and (9.8) with (9.10) for generalized harmonic analysis of irregular functions. In other words, he argued, the trigonometric integral of an irregular function's correlation in (9.10) adequately characterized its spectrum. For instance, he demonstrated that when $f(t)$ was the sum of a series of sinusoids $f(t) = \sum_{n=1}^{N} a_n e^{i\omega_n t}$, $S(\omega)$ resembled a train of stairs that jumped stepwise at $\omega_1 <$

[53] For Wiener's Göttingen year, see Wiener, *A Life in Cybernetics* (2018), 303–312.
[54] Wiener, "Harmonic analysis of irregular motion (second paper)," 185–189.
[55] Norbert Wiener, "The spectrum of an arbitrary function," *Proceedings of the London Mathematical Society*, 2:27 (1928), 484.
[56] Wiener, "Generalized harmonic analysis" (1930), 128.
[57] Ibid., 119.

$\omega_2 < \ldots < \omega_N$. From today's perspective, $S(\omega)$ could be understood as f's cumulative power spectrum that represented its power at frequencies lower than ω. The derivative of $S(\omega)$, $S'(\omega)$, equaled the correlation's Fourier transform $S'(\omega) \equiv \frac{1}{2\pi}\int_{-\infty}^{\infty} \varphi(t)\, e^{i\omega t} dt$, and represented f's power spectrum density. For $f(t) = \sum_{n=1}^{N} a_n e^{i\omega_n t}$, $S'(\omega)$ was a train of impulses at $\omega_1 < \omega_2 < \ldots < \omega_N$.[58]

As early as 1925 in Göttingen, Wiener entertained the idea of using the correlation function for generalized harmonic analysis. In a letter to Bertha, he wrote enthusiastically:[59]

> Schuster—the periodogram & white light man—is interested in my junk. I have also got in touch with a statistician who has done a very similar study independently, but has not published or used it. I lunch with him today. When in London, I am going to look up a guy at the Air Ministry, who is the periodogram expert, and see what practical man's opinion of my generalized periodogram is.

The expert at the British Air Ministry was not the only "practical man" that Wiener consulted. He also wrote to Edmund Wilson, a professor at the School of Public Health, Harvard University. Wilson did not think Wiener's idea too useful, for the available public-health data were not abundant enough to undertake a Wienerian periodical analysis with a meaningful resolution. Wilson also discouraged Wiener from finding doctoral students to work on this project. Instead, he told Wiener that the best place to develop it should be MIT or Göttingen.[60]

Wiener's mathematical treatment of random noise got noticed within the general technical communities. In 1929, George Kenrick at the Department of Electrical Engineering, University of Pennsylvania, utilized Wiener's theory to examine the spectrum of radio statics. Kenrick provided a more explicit visual illustration of how Wiener's concepts of generalized harmonic analysis worked. Wiener's cumulative spectrum $S(\omega)$ (which Kenrick denoted as $\theta(\omega)$) had a sharp jump at $\omega = \omega_0$ when the irregular signal had a frequency component at $\omega = \omega_0$. When the signal comprised a discrete set of frequency components, its correlation $\varphi(t)$ varied periodically and the corresponding $S(\omega)$ comprised a number of "steps" (Figure 9.1).

[58] Ibid., 132–171.

[59] Letter from Norbert Wiener to Bertha Wiener, August 28, 1925, Box 2, Folder 27, "Correspondence August-December 1925," Wiener Collection.

[60] Letter, Edmund Wilson to Norbert Wiener, October 6, 1925, Box 2, Folder 27, "Correspondence August-December 1925," Wiener Collection.

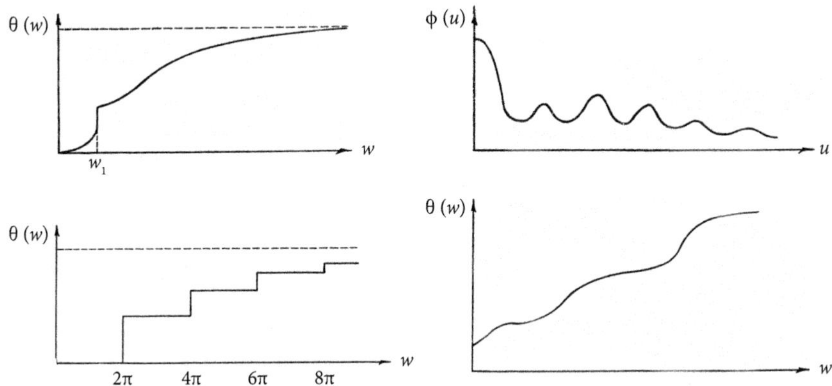

Figure 9.1 George Kenrick's illustration of the cumulative spectrum and correlation in Wiener's generalized harmonic analysis. George Kenrick, "The analysis of irregular motions with applications to the energy-frequency spectrum of static and telegraphic signal," *Philosophical Magazine* (Series 7), 7:41 (1929), 180 (Figure 1), 181 (Figure 2), 183 (Figure 3).

Moreover, Kenrick found Wiener's method useful in the spectral analysis for radio atmospherics. The prevailing approach to statics among electrical engineers at the time, as Kenrick acknowledged, was Carson's approach of modeling with a flat and continuous spectrum. Instead of following Carson, Kenrick inspected the plausible correlation function of statics, which were assumed to be a train of randomly arriving pulses. This assumption made statics a Poisson process, which generated an exponentially decayed correlation and a power spectrum $S'(\omega)$ decreasing monotonically with ω. Kenrick argued that such a pattern was more consistent (than the flat spectrum) with Louis Austin's measured data for atmospherics' power spectrum (chapter 4).[61]

Kenrick's Wienerian approach to noise in radio and telephony remained a minority at the time. Carson was clearly aware of this method, since he was listed as the communicator of Kenrick's paper in *Philosophical Magazine*. But throughout the 1930s, electrical engineers adopted primarily the Carsonian approach to treat noise as an entity with a continuous and flat spectrum, or treated it as an even simpler entity: a sinusoid. In 1934, Aleksandr Khinchin in Moscow came up with a similar theory of spectral analysis via the correlation function, but in terms of a formulation from Kolmogorov's newly developed axiomatic system of probability. Instead of a mean over time as Wiener chose,

[61] George Kenrick, "The analysis of irregular motions with applications to the energy-frequency spectrum of static and telegraphic signal," *Philosophical Magazine* (Series 7), 7:41 (1929), 176–196.

Khinchin's mean was an ensemble average over the sample space of a stochastic process.[62] Wiener's generalized harmonic analysis of irregular waveforms using correlation in (9.9) and (9.10) was placed together with Khinchin's work and given the name of "Wiener–Khinchin theorem." But this theorem was not incorporated into the technical canons of those dealing with random noise in physics and engineering until World War II.

Heterogeneity of a Theoretical Regime

In the 1920s–30s, three technical communities developed theoretical understandings, explanations, representations, and predictions for random fluctuations. Physicists inherited the intellectual tradition of statistical mechanics, specifically, Einstein's, Smoluchowski's, and Langevin's works on Brownian motion. They treated the subject as problems in dynamics, and focused on finding the governing equations of irregular movements under various conditions, ways to solve these equations, construction of the relevant physical entities' temporal evolution in coarse and fine scales, and irregularities' effects on metrological precision. The ensemble average, ergodicity, Fokker–Planck equation, Langevin equation, and equipartition constituted their repertoire.

Electrical engineers were preoccupied with the impacts of random noise on the receiving quality in telephony and radio. Their aim was to assess and reduce such impacts. To meet this goal, they understood fluctuations in the frequency domain, and modeled random noise either as an unwanted signal with a flat spectrum or as a sinusoid with an uncertain amplitude, phase, and frequency. Modeling of this kind helped gain progress in examining the frequency response of the signal-to-noise ratio in various frequency-selective electrical circuits and systems, including an explanation of newly developed FM radio's daunting capacity to suppress atmospherics.

Mathematicians were drawn into the studies of random noise for their potential to shed light on the nature of stochastic processes. Using Brownian motion as a template, they examined the abstract structure of irregular functions and represented them in an infinite-dimensional space. They also aimed at "clearing up" the status of various mathematical and empirical knowledge physicists and engineers held about random noise, in an attempt to build a more rigorous logical foundation for relevant theoretical investigations. This

[62] Alexander Khintchine, "Korrelationstheorie der stationären stochastischen Prozesse," *Mathematische Annalen*, 109:1 (1934), 604–615.

pursuit of a more rigorous basis applied to the case of pinning down the meaning of spectral analysis for haphazard motions.

In the 1920s–30s, the three different communities studying theoretical problems of Brownian motions, electronic noise, radio statics, and random fluctuations at large were not isolated from one another. Their subject matters overlapped. They were aware of the other communities' major works. Cross-community interactions occurred at individual levels. Mutual borrowing of one another's research products could be seen from time to time. Yet, a fuller and more systemic integration of the physical, engineering, and mathematical approaches to noise did not happen until World War II, when researchers from different intellectual traditions and epistemic cultures came to work on the problems of command, control, communication, and computation. The demand from the information warfare related to radar detection, antiaircraft gunfire control, and wireless telephony led to the convergence of the distinct theoretical approaches to noise. That development will be the subject of part IV.

PART IV

NOISE, SIGNAL, INFORMATION, AND WAR

10

Noise in Radar Detection

Ontology of the Ambience

By the late 1930s, the notion of noise as random fluctuations was established among scientists and technologists. They developed three approaches to understanding, representing, and computing it. Rooted in statistical mechanics and taking Brownian motion as a prototype, physicists focused on the temporal evolution of noise, and employed the methods of the Langevin equation, Fokker–Planck equation, and random walks. Concerned with the fidelity of telephony and radio, electrical engineers treated various disturbances as noise with a flat spectrum or a sinusoidal waveform. Inspired by progresses in measure theory and functional theory, mathematicians explored noise as stochastic processes with differential space and generalized harmonic analysis.

This theoretical regime of noise underwent a profound change in the following decade. During the 1940s, physicists', engineers', and mathematicians' disparate inquiries into random fluctuations converged toward the same aims and imperatives. Driven by World War II, these researchers looked closely into the implications of noise for the performance and improvement of military technologies in communication, control, and command in light of their understanding of noise. No matter whether they were interested in suspended particles in a fluid, shot and thermal effects in an electronic circuit, wobbling of a galvanometer, atmospherics plaguing radio broadcasting, differentiability of a Brownian motion, or the spectrum of a random curve, they entered defense research in the 1940s and applied their knowledge about noise to the problems in radar detection, gunfire directing, jamming and anti-jamming, and encrypted wireless telephony.

Whereas interwar research dealt with the nature, causes, and properties of noise, physicists, engineers, and mathematicians under the shadow of World War II paid primary attention to the behavior of systems, machines, and devices in noisy circumstances. This turn to military technologies opened a new direction for theoretical developments that linked the concepts of noise and related techniques or methods to the construal and handling of intelligence, computing, and information. The need to discern targets from background noise on a radar screen called for a theory of signal detection. The motivation to reduce unwanted fluctuations when a gunfire director tracked an enemy airplane led to a theory of estimation and filtering for stochastic processes. The considerations of the general character and fundamental limit of encryption or communication contaminated by disturbances prompted a conceptualization of information. In this novel context, the scrutiny of noise became part of the desiderata for making "smart" war machines.

The boom of research on intelligence, computing, and information—and their social imageries and cultural discourses—around the 1940s–60s was one of the most important episodes in twentieth-century science and technology. Historians have examined closely the origin of this development, especially its crucial connections to military projects during World War II. In a survey of the scientific conceptualization of information in this period, William Aspray reviewed Claude Shannon's information theory, Norbert Wiener's cybernetics, Warren McCulloch and Walter Pitts's artificial neural network, Alan Turing's theory of computation, and John von Neumann's theory of cellular automata.[1] To this list we can also add operations research, systems analysis, game theory, and more. Most of these fields were linked to wartime research and development—the atomic bomb, microwave radar, secret telecommunication, cryptography, servomechanical control, electronic computers, planning of bombing campaigns—that laid out the institutional and managerial backbones for, as Peter Galison and Thomas Hughes showed, the interdisciplinary "big science" and large-scale technology after the war.[2] Focusing on information sciences in a narrower sense, Ronald Kline traced how Wiener's cybernetics and Shannon's information theory during and immediately after World War II unfolded a "cybernetic moment" that sparked diverse discourses and speculations on the connections between humans and machines, which in turn led to today's mainstream narratives about the "information age."[3]

To many scholars in the history of cybernetics, computing, and information, the human–machine relationship constitutes the central issue. Steve Heims's study of the Macy group in the postwar US and Andrew Pickering's exploration of their British counterpart examined those researchers' deliberations and claims about the nature of mind, brain, and intelligence in light of new

[1] Aspray, "Scientific conceptualization of information" (1985), 117–140.

[2] Galison, *Image and Logic* (1997), 239–312; Thomas Hughes, *Rescuing Prometheus: Four Monumental Projects that Changed the World* (New York: Vintage Books, 1998), chapters 1–3.

[3] Kline, *Cybernetic Moment* (2015), 9–101.

information processors. David Mindell's work on the multi-threaded development of feedback control technology focused on human–machine interactions in engineering design.[4]

The preoccupation with the human–machine relationship in the early development of information sciences prompted theoretical discussions. In *Cyborg Manifesto*, science studies scholar Donna Haraway declared that the cybernetic picture of humans as black boxes with inputs and outputs had the potential of overcoming the prior cultural biases about human nature, and thus could help achieve more inclusive and progressive politics. Delving into Shannon's analysis of the English language, literary scholar Lydia Liu argued that information theory facilitated what she dubbed a "Freudian robot," a new type of human–machine relationship in which human actions—especially those revealing subconsciousness—were understood as random processes following fixed statistical regularities.[5] Galison examined Wiener's path from the anti-aircraft gunfire control project to the proposal of cybernetics as a worldview. Engaging Haraway's discourse, Galison reminded us of the military genealogy of the Wienerian cybernetic enterprise and referred to its underlying model about human subjectivity as an "ontology of the enemy." In Galison's words, this ontology:[6]

> was a vision in which the enemy pilot was so merged with machinery that (his) human-nonhuman status was blurred. In fighting this cybernetic enemy, Wiener and his team began to conceive of the Allied antiaircraft operators as resembling the foe, and it was a short step from this elision of the human and nonhuman in the ally to a blurring of the human-machine boundary in general. The servomechanical enemy became, in the cybernetic vision of the 1940s, the prototype for human psychology and ultimately, for all of human nature.

This scholarship on early cybernetics conveys a *bilateral* relationship with *humans* on one side and *machines* on the other. Yet, part IV of this book concentrates on an underexplored aspect of wartime and postwar information sciences and technologies that implicated a *trilateral* relationship, with the

[4] Heim, *Cybernetic Group* (1991); Andrew Pickering, *The Cybernetic Brain: Sketches of Another Future* (Chicago, IL: University of Chicago Press, 2011); Mindell, *Between Human and Machine* (2004).

[5] Donna Haraway, "A cyborg manifesto: Science, technology, and socialist-feminism in the late twentieth century," in *Simians, Cyborgs and Women: The Invention of Nature* (New York: Routledge, 1991), 169–181; Lydia Liu, *The Freudian Robot: Digital Media and the Future of the Unconscious* (Chicago, IL: University of Chicago Press, 2010).

[6] Galison, "Ontology of the enemy" (1994), 233.

environment as the third party. The imperatives of the studies on radar target detection, servomechanical estimation and filtering, and transmission of information were not just about grasping the quantitative and formal regularities of the machines-operating enemies or allies. They were about grasping such regularities of enemies or allies acting as machine operators *in highly indeterministic, chaotic, stochastic, and haphazard surroundings*. The addressed subjects—friends or foes—might use what Galison named "Manichean" tactics via active yet irregular maneuvers or produce unconsciously random messages as Liu's Freudian robot did. But a grave challenge for target detection, predicting and filtering trajectories, and characterizing information transmission was to cope with disturbances, interferences, and fluctuations of various kinds from outside the subjects that contaminated signal passage. To researchers of the 1940s, noise was a theoretical embodiment of these environmental factors. We may call the noise-centered theoretical treatments of information, communication, and control a manifestation of an "ontology of the ambience."

The ontology of the ambience in mid-twentieth-century information sciences and technologies attested to a way of viewing and handling uncertainty and indeterminacy with specific features. None of them should be taken for granted. First, with only few exceptions, the effects of the uncertain environment and machines were treated as *additive*. This was precisely where noise—an unwanted entity added on top of wanted signals in the original context of sound reproduction—became a powerful notion in the new context of weapons research.

Second, most researchers concentrated on the cases when noise followed a normal distribution. While they acknowledged that the probability density of random fluctuations in radars, gunfire control, and secret communication could take various forms,[7] Gaussian noise constituted the basis of their inquiries. The longstanding law of large numbers—which was used to legitimize the assumption of the normal distribution in statistical analysis since Siméon-Denis Poisson and Adolphe Quetelet in the mid-nineteenth century—provided only a partial justification, for wartime scientists and engineers were aware of certain non-Gaussian noise. A more important reason for the tenacity of Gaussian noise in theoretical treatments had to do with the availability

[7] For instance, in a letter to Uhlenbeck on January 4, 1944, Henry Hurwitz and Mark Kac discussed the accuracy of the Gaussian approximation to the synthetic total of n identical and independent random variables. They found that the approximation was satisfactory when $n > 4$ and presented a formulation for the deviations from the normal distribution. Such deviations were more significant at the "tail" of the distribution curve; i.e. when the values under consideration were very large. Box 2, Folder "Correspondence, Mark Kac, 1943-45," Uhlenbeck Papers.

of tools, as the prewar analytical apparatuses from the studies of Brownian motion and electronic noise in physics, mathematics, and engineering could be readily applied.

Third, the noise under scrutiny in the war was mostly continuous waveforms. Even for individuals like Shannon who were interested in discrete scenarios, dealing with continuous-time noise in signal transmission was inevitable. The handling of environmental uncertainties as continuous entities required a different set of techniques (sampling and filtering) than the conventional grappling with statistical errors (regression and hypothesis testing). Once again, this feature revealed the close connection between the informational conceptualization of noise and technology of sounds, which telephone and radio engineers had treated as continuous waveforms.

Last, in information sciences and technologies, noise was perceived *both* as a fundamental limit for signal transmission *and* as a resource and opportunity. It was not a surprise that uncertainty degraded the performance of technological systems. What became unexpected from information theory was the existential nature of this limit: In principle, signals could be completely recovered if the rate of information transmission did not exceed the capacity posed by noise on the communication channel. But there was no possibility of recovering the signals if the rate exceeded the capacity. To some, the random characters of noise turned out to be an asset. When messages were encoded with noise-like patterns, the transmitted signals looked sufficiently irregular for enemy code breakers to be unable to breach. After the war, scientists and engineers found that noise-modulation of signals could extend their spectra, which ironically became an effective means of reducing the plight of environmental noise on the quality of communication. This appropriation of noise resulted in the invention of spread-spectrum communication, a Cold War technology of sound reproduction that is built into today's cell phones.

In part IV, we examine the emergence of informational noise from military research and development circa World War II. Chapter 10 concerns Uhlenbeck's work for MIT Radiation Laboratory on radar target detection with his student Ming Chen Wang and his colleague Mark Kac. Uhlenbeck et al.'s task was to develop theoretical models to assess the probability of radar detection polluted by electronic noise, clutter from environmental scatterers, and jamming. The task prompted them to explore new mathematical problems about Gaussian processes.

In chapter 11, we look into Wiener's work at MIT with his assistant Julian Bigelow to build an antiaircraft gunfire director. Kline's and Galison's examinations of this episode stressed the relevance of Wiener's wartime research to his cybernetic worldview. We inspect Wiener's techniques of handling noise. In particular, we examine his development of a theory for stochastic estimating and filtering, leading to the so-called "Wiener filtering."

Chapter 12 discusses Shannon's development of information theory at Bell Labs. Whereas the existing accounts of this episode focused on conceptualization of information, we concentrate on Shannon's conceptualization of noise. We explore how his work on cryptography helped him forge a similarity between random keys and noise. We also inspect his geometric representations of noise that played a heuristic role in enacting the channel coding theorem.

In chapter 13, we review the launch of spread-spectrum communication through the information-theoretic framework, and inventions in early noise-modulation technology, including the ITT engineer Mortimer Rogoff's noise wheel and the NOMAC system developed at MIT Lincoln Laboratories.

The stories in part IV occurred at a few institutions of the American military–industrial–academic complex: MIT, Radiation Lab, Bell Labs, and Lincoln Labs. By no means does this suggest that the US East Coast was the only region that mattered for informational conceptualization of noise—Kolmogorov's and Khinchin's work in the USSR and the development of spread-spectrum systems in Continental Europe are notable counterexamples. Yet, MIT and Bell Labs were undeniable epicenters for mid-twentieth-century information sciences and technologies, and these episodes did represent the gist of that development. In these stories, moreover, we will not delve into the full-fledged military R&D but concentrate on the treatments of noise.

Uhlenbeck, Kac, and Wang at MIT Radiation Laboratory

After venturing into the problems of random fluctuations in the early 1930s, Uhlenbeck studied condensations, transport phenomena, radioactivity, electron–positron combinations, and cosmic rays. In 1935, he moved back to the Netherlands to take over Hendrik Anthony Kramers's professorship in physics at the University of Utrecht, when Kramers filled in Ehrenfest's position at Leyden after Ehrenfest's death. In 1938, Uhlenbeck visited Columbia University for a semester. As the Nazis' expansion in Europe became inevitable, going back to his homeland was a less viable option. Uhlenbeck decided not to return to Utrecht but to resume his position in Michigan.[8]

[8] Ford, "Uhlenbeck" (2009), 14–16.

What happened in the following years unfortunately confirmed his worry. In March 1938, Germany annexed Austria. Czechoslovakia fell into Hitler's control in March 1939. In September, the Wehrmacht invaded Poland. Denmark and Norway were occupied in April 1940. In May, Germany launched an offensive against France, and invaded the Netherlands, Belgium, and Luxembourg. By June, France capitulated. The Battle of Britain commenced in July.[9]

At Ann Arbor, Uhlenbeck was deeply saddened by the horrible news. In a letter to a Professor Stewart, he wrote:

> I have been feeling so badly depressed the last month or so, that I only managed to do my daily duties. I am trying hard to get over it, but it is difficult to get rid of the feeling of impotent hate; the news remains bad, and when we hear from Holland my wife and I are in the dumps for days. When Germany wins, it surely will be the end of Holland.

Not a surprise, he continued, "The Universities of Leiden and Delft have been closed already and all the Jews have been kicked out." The Jewish scientists under purge included Uhlenbeck's research partner Leonard Ornstein, coauthor of the 1930 Brownian motion paper. Ornstein was dismissed from the University of Utrecht in November 1940. Uhlenbeck indicated that Ornstein hoped to come to the US, and inquired about possible opportunities for him. Uhlenbeck tried to make arrangements for Ornstein through the Rockefeller Foundation; but they were unable to do something since Ornstein was "already almost sixty years old." After the dismissal, Ornstein secluded himself at home and died six months later. In addition, a student of Uhlenbeck's, van Lier, committed suicide with his whole family. To carry on, Uhlenbeck felt obliged to follow the words of William the Silent, leader of the Dutch revolt against Spain toward the nation's independence in the sixteenth century: "there is no need of hope to do one's duty, and there is no need of success to persevere."[10]

Uhlenbeck eventually overcame his depression, as new opportunities came up for him to do his duty in the war against the Axis. Anticipating full-fledged military conflicts with Germany and Japan, the US President Franklin Roosevelt followed the advice of Vannevar Bush, head of the Carnegie Institution of Washington and professor of electrical engineering at MIT, to enact the National Defense Research Committee (NDRC) in June 1940. Bush was

[9] John Keegan, *The Second World War* (London: Pimlico, 1989), 23–72.

[10] Uhlenbeck to Stewart, undated but likely 1940, in Box 1, Folder "Correspondence, 1938-44," Uhlenbeck Papers.

appointed director. As the largest wartime establishment for military research and development, NDRC marked a turning point for American science. Bush set up four divisions within NDRC: Division A on armor and ordinance headed by Richard Tolman at California Institute of Technology; Division B on bombs, fuels, gases, and chemistry headed by Harvard President James Conant; Division C on communication and transportation under Bell Labs director Frank Jewett; and Division D on radar, fire control, and instruments under MIT President Karl Compton.[11]

A major problem that NDRC aimed to tackle was air defense. While the use of airplanes on battlefields was sporadic in World War I, the military conflicts in Europe and Asia in the 1930s showed an increasing significance of airpower. In the Battle of Britain, the German Luftwaffe mobilized over a thousand bombers and fighters to launch attacks across the English Channel; and they were countered by the British Royal Air Force with comparable capacity.[12] Antiaircraft tasks became a pressing issue in preparing for the upcoming war.

In their defense against German air attacks, the British military demonstrated a powerful technology that utilized microwaves or radio waves to detect and locate enemy aircraft. Gaining the name of radar (**Ra**dio **D**etection **a**nd **R**anging), this new technology caught the attention of the American armed forces and military researchers. Soon after the launch of NDRC, Bush appointed a Microwave Committee. In September 1940, a British technical mission headed by physicist Sir Henry Tizard visited the US. The "Tizard Mission" presented Britain's cutting-edge radar technology, including the cavity magnetron they developed for generating high-power microwave pulses. This presentation prompted American researchers to take more proactive steps toward the development of radars. In October, Bush set up a laboratory for microwave research under the tutelage of NDRC. Instead of making the lab an institute of the armed forces like the Naval Research Laboratory, the NDRC leadership decided to build it within a university. They chose MIT. Lee DuBridge, a nuclear physicist from the University of Rochester, assumed directorship. To keep confidentiality, the establishment was given the

[11] Mindell, *Between Human and Machine* (2002), 185–188. For the activities of NDRC and its subsequent organization the Office of Scientific Research and Development (OSRD), see Larry Owens, "The counterproductive management of science in the Second World War: Vannevar Bush and the Office of Scientific Research and Development," *The Business History Review*, 68:4 (1994), 515–576; Daniel Kevles, *The Physicists: The History of a Scientific Community in Modern America* (New York: Knopf, 1977), 287–323.

[12] Liddell Hart, *A History of the Second World War* (London: Weidenfeld Nicolson, 1970), chapter 8.

misleading name of "Radiation Laboratory" to convey the impression of an institution conducting "pure" research in nuclear physics.[13]

MIT Radiation Laboratory ("Rad Lab") was the biggest research laboratory under NDRC. Beside the US Army's Manhattan Project, it employed the largest number of scientists, and embodied physicists' great influence on the course of the war. Throughout World War II, Rad Lab designed all the ground-based, airborne, and shipborne systems for the US Army, and a significant portion of those for the US Navy, for aircraft interception, anti-submarine campaigns, shore defense, and navigation. The lab's organizational structure reflected its mission: It had divisions on radar transmitters, radar receivers, beacons, gunfire controls, airborne systems, ground and ship systems, and the LORAN system of navigation.[14]

Although Rad Lab was dedicated to the development and design of radar systems and components, it administered a contingent on theoretical studies. Division 4, headed by the atomic physicist at Columbia University and Nobel laureate Isidor Isaac Rabi, was devoted to "research." Under this vague category, distinct groups were formed to investigate scientific and technical problems with theoretical characters. These included magnetron physics, electromagnetic wave propagation, behavior of microwaves in waveguides or other electronic components, and systemic implications of modulation schemes. Rabi recruited a sizable number of physicists to this division. For example, Julian Schwinger, a would-be founder of quantum electrodynamics and then a newly graduated Ph.D. from Columbia, worked on the theory of microwave circuits. A group in Division 4 looked into how various kinds of noise degraded the efficacy of radar target detection.[15]

Re-enter Uhlenbeck's collaborator Samuel Goudsmit. Throughout the 1930s, Goudsmit stayed at the University of Michigan and undertook research in atomic and nuclear physics. As with many other physicists at the time, the turmoil in Europe interrupted his career. As a Jew, Goudsmit held a strong anti-Nazi position from early on. (After Germany surrendered in 1945, he visited his hometown The Hague and found that his family apartment was in a

[13] Mindell, *Between Human and Machine* (2002), 244–245; Theodore A. Saad, "The story of the MIT Radiation Laboratory," *IEEE Aerospace and Electronic Systems Magazine*, 5:10 (1990), 46; Warren Flock, "The Radiation Laboratory: Fifty years later," *IEEE Antennas and Propagation Magazine*, 33:5 (1991), 43–44.

[14] Kevles, *The Physicists* (1977), 302–323; Galison, *Image and Logic* (1997), 243–254; Saad, "Story of the MIT Radiation Laboratory" (1990), 46–51; Flock, "Radiation Laboratory," (1991), 43–48. For a comprehensive survey of the MIT Radiation Laboratory's undertakings, see Guerlac, *Radar in World War II* (1987).

[15] Saad, "Story of the MIT Radiation Laboratory" (1990), 44; Galison, *Image and Logic* (1997), 820–827.

state of wreckage and his parents had died in a concentration camp.) In 1940, he left Ann Arbor for Boston, with a hope to contribute directly to the war preparations. In 1941, he joined Rad Lab's Division 4.[16] He was assigned to take charge of the noise group.

Knowing of Uhlenbeck's experience and expertise on noise, Goudsmit sought his advice. In a letter on March 12, 1942, Goudsmit laid out the scope of the noise problems that Rad Lab researchers encountered:

> the noise problem is getting more and more important because the apparatus used at present is so perfect that the noise level is the limit of its performance. There are two principal problems in connection with noise. First, there is the question of what causes the noise. Second, given the noise, how can one get rid of the discrimination between signal and noise.

Goudsmit pointed out that:

> about the second problem there are various learned statements and reports by Norbert Wiener and his disciples. However, they have not yet lead [*sic*] to much because some of that work cannot be understood by ordinary human beings and the part that can be understood is either too complicated for application or turns out to be rather trivial.

Here, Goudsmit referred to the curve-smoothing problem. He indicated that Uhlenbeck's work on infrared spectroscopy at Ann Arbor might be relevant to tackle it. The registration curves of the band spectra usually had "a lot of brownian motion which sometimes makes the identification of lines uncertain." What Goudsmit had in mind was that "a recording system with a mechanical or electrical time lag" would take the time average of recorded data that could smooth a registered curve and remove random fluctuations. Knowing the properties of "brownian motion" could help determine the appropriate value of the time lag, i.e. the window of average. In Goudsmit's view, Wiener's mathematically complicated optimum filter (which we will see in chapter 11) was unnecessary. Rather, he believed that "for practical purposes any system with an exponential decay will be good enough."[17]

[16] Benjamin Bederson, "Samuel Abraham Goudsmit," *Biographical Memoirs of the National Academy of Sciences*, 90 (2008), 11–12.

[17] Letter, Goudsmit to Uhlenbeck, March 12, 1942, Box 1, Folder "Correspondence 1938-1944," Uhlenbeck Papers.

Furthermore, Goudsmit sought Uhlenbeck's help with a concrete problem for radar detection at Rad Lab: He considered a sine-wave radar signal superimposed on random noise in voltage. At a receiver, this voltage passed through a rectifier made of a diode in series with a resistor-capacitor time-delay network (Figure 10.1). The diode was a semiconductor or vacuum-tube device that permitted passage of electric current when the input voltage was positive, while the resistor-capacitor network connected the peaks of the positive voltage lumps. Without noise, the circuit retrieved the signal amplitude. With noise, the circuit output was polluted by random fluctuations. Goudsmit's question was this: What were the statistical properties of the output from this radar signal detector under the designated condition? As he stated, "what one would like most to have are its Fourier components and its probability distribution." Goudsmit asked Uhlenbeck to solve this problem. He ended the letter with an encouragement: "The labs here and elsewhere are doing marvelous work. Thus don't despair, we'll lick the Japs and the verdomde rotmoffen ['damned rotten krauts', Dutch derogatory reference to German soldiers]."[18]

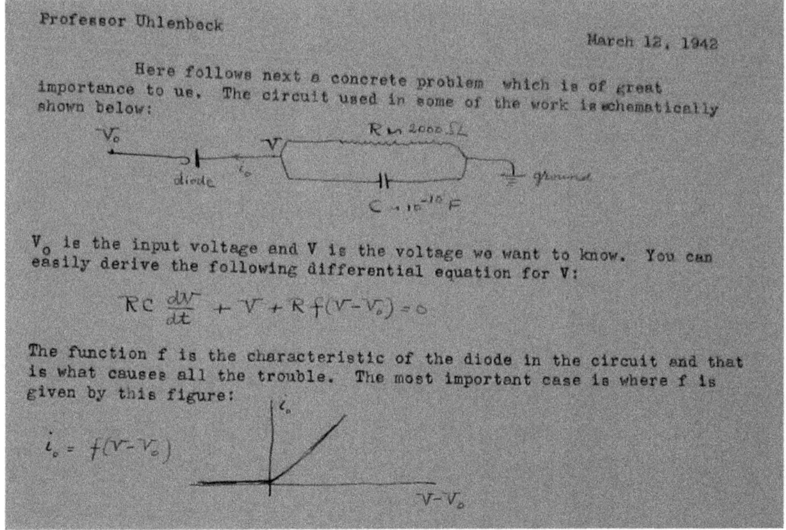

Professor Uhlenbeck

March 12, 1942

Here follows next a concrete problem which is of great importance to us. The circuit used in some of the work is schematically shown below:

V_0 is the input voltage and V is the voltage we want to know. You can easily derive the following differential equation for V:

$$RC\frac{dV}{dt} + V + Rf(V - V_0) = 0$$

The function f is the characteristic of the diode in the circuit and that is what causes all the trouble. The most important case is where f is given by this figure:

Figure 10.1 Goudsmit's sketch of a diode radar signal detector. He asked Uhlenbeck to assess the effect of noise on its signal detection. Letter, Samuel Goudsmit to George Uhlenbeck, March 12, 1942, Box 1, Folder "Correspondence 1938-1944," Uhlenbeck Papers.

[18] Ibid.

The two directions sketched in Goudsmit's letter represented the major framework of Rad Lab's research on radar noise, and the specific question he described exemplified the problems that the researchers encountered in the matters of noise, i.e. the statistical characteristics of signal plus noise (or noise alone) at the output of various radar detecting devices. In the following year, Uhlenbeck was increasingly involved in this undertaking and other wartime research. Opportunities for formal participation in defense projects came in a group. In March 1943, he accepted an invitation by O.S. Duffendack and E.S. Barker at the University of Michigan to take part in an NDRC project on spectroscopy. In April, Alfred Loomis, head of the NDRC Microwave Committee, wanted him to come to Rad Lab to succeed the German Jewish immigrant physicist Hans Bethe's leadership for theoretical studies. Uhlenbeck accepted Loomis's offer and agreed to spend half of his time in Ann Arbor and the other half in Cambridge, Massachusetts. Then an urgent request came from Dupont Company in Wilmington, Delaware, asking him to work on a project related to the atomic bomb. In June, Uhlenbeck sought Bush's advice on his choice between Rad Lab and Dupont. Bush encouraged him to continue with his original plan to work at Cambridge and Ann Arbor.[19] Uhlenbeck followed this advice and spent time alternately every two months at Rad Lab and Michigan until the war was over.

When Uhlenbeck started to direct the theoretical group in Rad Lab's Division 4, the lab's leadership in theoretical studies was undergoing a transition. Bethe responded to J. Robert Oppenheimer's call and joined the Manhattan Project. While Goudsmit continued to work at Rad Lab, his focus shifted to liaison and connection with the scientific communities in Allied Europe. In May 1944, he was assigned as the scientific head of the US Army's "Alsos Mission," which aimed to uncover the German progress on atomic weapons from the information gathered in European lands liberated from Nazi control.[20] Uhlenbeck had to take charge of the noise research.

To direct the investigations into noise issues in radar detection, Uhlenbeck relied on a few junior researchers. Outside the laboratory, his major collaborator was James Lawson, an associate of General Electric's Research Laboratory at Schenectady, New York. The contribution of Lawson's group to the studies of noise was experimental, as they conducted trials to measure the noise characteristics of radar signals under various conditions. Theoretically, Uhlenbeck's help came primarily from the mathematician Mark Kac at Cornell University

[19] Letters, Uhlenbeck to Vannevar Bush, undated but should be 1943; Bush to Uhlenbeck, June 7, 1943; both in Box 1, Folder "Correspondence 1938-1944," Uhlenbeck Papers.

[20] Benderson, "Goudsmit" (2008), 12–13.

and his Ph.D. student Ming Chen Wang, who was then a staff member at Rad Lab's Division 4.

In his book on the global history of the twentieth century, Eric Hobsbawm called the period from the beginning of World War I to the aftermath of World War II "an Age of Catastrophe." He lamented that:[21]

> For forty years [the world] stumbled from one calamity to another. There were times when even intelligent conservatives would not take bets on its survival.

Kac's and Wang's life experiences testified to Hobsbawm's comment about the calamities of the twentieth century. Mark Kac was born at the Polish city of Krzemieniec (now in Ukraine) in 1914. Then a territory of the Russian Empire, Krzemieniec was a Jewish town surrounded by Polish populations. Kac's father was a teacher who received advanced degrees in philosophy and history from Leipzig and Moscow. His mother's family had operated a mercantile business in the region for centuries. Showing mathematical talent from early on, Kac was cultivated by his father toward an academic career. During the interwar period, Poland enjoyed political independence and reunited the regions previously under Russian, German, and Austrian control into a single nation. Kac attended the John Casimir University of Lwów, with a major in mathematics.[22]

Kac was too young to witness Smoluchowski's teaching at Lwów. But Smoluchowski and a few scholars in the 1900s–30s created a research tradition of Polish physics and mathematics at Warsaw, Kraków, and Lwów. Kac benefited from this scientific development. At the University of Lwów, he worked as an assistant for the Göttingen-trained mathematician Hugo Steinhaus on probabilistic independence associated with normal distributions. He received a Ph.D. in 1937.[23]

When Kac was a student at Lwów, the situation in Europe went into a downward spiral. Across Poland's western border, Hitler rose in Germany and the Nazis posed an expanding threat to its neighboring states. Across the eastern border, Stalin launched purges against the "class enemies" in the Soviet Union. These gloomy events prompted Kac to "find a way to get out of Poland." In 1938, he obtained a fellowship that allowed him to visit Johns Hopkins University in Baltimore. In the next year, he moved to Cornell University in Ithaca,

[21] Eric Hobsbawm, *The Age of Extremes: A History of the World, 1914–1991* (New York: Vintage Books, 1994), 7.

[22] Henry McKean, "Mark Kac," *Biographical Memoirs of the National Academy of Sciences*, 59 (1990), 215.

[23] Ibid., 215–216.

and served as a lecturer at the university's Department of Mathematics, where he interacted with Eastern European refugee mathematicians who specialized in probability theory including Paul Erdös from Hungary and William Feller from Croatia.[24]

In America, Kac tried to keep an eye on what happened in Krzemieniec. He read from newspapers in late 1939 that German troops bombarded his hometown in their invasion of Poland.[25] Messages from the country became scarce and scattered during occupation. He was not optimistic about the fate of his families, relatives, and friends, but did not know exactly what they were experiencing. After V-E Day, he attempted to get information about his family from the Soviet Embassy in Washington, but to no avail. On his first postwar trip to Poland, he was informed by a cousin that the Jewish population in Krzemieniec was exterminated in 1942–43. Kac admitted that he never recovered fully from the trauma.[26] Around the time when this atrocity occurred, Kac got to know Uhlenbeck and began to work with him as a consultant for noise studies at Rad Lab.[27]

Ming Chen Wang (Wang Mingzhen, 王明貞) was another foreigner driven by circumstances toward military research at Rad Lab. Wang stands out in the history of noise as perhaps the only Asian woman who made a published contribution to its theoretical studies. Historians have shown women's active but unacknowledged roles in science and technology throughout history. In physics, mathematics, and engineering in the first half of the twentieth century, women appeared more frequently as human computers, scanners for processing laboratory images, or workers in manufacturing. While universities had been open to female students as early as the 1870s, their prospect in academic career was gloomy.[28] Geertruida de Haas-Lorentz, another female contributor to the theoretical studies of noise, taught as a university lecturer, despite her Ph.D., numerous publications, and being a daughter of Europe's leading physicist and the wife of a university professor. For those women of science from the Global South aspiring to graduate study or even career development in the West, the traditional culture and political instability of their native society imposed additional barriers.

[24] Ibid., 216–217; Mark Kac, *Enigmas of Chance: An Autobiography* (New York: Harper & Row, 1985), 41–42.

[25] Kac, *Enigmas of Chance* (1985), 88.

[26] Ibid., 106.

[27] Ibid., 109.

[28] Jenna Tonn, "Gender," *Encyclopedia of the History of Science* (March 2019), online journal from Carnegie Mellon University, https://lps.library.cmu.edu/ETHOS/article/id/20/ (accessed October 12, 2022).

Wang was born eight years earlier than Kac in Suzhou of southeast China. She came from a family of intellectuals. Her father was a self-taught engineer and mathematician serving at the Qing court, Republican government, and industrial firms. Her siblings included a gynecologist, a theoretical physicist, and three engineering scientists; three of them studied in the US or UK; and two were future academicians of the Chinese Academy of Science. As a straight-A student, Wang showed her aptitude in physics and mathematics from early on. Even so, being a girl in early-twentieth-century China meant she encountered obstacles to accessing higher education. Her father and stepmother did not support her intention to attend college. Rather, they arranged a marriage for her. With her gynecologist sister's sponsorship, she entered Ginling College, an American missionary institution in Nanjing. Two years later, she switched to another Anglo-American missionary college, Yenching University in Peiping (today's Beijing), with a major in physics.[29]

Wang wanted to pursue an advanced degree abroad after graduation. This was not easy. She persuaded her parents to nullify the prearranged marriage. Then she applied to the graduate school at the University of Michigan, which offered her admission and a fellowship. But she could not afford the ticket for the transpacific liner. These financial constraints forced her to give up the American opportunity and stay in China for graduate study. In 1932, she obtained an M.S. in physics from Yenching. To secure sufficient money for Ph.D. education abroad, she took three exams for the Boxer Indemnity Scholarship. She passed the last one with the highest score. Yet, the superintendent considered it a "waste" to send a female overseas to study physics, and instead gave the spot to a male student. Wang carried on and accepted an offer to teach physics and mathematics at Ginling College.[30]

The year 1937 marked a turning point both for Wang and for her nation. In July, a conflict broke out between Japanese and Chinese troops at the Marco Polo Bridge near Peiping. This incident soon escalated into a full-fledged war. The Japanese Imperial Army deployed about 300,000 combatants to invade Shanghai, countered by the Chinese National Revolutionary Army with twice as many soldiers. When the bloody battle of Shanghai was going on, Wang was teaching fall-semester courses at Ginling College and helped with the civilian mobilization for the war. Her brother advised her to leave the nation's capital as soon as possible, advice which she followed and she fled to Wuhan. In

[29] Ming Chen Wang, "Zhuanshun jiushi zai" (「轉瞬九十載」), *Wuli* (《物理》), 35:3 (2006), 174–176.

[30] Ibid., 176–177.

less than two months, the Japanese troops occupied Nanjing, and massacred over 20,000 civilians. In Wuhan, Wang met the President of Ginling College Wu Yifang and expressed again her hope of pursuing a doctoral degree overseas. Wu wrote a letter of recommendation to the University of Michigan, and helped Wang get admission and a fellowship. In 1938, Wang arrived at Ann Arbor.[31]

In the Department of Physics at the University of Michigan, Wang spent her first two years taking courses. Her performance impressed Goudsmit, who intended to supervise her dissertation. But this plan could not continue when he moved to Boston for defense research. Before his departure, Goudsmit transferred his supervisory responsibility to Uhlenbeck. Uhlenbeck and Wang chose a dissertation topic on statistical physics. In the following years, she worked with him on various solutions to the Boltzmann equation. Wang finished her dissertation in 1942.[32]

Wang's collaboration with Uhlenbeck went beyond the dissertation. When he supervised her thesis, he found her mathematically capable, and invited her to participate in his own research on statistical mechanics. After Wang obtained her Ph.D. from the University of Michigan in 1942, the blockage of transoceanic transportation due to the Pacific War made it difficult for her to return to China. She decided to stay temporarily at a friend's home in Boston. There she met Goudsmit again, who was affiliated with the theoretical group at Rad Lab. Goudsmit asked Wang to join it; she agreed. In 1943, Wang started to work as a human "computer" in Division 4. After Goudsmit left for Europe and Uhlenbeck took leadership of the theoretical group, Wang was involved more in research. She became Uhlenbeck's major aide for studies of noise.[33]

Radar Noise and Threshold Signals

In 1943–45, Uhlenbeck's theoretical group at Rad Lab examined the effects of noise on the performance of radar systems' target detection. Under the directive of the Office of Scientific Research and Development, NDRC, the lab after 1945 published a number of reports known as the "Radiation Laboratory Series" that summarized their major findings and outcomes during the war. The noise group's deliverable in this series was a monograph titled *Threshold*

[31] Ibid., 177.

[32] Ibid., 177–178. Ming Chen Wang, *A Study of Various Solutions of the Boltzmann Equation*, Ph.D. dissertation, University of Michigan, 1942.

[33] Wang, "Zhuanshun jiushi zai" (2006), 178.

Signals. It demonstrated their major problems with noise and the solutions.[34] *Threshold Signals* constitutes a primary source to look into Rad Lab's wartime work on noise.

To Uhlenbeck and his co-workers, the fundamental question was how noise limited the detection of signals in radars. In a radar system, a transmitter generated and sent high-power, high-frequency electromagnetic waves. The waves propagated over space, hit wanted targets and unwanted objects, and scattered back to a receiver. The receiver converted the returning waves into baseband signals and displayed them on an electronic imaging device, such as a one-dimensional A-scope that plotted the returning signal intensity over scanning time like an oscilloscope, or a Plan Position Indicator (PPI) that plotted the returning signal intensity as brightness on a two-dimensional plane spanned by range and azimuth. Ideally, an enemy target—a Junkers bomber, a V-1 rocket, a U-boat, a Zero fighter, or a Kamikaze attacker—was displayed as a clear pulse on an A-scope or a bright spot on a PPI; and a radar operator could use this visual information to locate the threatening object. Realistically, the radar signal to the indicator was contaminated by numerous disturbances. An important question was to determine the minimum signal intensity at which a target amidst noise could be detected. This minimum detectable signal was what Uhlenbeck et al. called the "signal threshold."[35]

Since a human operator was involved in judging whether a target was present in a radar return, the threshold signal must depend on psycho-physiological factors, such as the minimum visible brightness, the visually discernible contrast between signal and background, and the minimum lasting time of a bright spot. To Rad Lab researchers, while these factors had practical significances, they could not be subjected to a theoretical analysis, but could only be determined experimentally. Rather, they believed that:

> Only when the function of the observer is reduced to *measuring* or *counting* will the question of the minimum detectable signal become a definite statistical problem for which a theoretical analysis can be attempted. The success of such an analysis has shown that for a rather wide range of experimental conditions, the essential limitation for the detectability of a signal is due to the statistical nature of the problem.

[34] James Lawson and George Uhlenbeck (eds.), *Threshold Signals* (New York: McGraw-Hill, 1950).
[35] Ibid., 150.

Thus, their theoretical analysis was restricted to the "statistical limit" for signal detection.[36]

To Lawson, Uhlenbeck, and their collaborators at Rad Lab, a useful analogy to the problem of radar signal detectability was the employment of a sensitive mirror galvanometer for detecting a weak electric current. As we have seen in chapter 7, this was a problem that Ising, Uhlenbeck, Goudsmit, van Lear, and other statistical physicists had tackled in the 1920s–30s. Due to thermal agitations, the galvanometer's mirror underwent a small but recognizable Brownian motion. When the electric current was weaker than Brownian motion, it was overwhelmed by random jitter and the signal could not be detected. From a statistical perspective, however, this claim was true only when the duration of observation was limited. When that duration was sufficiently long, Brownian motion could be smoothed out and the signal (no matter how small it was) could stand out from averaging. In Uhlenbeck's handwritten draft report to NDRC circa 1945, he went further to point out that 100% correctness with sufficiently long time of observation in radar detection represented a situation that:[37]

> is fundamentally different from the limitation of measurements due to the uncertainty principle in quantum mechanics where each measurement affects in an uncontrollable way the to be measured object.

In addition to increasing the time of observation (which had practical difficulties in radar applications), one could raise the probability of correct detection by increasing the signal strength.[38]

Under this consideration, Uhlenbeck et al. defined a "betting curve" for the probability of successful detection versus signal strength. This betting curve was in an S-shape (Figure 10.2). Its character fit both theoretical intuition and experimental results from Lawson's team. Moreover, the minimum detectable signal was defined as the signal strength "for which the probability of guessing right is, let us say, 90 per cent." This threshold signal decreased with the time of observation; or, more precisely, the time of integration (viz. averaging) of the signal. The *sine qua non* for the analysis of radar detection was to construct theoretically or experimentally a radar system's betting curves and the associated threshold signals.[39]

[36] Ibid., 149 (italics in the original).
[37] Draft report to NDRC, circa 1945, in Box 7, Folder "Topical, Defense Research, 1944-46," Uhlenbeck Papers.
[38] Lawson and Uhlenbeck, *Threshold Signals* (1950), 150.
[39] Ibid., 150–151.

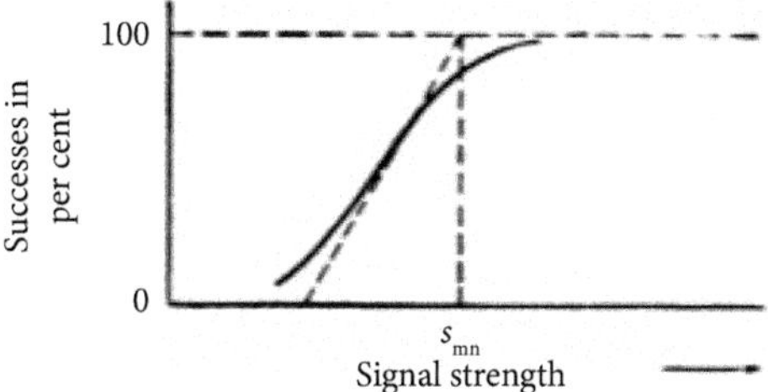

Figure 10.2 Lawson and Uhlenbeck's "betting curve" for characterizing radar detection. The horizontal axis represents the radar signal intensity, and the vertical axis represents the probability of correct detection. James Lawson and George Uhlenbeck (eds.), *Threshold Signals* (New York: McGraw-Hill, 1950), 151, Figure 7.2.

The analogy between radar detection and galvanometry suggested the heuristic value of Brownian motion to the Rad Lab researchers' treatment of radar noise. This analogy nonetheless implied a strong assumption about the nature of radar noise. The disturbances that fluctuated a galvanometer's mirror, as Ising, Uhlenbeck, and Goudsmit had shown, were due to thermal agitations. Thus, their mathematical properties were the same as those of a Brownian motion $x(t)$, in which $x(s) - x(u)$ was a random variable with a Gaussian distribution, a variance proportional to $s - u$, and statistical independence from $x(v) - x(w)$ when $[u,s]$ did not overlap with $[w,v]$.

Like the jittering galvanometer mirror, a radar screen had a non-zero irregular background when the signal was absent (Figure 10.3).[40] But since a radar was a much more complex system than a galvanometer, radar noise was much more involved than thermal agitation. To address what Goudsmit referred to as "the question of what causes the noise" in 1942, the Rad Lab researchers catalogued a byzantine list of plausible sources of disturbances. Thermal agitations undergoing Brownian motion were, in Lawson and Uhlenbeck et al.'s classification, a kind of "internal noise." The class of internal noise also included the shot effect—the discrete electronic discharges in a thermionic tube—and the flicker effect—the irregular but correlated emissions of discrete electronic or ionic charges (chapter 6). Moreover, the reception and processing of signals at different stages of a radar receiver introduced various types of "receiver noise." This included "antenna noise" as intake of unwanted background from cosmic

[40] Ibid., 33.

Figure 10.3 Noisy background on a radar display. Lawson and Uhlenbeck (eds.), *Threshold Signals* (1950), 33, Figure 3.1.

rays or environmental thermal radiation; "converter noise" and "local oscillator noise" caused by instabilities of the crystal electronic oscillator in a superheterodyne receiver's radio-frequency-to-intermediate-frequency conversion block; and "intermediate-frequency noise" as a cascade of noise produced by electronic components in the receiver's intermediate-frequency (IF) block.[41]

In addition to "internal noise" and "receiver noise," Lawson and Uhlenbeck et al. also listed "clutter" as "external noise," which referred to transmitted electromagnetic waves scattered and reflected by objects other than enemy targets from the environment. These objects included "rain, 'window' or 'chaff' [both referring to a large collection of thin metallic strips used as a jamming device], vegetation, and the surface of the sea."[42] On an A-scope or PPI, clutter created

[41] Ibid., 98–123.
[42] Ibid., 124.

similar types of noisy background as did electronic noise. There were also different kinds of electromagnetic radiation that enemies sent out to interfere with a radar's normal operations. The interferences included a simple continuous wave, a train of radio-frequency pulses, amplified thermal noise, a carrier amplitude-modulated by thermal noise, and a carrier frequency-modulated by the noise.[43]

These distinct kinds of radar noise were not Brownian motions. Yet, the mathematical configuration of Brownian motion provided the Rad Lab researchers with a useful platform for representing radar noise. A case in point was clutter. Lawson and Uhlenbeck et al. modeled clutter with N unwanted scatterers at a radar return:

$$u(t) = \sum_{k=1}^{N} \left[x_k \cos(2\pi_0 t) + y_k \sin(2\pi f_0 t)\right], \tag{10.1}$$

where (x_k,y_k) represented the scattering amplitude and phase from the kth scatterer. The radar return in (10.1) could be further reduced to a sum of a sine and a cosine term:

$$u(t) = X\cos(2\pi f_0 t) + Y\sin(2\pi f_0 t), \tag{10.2}$$

where

$$X = \sum_{k=1}^{N} x_k, \text{ and } Y = \sum_{k=1}^{N} y_k. \tag{10.3}$$

Assuming that $\{x_k,y_k\}$ from $k = 1$ to $k = N$ at any given instant t were mutually independent random variables with Gaussian distributions, Lawson and Uhlenbeck et al. conceived the radar return (X,Y) in (10.3) as a two-dimensional random walk (cf. Smoluchowski's random-walk representation of Brownian motion in chapter 5). From algebraic manipulations with Gaussian distributions, Lawson and Uhlenbeck et al. obtained clutter's first-order probability density function:

$$W_1(I)\,dI = \frac{dI}{I_0} e^{-I/I_0}, \tag{10.4}$$

where $I(t) = X(t)^2 + Y(t)^2$, and I_0 was the average power returned from the N scatterers. The expression in (10.4) was a Rayleigh distribution for I. To calculate the clutter's second-order probability density function at two instants t_1

[43] Ibid., 143–148.

and t_2, Lawson and Uhlenbeck et al. utilized a further assumption that stipulated the joint probability of $x_k(t_1)$, $y_k(t_1)$, and $x_k(t_2)$, $y_k(t_2)$ in terms of a scattering function. After manipulating $(x_k(t_1), x_k(t_2))$ and $(y_k(t_1), y_k(t_2))$, they obtained the second-order probability density for the amplitudes I_1 and I_2 as a two-dimensional Gaussian distribution for X_1, Y_1, X_2, and Y_2.[44]

Finding Novel Statistical Properties of Noise and Detection

The radar research gave Uhlenbeck's group a strong incentive to explore certain novel statistical properties of random fluctuations. These properties had not caught attention among interwar researchers on Brownian motion, for they did not seem to contribute much to understanding the dynamic behavior of Brownian particles. In the new context of radar detection, however, such properties became essential. In 1943–45, Uhlenbeck and Kac corresponded on the statistics of noise and detection. They started with a familiar topic: spectrum of radar noise. In a letter to Uhlenbeck in September 1943, Kac and his colleague Henry Hurwitz at Cornell treated radar noise displayed on a PPI or A-scope as a train of randomly arriving pulses. They affirmed that the noise's power spectrum was proportional to the magnitude square of an individual pulse's Fourier transform, if the pulses did not overlap. This was the Campbell theorem that interwar researchers had established in their characterization of electronic noise (chapter 6). Hurwitz and Kac also asked Uhlenbeck to provide Wang's calculations of an individual pulse's power spectrum in the overdamping and underdamping cases.[45]

In the following months, Kac and Uhlenbeck ventured into an unfamiliar terrain. One problem under their scrutiny concerned a random fluctuation's number of zeros. For a waveform $x(t)$, a zero was a value t_i so that $x(t_i) = 0$. If $x(t)$ was regular, then the locations of all zeros could be determined in principle. But since $x(t)$ was noise, the quantitative attributes of its zeros became random variables, including the number of zeros from $t = a$ to $t = b$, the separation between two consecutive zeros, etc. In 1942–45, Uhlenbeck's theoretical team explored the statistical features of these random variables. The mathematics of this work was intricate. In a letter to Uhlenbeck dated November 30, 1943, Kac and Hurwitz showed that when $(b - a)$ was small, the average number of zeros within the interval was proportional to $(b - a)$, while the average

[44] Ibid., 125–128.

[45] Kac and Hurwitz to Uhlenbeck, September 28, 1943, in Box 2, Folder "Correspondence, Mark Kac, 1943-45," Uhlenbeck Papers.

number of pairs of zeros was proportional to $(b - a)^2$. They obtained these results through a formula that specified the probability density of the separation between two consecutive zeros. This formula was expressed as a function of the power spectrum of $x(t)$. Their probability density did not apply to an agitating harmonic oscillator, into which Wang was investigating.[46] On January 4, 1945, Uhlenbeck wrote to Kac and presented a formula Wang derived for the latter case.[47]

Another problem tackled by the Uhlenbeck circle concerned the probability density of radar noise at the output of a video amplifier, the last electronic stage before radar display. In March 1944, Uhlenbeck and Kac considered IF random noise passing through a square rectifier. They represented IF noise as $x(t)\cos(\omega_{IF}t) + y(t)\sin(\omega_{IF}t)$, where $x(t)$ and $y(t)$ were mutually independent random processes. When this IF noise passed through a square rectifier, the IF noise's envelope square was retrieved as $x(t)^2 + y(t)^2$. Before entering the radar display, this waveform was directed into a video amplifier that effectively took an average over a period τ. The output from the video amplifier was $\int_0^\tau \left[x(t)^2 + y(t)^2\right] dt$. Uhlenbeck and Kac aimed to find the probability distribution of this quantity.[48]

To compute the probability density of $\int_0^\tau \left[x(t)^2 + y(t)^2\right] dt$, Kac approximated it with a sum $(\Delta/2)\sum_{j=1}^{2n}\left[x^2(t_j) + y^2(t_j)\right]$ where $t_j = (2j-1)\Delta/4$. As usual, he took IF noise's components $x(t)$ and $y(t)$ as stationary Gaussian. From probability theory, the joint probability density of $x(t_1), x(t_2), \ldots, x(t_{2n})$ was a $2n$-dimensional normal distribution

$$\frac{1}{\pi^n\sqrt{D}}\exp\left[-\sum_{r=1}^{n}\sum_{s=1}^{n}\alpha_{rs}x(t_r)x(t_s)\right],$$

where the quantity D and the matrix $[\alpha_{rs}]$ were the determinant and inverse of the correlation matrix $\boldsymbol{F}$, respectively. Each entry of the correlation matrix $\boldsymbol{F}$ was defined as $F_{pq} \equiv \varphi(t_p - t_q)$, where $\varphi(t) = \int_{-\infty}^{\infty} A(\omega)\cos(\omega t)\,d\omega$ was the real part of the complex envelope's correlation function, and $A(\omega)$ was the envelope's power spectrum. Since $y(t)$ was identical to and independent of $x(t)$, $y(t_1), y(t_2), \ldots, y(t_{2n})$ had the same form of joint probability density

[46] Kac and Hurwitz to Uhlenbeck, November 30, 1943, in Box 2, Folder "Correspondence, Mark Kac, 1943-45," Uhlenbeck Papers.

[47] Uhlenbeck to Kac, January 4, 1945, in Box 2, Folder "Correspondence, Mark Kac, 1943-45," Uhlenbeck Papers.

[48] Kac to Uhlenbeck, March 21, 1944, in Box 2, Folder "Correspondence, Mark Kac, 1943-45," Uhlenbeck Papers.

as $x(t_1), x(t_2), \ldots, x(t_{2n})$, and the total joint probability density was the product of both. Using the method of characteristic functions, Kac expressed the probability density of $(\Delta/2)\sum_{j=1}^{2n}\left[x^2(t_j)+y^2(t_j)\right]$ as the Fourier transform of its characteristic function, which took the form of a hyper-dimensional integral of $\exp\left\{(\Delta/2)\sum_{j=1}^{2n}\left[x^2(t_j)+y^2(t_j)\right]\right\}$ weighted by the probability density $\frac{1}{\pi^{2n}D}\exp\left[-\sum_{r=1}^{n}\sum_{s=1}^{n}\alpha_{rs}x(t_r)x(t_s)-\sum_{r=1}^{n}\sum_{s=1}^{n}\alpha_{rs}y(t_r)y(t_s)\right]$. Through diagonalization of the matrix $[\alpha_{rs}]$ and Cauchy's theorem of residues on the complex plane, Kac carried out the integral and obtained the probability density function of $(\Delta/2)\sum_{j=1}^{2n}\left[x^2(t_j)+y^2(t_j)\right]$:

$$Pr\left\{\beta < (\Delta/2)\sum_{j=1}^{2n}\left[x^2(t_j)+y^2(t_j)\right] < \beta + d\beta\right\} = \frac{2}{\Delta}\frac{d\beta}{D}\sum_{s=1}^{2n}\frac{\exp\left(-\frac{2}{\Delta}\frac{\beta}{\lambda_s}\right)}{\prod_{j=1}^{2n}\left(\frac{1}{\lambda_j}-\frac{1}{\lambda_s}\right)}, \tag{10.5}$$

where $\{\lambda_j\}$ were the eigenvalues of the correlation matrix $\boldsymbol{F}$ that satisfied $\boldsymbol{F}\cdot\boldsymbol{x} = \lambda_j\boldsymbol{x}$.[49]

Kac's mathematical tour de force did not stop here. To compute numerically the probability at the output of a video amplifier, Kac needed to calculate the eigenvalues $\{\lambda_j\}$ of radar noise's correlation matrix. This was far from easy. To simplify, Kac assumed that the correlation function $\varphi(t_p - t_q)$ could be approximated as non-zero only when $q = p - 1$, p, or $p + 1$. That is, $\varphi(t_p - t_q)$ vanished when two instants were separated by more than one step of $\Delta/2$. While this "nearest-neighbor" approximation was inaccurate, it greatly reduced the complexity of calculation, for the correlation matrix became a simple form:

$$F = \begin{bmatrix} 1 & \mu & 0 & 0 & . & . \\ \mu & 1 & \mu & 0 & . & . \\ 0 & \mu & 1 & \mu & . & . \\ 0 & 0 & \mu & 1 & . & . \\ . & . & . & . & . & . \\ . & . & . & . & . & . \end{bmatrix},$$

[49] Ibid.

where all the diagonal entries equaled one, all the immediate off-diagonal entries $\mu = \varphi(\Delta/2)$, and all the other entries were zero. Under this approximation, Kac obtained (with Hurwitz's help) $\lambda_j = 1 + 2\mu\cos[\pi j/(2n+1)]$, and the determinant D as a rational function of $\sqrt{1-4\mu^2}$.[50]

Kac was not satisfied with his nearest-neighbor approximation. Within a few months, he realized that he could find the eigenvalues without this assumption. In a confidential report to Uhlenbeck on July 6, he tackled the problem directly. If he did not discretize the stochastic processes $x(t)$ or $y(t)$, then its correlation was a continuous-time function, not a discrete matrix. In that case, the eigenvalue problem was formulated in terms of an integral equation:

$$\int_0^T \varphi(s-t)f(t)\,dt = \lambda f(s).$$

Kac assumed radar noise's correlation $\varphi(t) = e^{-\alpha|t|}$, and solved the integral equation. After an elaborate procedure, he obtained an analytical solution for all the eigenvalues. He left the numerical computations to Rad Lab assistants but checked the asymptotic case to make sure the results were qualitatively reasonable.[51]

The probability distribution of a random process's number of zeros and its output from a square envelope retriever and an integrator epitomized a new class of theoretical problems in wartime research on noise—those in the class also included the distribution of maxima and the output from a linear rectifier. Such problems rarely led to observable effects in the studies of suspended particles' Brownian motions, critical opalescence, a fluid's density variation, and an electrical device's current fluctuation. But they were crucial for radar signal detection amid noisy background, which was the aim of Uhlenbeck's group. The solutions to these problems could help compute the probability of correct target detection, and could inform how this probability varied with the threshold level, signal-to-noise ratio, the bandwidth and shape of frequency response of the IF filter or video amplifier, and other parameters of a radar system.

Take a look at the probability distribution of a random waveform's number of zeros. With a noisy background, the waveform on a radar display was jittering no matter whether a target signal was present or not. If a target showed up, the waveform would level up, resulting in a different number of zeros than the

[50] Ibid.

[51] Kac to Uhlenbeck, July 6, 1944, in Box 2, Folder "Correspondence, Mark Kac, 1943-45," Uhlenbeck Papers.

case when only noise existed. When the target signal strength was much higher than the threshold, signal-plus-noise would have much fewer zeros than noise-alone would. This implied that signal-plus-noise and noise-alone had different probability distributions for the number of zeros (or the period between two adjacent zeros, or the number of maxima, etc.). Thus, the probability of this quantity could serve as an index for target detection. As Kac and Hurwitz stated in their letter to Uhlenbeck on November 30, 1943, the next task, which was more difficult, was to state:[52]

> mathematically the criterion which determines whether a curve corresponding to signal + noise can be distinguished from one due to noise alone. We tried to give a general formulation by assigning to every possible curve a number Θ which tells how much this curve looks as though it contained a signal. Then if there are n sweeps there will be a collection of n numbers Θ_i characterizing the n curves, and from these Θ's one determines by the use of Bayes's rule the probability that a signal is present.

The number of zeros, or another zeros-related quantity, could be a candidate for this Θ.

The probability of a random waveform passing a video amplifier bore an even more direct connection to computing the probability of correct detection. Following the convention of American radar engineers, Uhlenbeck et al. focused on the "learning curve" or "betting curve," which plotted a radar system's percentage of correct target detection as a function of the target signal strength. The learning curve was a basis for assessing a radar's performance in target detection. From empirical results, this curve took an S-shape: the rate of successful detection was low when the target signal was much weaker than background noise; the rate increased with signal strength; the increase was more rapid when the signal strength was comparable to noise intensity; then the increase slowed down until the rate approached 100% (Figure 10.2). Uhlenbeck's group aimed to come up with a detection criterion that was "optimum" (giving the largest probability of detection), and to find the corresponding learning curve.[53]

In practice, the criterion for target detection in radar operations on battlefields was a simple threshold: In a set of multiple sweeps generating N deflections $r_1, r_2, \ldots, r_N$, their average $y = (1/N)\sum_{i=1}^{N} r_i$ was considered and the signal was just detectable if the shift of y due to the signal was in the same order

[52] Kac and Hurwitz to Uhlenbeck, November 30, 1943, Uhlenbeck Papers.
[53] Lawson and Uhlenbeck, *Threshold Signals* (1950), 149–151.

of magnitude as the standard deviation of y when only noise was present:[54]

$$\frac{\overline{y_{S+N}} - \overline{y_N}}{\sqrt{\overline{y_N^2} - \left(\overline{y_N}\right)^2}} = k, \tag{10.6}$$

where $\bar{y}$ was the ensemble average of y, and $k \sim 1$ was an empirical coefficient.

Theoretically, however, a more rigorous criterion entailing some kind of optimality was called for. Here Uhlenbeck et al. introduced an "ideal observer," who did not suffer physio-psychological biases as a real human being did and could always come up with the decision that gave the highest probability of detection. For an ideal observer, therefore, the problem of radar target detection could be formulated as a problem of optimization. Uhlenbeck et al. supposed that an ideal observer made N independent observations of deflection $r_1, r_2, \ldots, r_N$. They also let $P(o, r)$ and $P(s,r)$ ($P(r|o)$ and $P(r|s)$ in today's Bayesian notation) be the probabilities of r without and with a target signal. Then the probabilities of finding $r_1, r_2, \ldots, r_N$ with noise only and with signal plus noise were $P(o,r_1)\,P(o,r_2)\ldots P(o,r_N)$ and $P(s,r_1)\,P(s,r_2)\ldots P(s,r_N)$. Note that all possible values of $(r_1, r_2, \ldots, r_N)$ formed an abstract N-dimensional observation space. A detectability criterion was thus a division of this space into two regions: when the hyper-dimensional point $(r_1, r_2, \ldots, r_N)$ was in the "on region," the observer declared target detection; when the point was in the "off region," the observer declared no detection. The probability of correct detection was[55]

$$P = \frac{1}{2}\left[\int .. \int_{on} dr_1 \ldots dr_N P(s,r_1)\ldots P(s,r_N) + \int .. \int_{off} dr_1 \ldots dr_N P(s,r_1)\ldots P(s,r_N)\right]. \tag{10.7}$$

To maximize P in (10.7), Uhlenbeck et al. showed that the on and off regions were the following:

$$\begin{array}{ll} P(s,r_1)\,P(s,r_2)\ldots P(s,r_N) > P(o,r_1)\,P(o,r_2)\ldots P(o,r_N) & \text{on region} \\ P(s,r_1)\,P(s,r_2)\ldots P(s,r_N) < P(o,r_1)\,P(o,r_2)\ldots P(o,r_N) & \text{off region} \end{array}. \tag{10.8}$$

When a radar receiver's final stage before visual display was a linear rectifier, $r = \sqrt{x^2 + y^2}$, where x and y as the cosine and sine amplitudes of the

[54] Ibid., 163.
[55] Ibid., 168.

IF waveform were mutually independent Gaussian. Under this assumption, the probability density for x and y without signal was $(1/2\pi W)\, e^{-(x^2+y^2)/2W}$, and that with signal ($\alpha$ at x and β at y) was $(1/2\pi W)\, e^{-[(x-\alpha)^2+(y-\beta)^2]/2W}$. Expressing $x = r\cos\theta$ and $y = r\sin\theta$ and integrating the probability densities over θ from 0 to 2π, Uhlenbeck et al. obtained $P(o,r)$ and $P(s,r)$:

$$P(o,r) = \frac{r}{W}e^{-\frac{r^2}{2W}},\ P(s,r) = \frac{r}{W}e^{-\frac{r^2+S^2}{2W}} I_0\left(\frac{rS}{W}\right) \tag{10.9}$$

where S was signal intensity, W was noise's standard deviation, and I_0 was the modified 0th-order Bessel function of the first kind. With involved yet conceptually straightforward manipulations, Uhlenbeck et al. obtained from (10.7)–(10.9) the optimum probability of correct detection:[56]

$$P_{\max} = \frac{1}{2}\left[1 + Erf\left(\frac{\sqrt{N}}{2\sqrt{2}}z\right)\right], \tag{10.10}$$

where

$$Erf(x) = \frac{2}{\sqrt{\pi}}\int_0^x e^{-t^2}dt,\ \text{and}\ z = \frac{S^2}{2W}.$$

Equation (10.10) produced a desirable learning curve, as the optimum probability of correct detection was an error function of the signal-to-noise ratio z.

The above demonstration in *Threshold Signals* was a theoretical construction of learning curves for radar detection. But it was not the only scenario that Uhlenbeck's group considered. Kac's effort to find the probability distribution of waveforms after a square detector and a video amplifier $r = \int [x^2 + y^2]\, dt$ marked an attempt toward constructing another plausible learning curve (and hence another detectability criterion) under a different radar baseband block. In fact, Uhlenbeck and his associates even considered the learning curve for a non-ideal observer who employed a detectability criterion based on the time derivative of the video waveform, owing to a typical human visual illusion of relating the brightness of a light spot with its duration on a radar screen. In this case, once again, the Rad Lab researchers were obliged to deal with mathematical properties of random processes that made little sense to interwar physicists of Brownian motion.[57]

[56] Ibid., 169–172.
[57] Kac and Hurwitz to Uhlenbeck, January 17, 1944, in Box 2, Folder "Correspondence, Mark Kac, 1943-45," Uhlenbeck Papers.

S.O. Rice and Parallel Work at Bell Labs

Uhlenbeck and his associates at MIT Rad Lab were not the only researchers to delve into advanced mathematical properties of random noise. During World War II, there was at least another line of research on stochastic processes' statistical characteristics such as the probability distribution of zeros or maxima and their behavior after passing a nonlinear device. This research was done at AT&T Bell Labs by Stephen O. Rice. Its purpose was not to pave a theoretical ground for radar detection, but to solve theoretical problems in telecommunication.

A native of Shedds, Oregon, Stephen O. Rice received a B.S. in electrical engineering from Oregon State College in Corvallis in 1929. In 1930, he joined the Mathematical Research Department of Bell Labs as a consultant on transmission engineering.[58] Like Carson, Fry, and Nyquist, Rice was a theorist, a rare breed in the engineering profession in the early twentieth century but one which was gaining a more prominent presence in corporate research. His focus at AT&T was on communication theory. As one of the company's in-house theorists, Rice worked on problems related to the efficiency of modulation schemes, properties of shot and thermal noise and their effects on telephone and radio circuits, and the rate of telegraphic message transmission. His works were highly mathematical, and their connections to actual technological practice might not seem obvious.[59]

Rice began to pay attention to mathematical properties of random noise in the late 1930s. In the early 1940s, he produced a series of internal technical reports for Bell Labs on this subject. On April 10, 1940, he presented formulas for the average of a Gaussian random process's absolute value $\overline{|x|}$ when the fluctuation was distributed symmetrically about zero.[60] On October 3, 1941, he wrote about square-law detection of carrier plus noise.[61] In 1943, he gave a

[58] David Slepian, "Stephen O. Rice 1907-1986," *Memorial Tributes: National Academy of Engineering*, Volume 4, available at https://www.nae.edu/189134/STEPHEN-O-RICE-19071986?layoutChange=Print (accessed April 8, 2022).

[59] For example, see Stephen O. Rice, "Development and Application of Contour Integrals for the Product of Two Bessel Functions," Technical Report (October 27, 1934), in Legacy No. 622-08-01, Folder 03, AT&T Archives.

[60] Stephen O. Rice, "Formulas for <|x|> When x Is Distributed Symmetrically about Zero," Technical Report (April 10, 1940), in File Case No. 36760-11, Vol. A, "Probability Theory," Reel No. FC5184, AT&T Archives.

[61] Stephen O. Rice, "Square Law Detection of Carrier Plus Noise in Pulses," Technical Report (October 3, 1941), in File Case No. 36760-1, Vol. A, "1935-1945, Transmission Theory & Math Work," Reel No. FC3108, AT&T Archives.

report on the behavior of filtered thermal noise, in particular the fluctuations of average current and energy as a function of the averaging interval.[62]

These reports were outcomes of a major project Rice had been undertaking. By 1943, his mathematical treatments of random noise had become well known among American radar researchers. Uhlenbeck's laboratory notebook in January 1942 summarized "Rice's method" for finding the distribution of random noise's zeros.[63] Hurwitz and Kac's letter to Uhlenbeck on September 28, 1943 indicated that they obtained the same formula as Rice's for the spectrum of a random sequence of non-overlapping pulses.[64]

Upstream the Charles River from MIT Rad Lab, the staff at Harvard's Radio Research Laboratory (RRL) also became familiar with Rice's work. Formed under NDRC as a war research establishment, Harvard RRL was devoted to developing electronic countermeasures to enemy radar. In contrast to the Rad Lab researchers who treated noise as a plight to radar operations, the investigators at RRL viewed noise as a useful resource for degrading enemy radar performance. David Middleton, then a research assistant for the head of RRL's Research Group, physicist John Van Vleck, recalled later that the intention to employ random noise for countermeasures led the RRL researchers to "such mathematical questions as how to describe 'random noise' analytically and how one might then predict the behavior of such noise interacting with radar signals, for example to weaken or destroy their usefulness to an enemy." Van Vleck's team started exploring noise in early 1943. While struggling over these problems, they were told about "some new interesting studies on this subject going on at Bell Laboratories" of Rice's. In February, Van Vleck and Middleton made a classified visit to Bell Labs in New York. They met Rice, and found him "most helpful," as it "turned out he had just completed a long manuscript on our subject of interest." Rice generously offered them "a carefully guarded copy" of this manuscript. Van Vleck and Middleton returned to Boston "with renewed momentum and success" to attack their problems.[65]

[62] Stephen O. Rice, "Filtered Thermal Noise—Fluctuation of Energy as a Function of Interval Length," Technical Report (March 6, 1942), and "Filtered Thermal Noise—Fluctuation of Averaging Current as a Function of Interval Length," Technical Report (September 3, 1943), both in File Case No. 36760-11, Vol. A, "Probability Theory," Reel No. FC5184, AT&T Archives.

[63] Notebook 1932-1971, VIII, in Box 4, Uhlenbeck Papers.

[64] Hurwitz and Kac to Uhlenbeck, September 28, 1943, in Box 2, Folder "Correspondence, Mark Kac, 1943-45," Uhlenbeck Papers.

[65] David Middleton, "S.O. Rice and the theory of random noise: Some personal recollections," *IEEE Transactions on Information Theory*, 34:6 (1988), 1367–1368.

Rice's long manuscript became a paper published in *The Bell System Technical Journal.* Titled "Mathematical analysis of random noise," the article was printed in two installments in July 1944 and January 1945. It was a major publication on random noise during World War II. With a total length of 163 pages, the size of a monograph, "Mathematical analysis" comprised an encyclopedic synthesis of the existing body of mathematical knowledge about random fluctuations (in Parts I and II) and Rice's original research that furthered it (in Parts III and IV). Part I concerned shot noise as a sequence of identical, independent, and randomly arriving pulses. It summarized Campbell's theorem that linked the mean square of shot current to the square integral of an individual pulse, and the current's asymptotic convergence to a normal distribution at a high rate of arrivals. Part II discussed power spectra and correlation functions. It presented the tidy-up engineering version of the Wiener–Khinchin theorem that defined a stationary stochastic process's correlation and asserted that its Fourier transform was the process's power spectra. Rice also employed the theorem to derive the power spectra in a few examples of practical significance, such as a train of random telegraph signals.[66]

In Part III, Rice presented his novel findings about the statistical properties of noise current. He began with the premise that noise was normally distributed. Using the method of characteristic functions, he deduced the following:

- a noise current's average number of zeros per second;
- the probability distribution of the interval between two zeros;
- the distribution of the noise's maxima;
- the probability density of the envelope of the noise after passing a band-pass filter;
- the statistical behavior of a time average of the square noise current;
- the distribution of the noise current plus a sinusoidal current.

Part IV inspected noise through nonlinear devices. Rice presented results for a noise current's power spectrum, probability density, and other statistical properties when it passed through a square-law detector, a linear rectifier, a biased linear rectifier, or a square-law detector or linear rectifier and an

[66] Stephen O. Rice, "Mathematical analysis of random noise," *The Bell System Technical Journal*, 23 (July 1944), 282–332.

audio-frequency filter. He also undertook these calculations when noise was accompanied by a sinusoidal signal current.[67]

Rice's work on random noise exerted a crucial influence on American wartime radar detection. The major mathematical results in his 1944–45 paper—statistical characteristics of random noise's zeros, intervals between successive zeros, and maxima; the probability distribution of noise after passing through nonlinear devices and filters—were the core issues for those attempting to develop a statistical theory of radar detection amidst random noise. In their correspondence, Uhlenbeck and Kac kept referring to "Rice's formulas" as a basis for comparison with their own derivations, Wang's calculations, Wiener's equations, or experimental data.[68] Middleton contended that Rice's legacies to the studies of random noise included the method of characteristic functions, the probability distribution for the amplitude of noise without and with a signal component (i.e. (10.9), which Middleton named the "Ricean PDF"), and zero-crossing and maxima distributions.[69]

Yet, even though Rice examined similar mathematical problems of noise to those concerning Rad Lab and RRL researchers, he did not work on radar research. He was never assigned to missions related to target detection. He was not affiliated with Rad Lab or RRL. His task at Bell Labs during the war remained theoretical studies of telecommunication problems. That configured his formulation of the theory of random noise. A sequence of randomly arriving pulses resembled sporadic telegraphic messages; zero-crossing of a fluctuation was related to demodulation of an FM signal; passing a stochastic waveform through a nonlinear detector was a typical modeling of what was happening at the baseband stage of a radio receiver, etc.

Likely, Rice's distance from the radar people made his work accessible to a wider scientific audience. Unlike Uhlenbeck, Kac, Wang, Hurwitz, Lawson, Van Vleck, and Middleton, Rice did not work on a classified project. The technological problem situation of his research on random noise could still be civilian. The lack of military urgency gave him room to develop a more thorough treatment of noise. More importantly, he was still able to publish his findings. Rice's analysis became part of the disciplinary matrix for theoretical studies of random noise.

[67] Stephen O. Rice, "Mathematical analysis of random noise," *The Bell System Technical Journal*, 24 (January 1945), 46–156.

[68] Kac, "On the distribution of the average noise current in receivers," a report in August 1944; Kac to Uhlenbeck, December 27, 1944; Kac to Uhlenbeck, January 18, 1945, all in Box 2, Folder "Correspondence, Mark Kac, 1943-45," Uhlenbeck Papers.

[69] Middleton, "S.O. Rice and the theory of random noise," (1988), 1368–1370.

Making Marks on Studies of Random Fluctuations

The work on radar target detection amidst noise at Rad Lab during World War II brought a profound impact on theoretical studies of random fluctuations. Toward the end of the war, physicists and mathematicians gradually returned to the old problems of Brownian motion, stochastic processes, and statistical mechanics. The schemes and results developed in the treatments of threshold signals, probability of correct detection, and properties of noise nonetheless left methodological marks on the field. Noise became a label for random fluctuations.

A salient trend emerging during the war was a prioritization of spectral analysis over handling the Fokker–Planck equation for solving problems that involved stochastic processes. To interwar physicists, the Fokker–Planck equation had a particular significance: It served as the governing formulation for a stochastic process's probability density and was supposed to tackle the fundamental entities in statistical mechanics. The formulation of a stochastic system in terms of its Langevin equation and the expression of its solution in the frequency domain had a much higher practical value for researchers dealing with an electrical system like radar. The Langevin equation of a mechanical system with thermal agitations or shot effect could easily find a counterpart in the governing equation of an electrical network in a radar receiver. The various types of radar noise were often characterized experimentally with their measured power spectra. Understandably, Uhlenbeck and his associates at Rad Lab, Van Vleck's group at RRL, and Rice at Bell Labs focused on the Fourier transform of random noise, even when they tried to find the probability density functions of the quantities connected to such noise. Throughout Lawson and Uhlenbeck's *Threshold Signals*, the Fokker–Planck equation did not even get mentioned.

This move away from the Fokker–Planck equation (and the style of interwar statistical mechanics it represented) was not restricted to radar researchers only. Joseph Doob, a mathematician at the University of Illinois Urbana-Champaign who pioneered stochastic process theory, expressed explicitly this propensity. In his letter to Uhlenbeck on July 10, 1942, Doob provided an account of a recent paper he had published in *Annals of Mathematics*. The article developed a full-fledged probabilistic theory of a stochastic process that Ornstein and Uhlenbeck mentioned in their 1930 paper on Brownian motion—what Doob dubbed as the "Ornstein-Uhlenbeck process" $x(t)$, where $x(t + s) - x(s)$ was Gaussian with zero mean and variance $(\alpha/\beta)(e^{-\beta|t|} - 1 + \beta|t|)$.[70] Doob developed his theory without using the Fokker–Planck equation. As he told Uhlenbeck:[71]

> I neglect the F.P. equation, perhaps because as a 100% pure mathematician I am quite unable to solve any given differential equation. But it is true that up until recently probability theory has stressed the probability distributions, determined say by the F.P. equations, over the actual path functions of the motion, which I prefer, so I lean in the other direction.

Doob's emphasis on the "actual path functions of the motion" aligned with the general approaches among postwar mathematicians of random processes.[72]

Another mark of the wartime radar research on the studies of random fluctuations had to do with reassessing the legacies of "pure mathematics" epitomized by Wiener's and a few others' works. Throughout the war, the Rad Lab community interacted with Wiener and his protégés—they were on the same MIT campus anyway. Uhlenbeck and his associates were aware of Wiener's theory of random processes, and sometimes worked on similar problems such as prediction. Yet, they did not seem to embrace wholeheartedly Wiener's approach. To them, the eccentric child prodigy's method of handling stochastic processes invoked measure theory and formulated everything in a constructed functional space. Both were unnecessarily complicated and useless mathematical contrivance to the radar researchers, who were primarily concerned with computable numbers and empirically testable outcomes. In a letter to the head of a US naval laboratory in 1946, Uhlenbeck indicated that:[73]

> I think that the study of electrical noise not only is a beautiful application of the theory of random processes, but also allows an experimental approach to problems which are very hard to handle theoretically so that perhaps it would be useful to include it in a general project on random processes.

Samuel Goudsmit also held a down-to-earth attitude toward theories of noise. He wrote to Uhlenbeck in March 1942 that Wiener and his disciples had not led to much because "some of that work cannot be understood by ordinary human beings and the part that can be understood is either too complicated

[70] Joseph Doob, "The Brownian movement and stochastic equations," *Annals of Mathematics*, 43:2 (1942), 351–369.

[71] Joseph R. Doob to Uhlenbeck, July 10, 1942, in Box 1, Folder "Correspondence 1938-44," Uhlenbeck Papers.

[72] Davis and Etheridge, *Bachelier's* Theory of Speculation (2006), xiii–xv.

[73] Uhlenbeck to Carl Eckert, September 9, 1946, in Box 1, Folder "Correspondence 1945-46," Uhlenbeck Papers.

for application or turns out to be rather trivial." Wiener's optimum statistical filter, Goudsmit believed, was too complex to be useful; rather, "for practical purposes any system with an exponential decay will be good enough." The Rad Lab staff had circulated a story about an exchange between Goudsmit and Wiener. "You can keep your Hilbert space," Goudsmit reportedly told Wiener, "I want the answer in volts."[74] Even Doob, who was more sympathetic to the Wienerian approach of functional space and measure theory, admitted that "the main trouble is Wiener's incomprehensibility and the fact that he rarely distinguishes sufficiently clearly between the harmonic analysis of individual functions and that of a stochastic process."[75]

Kac had an even blunter view about this issue. In his letter to Uhlenbeck months after the end of the war, he talked about giving a reprint of Uhlenbeck's paper to his Cornell colleague William Feller, who was a protagonist of the highly abstract mathematical approach to stochastic processes. Kac quoted Feller's recent publication and added with sarcasm:[76]

> Let me quote from a recent paper by Feller: "A third approach is that from the classical time series problem, or, in modern language, from the measure theory in functional space (Wiener, Doob): this approach would lead to the theory of random noises (note the plural!*) which now occupies so many minds (c.f. Doob, 1944)". For your information Doob 1944 is the mysterious paper in Ann. of Math. Stat. Feller's paper just appeared in the Bull. Amer. Math. Soc. November 1945, p. 802. So we better sit down and wait until the "random noises" are safely placed in functional space so that no one can understand what is what.

In his reminiscence four decades later, Kac reflected on the status of the theory of random noise when it was just about to develop into a more general theory of stochastic processes in the wartime and postwar period. As he stated:[77]

> The theory suffers from a slight case of schizophrenia because it is shared, in an uneasy partnership, by the mathematicians on the one hand and by the physicists and communication engineers on the other. In 1943 the mathematical theory was in its infancy and it was beset by subtleties which the physicists

[74] Kac, *Enigmas of Chance* (1985), 110.
[75] Doob to Uhlenbeck, February 26, 1946, in Box 1, Folder "Correspondence 1945-46," Uhlenbeck Papers.
[76] Kac to Uhlenbeck, December 11, 1945, in Box 2, Folder "Correspondence, Mark Kac, 1943-45," Uhlenbeck Papers.
[77] Kac, *Enigmas of Chance* (1985), 110.

> were not inclined to take seriously. The mathematicians looked for rigorous, unambiguous concepts and definitions; the physicists wanted formulas and numbers.

Although Kac was a mathematician by training and profession, he admitted that he was not interested in many pure mathematical subtleties about the foundations of random processes. Rather, he was more concerned with the "superstructure." In a different metaphor, he pondered, his Rad Lab experience widened his scientific outlook and brought him "God-given" and not just "man-made" problems. These meant the practical problems in the real world that demanded solutions from mathematicians, not the problems mathematicians constructed themselves for conceptual clarity or logical foundation building. Such a dichotomy was similar to Henri Poincaré's distinction between the problems "qui se posent" (that are posed) and the problems "qu'on se pose" (that we pose).[78]

The wartime radar researchers' negative attitude toward the approach of functional space and measure theory to random noise certainly did not dispel the seeking of mathematical foundations from the studies of stochastic processes. Yet, the physicists seasoned with radar experiences attempted to frame the postwar research on random fluctuations within their insights on noise in electrical systems. In mid-1945, Wang and Uhlenbeck published an article in *Review of Modern Physics*. Titled "On the theory of the Brownian motion II," it was meant to be a sequel to the piece that Uhlenbeck and Ornstein had published in 1930 under the same title.[79] Like the Uhlenbeck–Ornstein paper, Wang and Uhlenbeck's article was a combination of a general review and original research. The topics of the two works were similar, too: finding the probability distribution of the speed or displacement of a free particle in a viscous liquid, a particle bounded by a simple harmonic force, and one under the influence of random thermal agitations. Fifteen years apart and with outcomes of intense research on radar noise, however, Wang and Uhlenbeck's paper in 1945 had a very different methodological scope from the Uhlenbeck–Ornstein paper in 1930. The Dutch physicists, living pretty much in the "time-domain" tradition of interwar statistical mechanics, dealt with the dynamics of suspended particles with the derivation of the Fokker–Planck equation and the manipulations of the Langevin equation for computing

[78] Ibid., 111.

[79] Ming Chen Wang and George Uhlenbeck, "On the theory of the Brownian motion II," *Review of Modern Physics*, 17:2&3 (April–July 1945), 323–342.

different statistical moments of stochastic quantities.[80] In contrast, the two Rad Lab veterans introduced various new methods and perspectives for studies of random fluctuations.

In light of Kolmogorov's and Wiener's endeavors to give a rigorous formulation of random functions in the 1930s, Wang and Uhlenbeck started with a definition and connotations of a general stochastic process $y(t)$, which was specified by its first-order probability density $P_1(y, t)$ for $y = y(t)$, second-order joint probability density $P_2(y_1,t_1;y_2,t_2)$ for $y_1 = y(t_1)$ and $y_2 = y(t_2)$, third-order joint probability density $P_3(y_1,t_1;y_2,t_2;y_3,t_3)$ for $y_1 = y(t_1)$, $y_2 = y(t_2)$, and $y_3 = y(t_3)$, etc. They also specified a purely random process when $P_2(y_1t_1;y_2t_2) = P_1(y_1,t_1)P_1(y_2,t_2)$, a stationary process when $P_2(y_1,t_1;y_2,t_2)$ depended only on $t = t_2 - t_1$, and a Markov process when the conditional probability density for $y_n = y(t_n)$ given $y_1 = y(t_1)$, $y_2 = y(t_2)$, ..., $y_{n-1} = y(t_{n-1})$ for $t_1 < t_2 < \ldots < t_{n-1}$ only depended on $y_{n-1} = y(t_{n-1})$ (i.e. $P_n(y_n,t_n|y_1,t_1;y_2,t_2; \ldots;y_{n-1},t_{n-1}) = P_2(y_n,t_n|y_{n-1},t_{n-1})$). Wang and Uhlenbeck contended that the Markov process correctly characterized a Brownian motion since thermal agitations were independent of one another at distinct instants. The Markov process led to what they called the Smoluchowski equation (7.4) that governed the second-order conditional probability. Using the properties of Markov processes, they summarized some major results of random walks as a solution of the difference equation governing the discrete conditional probability density.[81]

Moreover, Wang and Uhlenbeck recapitulated the relationship that a random process's power spectrum was the Fourier transform of its correlation. Based on this Wiener–Khinchin theorem, they represented a stochastic process in terms of its harmonic components. When the process was Gaussian, the coefficients of its harmonic components were normally distributed random variables. They—and therefore the stochastic process—could be fully characterized with its correlation matrix at different instants. Thus, the process's spectrum, or equivalently correlation matrix, could serve as the basis for computing its statistical properties. Following their wartime denotation at Rad Lab, Wang and Uhlenbeck called this approach the "method of Rice."[82]

Contrasting the method of Rice was the "method of Fokker-Planck." As statistical physicists, Wang and Uhlenbeck did not discard the Fokker–Planck equation for a Brownian motion's probability density. They rederived it from the Smoluchowski equation and random walks. Yet, the new findings about the spectrum and correlation matrix from "the method of Rice" greatly facilitated

[80] Uhlenbeck and Ornstein, "On the theory of the Brownian motion" (1930), 823–840.
[81] Wang and Uhlenbeck, "On the theory of the Brownian motion II," (1945), 323–328.
[82] Ibid., 326, 328–330.

the solution of the Fokker–Planck equation. In Wang and Uhlenbeck's discussions on the Brownian motion of a free particle and one bounded by a simple harmonic force (the problems that Uhlenbeck and Ornstein had tackled), the results computed from the corresponding power spectrum and correlation matrix were always presented first; the solutions of the Fokker–Planck equation came afterward as a hindsight.[83]

For instance, in tackling the problem of a simple harmonic oscillator,

$$\frac{d^2y}{dt^2} + \beta\frac{dy}{dt} + \omega_0{}^2 = F(t), \tag{10.11}$$

where $F(t)$ was a random force, Wang and Uhlenbeck characterized the system in terms of y and its derivative $p = dy/dt$. Instead of a fixed trajectory represented by a curve on the typical y-p phase plane, however, y and p were random processes represented by an uncertain region on the plane, with its center at the mean and size comparable to the variance. Here Wang and Uhlenbeck employed the method of Rice to obtain the variances of p and y as functions of time. They found that $(p-\bar{p})^2 \cong 2Dt$, $(y-\bar{y})^2 \cong (2/3)\,Dt^3$, and $(p-\bar{p})(y-\bar{y}) \cong Dt^2$ for small t. This implied that p and y on the phase space spread with different rates over time—p spread much faster than y when t was small. Starting at a point (p_0,y_0), therefore, the uncertain region of the stochastic simple harmonic oscillator moved along an inward spiral toward the origin, like the behavior of a deterministic harmonic oscillator. Yet, the uncertain region changed shape and size as it moved along the spiral: it would become an ellipse elongated much more saliently along p than y due to the former's much greater rate of spread at small t, but then the radii along both directions spread toward an equal size, making the uncertain region an expanding circle. At the end of the process when t went to infinity, the center of the uncertain region reached the origin, and $P(p, y,t|p_0,y_0)$ became the Maxwell–Boltzmann distribution (Figure 10.4). The conclusion from this examination of a fluctuating harmonic oscillator was not surprising. But Wang and Uhlenbeck's treatment of it—marked by the method of Rice based on spectrum and correlation and the representation of noise as an uncertain region in the phase space—was novel.[84]

Next, Wang and Uhlenbeck ventured into a new topic, Brownian motion of a system of mutually coupled harmonic oscillators. This corresponded to a scenario of multiple particles, each of which was subject to a velocity-dependent

[83] Ibid., 332–335.
[84] Ibid., 332–335.

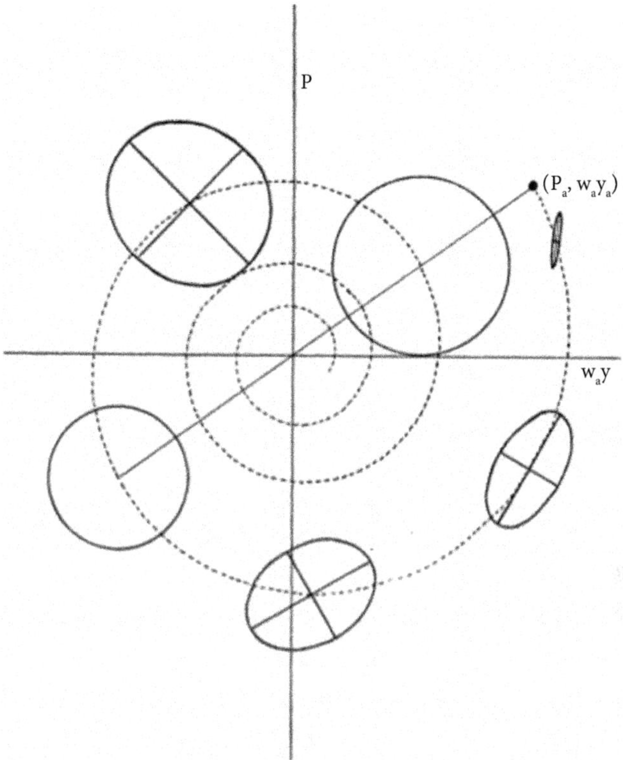

Figure 10.4 Wang and Uhlenbeck's illustration of a Brownian harmonic oscillator's temporal evolution in the phase space. The circular or elliptical areas represent the regions of uncertainty due to a stochastic force. Reprinted Figure 1 with permission from Ming Chen Wang and George Uhlenbeck, "On the theory of the Brownian motion II," *Review of Modern Physics*, 17:2&3 (April–July 1945), 336.

friction, an elastic restoration, a random thermal agitation, and similar forces coupled from other particles. The resultant Langevin equation of motion became a system of dynamic equations:[85]

$$\frac{d^2y}{dt^2} + \beta\frac{dy}{dt} + \omega_0{}^2\sum_{j=1}^{n}\left(L_{ij}\frac{d^2y_j}{dt^2} + R_{ij}\frac{dy_j}{dt} + G_{ij}y_j\right) = \sum_{j=1}^{n} E_{ij}. \qquad (10.12)$$

Wang had been dealing with the problem of coupled harmonic oscillators since she worked for Uhlenbeck at the University of Michigan and MIT

[85] Ibid., 335.

Rad Lab during World War II. To her, this dynamic problem was significant because she could find a pragmatic counterpart in electrical circuit theory for applications in radars or telecommunication. In other words, equation (10.12) could be used to characterize an n-branch electrical network, in which the inertia terms became inductive currents, the friction terms became resistive currents, the restoration terms became capacitive currents, and the molecular agitations became thermal noise. To deal with the problem, Wang and Uhlenbeck invoked again the correlation matrix for the multidimensional stochastic process in (10.12) and solved the matrix equation via the common techniques of diagonalization and eigen-problem formulation in linear algebra.[86]

The Wang–Uhlenbeck paper in 1945 was rooted in Wang's doctoral dissertation and her work at Rad Lab. The problems of single and coupled harmonic oscillators were already under her scrutiny for their practical implications for radar detection as well as their theoretical implications for the transport phenomena. The paper became a crucial reference for the studies of random fluctuations shortly after its publication. Wang and Uhlenbeck were aware that this work was not only a report of their original findings, but also a synopsis of what they deemed the framing intellectual elements of the field. They expressed in a subtle yet clear manner their view of the "pure" mathematical approach to Brownian motions, as they stated in a footnote:[87]

> The authors are aware of the fact that in the mathematical literature (especially in papers by N. Wiener, J.L. Doob, and others [...) the] notion of a random (or stochastic) process has been defined in a much more refined way. This allows for instance to determine in certain cases the probability that the random function $y(t)$ is of bounded variations, or continuous or differentiable, etc. However, it seems to us that these investigations have not helped in the solution of problems of direct physical interest, and we will, therefore, not try to give an account of them.

Instead, they articulated what they believed to be the major problems that called for solutions. At the end of their paper, they listed such "unsolved problems": the probabilistic distribution of particles subject to a gravitational force in addition to thermal agitations, the statistics of the time at which a random process first passed a threshold level, the probability density of the duration within which a random process returned back to its original value, the distribution of a fluctuation's average value, and the statistics of a random function's

[86] Ibid., 325–328.
[87] Ibid., 324.

maximum value for a given interval.[88] As a junior scientist who co-produced this central piece after four years working with her mentor in Ann Arbor and Cambridge, Wang would likely open up her career via tackling some of these problems she laid out with Uhlenbeck and carve out an avenue for herself in statistical physics.

But this avenue did not open to her. In 1946, Wang returned to China, stayed briefly, got married, and came back to the US the next year to take a research position in California for the navy through Uhlenbeck's help. Two years later, the Chinese Communist Party overturned the Nationalist government and established the People's Republic. Wang began to conceive the idea of moving back to China permanently when the Korean War broke out in 1952. Because of her working experience at Rad Lab, getting the immigration authorities' approval for doing so was difficult. Eventually, she received permission to leave the US, and returned to China with her husband in 1955.[89] Uhlenbeck lost touch with his student on the other side of the Iron Curtain, until June 1980, when he established contact through a Chinese visiting scholar in the US. Wang wrote a warm but sad letter to Uhlenbeck. She said:

> I was very touched to hear that you had inquired about me several times, but I am ashamed of myself for not able to accomplish any work worthy of reporting you. Perhaps you would like to know a little about my situation since I got back. I had been teaching physics here at Tsing Hwa [Tsinghua University] since then.

She was assigned to the sub-group teaching theoretical physics. She taught statistical mechanics and thermodynamics most of the time. While this arrangement eventually settled her down and gave her an opportunity to continue an academic career after decades of wars, atrocities, and turmoil, another characteristically Hobsbawmian catastrophe came to her. Wang continued:

> When our Cultural Revolution began in 1966, all the schools became a world of chaos and most of the teachers were accused of being guilty. My husband and I were put into the jail by the "gang of four" over night. Fortunately, we were both strong enough mentally as well as physically to go through the year of nightmare-like life.

[88] Ibid., 338–339.
[89] Wang, "Zhuanshun jiushi zai" (2006), 179.

Although she was released and was given back her professorial title, she admitted that "six years of prison life do leave a scar on me," including severe insomnia. She ended up not producing any further research on Brownian motion or random noise. The unfinished problems in the 1945 paper remained her road-not-taken.[90]

The Wang–Uhlenbeck paper epitomized postwar physicists' efforts to chart the intellectual landscape for the studies of stochastic processes in light of the wartime technological research on random noise in radar and telecommunication systems. Under this framework, the considerations and perusals of electrical noise, like statistical physics of Brownian motion, left significant marks on the theories of random fluctuations in terms of linguistic or lexicographical expressions, conceptual metaphors, and analytical or computational techniques. Their reservation about what they deemed the "pure" mathematical pursuits of the foundational issues in stochastic processes—represented by Wiener's approach via the measure theory and functional space—was also salient.

Yet, this charting and framing were hardly acceptable to those mathematicians who sided with Wiener. In 1946, Wiener's students Norman Levinson, Walter Pitts, and W.F. Whitmore wrote a letter to the editor of the journal *Science*, reminding its audience of the contributions their mentor had made to the studies of random noise well before engineers and physicists stepped into the field.[91] The problems that instigated a rapidly growing interest, they asserted, were the ones that dealt not "with a single numerical random variable or a finite collection of them, but instead with an infinite sequence of numbers given at random, or a continuous random function." Such random functions could include "the coordinates of a particle in the Brownian motion, or the path of a molecule in a gas, or the 'noise' potential in electrical machines." Levinson et al. made a clear priority claim about the treatments of such problems. According to them:

> The recent work in this field provides an excellent example of the impediment which the common relation between allied disciplines may place in the way of scientific advance. A large fraction of the methods and ideas of this subject has been rediscovered since 1940 by physicists and electrical engineers, almost wholly unaware that the same problems had been raised and solved in

[90] Wang to Uhlenbeck, June 28, 1980, in Box 2, Folder "Ming-Chen Wang, Mrs. Chang and Children, 1978-80," Uhlenbeck Papers.

[91] Norman Levinson, Walter Pitts, and W.F. Whitmore, "Recent contributions to the theory of random functions," *Science*, 103:2670 (1946), 283–284.

> the mathematical literature a decade before. What differences exist in treatment or proof are either notational, or minor ones dictated by the traditional oppositions between the physicist's ability to make physical intuition bolster an heuristic argument and the mathematician's demand for maximum rigor and generality.

Levinson et al. elaborated on what they meant by particulars. They indicated that in a series of papers published in the 1930s:

> N. Wiener develops a theory of Gaussianly distributed random functions, both in the wholly independent case with a "white" spectrum and the more general one with an arbitrary "power spectrum." He derives a general formula, in the form of a definite integral, for calculating the average of any function or functional of one or two such random functions—or any number, by obvious extension.

Moreover, they noted:

> The discussion proceeds from a theory of formal Fourier series with Gaussianly distributed coefficients. This is called the "Method of Rice" by M.C. Wang and G.E. Uhlenbeck in view of the extensive use made of it in S.O. Rice's review of 1944.

Levinson et al. contended that:

> All of the fundamental methods presented by Rice, except for the discussion of the shot effect, are to be found in this work of Wiener. A large part of the special results may be obtained easily by substituting in Wiener's general formula.

They went on by stating that Rice's various formulas for Gaussian noise after nonlinear devices and the fundamental equation of the correlation function either appeared in Wiener's work in different forms or could be deduced from it. Levinson et al. did acknowledge the "real advance" of the recent works on random noise such as Rice's and Kac's. However, "on precisely this account it is a greater pity that this further work could not have commenced when the basis was first obtained by the pure mathematician, in which case we should be a decade further today."

The position of Levinson et al. was clear: the "pure" mathematicians did more than what the physicists and engineers had thought. In fact, they argued,

Wiener and his collaborators had already obtained the major results for harmonic analysis of random fluctuations almost a decade before Uhlenbeck, Kac, Wang, and Rice delved into the subject. What delayed the scientific progress was the physicist and engineering communities' failure to notice the mathematicians' outcomes, Levinson et al. held, not the mathematicians' inability to produce useful results.

We do not have to take a side on the deliberations over priority between physicists/engineers and mathematicians. But Levinson et al.'s 1946 piece in *Science* reminds us of Wiener's role in the studies of random noise. Despite being downplayed by the likes of Kac and Goudsmit, in fact, that role was active not only in the 1930s but also during the war. In chapter 11, we will see Wiener's work on random noise through his involvement with stochastic prediction and filtering for antiaircraft gun control. This work would play a pivotal part in his development of cybernetics, a direction that the radar-detection circle did not imagine in the first place.

11
Filtering Noise for Antiaircraft Gunfire Control

While Uhlenbeck and his collaborators affiliated with Radiation Laboratory were improving radar target detection in noisy environments, another team across the MIT campus was coping with noise in a different war project. Under NDRC, Norbert Wiener and his assistant Julian Bigelow worked on the design of an antiaircraft gunfire control system (what Wiener called AA) for the US Army. Wiener's project was much smaller in scale than the radar business at Rad Lab. The product he and Bigelow delivered was not used in any battlefield. Yet, historians including Galison and Kline studied closely Wiener's wartime AA project, because it led to the formation of Wiener's cybernetics and facilitated his conceptualization of information.[1] In this chapter, we revisit Wiener's AA project from a different perspective: we will see that this undertaking marked another breakthrough in theoretical studies of noise.

Like Uhlenbeck, Kac, and Wang, Wiener and Bigelow employed the language of communication engineering and used the notion of noise to characterize the ambient factors that caused errors, fluctuations, and deviations in technological systems. To represent the errors of tracking an enemy airplane's position due to a gunner's oversights or devices' dysfunction, Wiener used additive noise modeled as a random stationary time series with a Gaussian distribution and a simple spectrum. This was similar to the flickers, clutter, and fluctuating bright spots that Uhlenbeck et al. tried to remove from a radar screen. In fact, Wiener's stochastic model of noise in AA had already existed in the interwar works on Brownian motion and electronic noise. The originality of Wiener's approach was a new way to *treat* such noise.

Uhlenbeck et al. aimed to optimize radar target detection amidst random noise—determining whether a target was present or not given statistical data from radar returns. In AA, in contrast, Wiener and Bigelow aimed to reproduce the enemy target trajectory polluted by random noise in observing and

[1] Galison, "Ontology of the enemy" (1994); Kline, *Cybernetic Moment* (2005), 18–21.

tracking devices. In other words, what they dealt with was a prediction problem (when the reproduced trace led the actual path) or a filtering problem (when the reproduced trace lagged or synchronized with the actual path). Removing unwanted components of a waveform had been a classical problem in telephone and radio engineering. The problem had been tackled through filter design that focused on specifying and implementing a proper frequency response for functional blocks. Wiener's optimum predictor was later known as the "Wiener filter." Drawing upon studies of stochastic noise, however, the Wiener filter differed fundamentally from typical filters in radio or telephony. Instead of selecting a designated bandwidth, the Wiener filter utilized statistical properties of signal and noise—especially their correlation functions—to minimize the latter. In so doing, Wiener turned noise into unwanted waveforms that could be manipulated and filtered through generalized harmonic analysis.

Wiener at the Dawn of the War

Throughout the 1920s–30s, Wiener made a reputation as a mathematician. He visited academic institutions in Göttingen, Cambridge, and Peiping, which facilitated his building of an international scholarly network. He also developed a collaborative research pattern, pivoting on correspondence with senior mathematicians (G.H. Hardy in Cambridge and Paul Lévy in Paris) and working directly with junior colleagues (Eberhard Hopf and Raymond Paley), graduate students (Yuk Wing Lee, Norman Levinson, and Ikehara Shikao) at MIT, and others (Aurel Wintner). Wiener's work on generalized harmonic analysis resulted in broader analysis of Tauberian theorems for series convergence and functional spaces. He also continued the studies of potential theory, which involved the exploration of various differential and integral equations. Meanwhile, his earlier interest in Brownian motion led him to examine measure theory of random processes, ergodic theory, and the mathematical foundation of statistical mechanics.[2]

By the late 1930s, Wiener's major focus in mathematical research was on the applications of measure theory, the Lebesgue integral, and ergodic theory to statistical mechanics. This subject was a product of a research program in mathematical probability that Wiener had helped initiate. In 1921–23, he constructed differential space as a formal representation of Brownian motion. In

[2] Wiener, *Life in Cybernetics* (2018), 273–387.

1931, Kolmogorov published his axiomatization of probability as a Lebesgue measure function in a state space. Two years later, Birkhoff at Harvard and von Neumann at Princeton presented the ergodic theorem that ensured the existence—in the limiting case of infinite iterations—of the sum of a measurable function mapped recursively by a measure-preserving transformation. As a corollary, this sum was almost always equal to the function's integral over the state space. Birkhoff's and von Neumann's ergodic theorem gave a measure-theoretic basis of a hypothesis that Boltzmann had entertained in the late nineteenth century: the time average of a fluctuating quantity in statistical mechanics equaled its average over all possible physical configurations (later called the ensemble average).[3]

Wiener was heavily influenced by this line of research. In 1938, he entertained the notion of "homogeneous chaos," attempting to formulate a statistical mechanical system in a probability space with a Lebesgue measure. In his view, statistical physics since Gibbs had dealt with a system of many entities in terms of their statistical averages in a typical phase space. From the recent development in mathematical probability, however, Wiener contended that it was time to represent such statistical averages in a separate and abstract probability space with properties stipulated by the ergodic theorem. Wiener characterized this probability space as a "homogeneous chaos" (not the "Lorenz-butterfly" type of chaos intrinsic to systems sensitive to initial conditions), where the system was sufficiently random and distributed widely. To him, Brownian motion was the only kind of "homogeneous chaos" that had received adequate mathematical treatments. He viewed the Gibbsian statistical mechanics as another kind of homogeneous chaos and aimed to tame it with the toolkit of probability space.[4]

This ethereal mathematical research was conducted at a turbulent time. In the late 1930s, the clouds of atrocities and wars loomed large. Wiener lived and worked in a safe place. Unlike Uhlenbeck, Goudsmit, Kac, and Wang, he did not have family members trapped in war-torn or dictatorially occupied homelands. Yet, the expansion of the Nazis gave him emotional stress as a Jew. As he recalled:[5]

> In America there was a reaction in our favor from the atrocity and terror of the German situation, but this did not completely compensate for the knowledge

[3] Wiener et al., *Differential Space, Quantum Systems, and Prediction* (1966), 2–9.

[4] Norbert Wiener, "The homogeneous chaos," *American Journal of Mathematics*, 60:4 (1938), 897–936.

[5] Wiener, *Life in Cybernetics* (2018), 378.

> that somewhere in the world we were being threatened with extermination, and that Nazi anti-Semitism had provoked an echoed anti-Semitism in some American quarters.

Wiener had relatives and colleagues in Germany. The news from these acquaintances was getting worse. Meanwhile, his visit in China in 1934–35 gave him first-hand experience with the dangerous situation there. When the Sino-Japanese War broke out two years later, Tsing Hua University which he had visited was shut down. The host of his visit—his former student and collaborator Yuk Wing Lee who was then a faculty member at Tsing Hua—was forced to escape from Peiping and got stuck in the International Settlement in Shanghai. Wiener's Chinese friends asked him to support China's cause against Japan's invasion. Lee asked for his help with entering the US. At the same time, he saw more and more European intellectual refugees arriving at American academic institutions. Even from the viewpoint of an established scholar at a secure and peaceful ivory tower, the world was on the verge of breakdown.[6]

By the end of the 1930s, many American scientists and engineers in academia believed that the US would enter a world war and began to prepare for it. Wiener contemplated about the possible role he could play in the war, too. His first thought was computing. At a summer meeting of the American Mathematical Society in August 1940 at Dartmouth, Massachusetts, Wiener proposed the idea of building a digital automatic computer for ballistic calculations. At this time, ballistics were formulated as solutions to partial differential equations and were tackled with analog computers like Vannevar Bush's differential analyzers. Wiener suggested that the problems be discretized and handled by binary machines. After the meeting, Wiener ran through this idea with Bush, his former MIT colleague and now the head of NDRC. Bush asked Wiener to develop the concept further. But their exchange of letters in September and October did not lead to a concrete project on digital computers, for Bush decided the idea impractical for urgent military purposes.[7] Soon after, Wiener entertained the possibility of contributing to cryptography and pitched Warren Weaver, head of the Fire Control section at NDRC Division 4, a scheme of encryption. That thought did not materialize, either.[8]

[6] Ibid., 375–387.

[7] Ibid., 391–398; Letters, Wiener to Bush, September 21 and 23, 1940, and Bush to Wiener, September 24 and 25 and October 7, 1940, all in Box 4, Folder 58, Wiener Papers.

[8] Wiener, *Life in Cybernetics* (2018), 398–399; Warren Weaver to Samuel Caldwell, November 15, 1940, in Box 4, Folder 58, Wiener Papers.

Tackling the Antiaircraft Problem

In his exploration of a plausible contribution to the war efforts, Wiener eventually settled down with antiaircraft gunfire control. At the outbreak of the war in Europe, the American and British militaries were aware of the importance of air defense. The enactment of Rad Lab for the development of radars was an immediate response to that demand. As the fighting experiences from the Battle of Britain showed, radars provided tremendous help in bringing down enemy aircraft by pinpointing their positions. Yet, another crucial factor for successful air defense was a gunfire control system that enabled gunners to aim at, track, and shoot down inimical targets. Since it took time for shells and missiles to reach a target, aiming gunfire directly at it would be useless. Key to successful elimination of a target was the ability to *predict* its trajectory several seconds in advance. At the end of 1941, Wiener believed that his mathematical expertise could help cope with this problem of prediction.

To Wiener, predicting an enemy airplane's trajectory was a problem of extrapolation. The future position of a target was estimated from its past positions. This could be done via constructing a smooth function of time based on the data points of observed target positions in the past. The easiest construction was linear, i.e. a function as a weighted sum of the past data points. This predictor was a linear operator that took a target's past trajectory as input. It resembled a linear electrical circuit taking a time-varying electrical current as input. The only design complexity was that the output at time t from this linear network depended only on the input before t. This demand on "causality" required that the linear system's transfer function had to satisfy certain conditions. Once that was met, the output from this predictor was a smooth and linear extrapolation of observational data.[9]

Wiener's scheme required sophisticated computation. In late 1940, he mentioned the idea to Samuel Caldwell at MIT Center for Analysis, who was in charge of the Institute's high-capacity computing facility, the differential analyzer that Bush had built in the 1930s. Wiener and Caldwell probed Wiener's ideas on the differential analyzer and reached out to NDRC Division D for further support. In January 1941, Wiener's project received a grant from Division D. He recruited an MIT-trained IBM electrical engineer Julian Bigelow and hired a technician Paul Mooney and a woman computer "Miss Bernstein"

[9] Wiener, *Life in Cybernetics* (2018), 399–400.

to work on the AA.[10] Meanwhile, Bush appointed Wiener as a consultant for Division D in March 1941, and introduced him to the projects at Rad Lab.[11]

In the following months, Wiener and Bigelow turned the mathematical formulation of the extrapolator into an electrical network. Wiener was not a stranger to filter design. In 1928–30, he supervised Yuk Wing Lee's doctoral dissertation at the Department of Electrical Engineering on representing an electrical network's impedance or admittance with a special function known as Laguerre polynomials. Wiener's insight to help solve Lee's central technical problem was to impose the causal condition onto the network's impedance and admittance. The physical implementation of Laguerre polynomials with passive electrical components turned out to be simpler and cheaper than that based on the impedance or admittance's typical frequency response. In 1935–38, Wiener and Lee further developed this scheme and obtained patents on network synthesis, which were acquired by AT&T.[12] Even so, Wiener was not an electrical engineer. He needed Bigelow's hands-on talent to transform his statistical predictor-filter into electrical circuits of resistors, capacitors, and vacuum tubes.[13]

In summer 1941, Wiener and Bigelow finished a prototype predictor, and put it into preliminary trials. Data from actual flight trajectories were unavailable. In lieu of them, Bigelow designed a simulator to generate pseudo-data arguably resembling realistic flight paths: a white spotlight was cast on a wall and moved back and forth every fifteen seconds. A human operator mimicking a pilot used a joystick to control a color spotlight, aiming to track the white light. To make the experimental situation realistic, Bigelow made the white light's movement irregular and fast, and introduced a lag to the joystick. The recorded data of the color light's position over time were a simulacrum of an actual flight path. They could be used to check the efficacy of Wiener and Bigelow's predictor.[14]

The outcomes from the preliminary testing in mid-1941 were interesting, but not favorable to Wiener's scheme. The Wiener–Bigelow predictor could estimate accurately the position of the color spotlight seconds in advance when its trajectory was *smooth*. When the color light's trajectory was piecewise linear, the predictor output had significant oscillations right after the

[10] Ibid., 400; Galison, "Ontology of the enemy" (1994), 234–235; Kline, *Cybernetic Moment* (2015), 19.

[11] Letter, Bush to Wiener, March 10, 1941, in Box 4, Folder 59, Wiener Papers.

[12] Charles Therrien, "The Lee-Wiener legacy: A history of the statistical theory of communication," *IEEE Signal Processing Magazine*, 19:6 (2002), 33–35.

[13] Galison, "Ontology of the enemy" (1994), 235.

[14] Ibid., 236–237.

pseudo-pilot trajectory transitioned from one linear segment to another. The predictor was oversensitive. Wiener and Bigelow tried various means to improve its performance, but to no avail. In the end, Wiener concluded that the tradeoff between accuracy and stability was fundamental as long as the predictor was designed on the basis of smooth extrapolation, like the conjugate pair of position and momentum in Heisenberg's uncertainty principle.[15]

Upon a closer look, Wiener believed that the solution to the problem of oversensitivity was statistics. Instead of using the data set of a single trajectory, the predictor should use and average the data sets of *many* trajectories created by the same operator under the same condition. A good-enough predictor accomplishing both accuracy and stability could be obtained via minimizing the mean-square difference between the actual trajectory and its prediction. In this regard, Wiener's new predictor was a solution to find a trajectory that minimized its mean-square error with respect to all observed target paths.[16]

Wiener's conversion to statistical operations as a response to an immediate technical difficulty ended up grappling with profound issues in the general problem of prediction. Two types of uncertainties were involved in the estimate of an enemy airplane's position: the aircraft's zigzag movements when its pilot maneuvered to escape gunfire, and observational errors of the aiming operation due to the antiaircraft system or its operator.[17] Wiener's new approach to antiaircraft prediction treated these two uncertainties as random functions of time whose statistical properties did not change with the moment when the observation began. That is, both the actual enemy aircraft trajectory and observational errors were *stationary* stochastic processes. In addition, he supposed that observational errors were *additive* to the actual trajectory. He modeled both stochastic processes with Gaussian distributions.

As Wiener was developing his theory of statistical prediction in the second half of 1941, he became aware of its broader implications for other war projects. Wiener's modeling of the antiaircraft problem was contingent upon the notions and language of telecommunication engineering—with the enemy aircraft trajectory understood as "signal" and observational errors as "noise." The relationship between this "signal" and "noise" in Wiener's antiaircraft approach connoted the same ontology of the ambience that Uhlenbeck et al. adopted in their work on radar detection: environmental uncertainties were additive and stationary Gaussian noise. This ontology permitted Wiener to characterize the statistical attributes of both the enemy aircraft trajectory and

15 Wiener, *Life in Cybernetics* (2018), 401–402; Kline, *Cybernetic Moment* (2015), 20.
16 Wiener, *Life in Cybernetics* (2018), 402–403.
17 Galison, "Ontology of the enemy" (1994), 235.

observational errors with their correlation functions. That means, the statistical handling of the problem could be conducted via generalized harmonic analysis on correlation functions, which Wiener had developed in the 1930s.

Wiener expected an immediate applicability of his theory to Rad Lab researchers' work on radar noise reduction. After the Pearl Harbor attack, he was more active in selling his approach to Rad Lab. In a letter on February 22, 1942, Wiener indicated to Weaver that "there is quite a demand for instruction in our theory here in the Radiation Laboratory and in such places." His student Norman Levinson "would like to give a restricted course on the subject. Can I let him?"[18] But Wiener's advocacy did not proceed well. The course that he wanted Levinson to teach was "cut to six lessons." The accumulating frustration led Wiener to break away from Rad Lab. On March 22, 1942, he wrote to its director E.L. Bowles and offered his resignation:[19]

> My reason for resignation is that with the highly chaotic and anarchic regime of theoretical work in the radiation laboratory, my efforts there have proved unwelcome, a waste of time on my part, and in view of the general amateurishness of the theoretical staff.

Circuit theory, he continued, should be an integral part of training for theoretical researchers at Rad Lab. Thus:

> where new members of the staff of your Laboratory are recruited from the theoretical physicists or mathematicians of the country, or indeed anywhere except from among the ranks of communication engineers in the strictest and narrowest sense of the term, some adequate instruction in this field should be given. It is not something which a quantum physicist has any reason to know the slightest thing about, and to turn such an individual loose in your laboratory without special training, no matter what a big shot he may be in his own subject, is like ordering a corn-doctor to amputate a leg.

What the theoretical staff at Rad Lab should know, Wiener contended, included "the proper use of Fourier integrals and of operational notation." Moreover:

> for work in which noise is as important a limiting factor as in what you are doing, there should be a thorough understanding of the Brownian motion

[18] Letter, Wiener to Weaver, February 22, 1942, in Box 4, Folder 62, Wiener Papers.
[19] Letter, Wiener to Bowles, March 22, 1942, in Box 4, Folder 62, Wiener Papers.

> and that of generalized harmonic analysis which includes the theories of Fourier series and of Fourier integrals, but is far more general, and is peculiarly applicable to noise problems.

Wiener pointed out that all these subjects were fields "in which I have been laboring for close to a quarter of a century, and in which the fundamental theorems are largely my own." Yet, he found in frustration that at the laboratory "there has been very little disposal to use this knowledge in any large way." Instead, he was only asked to provide "answers to specific problems." To him, continuing to be involved in the laboratory was a "waste of time."

Galison interpreted Wiener's resignation as a clash of his view on the centrality of "suppressing noise and conveying information" with those of the "fundamental" physicists, like DuBridge, N.F. Ramsey, Schwinger, Rabi, and Edward Purcell at MIT Radiation Laboratory.[20] Such a conclusion is inevitable, if we interpret literally Wiener's bitter letter of resignation. As we have seen in chapter 10, however, some theorists at Rad Lab examined closely the effects of noise on radar detection. Uhlenbeck, Goudsmit, Wang, and their external affiliates Kac and Lawson undertook a research program on the characteristics of noise in radar systems and its impacts on radar performance. They had no less expertise than Wiener on Brownian motion, Fourier integrals and series, and generalized harmonic analysis. We have also seen the postwar tension between Uhlenbeck's camp and Wiener's camp about credits for the spectral analysis of stochastic processes—whether it should be termed as "the method of Rice" or a Wienerian theorem largely "of his own." Also, Goudsmit's and Kac's view of Wiener's approach to random processes as too abstract to be useful for military purposes was a sharp contrast to Wiener's view about the lack among those "big shots" in "fundamental" physics of knowledge in practical communication engineering. Wiener's dissatisfaction with Rad Lab had multiple connotations.

After leaving Rad Lab, Wiener speeded up with the development of his statistical antiaircraft gunfire director. The hardware was almost ready, and the upcoming task was to test the prototype. On April 23, he informed Weaver in writing: "I got all the computations completed for our test and that Bigelow is getting very close to finishing the apparatus."[21] On July 1, George Stibitz, former Bell Labs researcher and now a technical aide of NDRC D-2 (Division D, Section 2) on fire control, paid a visit to Wiener's laboratory. Stibitz registered

[20] Galison, "Ontology of the enemy" (1994), 240.
[21] Letter, Wiener to Weaver, April 23, 1942, in Box 4, Folder 62, Wiener Papers.

his witness of the impressive and even uncanny performance of the Wiener–Bigelow apparatus in predicting simulated flight trajectories, even though the leading time was simply one to two seconds.[22]

To Wiener and Bigelow, the future of their project depended on a successful demonstration of the statistical predictor's capacity in more realistic antiaircraft scenarios. Testing the apparatus in a theater of war was out of the question. But they could hope for a trial on tracking data from actual flights. In the following months, they visited Langley Field, Aberdeen Proving Ground, Frankford Arsenal, and Foxboro Instrument Company. At the Anti-Aircraft Board at Camp Davis, North Carolina, they obtained two tracking data sets on test flights labeled 303 and 304.[23]

Wiener and Bigelow employed their apparatus on the tracking data for flights 303 and 304 and assessed the rate of successful "hits." Their reference was another antiaircraft gunfire director, T-15, developed at Bell Labs. The Bell researcher Hendrik Bode designed the core mechanism for T-15's target prediction.[24] This mechanism involved geometric extrapolations. In a simple form, the Bode predictor estimated the position of an airplane in the future by treating the trajectory as a straight line after the last observation, computing the trajectory's derivative, and using it as the slope of the extrapolated linear segment. In a more elaborate form, the Bode predictor recorded the error between the estimated and actual positions and corrected this error with negative feedback. Wiener and Bigelow computed the performance of their statistical predictor (termed "Statistical"), the simple Bode predictor (termed "Bode"), and the Bode predictor with feedback correction based on the extrapolation of data within ten seconds before (termed "10 Sec. Bode"):

Track	(1) Bode	(2) 10 Sec. Bode	(3) Statistical
303	6 hits	22 hits	23 hits
304	35 hits	55 hits	49 hits

The conclusion from these results was clear: The Wiener–Bigelow predictor outperformed the simple Bode method but was inferior to the Bode method with feedback correction. Given that the statistical method involved much more complicated computations than the Bode method, choosing Bell Labs' geometrical predictor made more sense. In Wiener's report to Weaver on

22 Galison, "Ontology of the enemy" (1994), 243.
23 Ibid., 244.
24 Mindell, *Between Human and Machine* (2002), 243–244.

January 15, 1943, he considered this "winding up of his responsibility" for the AA project.[25] Weaver was unimpressed, too. At a meeting of NDRC Division 7 on Fire Control (formerly D-2) in January 1943, Wiener's contract was not renewed. His project was terminated. Bigelow left to join a statistical fire control group at Columbia University. Wiener's venture into the antiaircraft problem ended.[26]

Noise Filtering through Statistical Prediction

The significance of Wiener's antiaircraft project lay not in the weapons system it promised to deliver, but in its theoretical implications. It is well known that Wiener's exploration of the antiaircraft problem led him to conceptualize a new human–machine relation, in which actions of a human or even an organism were construed as results of a feedback control system that connected sensing, decision, and actuation through communication links. In 1943, Wiener, Bigelow, and Harvard physiologist Arturo Rosenblueth published "Behaviour, purpose, and teleology," marking Wiener's departure from detailed electromechanical control for a general understanding of seemingly purposeful behavior in terms of feedback loops.[27] In the following years, Wiener developed these ideas into a worldview and a research program on feedback mechanisms underlying biological, social, cognitive, and technological phenomena. His book *Cybernetics, Or Control and Communication in the Animal and the Machine* in 1948 summarized the major elements of his new worldview.[28] Along with his building of cybernetics, Wiener entertained the notion of information as a quantifiable entity, and came up with a similar formulation of information as Shannon did.[29] Galison, Kline, and Mindell have focused on the cybernetic ramifications of Wiener's antiaircraft work for the emergence of a new human-machine ontology, conceptualization of information, and development of feedback control technology.

This perspective nonetheless has sidelined a crucial theoretical aspect of Wiener's antiaircraft work that handled what he considered were intrinsically uncertain situations in information, communication, control, and computing in general, and the grappling with noise in particular. Random messages,

[25] Letter, Wiener to Weaver, January 15, 1943, in Box 4, Folder 64, Wiener Papers.
[26] Mindell, *Between Human and Machine* (2002), 280–282.
[27] Arturo Rosenblueth, Julian Bigelow, and Norbert Wiener, "Behaviour, purpose, and teleology," *Philosophy of Science*, 10:1 (1943), 18–24.
[28] Norbert Wiener, *Cybernetics, Or Control and Communication in the Animal and the Machine* (Cambridge, MA: MIT Press, 1948).
[29] Kline, *Cybernetic Moment* (2015), 21–23.

signals, and interferences were an essential character of Wiener's cybernetic worldview. Not a surprise, he opened his 1948 monograph with the reversible Newtonian time and irreversible Bergsonian time, Gibbs's statistical mechanical picture of the universe, and probabilistic, measure-theoretic treatments of the picture. Quantity of information, feedback and oscillation, computing machines and the nervous system, cognition, language, and forms of society—the well-known hallmarks of the cybernetic framework—came later.[30]

To understand the statistical elements of Wiener's worldview and its relationship to the theoretical tackling of noise in mid-century information sciences, we must look into another document that has been largely neglected in the historiography of cybernetics due to its obscure and difficult mathematics. In February 1942, a month after NDRC closed his antiaircraft contract, Wiener submitted a report to summarize the statistical theory of prediction that he and Bigelow had developed. Titled "Extrapolation, interpolation, and smoothing of stationary time series," this report gained a racist nickname "Yellow Peril" (in the background of the US's Pacific War against Japan) among NDRC engineers, for its yellow cover and challenging contents.[31] Four years after the end of World War II, the document was declassified and published as a monograph.[32]

The core problem that preoccupied Wiener in his antiaircraft project was prediction of an enemy airplane's position in the future based on recorded data of its trajectory in the past. This was a problem of extrapolation. As mentioned, Wiener and Bigelow found the geometric extrapolation inadequate and decided to pursue a statistical approach. Specifically, Wiener assumed the target trajectory was a stochastic process $f(t)$, which he called a continuous time series à la statisticians. He supposed that $f(t)$ was stationary, in the sense that its statistical properties did not depend on the initial time, since the behavior pattern of the enemy pilot was time-invariant. The autocorrelation of a stationary process $f(t)$ at two instants $\varphi(t_1, t_2) = \langle f(t_1) f^*(t_2) \rangle$ depended only on the time difference $t = t_2 - t_1$ (f^* was the complex conjugate of f). Invoking the ergodic theorem, the autocorrelation

$$\varphi(\tau) = \lim_{T\to\infty} \frac{1}{2T} \int_{-T}^{T} f(t+\tau) f^*(t)\, d \tag{11.1}$$

[30] Wiener, *Cybernetics* (1948).

[31] Kline, *Cybernetic Moment* (2015), 21.

[32] Norbert Wiener, *Extrapolation, Interpolation, and Smoothing of Stationary Time Series* (Cambridge, MA: MIT Press, and New York: John Wiley & Sons Inc., 1949).

should exist. When the stationary process was Gaussian, its statistics were fully specified by the autocorrelation in (11.1).[33]

To Wiener, a predictor was a mathematical operation so that given the values of f at present time t and before, the operation provided an estimate of f at a future time $t + \alpha$. In other words, the predictor took all values of $f(t - \tau)$ from $\tau = 0$ to ∞ as inputs and gave an estimate of $f(t + \alpha)$ as output. If the predictor was linear, then its output was a weighted sum of all its inputs, or an integral of its input over a weight function in the continuous case. The output from this linear predictor was a Stieltjes integral

$$\hat{f}(t + \alpha) = \int_{\tau=0}^{\infty} f(t - \tau)\, dK(\tau), \tag{11.2}$$

where K corresponded to the linear predictor's transfer function (more precisely, the time integral of its transfer function).[34]

An antiaircraft predictor could take any possible form of the transfer function K. What would be the "best" predictor among all of them? Here Wiener turned the determination of K into an optimization problem: K took the form that minimized the difference between the predictor output $\hat{f}(t + \alpha)$ and the function's actual value $f(t + \alpha)$ at $t + \alpha$. But what would be the definition of "minimum difference?" Wiener chose the mean-square error as the objective for optimization. In other words, the optimal K should minimize

$$\lim_{T \to \infty} \frac{1}{2T} \int_{-T}^{T} f(t + \alpha) - \int_{\tau=0}^{\infty} f(t - \tau)\, dK(\tau)^2 dt. \tag{11.3}$$

Here, the "error" $\hat{f}(t + \alpha) - f(t + \alpha)$ was squared and averaged over time. Expanding the objective in (11.3) and removing the term independent of K, Wiener obtained a new objective

$$-\mathrm{Re}\left\{\int_{0}^{\infty} \varphi(\alpha + \tau)\, dK^*(\tau)\right\} + \int_{0}^{\infty} dK^*(\tau) \int_{0}^{\infty} dK(\sigma)\, \varphi(\tau - \sigma). \tag{11.3'}$$

To minimize this functional, Wiener employed a familiar technique in the calculus of variation by replacing $K(\tau)$ with $K(\tau) + \varepsilon\delta K(\tau)$, differentiating (11.3') with respect to ε, equating the result to zero, and setting $\varepsilon = 0$. In so doing, the

[33] Ibid., 56–57.
[34] Ibid., 57.

K that minimized (11.3') satisfied:[35]

$$\varphi(\alpha + \tau) = \int_0^\infty dK(\sigma)\,\varphi(\tau - \sigma) \text{ for } \tau > 0. \tag{11.4}$$

Equation (11.4) did not give an explicit expression for K. Rather, it was an integral equation. Fortunately, Wiener and his junior colleague Eberhard Hopf at MIT had encountered similar integral equations when they worked on potential theory in the 1930s. A ready-made solution to this "Wiener–Hopf equation" was available.[36] A handy solution to (11.4) existed when the cumulated spectrum $\Lambda(\omega) = (1/2\pi)\int_0^\omega \Phi(\omega')\,d\omega'$ of the stochastic process $f(t)$ was continuous in frequency ω, where $\Phi(\omega) = \int_{-\infty}^\infty \varphi(t)\,e^{-i\omega t}dt$ was $f(t)$'s power spectrum. Moreover, the physical properties of $f(t)$ gave causality and positive $\varphi(0)$, which ensured that $\Phi(\omega)$ was real and positive for real ω. Under these conditions, Wiener could factorize $\Phi(\omega)$ into a product of a function and its complex conjugate: $\Phi(\omega) = |\Psi(\omega)|^2$. When Φ was a rational function of ω, he could regroup all its poles and zeros on the upper and lower halves of the complex plane to obtain Ψ. The solution of (11.4) was expressible with this Ψ. Wiener demonstrated that

$$k(\omega) = \frac{1}{2\pi\Psi(\omega)}\int_0^\omega e^{-i\omega t}dt \int_{-\infty}^\infty \Psi(u)\,e^{-iu(t+\alpha)}du \tag{11.5}$$

was the solution to (11.4) and thus the minimum mean-square-error predictor, where $k(\omega) = \int_0^\infty e^{-i\omega t}dK(t)$ was the predictor's transfer function in the frequency domain.[37]

After obtaining the general formula (11.5) for statistical prediction, Wiener tried it on several cases for the observed trajectory's power spectrum: when $\Phi(\omega)$ was $1/(1+\omega^2)$, $1/(1+\omega^4)$, $1/(1+\omega^2)^2$, $e^{-\omega^2}$, or $\sum_n b_n e^{-in\omega}$. In the first few cases, the simple form of the trajectory's spectrum gave a straightforward solution to the optimum predictor. For example, when $\Phi(\omega)$ was $1/(1+\omega^2)$, the optimum predictor was $k(\omega) = e^{-\alpha}$. When the trajectory's power spectrum was more complicated, the Wiener–Hopf procedure for solving (11.4) became more involved. But the manipulation of poles and zeros on the complex plane was similar to the procedure of filter design which Wiener had used in his collaboration with Lee in the 1930s.[38]

[35] Ibid., 57.
[36] Wiener, *Life in Cybernetics* (2018), 403.
[37] Wiener, *Extrapolation, Interpolation, and Smoothing of Stationary Time Series*, 62–64.
[38] Ibid., 65–71.

The statistical prediction that Wiener had developed so far handled only one kind of uncertainty in the antiaircraft problem: the target flight's irregular trajectory. In a real combat, there was another kind of uncertainty, as the gunner or the aiming device misaligned the target and introduced observational errors. To Wiener, this second kind of irregularity, which he dubbed "disturbance," could be represented with an additional random process $g(t)$. While the enemy airplane's actual path was $f(t)$, the path observed and registered at the antiaircraft apparatus was $f(t) + g(t)$. The task of the predictor was to take all the values of $f + g$ at and before time t as input and generate an estimate of $f(t + \alpha)$ as output.[39]

With this disturbance, the prediction problem involved not only estimating the value of f, but also removing the unwanted fluctuation g. When $\alpha < 0$, the estimate was retrospective, and the whole exercise of optimization became filtering. As Wiener pointed out, such a problem situation was similar to those in communication, in which the stochastic f represented a haphazard signal such as a random sequence of telegraphic messages while g represented noise. In this regard, Wiener's statistical procedure of minimizing the mean-square error turned out to be a method of eliminating noise and recovering signal. The Wiener predictor became the Wiener filter.[40]

Wiener's optimum linear statistical filter for removing additive random noise minimized the following objective (cf. (11.3)):

$$\lim_{T\to\infty}\frac{1}{2T}\int_{-T}^{T}\left|f(t+\alpha)-\int_{\tau=0}^{\infty}\left[f(t-\tau)+g(t-\tau)\right]dK(\tau)\right|^{2}dt. \qquad (11.6)$$

Expanding this objective led to an expression in terms of the signal's autocorrelation φ_{11}, the noise's autocorrelation φ_{22}, and the two entities' cross-correlation φ_{12}. Thus, (11.6) became

$$-\mathrm{Re}\left\{\int_{0}^{\infty}h(\tau)\,dK^{*}(\tau)\right\}+\int_{0}^{\infty}dK(\sigma)\int_{0}^{\infty}dK^{*}(\tau)\,\varphi(\tau-\sigma), \qquad (11.6')$$

where

$$\varphi(\tau)=\varphi_{11}(\tau)+\varphi_{12}(\tau)+\varphi_{12}^{*}(-\tau)+\varphi_{22}(\tau) \qquad (11.7)$$

[39] Ibid., 81–82.
[40] Ibid., 81–82.

and

$$h(\tau) = \varphi_{11}(\alpha + \tau) + \varphi_{12}(\alpha + \tau). \qquad (11.8)$$

Again, Wiener showed that the objective in (11.6') could be minimized when K satisfied the following integral equation:

$$h(\tau) = \int_0^\infty dK(\sigma)\,\varphi(\tau - \sigma) \text{ for } \tau > 0. \qquad (11.9)$$

Following a similar Wiener–Hopf procedure as Wiener adopted in solving (11.4), he obtained the solution of (11.9):

$$k(\omega) = \frac{1}{2\pi\Psi(\omega)} \int_0^\omega e^{-i\omega t} dt \int_{-\infty}^\infty \frac{H(u)}{\Psi^*(u)} e^{-iut} du \qquad (11.10)$$

where $\Phi(\omega) = |\Psi(\omega)|^2$ and $H(\omega)$ was the Fourier transform of $h(\tau)$.[41]

Equation (11.10) specified the predictor-filter that minimized the mean-square error given a stochastic signal (target trajectory) and noise (disturbances to observation). The exact form of this apparatus depended on $H(\omega)$ and $\Psi(\omega)$, which were contingent upon the spectral characters of the signal's and noise's autocorrelation and cross-correlation. To compute, Wiener assumed that the signal was statistically independent of noise, so that $\varphi_{12} = 0$. Like many of his predecessors in communication engineering, moreover, he focused on the cases "in which the noise input is due to a shot effect and has an equipartition of power in frequency." In other words, he worked on white noise, which had been dealt with frequently in the studies of Brownian motion, thermal fluctuations, and shot effects. Under these assumptions, $\Phi_{22}(\omega)$was a constant ε^2, $\Phi(\omega) = \Phi_{11}(\omega) + \varepsilon^2$, and $H(\omega) = e^{i\alpha\omega}\Phi_{11}(\omega)$. The predictor-filter's transfer function $k(\omega)$ could be calculated by factorizing this $\Phi(\omega)$ into $|\Psi(\omega)|^2$, and plugging $H(\omega)$ and $\Psi(\omega)$ into (11.10).[42]

Wiener was not the only mathematician to work on a statistical theory of prediction and filtering based on minimization of the mean-square error. When he finished the development of his theory by late 1941, he received a notice from William Feller about an article published by Kolmogorov in *Bulletin d'académie des sciences de l'URSS* in March 1941 on the extrapolation and interpolation of stationary time series. Wiener quickly wrote to Bush about

[41] Ibid., 82–86.
[42] Ibid., 91.

this. Wiener hoped that Bush could find a way to warn Kolmogorov not to publish on this militarily sensitive matter anymore but was worried that such a warning might hurt the Russian mathematician, given the way the Stalinist regime worked. In his report to NDRC, Wiener concluded that Kolmogorov's theory did not produce as useful results as his own. The Russian mathematician wanted to be rigorous and refrained from assuming the continuity of the target trajectory's cumulative spectrum $\Lambda(\omega)$. This made it difficult to derive the analytical form of the linear predictor-filter.[43]

The Wiener–Bigelow approach to random noise formed an interesting contrast to that of Uhlenbeck et al.'s. Both subscribed to the ontology of the ambience that treated noise as additive, stochastic, Gaussian, and stationary time series. Both were embedded in the American war projects on air defense using radars or optical devices. And both invoked generalized harmonic analysis, or the "method of Rice," that represented, computed, and manipulated random noise via its spectral characteristics. Because of their overlapping problem situations, we can even see the two camps competing over the legitimacy of reasoning and the credits of work.

Yet, the Wienerian predictor-filter had important differences from the Rad Lab researchers' tackling of noise. Whereas Uhlenbeck, Kac, Wang, and Goudsmit were concerned with the issues of target *detection* amidst a noisy environment, Wiener and Bigelow aimed to solve the problem of *estimation* under uncertain circumstances. The Rad Lab researchers devoted much effort to computing the probabilities of correct detection under various noise characters. Their hope was to use the outcomes from such computations as a guideline for setting up "threshold signals," the cutoff levels of radar return intensity that warranted a belief in the existence of a real target. Wiener and Bigelow endeavored to develop a predictor-filter that took the observed waveform as input and outputted an optimum estimate of the irregular signal while removing fluctuating noise. Wiener's approach did not add novel understanding about the physical nature or mathematical structure of random noise. Yet, his statistical theory of prediction provided a new means to *manipulate* noise. The theory stipulated a filtering mechanism to screen out random noise and to preserve the original signal, which now was also a stochastic process. This treatment of noise cast new light on the theoretical construal of telecommunication.

[43] Ibid., 59; Letter, Wiener to Bush, January 6, 1942, Box 4, Folder 62, Wiener Papers.

12
Information, Cryptography, and Noise

While Uhlenbeck's associates at Rad Lab were examining the effects of noise on radar target detection and Wiener's team at MIT were producing statistical filtering of noise for antiaircraft gunfire direction, another individual was also working on problems related to noise. Claude Shannon spent his wartime years at AT&T Bell Labs as an in-house mathematician. Although his involvements in military projects were not as substantial as Uhlenbeck's or Wiener's, throughout World War II Shannon participated in Bell Labs' defense research in antiaircraft gunfire control, secret telephony, and cryptography. Combining with his longstanding interest in the general process of communication, this war experience enabled him to develop a theory of information transmission. With the publication of a monumental seventy-nine-page paper "A mathematical theory of communication" in 1948, Shannon made himself a central figure of postwar information sciences, and he eventually became the "founding father" of information theory.[1]

As crucial ingredients of the postwar and Cold War cybernetic worldview, Shannon's information theory has not only influenced electrical engineers, but also attracted those interested in language, meaning, communication, media, and intelligence at large. For this reason, the origin of Shannon's information theory has caught the attention of historians of science and technology as well as media and literary scholars. And they have almost always concentrated on the conceptualization of information. Ronald Kline placed Shannon at a pivotal place in the intellectual and cultural history of the information discourses after World War II and traced the emergence of the notion of information as entropy from his theory. Lydia Liu examined Shannon's statistical analysis of the English language and especially his introduction of the space as the twenty-seventh letter and explored their implications for the nature of language and writing. Friedrich Kittler highlighted Shannon's wartime involvement in

[1] Claude Shannon, "A mathematical theory of communication," *The Bell System Technical Journal*, 27 (1948), 379–423, 623–656.

AT&T's vocoder-based secret telephony as a new form of media that converted speeches into digital signals.[2]

The concept of information indeed constituted the core of Shannon's theoretical breakthrough. In this chapter, however, we focus on his treatment of noise, the opposite of information, in the contexts of wartime and postwar research on random disturbances in command, control, communication, and computing. Shannon's construal of information and his first fundamental principle of information transmission—what was later named the "source coding theorem"—were independent of noise. Yet, his second fundamental principle with crucial pragmatic significances—the so-called "channel coding theorem" that stipulated the optimum rate of information transmission through a communication channel contaminated by stochastic noise—was built upon noise as a basic and unavoidable element of a communication process. This picture of noise-polluted communication did not differ intrinsically from the Rad Lab researchers' model of radar returns plagued by random fluctuations or Wiener and Bigelow's idea of a target flight trajectory offset by haphazard disturbances. Compared with Uhlenbeck and Wiener, what novelties did Shannon bring to the studies of noise in postwar information science and technology?

Shannon's tackling of noise in his information theory did not add new contents to the mathematical features of noise as such. To him, noise in most practical circumstances was similar to what Uhlenbeck et al. and Wiener et al. portrayed: an additive disturbance as a stationary random time series onto a stochastic signal. Shannon did not delve into the statistical properties of such noise and specific schemes of removing, reducing, or minimizing its effects as Uhlenbeck and Wiener did. To him, noise epitomized an essential uncertainty in the ontology of the ambience that he used to grasp what he believed was a fundamental limit of all processes of information transmission. As his Bell Labs colleague David Slepian remarked, "his theory differed from earlier treatments of communication in that it is a macroscopic theory in contrast to a microscopic theory. It is not concerned with the small scale details and circuitry of particular communication systems, but rather focuses attention on the large scale or gross aspects of communication devices."[3]

[2] Kline, *Cybernetic Moment* (2015); Liu, *Freudian Robot* (2010); Friedrich Kittler, *Gramophone, Film, Typewriter*, translated by Geoffrey Winthrop-Young and Michael Wutz (Palo Alto, CA: Stanford University Press, 1999).

[3] David Slepian, "Information theory" (circa 1953), in Legacy No. 43-10-03, Folder 01, AT&T Archives.

To tackle "macroscopic" features of communication processes, Shannon came up with several original representations to conceptualize noise. In his attempt to portray the generic flow of information, noise was construed as environmental effects in a telecommunication system or deliberate cryptographic scrambling that caused equivocation of original messages and was visualized in a "fan diagram." When he tried to give a geometric demonstration of the channel coding theorem, he invoked the "sphere-packing" model in which noise was pictured as spherical regions of uncertainty surrounding signal-points in a high-dimensional abstract space. When he applied the channel coding theorem to determine the optimum strategy for signal power allocation, noise appeared as the channel's spectral terrain that facilitated certain "water pouring"—designating more power to signals at frequencies where noise power was less. Shannon's "pictures" of noise formed an interesting contrast to Brownian motion, thermal agitations, shot current, stochastic time series, and the Lebesgue measure which scientists and technologists used to represent noise. By introducing new notions and tools, Shannon's approach extended the understanding of noise from its original physical, mathematical, and engineering contexts.

Shannon's Point of Entry

In 1916, Claude Shannon was born in Gaylord of northern Michigan. His father was a judge of probate, his mother was a teacher and once principal of a local high school. Shannon attended the University of Michigan, obtained a B.S. in both electrical engineering and mathematics in 1936, and moved to MIT to pursue a master's degree.

At MIT, he worked for Vannevar Bush on the differential analyzer. Bush's differential analyzer comprised electromechanical integrators and differentiators that could be wired to form circuitry to simulate physical or engineering systems. While the apparatus was an analog computer, it had a digital control module with vacuum tubes and electromagnetic switches. This module reminded Shannon of Boolean algebra he had learned in Michigan—a formal tool in mathematical logic developed by the British mathematician George Boole in the nineteenth century—for determining the truth-values of logical statements given the truth-values of their constituents. To Shannon, the switches' on-off resembled logical elements' true-false. Thus, an electromagnetic switching circuit could be translated into a logical statement. This connection made the design of binary switching circuits an application of

Boolean algebra. Based on this concept, Shannon completed a master's thesis at MIT in 1937.[4]

After receiving an M.S. in electrical engineering, Shannon decided to pursue a Ph.D. in mathematics at MIT. Bush suggested that Shannon apply his knowledge in symbolic algebra to population genetics. Shannon followed this advice and produced a dissertation on the mathematics of Mendelian and post-Mendelian theories of inheritance. Upon graduation in 1940, he received a National Research Fellowship—through Bush's arrangement—to work as a postdoctoral researcher at the Institute for Advanced Study (IAS) in Princeton. The project in Shannon's fellowship application was to further mathematical genetics. His supervisor at the IAS was the German émigré mathematician Hermann Weyl, a long-time professor at ETH Zurich and successor of Hilbert's chair at Göttingen who moved to Princeton to escape the Nazis. Weyl specialized in geometrical manifolds for general relativity, algebraic and topological representations of quantum fields, and number theory. None of these areas appealed to Shannon. Neither did he intend to continue with genetics. Instead, he began to entertain an analysis of some "fundamental properties of general systems for the transmission of intelligence, including telephony, radio, television, telegraphy, etc." Weyl granted him freedom to do so.[5]

In the previous summer, Shannon spent three months at Bell Labs working in the Mathematical Research Department headed by Thornton Fry. Soon after Shannon arrived at the IAS in fall 1940, Fry recommended to the NDRC Section on Fire Control that they recruit him. Warren Weaver reached out and offered Shannon a contract to analyze two antiaircraft gunfire directors based on geometrical extrapolation. As he completed the analysis and submitted a report to NDRC in spring 1941, he was considering his next step. At this time, the US was actively preparing for war with Germany and Japan. Years later, Shannon recalled that "I could smell the war coming along. It seems to me I would be safer working full-time for the war effort, safer against the draft, which I didn't exactly fancy." He also thought that staying home and doing military research was "to the best of my ability" to serve the American war effort. In fall 1941, he withdrew his fellowship at the IAS, and joined Fry's department in New York.[6]

[4] Kline, *Cybernetic Moment* (2015), 26–27; Ioan James, "Claude Elwood Shannon," *Biographical Memoirs of Fellows of the Royal Society*, 55 (2009), 259–260.

[5] Kline, *Cybernetic Moment* (2015), 28–29; James, "Shannon" (2009), 260; Robert Price, "Oral history: Claude E. Shannon," IEEE History Center, Interview #423, July 28, 1982, https://ethw.org/Oral-History:Claude_E._Shannon (accessed April 8, 2022).

[6] Kline, *Cybernetic Moment* (2015), 29–30; Mindell, *Between Human and Machine* (2002), 289–290; Price, "Shannon" (1982).

As a junior fellow with a fresh Ph.D. in mathematics, Shannon's primary duty at Bell Labs was to provide consultation and analysis when theoretical issues came up from AT&T's defense projects. Glimpsing through his wartime articles and reports, we find topics as diverse as self-checking relay computers, feedback systems with periodic loop closure, circuits for a pulse-coded Modulation (PCM) transmitter and receiver, optimum detection of pulses, power spectra of a certain ensemble of functions, and pulse counters.[7]

Shannon was involved in three military projects at Bell Labs from 1941 to 1945. First, he was a member of Hendrik Bode's team to develop an anti-aircraft gunfire director. Through this project, Shannon got familiar with Wiener's work on the modeling of noise and signal as stochastic stationary time series and statistical filtering of noise. When Shannon was still a student at MIT, he took a course on Fourier analysis with Wiener. The wartime anti-aircraft research prompted more exchanges. Shannon read Wiener's "Yellow Peril" shortly after it was submitted to NDRC in February 1942. Bigelow also recalled that for a while Shannon came up to MIT to talk with Wiener and himself "every couple of weeks." Through these interactions, Shannon learned about Wiener's measure-theoretical framework for stochastic processes, generalized harmonic analysis, their applications to model signals and noise, and the mean-square predictor-filter.[8]

Cryptography and Information Transmission

Shannon's second wartime project concerned a secret voice communication system. In 1940, the US Army Signal Corps conceived a long-distance radio telephone with enhanced security features. The Army contracted this project—known as SIGSALY—to AT&T. Within three years, Bell Labs produced a prototype. Named "X-System," the secret communication link employed a vocoder introduced in 1928 by the Bell engineer Homer Dudley that encoded a voice waveform with its components at distinct frequency bands. These signals were passed through a pulse-coded modulation (PCM) that sampled the signals at discrete times and digitized their amplitudes. Through the vocoder and PCM, the voice waveform was transformed into a sequence of numbers. To encrypt these numbers, the X-System combined

[7] These articles and reports can be seen in Box 8 "Articles and Scientific Papers (alphabetical) U-Z, Chronological File 1938-46," Folder 6 "1942-43," Folder 7 "1944 Mar-Aug," Folder 8 "C.A. 1944," Claude Shannon Papers, Library of Congress, Washington, D.C., US.

[8] Kline, *Cybernetic Moment* (2015), 30–31; Price, "Shannon" (1982).

them with a set of random keys through logical-algebraic operations from a Vernam cipher invented circa World War I. The random keys were obtained from digitizing electrical resistors' thermal noise recorded on phonographic discs. After successful testing of the prototype in 1942, Signal Corps commissioned Western Electric to make multiple sets, and deployed them for critical military communications, including the hotline between Roosevelt and Churchill.[9]

Shannon did not have the security clearance to participate directly in the development of the X-System. His role was to provide theoretical support. In 1943, he was asked to analyze the Vernam cipher. He submitted a report to prove the security of the Vernam system. He had vague ideas about the work on "speech scrambling" going on at Bell Labs. But he was not given details about the top-secret X-System.[10]

While working on the specific tasks assigned by Bell Labs, Shannon was contemplating a general theory of cryptography, his third wartime project. Shannon's experience with the X-System might have played a part in his cryptography. Another plausible factor was his interactions with Alan Turing, Britain's leading cryptographer who was involved in decoding the Enigma system used for German military communications. In early 1943, Turing was in the US to help assess the X-System. After his trip in Washington, he paid a visit to Bell Labs. During his two-month stay in New York, Turing had lengthy conversations with Shannon. These episodes occurred around the time when Shannon was constructing a theoretical framework for cryptography. Since this framework had crucial underpinnings in Shannon's doctrine of information transmission in 1948, Kline considered his wartime cryptography an important context and pretext for his development of information theory.[11]

The historical relevance of cryptography to information theory is undeniable. According to Shannon's recollection, however, their causal connections were different from their apparent chronological order. He insisted that he had brewed the notions of information, entropy, redundancy, and equivocation when he was in Princeton before the war, or even earlier. As he claimed, he worked out a lot of information-theoretical results from 1940 to 1945—e.g. "random sequences of letters" as their statistics approached that of

[9] For the X-System, see Millman, *History of Engineering and Science in the Bell System, Communications Science* (1984), 104, 405; M.D. Fagen (ed.), *A History of Engineering and Science in the Bell System: National Service in War and Peace (1925–1975)*, vol. 2 (New York: Bell Telephone Laboratories, 1978), 296–313.

[10] Kline, *Cybernetic Moment* (2015), 31–32; Price, "Shannon" (1982).

[11] Kline, *Cybernetic Moment* (2015), 31–32.

English, and their Markov chain analysis—and included them in his cryptography. These contents were "things that I did around the time at home, not office hours, so to speak." When asked whether this study was motivated by cryptography, Shannon stated:[12]

> That was, my first getting at that was Information Theory. And the cryptography, I used that as a way of legitimatizing it all, you know, if you understand what I mean. To make it sound like I'm working on decent things .... In fact. You might say that cryptography was there, and people were working on secrecy systems in Bell Labs, and it seemed to me that here this was very closely related. And I should go over to that too. And the other thing I was not yet ready to write up the Information Theory anyway. This you could write up anything in any shape, which I did.

In Shannon's own account, therefore, his cryptographic theory was a surrogate for the early part of his information theory, since he had not fully developed the latter and writing up ideas in the style of the former was more justified under the war-looming circumstances at the time. In fact, he believed that his original source of inspiration came from the Bell researcher Ralph Hartley's paper "Transmission of information" in 1928 that proposed to encode M telegraphic messages in terms of $\log_2 M$ binary digits.[13] Shannon read the Hartley paper when he was at Ann Arbor.[14]

From Shannon's reminiscence, moreover, his conversations with Turing were not about cryptography in any substantive way. Rather, they talked more about the theory of computability, specifically the "Turing machine" that the Englishman had developed in the 1930s. They:

> spent much time discussing the concepts of what's in the human brain. How the brain is built, how it works and what can be done with machines and whether you can do anything with machines that you can do with the human brain and so on.

Shannon mentioned to Turing several times about the notions of information theory. But Turing "didn't believe they were in the right direction."[15]

[12] Price, "Shannon" (1982).
[13] Ralph Hartley, "Transmission of information," *The Bell System Technical Journal*, 7 (1928), 535–563.
[14] Price, "Shannon" (1982).
[15] Ibid.

On September 1, 1945, two weeks after V-J Day, Shannon submitted a confidential technical memorandum to Bell Labs. Titled "A mathematical theory of cryptography," this report synopsized Shannon's theoretical work on cryptography during the war. At this point, he had not yet worked out the whole picture of his information theory. In 1949, he turned the non-confidential contents of his 1945 report into a publication "Communication theory of secrecy systems" under the framework of the newly developed information theory.[16]

Shannon's 1945 report contained important elements of his forthcoming theory of information transmission. First, he developed a model for messages in secret communication systems. In contrast to Uhlenbeck's continuous radar return waveforms or Wiener's enemy flight trajectories, each message in Shannon's theory comprised a discrete sequence of elements, like a train of telegraphic codes. Suppose a source could generate n possible sequences. Shannon assumed that each sequence i appeared stochastically with probability p_i. He defined a quantitative measure of the "uncertainty" of a generic message from this source:

$$H = -\sum_{i=1}^{n} p_i \log_2 p_i. \tag{12.1}$$

To Shannon, this definition was reasonable since it had the following properties: It was maximal when the probabilities of all messages were identical. The uncertainty of two statistically independent messages was the sum of their respective uncertainties. The uncertainty of two statistically dependent messages was the uncertainty of one message plus the conditional uncertainty of the other message given the previous one. The uncertainty of a message equaled the average number of binary digits required to encode it. Thus, the uncertainty H could also serve as a measure of the message's quantity of information, à la Hartley. Shannon noted that (12.1) had the same form as a statistical ensemble's entropy in Boltzmann's statistical mechanics.[17]

In telecommunications, messages were produced from a natural language. Shannon's next step was to characterize a language with a discrete stochastic process, and to explore its statistical configuration as a basis to compute its uncertainty. He modeled English with a series of Markov chains. In the zeroth-order approximation, a message from English was modeled as a sequence

[16] Claude Shannon, "A mathematical theory of cryptography," Technical Memoranda, AT&T Bell Telephone Laboratories, September 1, 1945; Claude Shannon, "Communication theory of secrecy systems," *The Bell System Technical Journal*, 28 (1949), 656–715.

[17] Shannon, "A mathematical theory of cryptography" (1945), 7–10.

in which all elements were statistically mutually independent. In the first-order approximation, the chance for the appearance of a letter at an element depended on its previous element. In the second-order approximation, that probability depended on the previous *two* elements, and so on. Following this procedure, Shannon extended the statistical characterization from letters to words and constructed the Markov chains for single words, double words, etc. In principle, these probabilities could be obtained from frequency analysis of English texts, and English's "uncertainty" (entropy) could be calculated from (12.1) given this statistical structure.[18]

Shannon noted that a typical source of messages did not fully utilize all possible combinations. In English, not all possible combinations of letters appeared with equal chances (the case of equal chances corresponded to the maximum uncertainty H_0). Rather, the set of all meaningful messages in English was smaller and had smaller uncertainty H. Shannon defined $D = H_0 - H$ as the "redundancy" of this source. The positive redundancy of English implied the possibility of compressing all meaningful messages into a more compact linguistic structure. He provided two examples to illustrate this idea. The linguist C.K. Ogden's "Basic English" boasted of accomplishing most communicative acts with only 850 words. From Shannon's perspective, this was a highly redundant system, for most expressions in English did not get used in Ogden's scheme. On the other hand, James Joyce's *Finnegans Wake* had a very low redundancy, for it utilized perhaps more expressions than most English texts.[19]

Having laid out the conceptual ground for a message source, Shannon explored the mathematical configuration of secret communication. A secrecy system comprised a transmitter (including a message source, a key source, and an encipherer) and a receiver as a decipherer (Figure 12.1). The message source generated a message M, whereas the key source generated an encrypted key K. M and K were combined algebraically at the encipherer to produce a cryptogram E. This cryptogram was a transformation T_K of the message M, or $E = T_K M$. Shannon supposed that the cryptogram E and the key K were sent faithfully to the decipherer without distortion or interference. The decipherer then performed the inverse operation of the encipherer based on the received E and K, $T^{-1}{}_K E$. This operation gave back M. Meanwhile, the cryptogram E could be intercepted by an enemy cryptoanalyst, who aimed to reconstruct the

[18] Ibid., 11–16.
[19] Ibid., 18–24.

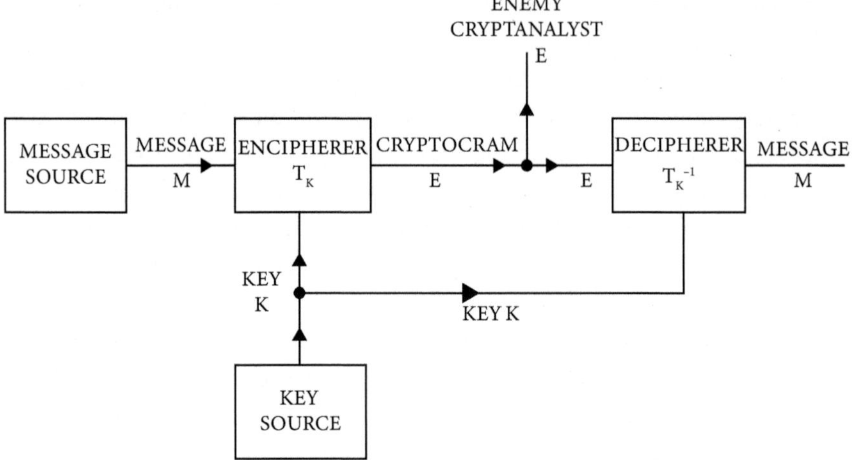

Figure 12.1 Shannon's block diagram of a secret communication system. Claude Shannon, "Communication theory of secrecy systems," *The Bell System Technical Journal*, 28 (1949), 661, Figure 1.
Courtesy of AT&T Archives and History Center.

original message *M* from the cryptogram *E* without knowing the encrypted key *K*.[20]

The aim of Shannon's analysis was to determine under which circumstances could a secrecy system achieve "perfect secrecy," when cryptoanalysis did not enhance the chance of deciphering the message. In his own words, "perfect secrecy" is "defined by requiring of a system that after a cryptogram is intercepted by the enemy the a posteriori probabilities of this cryptogram representing various messages be identically the same as the a priori probabilities of the same messages before the interception."[21]

For our purpose, the most relevant part of Shannon's cryptographic analysis was the similarity between the effect of encryption on a message and the effect of noise on a communicated signal. As mentioned, the encipherer transformed a message *M* via the encrypted key *K*, $E = T_K M$. For a given message *M*, when the key *K* was fixed, the cryptogram *E* was fixed. If *M* was transmitted multiple times, then the cryptoanalyst could have a good chance to reconstruct it from multiple appearances of the same cryptogram *E*. Yet, this was not the way encryption worked. Instead, each time *M* was transmitted, *a*

[20] Ibid., 24–26.
[21] Ibid., 5.

randomly selected key K was fed into the encipherer. For example, the German High Command reset the keys of the Enigma systems at the beginning of every day. As a result, the cryptogram *E* had random variations despite the same message *M* being repeated. That means, the same cryptogram could correspond to different messages, and vice versa. To Shannon, this uncertainty could be understood as the "equivocation" of the secrecy system due to the random encrypted key. Shannon made an explicit analogy between the random variation due to the irregular key and noise in telecommunication:[22]

> From the point of view of the cryptoanalysis, a secrecy system is almost identical with a noisy communication system. The message (transmitted signal) is operated on by a statistical element, the enciphering system, with its statistical chosen key. The result of this operation is the cryptogram (analogous to the perturbed signal) which is available for analysis.

He pointed out that:[23]

> The chief differences in the two cases are: first, that the operation of the enciphering transformation is generally of a more complex nature than the perturbing noise in a channel; and, second, the key for a secrecy system is usually chosen from a finite set of possibilities while the noise in a channel is more often continually introduced, in effect chosen from an infinite set.

Yet, the mathematical analysis of equivocation as conditional entropy was identical in both cases. By making an analogy between communication with noise and random encryption of messages, Shannon cast the notion of noise in information theory.

Information Theory and Representing Noise with Equivocation

Shannon's cryptographic report in 1945 introduced basic building blocks for information theory: uncertainty as entropy, message sources as Markov chains, redundancy, and equivocation. While he did not explicitly tackle noise, the irregular character of random keys in cryptography provided an analogous heuristic for him to think about the nature and effects of noise in communications. In the following years, he developed more elements of information

[22] Shannon, "Communication theory of secrecy systems" (1949), 685.
[23] Ibid., 685.

theory, including the two fundamental principles of information transmission, and the extension of results from discrete messages to continuous signals. In the final form of his theory, noise was conceptualized and visualized in various ways: as equivocation represented by fan diagrams, "spheres" of uncertainty, a spectral terrain for exercising a power-allocation strategy of "water pouring," and a resource for overcoming disturbances in communications.

Shannon's theory began to take a complete shape within two years after he submitted "A mathematical theory of cryptography" in 1945. In spring 1947, he gave a talk at Harvard University on white noise in telecommunications. On November 12, he presented a preliminary version of his theory, with the title of "Transmission of information," at the Institute of Radio Engineers (IRE) Annual Congress in New York City. Both speeches attracted requests for papers from the audiences.[24] In July and December 1948, Shannon published "A mathematical theory of communication" in two installments in *The Bell System Technical Journal*.[25] This paper was the founding document of information theory.

Shannon started his 1948 paper by reasserting the advantages of using a logarithmic measure for the quantity of information, after Hartley's and Nyquist's proposals in the 1920s.[26] This measure could index the number of binary digits—or "bits" following the Princeton mathematician John Tukey's notation—needed for representing possible messages from a source. In addition, Shannon provided a generic model of communication: An information source produced messages, which were converted at a transmitter into signals. The signals passed through a communication channel, which introduced noise contaminating the signals. The received signals were converted back to messages at a receiver. These recovered messages reached a destination and were interpreted there (Figure 12.2).[27]

Shannon considered several general situations in telecommunication. First, the communication system produced discrete messages, and the channel was noiseless. Under this condition, he reiterated the major concepts in his 1945 report: modeling of a message source with Markov chains, information or uncertainty as entropy, and redundancy. Shannon asserted that the quantity of information for a message source in (12.1) could be understood as the average

[24] Letters, Calvin Mooers to Shannon, December 29, 1947, and C.A. Smith to Shannon, November 24, 1948, both in Box 1 "Correspondence General 1938-55," Folder 1, Shannon Papers.

[25] Shannon, "A mathematical theory of communication," (1948).

[26] Hartley, "Transmission of information" (1928); Harry Nyquist, "Certain factors affecting telegraphic speed," *The Bell System Technical Journal*, 3 (1924), 324–346, and "Certain topics in telegraph transmission theory," *Transactions of the American Institute of Electrical Engineers*, 43 (1928), 412–422.

[27] Shannon, "A mathematical theory of communication" (1948), 379–382.

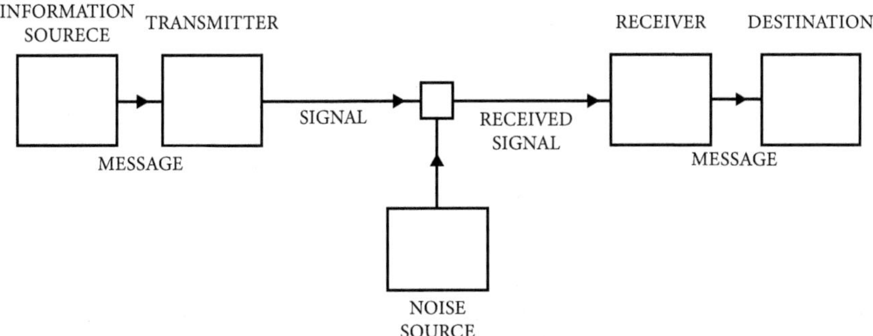

Figure 12.2 Shannon's block diagram of a communication system. Claude Shannon, "A mathematical theory of communication," *The Bell System Technical Journal*, 27 (1948), 381, Figure 1.
Courtesy of AT&T Archives and History Center.

number of bits needed to represent a symbol, should the symbols be encoded in an efficient manner to minimize the average length (e.g. by coding more frequent letters with fewer bits and less frequent ones with more bits). In addition, he defined a noiseless channel's capacity C as

$$C = \lim_{T\to\infty}\frac{\log_2 M(T)}{T}, \quad (12.2)$$

where T was a period of time and $M(T)$ was the number of all possible signals within T. The channel capacity took the unit of bits per second.[28]

With these preliminaries, Shannon presented his "fundamental theorem for a noiseless channel." Known later as the "source coding theorem," this principle stipulated that if a source had entropy H bits per symbol and a noiseless channel had capacity C bits per second, then it would be possible to encode without errors messages from the source in such a way that they could be transmitted at an average rate of C/H symbols per second or less. On the other hand, it would be impossible to encode without errors the messages in such a way that they could be transmitted at an average rate higher than C/H symbols per second. In other words, channel capacity constrained the maximum speed of information transmission.[29]

The mathematical proof of the source coding theorem was involved. But Shannon offered an intuitive grasping of it: Within a very long time T, the total number of bits transmitted over the channel would be roughly CT, and

[28] Ibid., 382–396.
[29] Ibid., 401.

the corresponding number of possible binary sequences would be roughly 2^{CT} (see (12.2)). Statistically, the collection of all sequences not belonging to those 2^{CT} ones would have a zero measure in probability space when T became infinite. Similarly, when the message source outputted a very large number N of symbols, the total number of output bits would be NH, and the corresponding number of possible binary sequences would be roughly 2^{NH}, where the entirety of exceptions had zero probability. At the maximum rate of signal transmission, all possible binary sequences from the source were equal to all possible binary sequences the channel could transmit. That is, $2^{CT} = 2^{NH}$. Thus, the maximum rate of transmission was $N/T = C/H$ symbols per second.[30]

Next, Shannon examined a more realistic situation in which signals were still discrete but the channel was noisy. Instead of imposing a specific model of noise as Uhlenbeck and Wiener did, Shannon introduced a notion of noise in the broadest sense. A noiseless channel took a signal x from the transmitter as input and gave the signal $y = x$ to the receiver as output. For a noisy channel, y did not equal x in general. Noise could deviate the channel output y from the channel input x. This deviation did not have to be additive as Uhlenbeck and Wiener had assumed. Neither did it have to take any specific form like white Gaussian. All one could know was that this deviation was stochastic in a similar way to how a random key shifted an encrypted message. The signal x at the channel input was stochastic due to the nature of any nontrivial message source. Yet, the signal y at the channel output for a given x was stochastic because of random noise. In other words, a fixed value of x did not correspond to a fixed value of y. Similarly, a fixed value of y could correspond to multiple values of x due to noise. This ambiguity brought difficulty for recovering the transmitted signal x from the received signal y. Shannon named such ambiguity *equivocation*—as he had done in his cryptographic report. Moreover, he gave the quantitative measure of equivocation as a kind of uncertainty, which was the average of the conditional entropy of x for a given y:

$$H_y(x) = -\sum_y \{p_y \sum_x p_{x|y} \log_2 p_{x|y}\}. \tag{12.3}$$

To Shannon, the non-zero value of equivocation due to noise implied that not all the quantity of information $H(x)$ transmitted from the message source could reach the receiver. Rather, the average number of bits per symbol at the receiver was reduced by the amount of the channel's equivocation, i.e. $R(x) = H(x) - H_y(x)$. Shannon further showed that this quantity could be

[30] Ibid., 402–404.

$R(x) = H(x) - H_y(x) = H(y) - H_x(y) = H(x) + H(y) - H(x,y)$.[31] Normalizing this quantity with the average number of transmitted symbols per second, this R could be construed as the rate of information transmission in the unit of bits per second. Shannon defined a noisy channel's C as the maximum rate of information transmission over all possible sources x:[32]

$$C = \max_x \{H(x) - H_y(x)\}. \tag{12.4}$$

With the definition of a noisy channel's capacity, Shannon presented his second fundamental principle of information transmission, which was later known as the "channel coding theorem." He considered a noisy channel with capacity C bits per second and a source producing messages with entropy H bits per second. If $H \leq C$, then there existed a coding scheme that could encode the messages and transmit them over the channel "with an arbitrarily small frequency of errors." If $H > C$, then the transmitted messages over the channel had at least an equivocation of $H - C$. In other words, the channel coding theorem set the limit for how best a communication system could do in terms of information transmission. When the source entropy was less than the channel capacity, error-free transmission was possible. That was impossible when the source entropy exceeded the channel capacity.[33]

Two peculiar features about the channel coding theorem are worth noting. First, the effect of noise was threshold-like, not gradual. In the problems that preoccupied Uhlenbeck and Wiener before and during World War II, thermal agitations or Brownian motion degraded the performance of a measuring device or a statistical predictor no matter how weak they were. By contrast, noise at a communication channel did not produce any errors when the rate of information transmission did not exceed the channel capacity. While noise did incur deviations of signals, original messages could still be accurately recovered presumably through a smart coding design. Only when the rate of information transmission exceeded the channel capacity did the detrimental effect of noise on communication become irrepairable. Second, even though Shannon's channel coding theorem asserted the presence of an optimum coding that could match the entropy of the message source with the channel capacity, he did not provide this "channel coding." The principle was a theorem of existence, not of construction.[34]

[31] Ibid., 407–409.
[32] Ibid., 410.
[33] Ibid., 411.
[34] For comments on these features, see Price, "Shannon" (1982).

Despite mathematical intricacies, Shannon's proof of the channel coding theorem was based on an intuitive concept about the effect of noise on communication. He considered a message source whose entropy $H(x)$ bits per second matched the channel capacity C in (12.4). All possible messages produced by the source within a long time T (in the form of sequences of bits) at the channel input belonged to one of two classes: the set of $2^{TH(x)}$ sequences with a high probability of appearance, and the set of all others. From the law of large numbers in statistics, he argued that when T became asymptotically infinite, the set of $2^{TH(x)}$ sequences had a probability close to one, and the total probability of all others was effectively zero. Under this large-number (i.e. long-duration) condition, one could safely assume that the source produced approximately $2^{TH(x)}$ signals. Similarly, at the channel output, the entropy of the signals was $H(y)$, and the number of possible sequences was approximately $2^{TH(y)}$ when T was long.[35]

Because of noise, the channel output y was not a regular transformation of the channel input x. Due to the random character of noise, different signals at the channel input could produce the same signal at the channel output. As mentioned, Shannon used the conditional entropy $H_y(x)$ to measure the uncertainty for such noise-generated equivocation. Invoking the same reasoning of large numbers, he contended that when T was long, each high-probability sequence at the channel output could be produced by about $2^{TH_y(x)}$ distinct sequences at the channel input.[36]

The non-zero conditional entropy $H_y(x)$ implied that not all $2^{TH(x)}$ sequences at the channel input were usable for effective communication. If two different sequences at the channel input fell within the same domain of equivocation for a certain sequence at the channel output, then these two sequences could not be used to represent different messages. The receiver would not be able to distinguish them from the same received signal. To avoid ambiguity and to make communication error-free, one had to select *only one* sequence from each domain of equivocation as a possible transmitted signal. Thus, the total number of possible sequences that could be transmitted through the channel without errors was $2^{TH(x)}/2^{TH_y(x)}$. Since the source matched the channel capacity, this possible number of sequences was $2^{T[H(x)-H_y(x)]} = 2^{TC}$, meaning that C bits per second could be transmitted through the channel. For another source with entropy R bits per second, if $R < C$, then it was plausible to allocate all 2^{TR} high-frequency messages to different domains of equivocation,

[35] Shannon, "A mathematical theory of communication" (1948), 411.
[36] Ibid., 411–412.

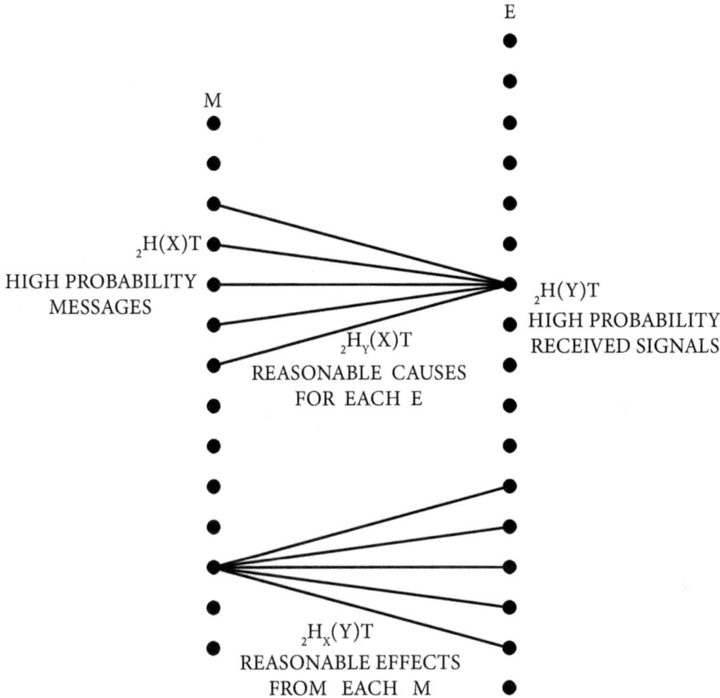

Figure 12.3 Shannon's "fan diagram" to represent equivocation of messages due to noise. Claude Shannon, "A mathematical theory of communication," *The Bell System Technical Journal*, 27 (1948), 412, Figure 10.
Courtesy of AT&T Archives and History Center.

and thus to attain error-free communication. If $R > C$, then such an allocation was impossible, and so was error-free communication.[37] Shannon used a scheme of visualization to represent this concept. Named "fan diagram" among later information theorists, this chart represented the effect of noise on communication of discrete messages in the framework of large numbers (Figure 12.3).[38]

Continuous Noise, Sphere Packing, and Water Pouring

After establishing the two fundamental principles for discrete messages, Shannon considered the case involving continuous signals. When communication shifted from transmission of discrete messages (e.g. telegraphic pulses) to

[37] Ibid., 412–414.
[38] Price, "Shannon" (1982).

continuous waveforms (e.g. telephonic voice), the signals took continuous values at continuous time. With respect to continuous signal values, Shannon extended the entropy formulation in (12.1) to a continuous integral

$$H = -\int_{-\infty}^{\infty} p(x)\log_2 p(x)\,dx. \tag{12.5}$$

A shortcoming of this extension was that the entropy in (12.5) was no longer a fixed quantity for a given source, but rather varied with the chosen coordinate system for x. But this variation did not matter much, for the *entropy difference* between two sources remained the same, regardless of their system of coordinates.[39]

To grapple with the signals changing over a continuous time, Shannon invoked the so-called "sampling theorem" first demonstrated by the British mathematician John M. Whittaker in 1935:[40] for a signal $f(t)$ band-limited by frequency W,

$$f(t) = \sum_{n=-\infty}^{\infty} f\left(\frac{n}{2W}\right)\frac{\sin[\pi(2Wt-n)]}{\pi(2Wt-n)}. \tag{12.6}$$

Specifically, (12.6) implied that within a unit duration, the continuous-time waveform $f(t)$ could be fully represented and accurately recovered by its $2W$ sample points $f(1/2W)$, $f(2/2W)$, ..., $f(2W/2W) = f(1)$. When $f(t)$ was restricted within a finite time window of period T, $2TW$ samples were necessary and sufficient to represent and recover the waveform. As Shannon indicated, this claim was consistent with Nyquist's observation in 1924 and the Hungarian-British engineer Dennis Gabor's 1946 discussion on communicating band- and time-restricted signals.[41]

The sampling theorem in (12.6) enabled the representation of a continuous signal with bandwidth W and duration T with $2TW$ samples, while the extension of the entropy formulation in (12.5) allowed the calculation of these $2TW$ samples' total entropy. Equipped with these tools,

[39] Shannon, "Mathematical theory of communication" (1948), 628–629.

[40] John M. Whittaker, *Interpolatory Function Theory* (Cambridge: Cambridge University Press, 1935), chapter 4.

[41] Nyquist, "Certain topics in telegraph transmission theory" (1928); Dennis Gabor, "Theory of communication," *Journal of the Institution of Electrical Engineers*, 93:3 (1946), 429–458. Gabor pointed out that a strict band-limited and duration-limited waveform did not exist, due to the nature of the relationship between a waveform in the time domain and its Fourier transform in the frequency domain. Realistically, the signal would be more like a "wave-packet" with a rough time range Δt and a rough frequency range Δf. Gabor proved that the product of the two ranges would exceed a constant $\Delta t \Delta f \geq 1/2$. He designated this relation as an "uncertainty principle" in signal processing, in analogy to the Heisenberg uncertainty principle involving momentum Δp and position Δx in quantum physics.

Shannon derived the capacity of a noisy channel permitting transmission of continuous signals. The continuous signal source's probability distribution was characterized by $p(x_1, x_2, \dots, x_{2TW}) \equiv p(x)$. At the output of the channel the received signals' probability distribution was characterized by $p(y_1, y_2, \dots, y_{2TW}) \equiv p(y)$. And their mutual conditional probability density was $p(x_1, x_2, \dots, x_{2TW}|y_1, y_2, \dots, y_{2TW}) \equiv p(x|y)$. These probability density functions determined the entropy of the channel input and output, and their mutual entropy:

$$
\begin{aligned}
H(x) &= -\int_{-\infty}^{\infty} p(x)\log_2 p(x)\,dx, \\
H(y) &= -\int_{-\infty}^{\infty} p(y)\log_2 p(y)\,dy, \\
H_y(x) &= -\int_{-\infty}^{\infty}\int_{-\infty}^{\infty} dydx\cdot p(x,y)\log_2 p(x|y).
\end{aligned}
\tag{12.7}
$$

The rate of information transmission for a given source x through this continuous channel was $R = H(x) - H_y(x)$, where $H(x)$ and $H_y(x)$ were calculated from (12.7). Like in the discrete condition, the channel capacity C was the maximization of R averaged over a long period T. From the Bayesian formulation of conditional probabilities, Shannon obtained[42]

$$
C = \lim_{T\to\infty} \max_{p(x)} \frac{1}{T}\int\int p(x,y)\log_2\frac{p(x,y)}{p(x)p(y)}dxdy.
$$

The capacity of a continuous channel depended on noise which caused equivocation of the received signals. To obtain a tractable formulation of C, Shannon worked on what he deemed a simple but important case: that noise was a white thermal noise with an average power N, the transmitted signals had an average power P, noise was additive onto signals, and signals and noise were statistically independent of each other. This modeling of channel noise invoked the same ontology of the ambience as Uhlenbeck's radar noise and Wiener's gunfire directing errors. In all these cases, the environmental disturbances appeared in the form of additive, stationary, white Gaussian fluctuations.[43]

[42] Ibid., 637.
[43] Ibid., 639.

Since noise was additive, the received signal y at the channel output was the sum of the transmitted signal x at the channel input and noise n: $y = x + n$. Shannon showed that the rate of information transmission at the channel output was the entropy of the received signal less the entropy of noise; i.e. $R = H(y) - H(n)$. The channel capacity was the maximization of R over the probability distribution of the message source $p(x)$. Because noise n was independent of the signal x at the channel input, $H(n)$ was independent of $p(x)$, meaning that the maximization was over only the first term $H(y)$:[44]

$$C = \lim_{T\to\infty} \frac{1}{T} \left\{ \max_{p(x)} H(y) - H(n) \right\}. \tag{12.8}$$

In (12.8), n was a short-handed expression of $2TW$ random variables $n_1, n_2, \ldots, n_{2TW}$ corresponding to the samples of noise waveform at $2TW$ instants within the time window T. Since noise was white, $n_1, n_2, \ldots, n_{2TW}$ should have identical and mutually independent probability distributions. Thus, $H(n) = H(n_1, n_2, \ldots, n_{2TW}) = 2TWH(n_1)$. Moreover, n_1 had a Gaussian distribution $p(n_1) = \left(1/\sqrt{2\pi N}\right) e^{-(n_1{}^2/N)}$, where N was the noise power. Plugging this expression into (12.8), Shannon obtained $H(n) = WT\log_2(2\pi eN)$.[45]

The computation of $\max_{p(x)} H(y)$ in (12.8) required an optimizing procedure over all the probability distribution of x that gave power P to x and power $P + N$ to y (the latter held since x was independent of n and $y = x + n$). Employing the technique of the Lagrange multipliers, Shannon demonstrated that the probability distribution $p(x)$ that gave the maximum entropy $H(y)$ had the same form as white Gaussian noise. This implied that $y = x + n$ also took the form of white noise with power $P + N$. In his own words:[46]

> The maximum entropy for the received signals occurs when they also form a white noise ensemble since this is the greatest possible entropy for a power P+N and can be obtained by a suitable choice of the ensemble of transmitted signals, namely if they form a white noise ensemble of power P.

Under this optimized condition, the entropy of the received signal y was $H(y) = WT\log_2[2\pi e(P + N)]$. Plugging the expressions of $H(y)$ and $H(n)$ into

[44] Ibid., 639.
[45] Ibid., 629–632.
[46] Ibid., 639.

(12.8), Shannon obtained the formula for the capacity of a continuous channel plagued by white Gaussian noise:[47]

$$C = W \cdot \log_2\left(\frac{P+N}{N}\right). \tag{12.9}$$

Equation (12.9) was a central result of Shannon's information theory. Here the effect of noise on transmission of information was fully manifested in the channel capacity. In this formula, the capacity was determined by the channel's bandwidth W and the signal-to-noise ratio P/N—the broader the bandwidth or the higher the signal-to-noise ratio, the larger the channel capacity. This formula depended on a particular form of noise, and thus was not the most general result for information transmission through a noisy channel. Yet, modeling noise with an additive white Gaussian fluctuation constituted the core of the ontology of the ambience shared by the wartime and postwar researchers on military information technologies. As Uhlenbeck's and Wiener's experiences showed, white Gaussian noise not only represented physical forms of disturbances frequently found in electronic systems, but also epitomized a mathematical entity with a rich set of technical tools and intellectual heritage.

Shannon's work added a new layer on the conceptual and technical aspects of this environmental ontology, for it implied a connection between what was communicated and what disturbed the communication. In his derivation of (12.9), the maximum rate of information transmission under white noise was achieved when the signal itself was like white noise. The similarity between the two implied the *usefulness* and *advantages* of stochastic and noise-like fluctuations in information processing. This insight was already present in Shannon's construal of random keys in cryptography. As we will see in chapter 13, it would also play a part in the development of spread-spectrum communication.

Shannon deduced the central formula (12.9) for the capacity of a continuous channel from the channel coding theorem that he had obtained through considering equivocation for discrete messages. But this was not the only route—and perhaps not even the first route—he traversed to reach that conclusion. Before publishing his article in July 1948, he gave a presentation at the IRE New York Section on November 12, 1947, and another one at the IRE National Convention on March 24, 1948. Around the time when "A mathematical theory of communication" was published, he submitted the manuscript of his presentations to the *Proceedings of the Institute of Radio Engineers.* The

[47] Ibid., 639–640.

latter appeared in January 1949, with the title "Communication in the presence of noise."[48] This paper was not a reiteration or synopsis of his full-fledged information theory in 1948. Rather, Shannon's 1949 work provided a geometric understanding of the channel capacity formula in (12.9). In this approach, signals were viewed as points in a high-dimensional space, noise was construed as spheres of uncertainty surrounding those points, and the channel capacity was interpreted as the maximum number of non-overlapping spheres that could fill the domain bounded by the power of signal plus noise.

Shannon's 1949 paper began with the transmission of continuous signals in the presence of white Gaussian noise. Employing the same Whittaker sampling theorem, he represented the continuous, band-limited, and duration-limited waveform $x(t)$ with $M = 2TW$ sampling points $x_1 = x(t_1), x_2 = x(t_2), \ldots, x_M = x(t_M)$. Therefore, a signal waveform $x(t)$ corresponded to a set of values $(x_1, x_2, x_3, \ldots, x_M)$. Shannon gave a geometrical interpretation of this correspondence: The function $x(t)$ was represented by a point in a high-dimensional space. The M-tuple $(x_1, x_2, x_3, \ldots, x_M)$ was the Cartesian coordinates of the point; and M was the dimension of the abstract space, or, in Shannon's words, the function's "degrees of freedom." This was often a very high dimension for a typical signal in electrical communications. For example, a five-megacycle television signal lasting for an hour was represented by a point in a space of $2WT = 3.6 \times 10^{10}$ dimensions.[49]

When a signal $x(t)$ passed through a noisy channel, its corresponding point in the signal space was shifted by noise. Since noise was stochastic, the noise-plagued signal would be best represented not by a fixed translation from the original point, but by a region of uncertainty surrounding it—each point in this region corresponded to an individual realization of the stochastic function, and the region represented the statistical ensemble of the stochastic function. For white noise with average power N, the noise waveform $n(t)$'s coordinates $n_1, n_2, \ldots, n_M$ in the signal space were identical and independent Gaussian random variables, with $\langle {n_1}^2 \rangle = \langle {n_2}^2 \rangle = \ldots = \langle {n_M}^2 \rangle = N$. According to the central limit theorem in probability theory, when M was large enough, most realizations in that ensemble should satisfy $|n_1|^2 + |n_2|^2 + \ldots + |n_M|^2 \leq MN$. That means, white Gaussian noise could be represented with a hyper-sphere of uncertainty with radius $\sqrt{MN}$ in the M-dimensional signal space. To maximize the rate of information transmission, Shannon assumed that the signal $x(t)$ was also a white Gaussian random process with average power P. Thus, the signal could

[48] Claude Shannon, "Communication in the presence of noise," *Proceedings of the Institute of Radio Engineers*, 37 (1949), 10–21.
[49] Ibid., 12–13.

also be represented with a hyper-sphere of uncertainty with radius $\sqrt{MP}$ in the signal space. Since $x(t)$ and $n(t)$ were mutually independent, the noise-corrupted signal $y(t) = x(t) + n(t)$ at the receiver could be represented with a hyper-sphere of uncertainty with radius $\sqrt{M(P+N)}$.[50]

This "billiard-ball" representation of signal and noise offered an immediate visual heuristic. The channel capacity was the maximum number of bits per unit time a channel could transmit without error. For error-free transmission, the transmitter must encode messages in a way so that their signals did not equivocate with one another—i.e. no two messages corresponded to the same signal. For a channel with white Gaussian noise at average power N, each message corrupted by noise could be represented by a hyper-sphere with its center at the signal point $(x_1, x_2, x_3, \ldots, x_M)$ and radius $\sqrt{MN}$. No equivocation between two messages meant that their spheres of uncertainty did not overlap. Since almost all noise-corrupted signals lay within the M-dimensional sphere of radius $\sqrt{M(P+N)}$, the question of channel capacity became: How many non-overlapping hyper-spheres of radius $\sqrt{MN}$ could be packed within the large hyper-sphere $\sqrt{M(P+N)}$? This number must be no greater than the ratio of the volume of the big hyper-sphere (radius $\sqrt{M(P+N)}$) over that of the small hyper-sphere (radius $\sqrt{MN}$). From the formula of an M-dimensional sphere, this ratio was $\left[\sqrt{(P+N)/N}\right]^M = \left[\sqrt{(P+N)/N}\right]^{2TW}$. Representing this number in terms of bits, the maximum number of bits transmitted over the channel was $\log_2\left[\sqrt{(P+N)/N}\right]^{2TW} = TW \cdot \log_2[(P+N)/N]$. The channel capacity, or maximum number of bits transmitted per unit time, was $C = W \cdot \log_2[(P+N)/N]$.[51]

This exercise reproduced (12.9), the central result of Shannon's 1948 paper. But here he reached the conclusion through a much more intuitive and visual reasoning that treated a signal's regions of uncertainty caused by white noise like billiard balls. This idea traced back to his days at the IAS when he worked with Weyl. As he recalled:[52]

> I remember the very first lecture I went to where Herman Weyl talked about, he gave, at Princeton .... Now, it had to do with how many things you could pack into a certain space, proving that—like how many spheres could you pack into another sphere, something of that sort, in a dimension, or whatever. Which is of course very closely related to all this subject. The bounds were

[50] Ibid., 13–14.
[51] Ibid., 15–16.
[52] Price, "Shannon" (1982).

often crude and they were based on, the spheres had to not intersect. But anyway it was oddly related to the sort of things I'd been thinking about.

According to Shannon's protégé Peter Elias at MIT, the notion of billiard-ball packing formed the conceptual origin of Shannon's information theory.[53] While Shannon himself had reservations about this view, the crucial early readers of his theory at Bell Labs focused on his capacity formula (12.9) and understood it in terms of sphere packing. For example, in Nyquist's note to Rice on July 30, 1948 that summarized Shannon's theory, the discussion started with (12.9), and proceeded with calculating the number of noise spheres within a signal region at various dimensions.[54] Similarly, on July 19, Fry asked Shannon a question about information transmission. Fry believed that Shannon did not consider a channel's distortion of both signal and noise. With distortion, Fry maintained, a region of uncertainty at the channel output would not spread equally along all dimensions as a hyper-sphere did, but would be more like a hyper-ellipsoid.[55]

The formula (12.9) for the capacity of a noisy channel was derived when noise was white. In actual communications, noise could be more complicated than Brownian motion, thermal agitations, or shot effect. Shannon was aware of this. He explored the condition in which the channel noise was still Gaussian and stationary, but not white. Instead, noise now had a non-flat power spectrum $N(f)$ within a bandwidth W. Since the power of noise varied with frequency, to match this channel the power of signals also had to vary with frequency with a spectrum $P(f)$ within the bandwidth W. Thus, (12.9) would be generalized:

$$C = \int_0^W \log_2 \left[1 + \frac{P(f)}{N(f)} \right] df, \tag{12.10}$$

where $P(f)$ was constrained by the signal power P:[56]

$$\int_0^W P(f)\, df = P. \tag{12.11}$$

[53] Ibid.

[54] Letter, July 30, 1948, Harry Nyquist to Stephen O. Rice, Legacy No. 43-10-03, Folder 01, AT&T Archives.

[55] Letter, July 19, 1948, Thornton Fry to Claude Shannon, in Box 1 "Correspondence General 1938-55," Folder 2, Shannon Papers.

[56] Shannon, "Communication in the presence of noise" (1949), 19.

To find the "channel coding," the signal coding that optimized the rate of information transmission, Shannon had to maximize the channel capacity C in (12.10) over all the possible signal power spectrum $P(f)$ under the signal power constraint in (12.11). This optimization could be accomplished by taking a Lagrange multiplier with respect to (12.11), incorporating it into (12.10), and employing the calculus of variation over $P(f)$. The solution to this optimization problem was $P(f) + N(f)$ = a constant K, where K was adjusted to satisfy (12.11).[57]

This solution had different meanings when the total signal power P was high or low. When P was high, a positive K existed. At frequencies where noise power $N(f)$ was low, the signal power $P(f)$ should be high. This was like filling water over an uneven terrain of $N(f)$ to the level K. The signal power $P(f)$ at frequency f was analogous to the depth of water at this position. When P was low, the relation $P(f) + N(f) = K$ could not hold, for K was so low that at certain frequencies that relation would lead to an unphysical outcome $P(f) = K - N(f) < 0$. In this case, Shannon advised, we should allocate signal power $P(f) = K - N(f)$ when $K - N(f) \geq 0$, but $P(f) = 0$ when $K - N(f) < 0$. Again, this strategy was similar to filling a terrain with water. The only difference from the previous case was that the water level K was so low that at some frequencies the terrain $N(f)$ stuck out from the water (Figure 12.4).[58]

Shannon's power-allocation strategy for communication under noise of a given power spectrum made intuitive sense: investing more signal power at frequencies where noise was weak, allocating less signal power at frequencies where noise was strong, and assigning no signal power at frequencies where noise strength exceeded a certain threshold. He reached this conclusion through an application of information theory. Later nicknamed "water pouring" among the first-generation students of information theory at MIT,[59] this strategy was Shannon's one of very few channel coding schemes.

For a channel infested by white noise, Shannon's theory of information transmission entailed an optimum channel coding with signals resembling a white noise. For a channel plagued by colored noise, the theory implied an optimum coding invoking the strategy of "water pouring." For more general types of noise, a channel coding might not be readily available. But Shannon affirmed that we could still estimate the channel capacity through the signal power P, the noise power N, and the power N_1 of the corresponding "entropy noise"—equivalent white Gaussian noise with the same entropy as the channel

[57] Ibid., 19.
[58] Ibid., 19–20.
[59] Price, "Shannon" (1982).

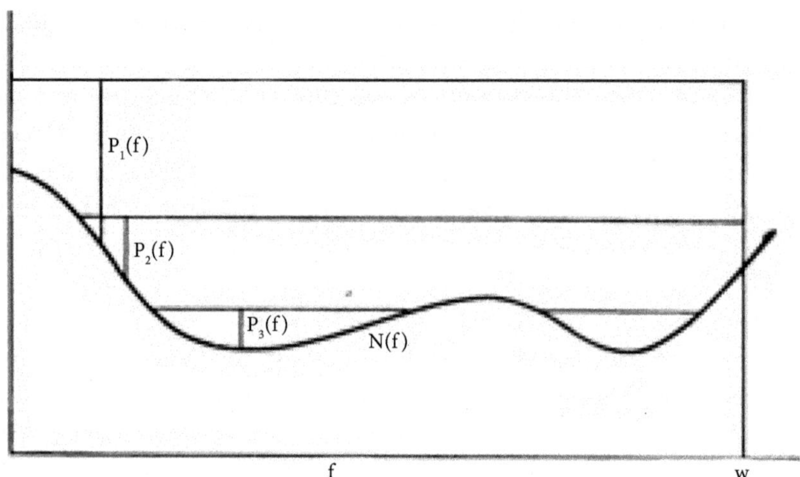

Figure 12.4 Shannon's "water filling" scheme to maximize the rate of signal transmission in a noisy channel. Claude Shannon, "Communication in the presence of noise," *Proceedings of the Institute of Radio Engineers*, 37 (1949), 19, Figure 8. Permission from IEEE.

noise. Shannon proved that the channel capacity for this general type of noise was bounded by $W \cdot \log_2 [(P + N_1)/N_1]$ and $W \cdot \log_2 [(P + N)/N_1]$.[60]

Shannon's handling of noise in information theory constituted an important element of wartime and postwar information sciences, like Uhlenbeck et al.'s treatment of noise in radar detection and Wiener's et al.'s grappling with observational errors in gunfire directing. These works shared the same ontology of the ambience that took noise as additive random Gaussian processes. Different from Uhlenbeck et al.'s preoccupation with metrological accuracy and Wiener's focus on filtering noise, however, Shannon paid attention to noise as channel effects, and what was possible or impossible under such effects.

Uhlenbeck, Wiener, and Shannon held different attitudes toward their research on noise. An established statistical physicist, Uhlenbeck was committed to the ontology of a world filled with Brownian-like random fluctuations. His and his close colleagues' personal experiences during the war gave him a sense of urgency for concentrating on immediately useful and deliverable analysis related to the American defense projects. A mathematician coming to terms with military technological development, Wiener was forming a

[60] Shannon, "Communication in the presence of noise" (1949), 20.

comprehensive worldview of cybernetics that took feedback control, communication, and computing as a general process underlying engineering, organic, and social systems. In contrast, Shannon's propensity was much more playful. He was not an architect of an overarching worldview, an advocate for a universal approach, a builder of a technological system, or a smart inventor. Rather, he was much more a problem solver, who chose to delve into problems that interested him. As he admitted:[61]

> Well, let me put it this way, that my mind wanders around and I will conceive of different things day and night. And the, like the science fiction writer or something like that. I'm thinking what if it were like this, or what is this problem, or is there an interesting problem of this type? And I'm not caring whether somebody's working on that or whether Washington would care one way or the other, or anything of that sort. It's usually, I just like to solve a problem. And I work on these all the time. And this problem of the best strategy against the worst noise and so on is just a thing that would occur to me. And in thinking about how would you handle this kind of noise or that kind of noise I would say, well, what would be the worst kind of noise?

Not a surprise, Shannon did not participate directly in AT&T's construction of communication systems, jump on the bandwagon of postwar information sciences, or get closely involved in the development of error-correction coding and compression coding in light of information theory. He preferred to look into the entropy of English, conceived an automatic chess player, designed a machine that could turn itself off, and explored ways to anticipate the variations of stock prices.

Yet, Shannon's information theory did provide a crucial pragmatic insight: Noise—or more precisely, noise-like patterns—was not only a detrimental uncertainty for engineers, physicists, and mathematicians to suppress and overcome. It could be a valuable resource for the betterment of information systems. In the derivation of the channel capacity under the influence of white noise in his 1948 paper, Shannon pointed out that the maximum rate of information transmission was attained when the signal resembled white noise. This finding traced back to 1947. In a letter to Shannon on December 29, a Calvin Mooers at the Zator Company in Cambridge, Massachusetts included

[61] Price, "Shannon" (1982).

his paper on a new theory of punch card selection for indexing and finding of filed information. Mooers told Shannon that:[62]

> because of the new and unusual use of random patterns, it should be of considerable interest to you. I recall your talk at Harvard last spring on the use of white noise spectral characteristics for the maximum transmission of information. Zatocoding [Mooers's filing scheme], as described in this paper, has much in common with this.

This utilization of "white noise spectral characteristics" had a conceptual affinity to Shannon's cryptographic investigation. The X-System's employment of prerecorded thermal noise to scramble voice signals testified to the security of random keys. Among signals at the same power level, those with the spectral patterns of white noise possessed the widest bandwidth. Shannon's idea of coding signals with white noise implied an artificial spreading of the signal spectrum.

A similar concept was developed when Shannon cowrote a paper with his Bell Labs colleagues Bernard Oliver and John Pierce. Titled "The philosophy of PCM," this piece was published in November 1948.[63] In light of the newly developed information theory, the three authors appraised PCM, a modulating scheme popular in AT&T's postwar development of digital voice communications. Treating the quantization errors in digitization as disturbances similar to white noise, Oliver, Pierce, and Shannon employed (12.9) to calculate the channel capacity for PCM. They compared the theoretical performance for PCM and FM, in terms of a simple signal that increased linearly with time. Oliver et al. discretized the time into small segments, and divided the usable signal bandwidth into four distinct frequency bands B_1, B_2, B_3, B_4. In PCM, there were four different types of pulses for each frequency B_1, B_2, B_3, B_4; whereas, in FM, the carrier wave was modulated at one of these four frequencies at each time step. For PCM, the ramp signal could be coded with a binary sequence of 000, 001, 010, 011, 100, 101, 110, 111, etc. at consecutive steps, which was translated into a transmitted sequence of pulses 0, B_1, B_2, B_1B_2, B_3, B_1B_3, B_2B_3, $B_1B_2B_3$, etc. For FM, the transmitted signal at each step could only take B_1, B_2, B_3, or B_4 (Figure 12.5). To Oliver et al., this contrast showed the advantage of PCM over FM:[64]

[62] Letter, December 29, 1947, Calvin Mooers to Claude Shannon, in Box 1 "Correspondence General 1938-55," Folder 2, Shannon Papers.

[63] Bernard Oliver, John Pierce, and Claude Shannon, "The philosophy of PCM," *Proceedings of the Institute of Radio Engineers*, 36 (1948), 1324–1331.

[64] Ibid., 1329.

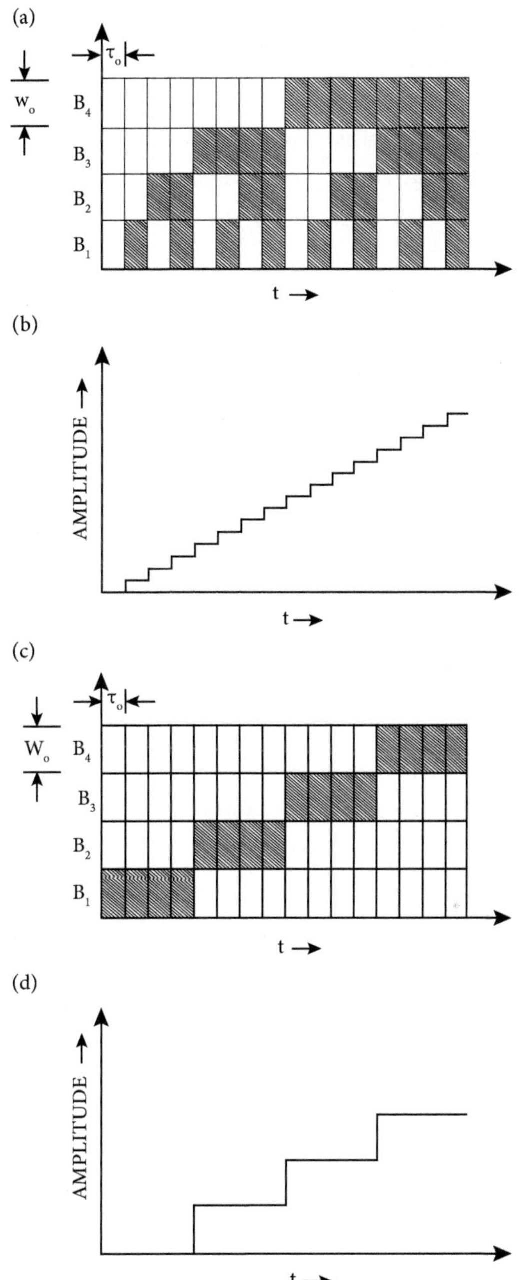

Figure 12.5 Oliver, Pierce, and Shannon's comparison of PCM with FM. The first panel (a) represents the band usage of PCM, and the second panel (b) represents its corresponding digitized signal in the time domain. The third (c) and fourth panels (d) illustrate the similar case for FM. Bernard Oliver, John Pierce, and Claude Shannon, "The philosophy of PCM," *Proceedings of the Institute of Radio Engineers*, 36 (1948), 1329, Figure 2.

Permission from IEEE.

> The trouble with the FM type of signal is that only a few of the possible signals which might be sent in the four bands B1-B4 are ever produced; all the others, those for which there is a signal in more than one band at a time, are wasted. Ideally, PCM takes advantage of every possible signal which can be transmitted over a given band of frequencies with pulses having discrete amplitudes.

From their perspective, PCM was better than FM in combating external disturbances because the former could on average utilize more parts of the entire signal spectrum B_1–B_4 at each time step, whereas FM utilized only one of the four bands at each time step. In other words, PCM's strength lay in its apparent scrambling or randomness in utilizing the spectrum resource.

To Shannon's interviewer Robert Price, a veteran of MIT Lincoln Laboratories, this line of thought epitomized a crucial conceptual trace for the beginning of spread-spectrum communication. Price found from his conversation with Shannon and other inquiries that around 1949, soon after the publication of the PCM paper, Shannon entertained the core notions of spread-spectrum communication and mentioned them to Oliver and Pierce: modulating signals with noise-like carriers to broaden their spectra, designing noise-like carriers so that they were statistically uncorrelated with one another ("orthogonal" in the geometrical jargon), and sending those signals from distinct transmitters. The use of random noise as signal carriers ensured that each signal could make use of the entire spectrum. Meanwhile, orthogonality between noise-like carriers protected communication links from mutual interference. Shannon called this scheme a "democratic way" of using the spectrum. Shannon did not put this idea in writing. Moreover, he was not aware that as he was entertaining the idea, similar noise-enabled spread-spectrum technologies were being developed at Lincoln Labs and elsewhere in the US.[65]

[65] Price, "Shannon" (1982).

13
Spread-Spectrum Communication

The spread-spectrum technology was one of the first applications of the wartime and postwar informational studies of noise. The term refers to telecommunication systems that place transmitted signals (sound waves, television patterns, or digital data) on carriers of much wider bandwidths. A design for fighting against environmental disturbances or artificial interferences, spread-spectrum communication is a core constituent of today's 4G and 5G wireless networks. In the academic and popular narratives about its birth, noise does not play a central part. Rather, what get highlighted are the needs of security, secrecy, and robustness against interception and jamming for military electronic communications in encrypted telephony, radar, sonar, torpedo and missile guidance, and aircraft navigation during World War II and the Cold War.

In their historical reviews, several authors have traced various origins of spread-spectrum communication to electronic warfare: the Swiss inventor Gustav Guanella's 1938 patent on noise-modulated radar and speech-privacy systems; the German company Telefunken's schema of combining voice with scrambled signals and the "Optiphone" that Erwin Rommel's troops used in North Africa; AT&T's "X-System"; the time-wobbling system of facsimile transmission invented in 1942 by E.M. Deloraine, Henri Busignies, and Louis de Rosa at the Paris laboratory of International Telephone and Telegraph (ITT) Company.[1] Taking these inventions as straightforward precursors of today's spread-spectrum systems may be problematic, since the inventors might not have had bandwidth extension in mind, and much of their work remained unknown or confidential for a long time, bearing no direct relationship with later developments. Yet, these communications systems shared the same core features that reflected engineers' general approach to interception

[1] Robert A. Scholtz, "The origin of spread-spectrum communications," *IEEE Transactions on Communications*, 30:5 (1982), 822–854; Mark K. Simon, Jim K. Omura, Robert A. Scholtz, and Barry K Levitt, *Spread Spectrum Communications*, vol. 1 (Rockville, MD: Computer Science Press, 1985), chapter 2, "The historical origins of spread-spectrum communications," 39–134; Robert Price, "Further notes and anecdotes on spread-spectrum origins," *IEEE Transactions on Communications*, 31:1 (1983), 85–97; William R. Bennett, "Secret telephony as a historical example of spread-spectrum communication," *IEEE Transactions on Communications*, 31:1 (1983), 98–104.

and jamming. Such features would indeed become intellectual resources for the technical designs of spread-spectrum systems.

A much better known episode in the prehistory of spread-spectrum technology is the invention of frequency hopping in 1941 by the movie star Hedy Lamarr and film music composer George Antheil. Lamarr's legendary tale as the "founding mother" of spread-spectrum technology has been made popular through substantial media coverage and the Pulitzer Prize winner Michael Rhodes's biographical account *Hedy's Folly*. The storyline goes like this: A young Austrian actress in an unhappy marriage, Lamarr (neé Hedwig Maria Kiesler) escaped from her authoritarian husband and the Nazi expansion in Europe, headed for the US, and made a career on the silver screen in Hollywood in the late 1930s. Lamarr supported the American war preparations against the Axis. As a handy tinkerer with electronics and machines since childhood, she believed that she could contribute to the cause with technological inventions and conceived a design for torpedo control.

At this time, torpedoes were often controlled remotely through radio signals, which suffered from jamming. To reduce interferences, Lamarr came up with an idea of making the wireless control signal jump randomly across a range of frequencies over time. This "frequency hopping" scrambled signals in the spectral domain and made them difficult to jam with narrow-band waves. To synchronize hopping at a transmitter and a receiver, Lamarr sought the help of her working partner Antheil, who devised a mechanism inspired by automatic player pianos to execute an irregular frequency shift in synchrony with preset machinery keys at both ends of communication. Lamarr and Antheil filed their patent in 1941.[2]

Lamarr and Antheil's invention of frequency hopping could have an interesting connection to the modern history of noise at large. As Bijsterveld pointed out, Antheil was an avant-garde music composer in the 1920s. Like the Italian futurists Luigi Russolo and Filippo Tomaso Marinetti, Antheil explored forms of music that were un-rhythmic or non-tonal and incorporated environmental noise of the industrial age. His *Ballet Mécanique* debuted in 1927 featured "ten pianos, a player piano, xylophones, electric bells, sirens, airplane-propellers, and percussions." He delved into the timbral characters of industrial cacophony and their potential use in music composition.[3] Yet, researchers have not found an explicit relationship between Antheil's thoughts

[2] Richard Rhodes, *Hedy's Folly: The Life and Breakthrough Invention of Hedy Lamarr, the Most Beautiful Woman in the World* (New York: Vintage Books, 2011), 133–193.

[3] Bijsterveld, *Mechanical Sound* (2008), 137–158.

on or experiments with noise and his work on frequency hopping. He did not seem to associate the random frequency shift Lamarr and he explored in torpedo control with the aircraft noise or street din he had appropriated for musical purposes. Rhodes stressed the proximity between Antheil's mechanical design in *Ballet Mécanique* and that in his torpedo control.[4] Although Lamarr and Antheil's invention effectively extended a torpedo control signal's bandwidth, they did not understand their design in terms of spread-spectrum.

In these antijamming and counter-intercepting communication systems, a transmitted signal was scrambled with a complicated, noise-like pattern. Scrambling was done by randomly shifting the signal's carrier frequency, modulating it by an irregular waveform, sampling it into pulses and wobbling their arriving times, or adding these pulses onto another complex train of pulses. In these cases, the signal "rode on" a noise-like carrier, became hardly intelligible to eavesdroppers, and prevented interception. Since the carrier did not have a regular pattern, many jamming techniques that deployed simple interfering waveforms could not tune to the carrier and hence caused little disturbance to the signal. In this sense, it was the *diversity* of the noise-like carrier that protected the signal. The apparently random signal should be unintelligible only to interceptors, not to recipients. Thus, a receiver should have a copy of the noise-like carrier and a machinery to decode the received signal.[5]

Two immediate questions followed. First, what was the proper receiver design to decipher noise-modulated signals? How to make a copy of the noisy carrier available at a receiver? What should the decoding machinery be? Second, how could a transmitter generate sufficiently irregular carriers? How noisy was "noisy enough?" After World War II, inventors and designers of scrambling communication systems began to treat these problems as those of noise, and attempted consciously to understand and solve them through the angles of the wartime and postwar informational studies of noise.

Wiener's, Shannon's, and Uhlenbeck's stochastic approaches to noise and communications brought novel insights to the designs of antijamming systems. First, they provided a perspective regarding why noise-like signals could counter interference. In the Wienerian theory, noise was a random process

[4] Rhodes, *Hedy's Folly* (2011), 154–169.

[5] For example, Guanella's system multiplied the signal with a noise-like waveform. The "X-System" added digital signal pulses with random numbers retrieved from prerecorded waveforms of thermal noise. Lamarr and Antheil employed random hopping of the carrier frequency. The ITT invention jittered the arriving times of signal pulses.

with a well-defined spectrum. When noise was "noisy enough" so that its autocorrelation between two different times was small, its spectrum was flat and wide. Thus, a signal modulated with such noise would spread its power over a broad bandwidth. Any narrow-band interference would affect only a small portion of the signal. Under the stochastic framework, a noise-modulated antijamming system became a "spread-spectrum" system.

Second, the statistical theory of information specified the decoding machinery at the antijamming receiver. The primary component of stochastic estimators, predictors, and filters was correlation. From the Wienerian perspective, a correlator provided a mechanism for decoding spread-spectrum signals. To recover a signal from a noise-modulated waveform $x(t)$, a receiver computed its correlation with the noisy carrier $y(t)$, $\int x(t)\, y(\tau - t)\, dt$, which could smooth out the fast-fluctuating carrier from the signal. Therefore, the theoretical and engineering work on correlators became relevant to spread-spectrum communication.

While the stochastic approach provided theoretical insights into spread-spectrum communications, their implementation relied on nuts-and-bolts technical ingenuity. The Wienerian doctrine informed very little about how to generate sufficiently noisy carriers, how to duplicate carriers at receivers, and how to make efficient correlators. In the 1940s, designers of secret communications systems had to generate random patterns by lengthy algebraic procedures or recording physical noise from tubes or resistors. An exact copy of a random pattern had to be prepared and delivered in a bulky medium like tapes to a receiver before communication. Synchronizing noise carriers at a transmitter and a receiver was challenging. Before digital computers, correlation could only be implemented with complicated and special-purpose analog circuits. These difficulties restricted the applications of the early spread-spectrum systems.[6]

A series of innovations in the 1940s–50s overcame the technical barriers and made spread-spectrum systems a widely used technology. The confrontation between the United States and the Soviet Union after Hiroshima and the outbreak of the Korean War boosted American research and development in military communications, control, and sensing. In this military–industrial context, engineers introduced more effective means for noise generation and correlation and broadened spread-spectrum from an antijamming tool into a counter-interference method.

[6] Scholtz, "Spread-spectrum" (1982), 824–829.

Mortimer Rogoff's Noise Wheel

Mortimer Rogoff was one of these pioneers. A radar operator in the US Navy during World War II with a B.S. in electrical engineering from Rensselaer Polytechnic Institute, Rogoff entered in 1946 the Federal Telecommunications Laboratory (FTL) that ITT had recently established in New Jersey. His first task at FTL was to work on a radio navigation system named NAVAGLOBE contracted by the Air Force. Designed to guide long-range bombers and missiles, NAVAGLOBE comprised an array of three transmitting antennas, whose radiation pattern enabled a receiver at a bomber or a missile to discern its exact direction. NAVAGLOBE drew upon the same principles as LORAN, the navigation system that MIT Rad Lab had developed for the US military during the war. Yet ITT's new system worked at the frequency range of 100 KHz, at which significant atmospheric noise occurred. NAVAGLOBE was seriously infected by atmospherics and did not work well.[7]

NAVAGLOBE offered Rogoff an opportunity to study noise, though. To solve the problem of atmospheric interference, ITT asked him to conduct an analysis of noise's effect on the navigation system. Based on Rice's "Mathematical analysis of random noise" in 1944–45, Rogoff calculated the changes of statistical properties of noise after passing through various circuits. His central measure to gauge a system's noise-resisting performance was the signal-to-noise ratio. Rogoff tried different signal detectors to see which one accomplished the higher signal-to-noise ratio at the output, and found that the square-law detector was better than the linear detector if the signal-to-noise ratio was larger than one. He designed and patented a square-law detector.[8]

A more impressive lesson to Rogoff concerned how noise became an engineering issue. "At this time," he later recalled, "I had the idea of pursuing noise as a friend, rather than as an enemy." His attention turned to secret communications. In such systems, artificial jamming resembled atmospheric interference—both were unwanted noise infecting wanted signals. Rogoff understood the function of noise modulation in the same way that ITT engineers pursued to fight atmospheric interference in NAVAGLOBE—both aimed at achieving a high signal-to-noise ratio. From this perspective, the correlator worked because it retained a good signal-to-noise ratio. Moreover, since antijamming was functionally identical to counter-interference for

[7] Interview with Mortimer Rogoff by David Mindell (November 28, 2003), 5–9.

[8] Ibid., 9–10; Mortimer Rogoff, "Spread-spectrum communications (for IEEE 1981 Pioneer Award)," *IEEE Transactions on Aerospace and Electronic Systems*, 18:1 (January 1982), 154–155.

enhancing the signal-to-noise ratio, one could use antijamming techniques to resist atmospherics.[9]

The design of an antijamming system's correlator and noise generator demanded more than a theoretical analysis of spread-spectrum. Here Rogoff's hobby in photography played a part. Working at his home garage in 1948–49, he experimented with a photoelectric implementation of noise modulation and demodulation. A crucial building block of his idea was a 4" × 5" sheet of photographic film whose transmissivity varied linearly in both horizontal and vertical directions. A unit-intensity light spot projected at the point (x,y) of the film (x denotes its horizontal coordinate and y its vertical coordinate) would penetrate it with intensity xy. This special film was used to make a correlator. Rogoff masked an oscilloscope with the film, deployed an electrical signal $x(t)$ and another time-reverted and -delayed signal $y(\tau - t)$ to the horizontal and vertical inputs of the oscilloscope, used a photodetector to convert the light intensity from the film into an electrical current $x(t)y(t - \tau)$, and passed this current through a resistor-capacitor integrator. The output of the integrator was the cross correlation of $x(t)$ and $y(t)$. This analog, photoelectric correlator was simpler and cheaper than many vacuum-tube correlators.[10]

The photoelectric correlator was only half of Rogoff's invention. A spread-spectrum system needed a noise generator, too. Again, the special photographic film played a part. To produce a random electrical signal, Rogoff plotted a dense sequence of irregular radial spikes in a circle on the film. The film was then mounted on a wheel. When rotated past a slit of light, the film modulated the intensity of the light beam with the irregular spikes. A noise-like electrical signal could be obtained by detecting with a photocell the irregularly modulated light beam. How "noisy" this signal was depended upon how random the radial spikes were. Contrasting with those who produced irregular patterns with algebraic procedures or recorded electronic noise, Rogoff used an easily accessible random-number generator: the phone book of Manhattan, New York. He randomly selected 1440 numbers not ending in 00 from the Manhattan telephone directory. For each number, he plotted radially the middle two of the last four digits every fourth of a degree along the circular perimeter. The product was known as a "noise wheel"[11] (Figure 13.1). The noise wheel was a handy way to generate and reproduce noise. It could be easily

[9] Rogoff, "Interview," 11–12.

[10] Scholtz, "Spread-spectrum" (1982), 833; Rogoff, "Interview," 16–19.

[11] Scholtz, "Spread-spectrum" (1982), 833; Rogoff, "Interview," 14–16.

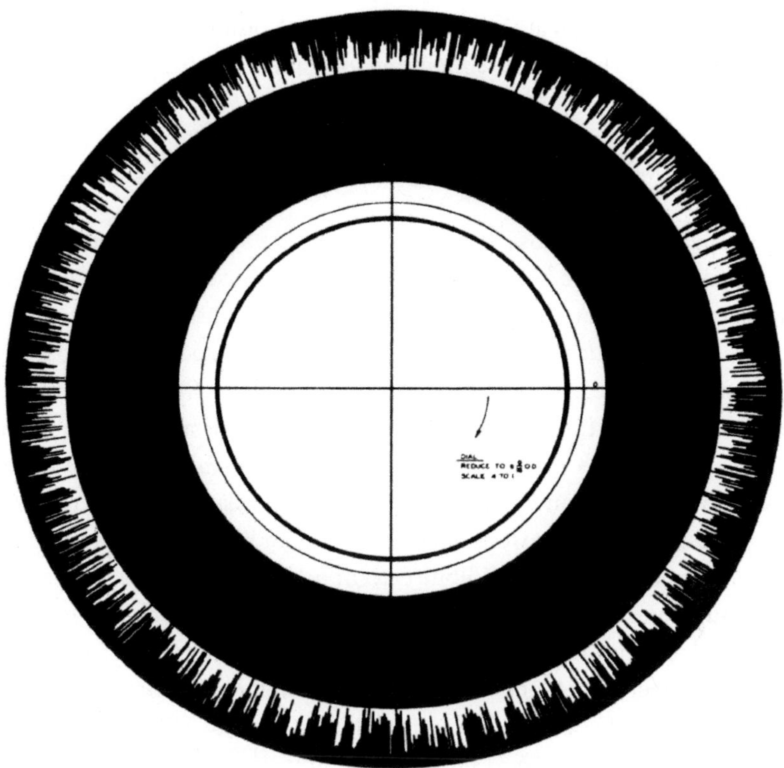

Figure 13.1 Rogoff's noise wheel. Robert A. Scholtz, "The origin of spread-spectrum communications," *IEEE Transactions on Communications*, 30:5 (1982), 833 Figure 6.
Permission from IEEE.

duplicated, delivered, and stored. A transmitter and a receiver could undergo spread-spectrum communication if they possessed identical noise wheels and synchronized the wheel motors.

Rogoff's work was soon noticed by de Rosa, who had co-invented time-wobbling encryption during the war and was still working at ITT on military communications. De Rosa incorporated Rogoff's inventions of the photoelectric correlator and noise wheel and helped turn them into a communication system. In 1952, ITT organized a long-range test of their spread-spectrum radio for the US Air Force. This transcontinental test between California and New Jersey was successful: A witness recalled that the spread-spectrum equipment worked well even under a severe magnetic storm that interrupted ordinary radio systems at the same frequency. Noise was useful not only in fighting jamming, but also in containing ionospheric disturbances.[12]

NOMAC at MIT Lincoln Laboratories

While Rogoff and de Rosa were working on a noise-modulated system at ITT, similar ideas were being developed in Massachusetts. In the 1940s, MIT was a center of communications science and information warfare technology. Wiener's statistical communications theory prevailed at the Department of Electrical Engineering. A generation of mathematically minded engineers were educated with the notions of generalized harmonic analysis and stochastic filtering. Wiener's former student and colleague Yuk Wing Lee devoted himself to the design and analysis of electrical correlators for control, detection, and estimation. This work offered training for junior faculty and students such as Jerome Wiesner and Wilbur Davenport, and nurtured an engineering culture surrounding correlation-based communication systems.[13] When Shannon introduced information theory in the late 1940s, young MIT researchers including Robert Fano and Peter Elias were among the first to greet and elaborate the new concept. These individuals did not focus exclusively on theory, though. MIT in the 1940s was the home of several crucial research establishments on radar, military radio, guidance, and navigation: Radiation Laboratory and its subsequent Research Laboratory of Electronics (RLE), Servomechanism Laboratory, Instrumentation Laboratory, etc. MIT researchers' stochastic-oriented engineering culture was shaped by their working experience at these laboratories.[14]

MIT continued to be a hub of the American military–industrial–academic complex as the Cold War began.[15] The anticipation of plausible Soviet attacks from the ocean, sky, and ether motivated further research on spread-spectrum technology at the Institute. In 1950, the Committee on Undersea Warfare of the US National Research Council called for an early warning system against possible invasions of Russian submarines. This led to the formation of a study group, "Project Hartwell," directed by Jerold Zacharias at RLE. An issue

[12] Scholtz, "Spread-spectrum" (1982), 834.

[13] Therrien, "The Lee-Wiener legacy" (2002), 33–44.

[14] For the MIT engineering culture at the laboratories during and immediately after World War II, see Karl Wildes and Nilo Lindgren, *A Century of Electrical Engineering and Computer Science at MIT, 1882–1982* (Cambridge, MA: MIT Press, 1986), 182–235.

[15] For MIT's role in the Cold War, see Stuart Leslie, *The Cold War and American Science: The Military-Industrial-Academic Complex at MIT and Stanford* (New York: Columbia University Press, 1993).

discussed at the group was covert fleet communications to prevent enemy submarines from triangulating warships' positions. De Rosa participated in Project Hartwell. He and researchers from Bell Labs proposed to implement covert fleet communications with spread-spectrum.[16]

The idea of spread-spectrum impressed another Hartwell participant. An Associate Director of RLE, Jerome Wiesner brought the idea back to MIT and discussed it with Fano and Davenport. This was timely to the RLE engineers. Fano had done work on radar detection during the war, got excited by information theory, and was entertaining the idea of a framework integrating Wiener's and Shannon's perspectives into engineering applications.[17] Davenport was a consultant of the US Army Signal Corps on how to counter the Soviet Union's high-power jamming of American long-range radio systems in Europe.[18] Wiesner's input addressed both needs. Led by Fano and Davenport, the graduate students at RLE began to investigate spread-spectrum and its applications to radar and radio. In 1951, MIT launched another study group, "Project Charles," for defense against possible long-range bomber attacks from the Soviet Union. This project eventually led to the first air defense system (Semi-Automatic Ground Environment, or SAGE) and the establishment of Lincoln Laboratories in Lexington, Massachusetts. The research on spread-spectrum was transferred from RLE to Lincoln Labs. Davenport's group became part of Lincoln Labs' Communications Division (Division 3).[19]

Division 3 developed a spread-spectrum system in 1950–52. Davenport, a small group of MIT graduate students, and a few technicians were the major developers, but Fano and Wiesner also played parts through their discussions with Davenport and their supervision of the students. The Lincoln Labs' spread-spectrum system was named NOMAC, the acronym for Noise Modulation And Correlation. As its name suggested, the system was built on the principle of noise modulation. Messages were converted into binary sequences and modulated by two noise-like carriers at a transmitter. At a receiver, incoming signals were correlated with the same random carriers to decipher the messages.

[16] Scholtz, "Spread-spectrum" (1982), 835.

[17] Robert Fano, *The Transmission of Information* (I and II), MIT RLE Technical Report No. 65 (March 1949) and No. 149 (February 1950).

[18] Wilbur Davenport, "The MIT Lincoln Laboratory F9C System (for IEEE 1981 Pioneer Award)," *IEEE Transactions on Aerospace and Electronic Systems*, 18:1 (January 1982), 157–158.

[19] For a history of the SAGE project and the establishment of Lincoln Laboratories, see Hughes, *Rescuing Prometheus* (1998), 15–67.

NOMAC differed from the wartime secret communications systems and ITT's noise-wheel sets in its implementation of correlation and noise generation. Its correlator was similar to the correlation networks that Lee and several students and alumni had worked on since the 1940s: an electronic circuit comprising a vacuum-tube multiplier and a low-pass filter. The correlation approach that constituted the core of the MIT communication engineering culture led to a technical solution quite different from Rogoff's gadget: Whereas the former implemented the correlator with an electronic network via general-purposed circuit design, the latter did so with a photoelectric device that was analog and special-purposed.[20]

Another innovative aspect of NOMAC lay in noise generation and synchronization. A young engineer Paul Green contributed to this innovation. A native of Chapel Hill with an M.S. in electrical engineering from North Carolina State College, Green entered MIT's Ph.D. program in 1949.[21] At MIT, he was under Davenport's supervision and was recruited into NOMAC's development at Lincoln Labs. Davenport offered Green a thesis topic that was a major challenge for the spread-spectrum systems: inventing a method to generate noise-like carriers at a transmitter *and* to convey the carrier to a receiver. Green came up with an idea. If the noise were generated properly, then one could repeat noise generation at the receiver without having to send the carrier via a separate channel or store the carrier.

Green's solution to noise generation was the new digital circuitry. His noise generator consisted of a stack of digital counters that produced trains of pulses with different periods but which were all synchronized by a standard-frequency oscillator. The periods of these pulses were not in simple ratios to one another. Therefore, when mixing them together, the result was a sequence of pulses with a highly irregular appearance. The waveforms produced in this manner were not truly random, but they were "noisy enough" to spread signal spectrum. Moreover, the digital-counter-enabled noise generator offered a solution to the problem of delivering noise-like carriers to the receiver: One only had to install an identical noise generator at the receiver and synchronize it with the digital circuit at the transmitter. No need to record the carrier waveforms in storage, and no need to send them via another wireless channel. To synchronize noise generation between the transmitter and receiver,

[20] Paul Green, *The Lincoln F9C Radioteletype System*, Lincoln Laboratories Technical Memorandum No. 61, May 14, 1954, 2, 28–41.

[21] David Hochfelder, "Oral history: Paul Green," IEEE History Center, Interview #373, October 15, 1999, 4, https://ethw.org/Oral-History:Paul_Green (accessed April 8, 2022).

Green used a track-and-lock controller that was the prototype of the modern phase-lock loop circuit.[22]

MIT Lincoln Labs' work on NOMAC not only contributed to the technical implementation of spread-spectrum systems. Davenport, Fano, Wiesner, and their graduate students also developed stochastic-based theoretical tools and concepts to analyze spread-spectrum communications. Bennett Basore, a B.S. from Oklahoma A&M College, a student of Davenport's, and a member of Division 3, wrote a doctoral dissertation in 1952 that compared the performance of the stored-carrier and transmitted-carrier spread-spectrum systems in the presence of a random Gaussian noise. Like Rogoff, Basore used the signal-to-noise ratio as the measure of performance. But his analysis was grounded much more on the statistical detection theory that Uhlenbeck's team at Rad Lab had helped develop during World War II. Moreover, Basore suggested that the new information theory might provide a novel meaning for spread-spectrum. According to Basore's interpretation of Shannon's 1949 paper "Communication in the presence of noise," Shannon "noted that the waveforms of the set of signals which would lead to the full utilization of the system capacity with arbitrarily low equivocation would be in all respects similar to white gaussian noise."[23] Thus, noise-like carriers could serve as a means not only to spread the signal spectrum, but also to optimize coding to achieve channel capacity.[24]

After Lincoln Labs finished the design of NOMAC in 1952, the US Army Signal Corps took over manufacturing and renamed it F9C. In 1954, the Signal Corps performed a field test of F9C between Davis, California, and Deal, New Jersey. The test went well for low data rates; but as the data rate increased, the received signals suffered a considerable distortion. The problem, the army engineers figured out, was due to multipath propagation. For long-range signal delivery at F9C's frequency range (130–150 kHz), radio waves propagated via multiple routes: They bounced back and forth between the earth and the ionosphere with different rounds, and they could split in the ionosphere. The details of multipath propagation were determined by the condition of the ionosphere. The outcome was pathological: A single waveform from the transmitter

[22] Green, *The Lincoln F9C Radioteletype System* (1954), 7–27.

[23] Bennett Basore, *Noise-like Signals and Their Detection by Correlation*, Sc.D. dissertation (Cambridge, MA: MIT, 1952), 5.

[24] This interpretation differed slightly from Shannon's original argument. Shannon's random coding in his derivation of the channel coding theorem exhibits a "statistical diversity" that exhausts all possible cases. By contrast, the noise-like carriers in spread-spectrum communications exhibit a "complexional diversity" and are sufficiently irregular but do not have to exhaust all statistical possibilities.

would arrive at the receiver multiple times, resulting in the superposition of several duplications with different time delays.[25]

Multipath propagation posed a challenge to spread-spectrum systems. While modulating signals with noise-like waveforms was to remedy a specific kind of uncertainty (i.e. additive noise), the multipath problem corresponded to another kind of uncertainty—duplication of signals with irregular multiple delays. The diversity offered by noise-like carriers was not sufficient. The Lincoln Labs researchers needed another kind of diversity to overcome the multipath problem.

Their approach may be epitomized by an episode Davenport recalled later. In the early days of NOMAC/F9C, a British delegation visited Division 3 and saw the system. The leader of the delegation expressed that the British group once considered making a quite similar noise-modulation system, but they gave up on the idea because they knew too well that multipath would be a problem. Davenport was not deterred from continuing the work on NOMAC/F9C, and he considered this a good example of "the value of not knowing too much."[26] In the context of fighting multipath, "not-knowing-too-much" characterized a technical scheme that treated different delay times as unknown and tried to match instead of knowing them.

This scheme was proposed by Paul Green and Robert Price, an A.B. from Princeton University and another graduate student/researcher of MIT/Lincoln Labs. Key to their scheme was a receiver comprising *a bank of* filters to match an incoming noise-like waveform delayed at thirty consecutive times. The outputs from the thirty matched filters were then multiplied by distinct weights and summed together. In principle, those matched filters whose delay times were closest to the multipath waveform's actual delays should have the strongest outputs, since they were the best "matches" to the waveform. This led to an arrangement in which the weights assigned to the matched filters were *adaptive* and corresponded to the matching results. The filters that matched better the multipath waveform were given higher weights. In so doing, maximum signal power could be retrieved from the bank of matched filters and the receiver did not have to utilize knowledge about ionospheric wave propagation. Green and Price coined the name "Rake" for this anti-multipath spread-spectrum system, since the filter bank looked like an agricultural tool.[27]

The NOMAC/F9C/Rake was not the only postwar spread-spectrum system. In the early 1950s, ITT, Sylvania Company, Bell Labs, and Caltech Jet Propulsion Laboratory were all developing spread-spectrum communications

[25] Scholtz, "Spread-spectrum" (1982), 839.
[26] Davenport, "The MIT Lincoln Laboratory F9C System" (1982), 159.
[27] Robert Price and Paul Green, *An Anti-Multipath Communication System*, Lincoln Laboratories Technical Memorandum No. 65, November 9, 1956.

systems. Yet the Lincoln Labs instrument was perhaps the first functional wideband spread-spectrum system and the first one discussed in published literature. By the late 1950s, Rake became a standard technique to deal with fading (i.e. multipath) problems, and noise modulation was a well-known counter-interference scheme. The spread-spectrum technology was incorporated into satellite communications in the 1960s. Borrowing key ingredients of satellite communications twenty years later, MIT-trained information theorists Irwin Jacobs and Andrew Viterbi developed spread-spectrum further into a means to increase the data transmission rate, commercialize the technology into today's Code Division Multiple Access (CDMA), and establish the company Qualcomm in San Diego that has continued to dominate wireless communications.

In the early history of spread-spectrum technology, noise was transformed from an environmental plight into something useful, as it was turned into a source of *diversity*. Since noise appeared in irregular or random forms, it could drive a system into a variety of different states that did not exhibit simple orders. Apparently, this made the system hard to control. At another level, however, such irregularity or randomness prevented the system from being entrenched at certain specific states that were defective or susceptible to the attacks of external interferences and disturbances. This was the principle underlying the early secret communications systems that "scrambled" signals. Even though signal scrambling later gained the meaning of "spread-spectrum" in the framework of stochastic theories, the idea was still to make the signal more robust by diversifying it into a wider bandwidth. The multipath phenomenon posed a problem to engineering design, but it also offered additional signal diversity with multiple, delayed duplications of transmitted waveforms. The MIT researchers solved the multipath problem by using the Rake architecture to capture such diversity. Although the spread-spectrum technology was born in the military–industrial–academic context of World War II and the Cold War that stressed determinacy and abhorred uncertainty, it was made possible, ironically, by appropriating noise and diversity.

Theories and Technologies of Noise under the Shadow of War

Theoretical studies of noise underwent a fundamental change during the Second World War and the years afterward. Before the war, physicists endeavored to characterize the properties of random fluctuations, mathematicians attempted to develop conceptual frameworks to represent stochastic processes, and engineers explored models to assess the impacts of noise on communications. These studies were conducted in the domain that followed the variations of random noise over time, or in the domain that inspected the power of random noise at various frequencies. The war provided a strong driving force toward the integration of these undertakings. Under the cause of military preparation, physicists, mathematicians, and engineers worked together on radar, gunfire control, and secret communication. In the new context, noise was not only an omnipresent entity in various realms that caught scientific attention. It became crucial elements of an "ontology of the ambience" that influenced the performance of defense information technologies, and thus important objects to manipulate, control, predict, and estimate.

Uhlenbeck's associates at MIT Rad Lab came up with a body of works that treated the effects of noise on radar target detection as those of Brownian motions on measuring instruments. Wiener's theory of statistical prediction and filtering amidst random noise for antiaircraft gunfire directing led to his cybernetic worldview. Shannon's construal of the rate of signal transmission across a noisy channel culminated in his information theory. These examinations and analyses eventually helped integrate the considerations and handling of noise into scientific and engineering treatments of uncertainties in general. An application of the new noise studies after the war was the development of spread-spectrum communication systems. In their development, engineers came up with novel technical designs for noise generation and synchronization, calculations of correlation, and adaptation to multipath interferences. These inventions together formed a technological repertoire for handling and applications of noise in its informational sense. With them, the transformation of noise was complete.

14
Conclusion

Throughout the long history before the end of the nineteenth century, noise appeared exclusively as annoying sounds. To the philosophers, scholars, and scientists of music, it was the opposite of musical tones, discordance, dissonance, disharmony, and irregular combinations of many different resonating keys. To the advocates for a tranquil life, noise was a public nuisance, environmental plight, sonic pollution, and an aural symptom of the urban civilization or the industrial life. In the first half of the twentieth century, the meaning of, understanding of, and grappling with noise underwent a radical transformation. This change started with the popularity of sound reproduction technologies. In the process of improving the sonic quality of their products, phonographic technologists extended the concept of noise to the material defects on recording media. On the other hand, electrical engineers invented methods and instruments to measure not only ambient din but also statics, crosstalk, and other interferences in telephony and radio, turning noise into abstract and quantifiable entities connected to various natural and human-made effects.

Meanwhile, statistical physicists at the turn of the century re-examined the phenomenon of Brownian motion—haphazard movements of suspended particles in a fluid—and came up with a theory to interpret it in terms of random molecular thermal collisions. In the 1920s, researchers at industrial laboratories discovered electronic noise—erratic deviations from signals in telecommunication systems—and conceived it as electrical currents' Brownian motions at various stages of radio and telephone links. The physical model of Brownian motion and the mathematical formulation of stochastic processes began to form a theoretical picture of noise as random fluctuations in general.

In the interwar period, physicists, engineers, and mathematicians developed this picture in different ways. Physicists tracked the evolution of random fluctuations over time via standard approaches in statistical mechanics. Engineers approximated frequency components not only of electronic noise but also of other disturbances in telecommunications with a "white" spectrum. Mathematicians attempted to formulate stochastic processes in an esoteric language of measure theory, functional space, and generalized harmonic analysis.

These different threads of developments merged circa World War II. The war imposed an imperative for physicists, engineers, and mathematicians to work together, in tandem, or in proximity on the military technologies of radar detection, antiaircraft gunfire control, and encrypted wireless telegraphy and telephony. The characterization, suppression, and filtering of random noise and the design of effective signal-transmission schemes within a noisy background became a crucial foundation not only for the military technologies but also for the new informational worldview after the war. By the late 1940s, noise constituted a central part in the new paradigm of cybernetics and information theory. Meanwhile, noise was turned into a fully informational concept that represented uncertainty and disturbances at large, with a repertoire of theoretical treatments of Gaussian random processes including threshold detection, filtering, estimation, and geometric representations.

After World War II, the new meaning of noise as generic fluctuations and deviations spread to wider audiences. A Google Ngram with respect to books published in English between 1800 and 2019 shows that the number of works containing the keywords "noise" and "information" or "noise" and "signal" launched into a steep rise from almost null around the mid-1940s. For "noise" and "information," the number reached a peak around the mid-1970s (pinnacle of the "information hype") and oscillated about a plateau afterwards. For "noise" and "signal," the peak arrived around the early 1960s and the oscillation about the plateau was milder. For reference, the number of books containing the keyword "noise" was already significant at the height of the Industrial Revolution in the 1800s. That figure declined slightly over the nineteenth century and the first decades of the twentieth century until the 1940s, when it grew rapidly like the previous two diagrams (Figure 14.1). If we treat these Ngrams as a macroscopic marker, then the association of "noise" with "information" or "signal" was basically unknown before World War II but became more and more frequent by the 1950s–70s.

Noise since the Mid-Twentieth Century: Ontology, Epistemology, Technology

The historical process in this book transformed noise from unpleasant sensory experiences into a set of general and heterogeneous concepts, theories, and techniques shared by specialists working on various subjects from sound reproduction and statistical physics to mathematization of probability and military information technology. These developments paved the way for an

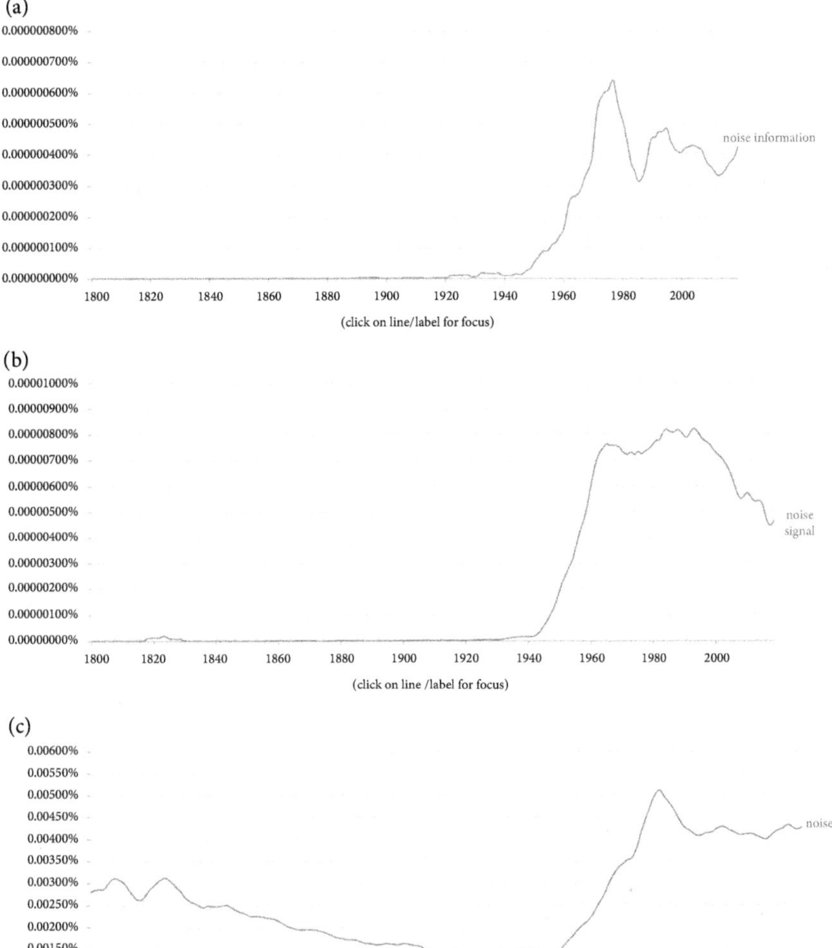

Figure 14.1 Google N-grams for books published in English from 1800 to 2019 containing the keywords of "noise" and "information" (panel (a): https://books.google.com/ngrams/graph?content=noise+information&year_start=1800&year_end=2019&corpus=26&smoothing=3&direct_url=t1%3B%2Cnoise%20information%3B%2Cc0#t1%3B%2Cnoise%20information%3B%2Cc0), "noise" and "signal" (panel (b): https://books.google.com/ngrams/graph?content=noise+signal&year_start=1800&year_end=2019&corpus=26&smoothing=3&direct_url=t1%3B%2Cnoise%20signal%3B%2Cc0), and "noise" (panel (c): https://books.google.com/ngrams/graph?content=noise&year_start=1800&year_end=2019&corpus=26&smoothing=3&direct_url=t1%3B%2Cnoise%3B%2Cc0) (accessed April 8, 2022).

even more encompassing and overarching picture. Since the mid-twentieth century, noise has reached even broader circles in science and technology and gained a reputation as something related to the existential conditions of the modern world. To some, noise represented fundamental fluctuations and indeterminacies that permeated nature and the universe. From time to time, these fluctuations and indeterminacies moved from background to foreground and were construed as a basis to form crucial order or to constitute surprisingly novel effects. To others, noise epitomized the intrinsic uncertainty that posed a limit on the exactitude of empirical knowledge but underscored the wise methods, strategies, and attitudes for retrieving useful and sensible conclusions from data with a fuzzy background. Under certain other circumstances, noise was an important part of the technological backbone for the contemporary computing form of life—be it electronic hardware, signal processing software, or mathematical tools. In other words, noise has had implications in ontology, epistemology, and technology.

Ontology

An immediate sequel of prewar noise studies was an increasing awareness that Brownian motion and electronic noise were members of a much larger class of phenomena. Smoluchowski placed irregular movements of suspended particles together with random variations of concentration in a fluid and critical opalescence. Langevin contemplated the fluctuating nature of all physical effects. Uhlenbeck, Goudsmit, and Ising explored the constraints such fluctuations incurred on the precision of galvanometers and other measuring instruments. The postwar researchers moved one step further. In 1951, the American statistical physicists Herbert Callen and Theodore Welton at the University of Pennsylvania published an article in *Physical Review* on the fluctuations of an arbitrary irreversible system. They contended that the dissipation, friction, damping, or another form of energy loss that characterized any irreversible physical system was an outcome of random molecular actions and corresponded to fluctuations of observable quantities. To them, these fluctuations were generalized noise in irreversible systems, like thermal noise in a resistor. Moreover, Callen and Welton established a formulation, known as the "fluctuation-dissipation theorem," that governed the relationship between an irreversible system's generalized noise and its energy loss due to dissipation. Einstein's equation connecting the mean-square displacement of Brownian particles with the fluid's Stokes frictional coefficient and Nyquist's theory relating the power of the thermal electronic noise to a circuit's resistance were two special cases of this generalization. Callen and Welton's formulation covered a wider range of effects: fluctuations of barometric pressure in a gas connected to its acoustic radiation resistance, random variations of electric field in a cavity corresponding to its radiation resistance, etc.[1] Throughout the second half of the twentieth century, this generalized noise and the fluctuation–dissipation relation were considered an omnipresent feature of irreversible thermodynamic systems' responses to external perturbations. Callen and Welton's generalization of Nyquist's electronic-noise theory was extended and employed in a broader range of problem situations from fluid turbulences and climate modeling to cellular biochemistry and phase transitions of granular materials.[2]

These considerations of random noise in the contexts of thermodynamics and statistical mechanics gained a new significance as they were integrated into the powerful scientific discourse by the third quarter of the twentieth century that orbited around the notions of systems, order, and complexity. Evelyn Fox Keller has traced the complicated origin and early development of the concept of "self-organization"—formation of regular patterns out of disordered initial states because of the specific features of dynamic systems—in mathematics, physics, chemistry, and theoretical biology in the 1960s–90s. A prominent figure in the first stage of what Keller viewed as a paradigm shift was a Brussels-based Russian-Belgian physical chemist, Ilya Prigogine. Prigogine studied the thermodynamics of states far from equilibrium. He found that in a dissipative system that exchanged energy and material with its surroundings, small-scale random fluctuations could be transformed into large-scale patterns when pertinent mathematical conditions about systemic stability were met. Prigogine and his contemporaries Manfred Eigen, Herman Haken, and Grégoire Nicolis used this concept of "dissipative structure" to explain not only the formation of circular patterns out of turbulence around the boundaries of a fluid (known as "Bénard cells") and the autocatalysis of multiple chemical substances in a solution, but also more involved phenomena such as the homeostasis, regeneration, and functional division of a living organism.[3]

Along a similar line of thought, meteorologists, mathematicians, ecologists, and physicists including Edward Lorenz, Stephen Smale, Benoit Mandelbrot,

[1] Callen and Welton, "Irreversibility and generalized noise" (1951), 34–40.

[2] For example, see Umberto M.B. Marconi, Andrea Puglisi, Lamberto Rondoni, and Angelo Vulpiani, "Fluctuation-dissipation: Response theory in statistical physics," *Physical Reports*, 461 (2008), 111–273.

[3] Evelyn Fox Keller, "Organisms, machines, and thunderstorms: A history of self-organization, part II: complexity, emergence, and stable attractors," *Historical Studies in the Natural Sciences*, 39:1 (2009), 1–31. For the prehistory of this development, see Evelyn Fox Keller, "Organisms, machines, and thunderstorms: A history of self-organization, part I," *Historical Studies in the Natural Sciences*, 38:1 (2008), 45–75.

Robert May, and Mitchell Feigenbaum during the same period noticed the peculiar behavior of certain nonlinear dynamical systems—sensitivity to initial conditions, involution into strange attractors in a phase space, period doubling, fractal geometry, etc.—and developed a theory of chaos.[4] In the narratives of self-organization and chaos, regularities could emerge out of a sea of disorder. The role of random fluctuations in this type of pattern formation was to make a dynamical system's state wander within a sufficiently diverse region in the phase space, so that the stable or robust configurations had a higher chance to be selected from the system's evolution. Under the complex-system discourse, noise became a crucial seed of emerging order.

Self-organization, chaos, and nonlinear dynamics were not the only narratives that modern scientists invoked to explore the ontological implications of noise. The revolution in molecular biology in the 1950s–60s triggered a new kind of imagination about the relationship between life and random fluctuations. The discovery of DNA as genetic material makeup and the unveiling of its mechanism of reproduction entailed a highly universal and deterministic picture of life: The sequences encoded on DNA molecules are translated into proteins that work together to make cells functional units; and this process applies to all forms of organisms. But individual variations of biological traits across a population had been well known among naturalists for a long time. Here, the postwar informational worldview offered an account. As historian Lily Kay indicated, cybernetics, information theory, and other wartime information sciences opened a "discursive space" that represented life and heredity in terms of the operations of information and control, the base-pair sequences on DNA molecules as codes and programs, and the genetic makeup of organisms as the "book of life."[5] Under this framework, individual variations, mutations, and functional defects were understood as "noise" in the process of genetic replications *qua* information transmission.

For example, the British geneticist John B.S. Haldane at University College London, Wiener's friend and one of the earliest biologists to embrace the cybernetic doctrine, remarked in 1948 that:

> I suspect that a large amount of an animal or plant is redundant because it has to take some trouble to get accurately reproduced, and there is a lot of noise around. A mutation seems to be a bit of noise which gets incorporated

[4] For a historical review of the development of chaos theory, see Gleick, *Chaos* (1987).

[5] Lily Kay, *Who Wrote the Book of Life: A History of the Genetic Code* (Stanford, CA: Stanford University Press, 2000); and "Who wrote the book of life: Information and the transformation of molecular biology, 1945–1955," *Science in Context*, 8:4 (1995), 609–634.

> into a message. If I could see heredity in terms of message and noise I could get somewhere.[6]

Jacques Monod, Nobel Laureate and director of the Pasteur Institute in Paris who was renowned for his findings in the genetic control of enzymes, made a similar comment two decades later. Monod was a central figure in Kay's story, especially with respect to the "Pasteur connection" surrounding the discovery of mRNA, *cybernétique enzymatique*, and *gène informateur*. In his well-selling book *Chance and Necessity* on the "natural philosophy" of modern biology in 1971, Monod affirmed the adequacy of mathematical, physical, and chemical principles as the basis for the invariance and regularity of biological phenomena. But he pointed out that perturbations from the normal could persist and constitute new effects on living organisms. Such effects due to "noise"—random fluctuations—were salient in the process of evolution. As he wrote:[7]

> But where Bergson saw the most glaring proof that the "principle of life" is evolution itself, modern biology recognizes, instead, that all the properties of living beings rest on *a fundamental mechanism of molecular invariance*. For modern theory *evolution is not a property of living beings*, since it stems from the very *imperfections* of the conservative mechanism which indeed constitutes their unique privilege. And so one may say that the same source of fortuitous perturbations, of "noise," which in a nonliving (i.e., non-replicative) system would lead little by little to the disintegration of all structure, is the progenitor of evolution in the biosphere and accounts for its unrestricted liberty of creation, thanks to the recreative structure of DNA: that registry of chance, that tone-leaf conservatory where noise is preserved along with music.

To Monod, therefore, noise that engendered random errors and made all worldly phenomena—non-living and living alike—chance events would constitute a legitimate driving force for evolution in lieu of the ghosts of the Aristotelian teleology or Bergsonian vitalism.

Kay criticized the postwar informational and scriptural representations of life and heredity and showed that they "were not an outcome of the internal cognitive momentum of molecular biology; they were not the logical

[6] Quoted from Kay, *Who Wrote the Book of Life* (2000), 87.
[7] Jacques Monod, *Chance and Necessity: An Essay on the Natural Philosophy of Modern Biology*, translated by Austryn Wainhouse (New York: Vintage Books, 1971), 116–117 (italics in the original).

necessity of the unraveling of the base-pairing of the DNA double-helix." Rather, these biological representations were transported from "cybernetics, information theory, electronic computing, and control and communication systems—technosciences that were deeply embedded with the military experiences of World War II and the Cold War." In her view, the informational models, metaphors, and languages in molecular biology prevailed not because they "worked in the narrow epistemic sense," but because they "positioned molecular biology within postwar discourse and culture, perhaps within the transition to a postmodern information-based society."[8]

Kay's criticism makes sense. Indeed, thinking of variations at the molecular, cellular, physiological, or evolutionary levels in terms of random fluctuations had very limited effects on the actual laboratory research in molecular biology of the 1940s–70s. The development in life sciences in the following decades nonetheless changed the relationship between life sciences and the information-based technoscience. Genomics, systems biology, and bioinformatics introduced in the early 2000s looked into the mutually interacting genes and their "pathways" or "circuits" toward enzyme production, regulation of the cellular environment, and responses to external alterations as a highly complicated system of control. In this new context, information, redundancy, and feedback no longer remained only as cultural discourse; they were turned into ideas and notions integrated into biologists' laboratory research design, data processing and interpretations, and computer simulations for theoretical modeling. Along the same line of thought, noise became something more than a metaphor to molecular biologists.

Recently, they have started to explore the functional roles of noise—as random fluctuations of chemical concentrations within cells—in genetic circuits. As researchers have claimed, "noise, far from just a nuisance, has begun to be appreciated for its essential role in key cellular activities." For instance, random fluctuations can constitute a stochastic control mechanism to coordinate the expression of multiple genes. Noise also enables probabilistic differentiation strategies that engender diverse traits across a cell population. In a longer timescale, noise may facilitate evolutionary transitions.[9] In contrast to Haldane's and Monod's intuitive link of noise with mutations, diversity, and evolution, biologists of the early twenty-first century endeavored to pin down the specific mechanisms through which stochasticity enters the

[8] Kay, "Who wrote the book of life" (1995), 609.

[9] Avigdor Eldar and Michael Elowitz, "Functional roles for noise in genetic circuits," *Nature*, 467:9 (2010), 167–173.

cellular biochemistry of gene regulation, functional division, and evolutionary dynamics.

Another novel narrative about the physical origin of noise also appeared after World War II. Before the mid-twentieth century, random noise was generally conceived as the outcomes of atoms' or molecules' irregular motions. With the advancement of quantum physics, a different notion of fluctuations came to the fore. At the debut of quantum mechanics in the 1920s, Werner Heisenberg proposed the uncertainty principle concerning an intrinsic ambiguity of two canonically conjugate physical quantities (e.g. position and momentum). An implication of this principle was the sudden creation of quantized virtual particles out of a "vacuum" quantum state (i.e. nothing) and their subsequent annihilation. In the 1930s–40s, the assertion of such virtual particles became a crucial basis for the development of quantum electrodynamics that governed electromagnetic interactions between particles.[10] Within the framework of quantum electrodynamics and the quantum field theories for other elementary forces, therefore, a quantum state of a physical system would undergo incessant fluctuations due to creations and annihilations of virtual particles out of "the vacuum."[11]

Although the use of virtual particles was key to quantum field theorists' success with predicting the results from atomic spectroscopic experiments in the late 1940s, quantum fluctuations did not associate with the notions of noise until the invention of the maser and laser in the 1950s–60s. The materialization of microwave and light amplification by stimulated emissions was a primary breakthrough of early Cold War atomic physics. The quantum mechanical processes of stimulated emissions from higher atomic states to lower ones promised the production of highly coherent and focused electromagnetic waves. But quantum fluctuations engendered spontaneous emissions that degraded the coherency, spectral sharpness, and spatial focus of masers and lasers. To atomic physicists, fluctuations out of the vacuum formed a type of noise that affected the stability and performance of opto-electronic instruments. Historian Joan Lisa Bromberg followed the theoretical understandings and practical treatments of quantum noise among pioneering maser researchers.[12] With widespread applications of masers and lasers

[10] David Kaiser, *Drawing Theories Apart: The Dispersion of Feynman Diagrams in Postwar Physics* (Chicago, IL: University of Chicago Press, 2005), 27–42.

[11] For a comprehensive examination of the notion of the vacuum in quantum physics and general relativity, see Aaron Sidney Wright, *More than Nothing: A History of the Vacuum in Theoretical Physics, 1925–1980* (Oxford: Oxford University Press, forthcoming).

[12] Joan Lisa Bromberg, "Coherence and noise in the era of the maser," *Perspectives on Science*, 24:1 (2016), 93–111.

since the 1970s, quantum fluctuations have been a salient source of noise that no optoelectronic researchers and engineers could afford to ignore. When quantum computing and communication utilizing entangled states became realistic by the 2000s, quantum noise was posed again as an essential constraint to what was boasted as information technology of the next generation that could execute tasks intractable to traditional computers and communication frameworks.[13]

Epistemology

To many postwar researchers outside physical sciences, noise referred not necessarily to fluctuations of specific causes, but to any sorts of uncertainty that could affect the accuracy of observed data. This notion of uncertainty had existed for a long time. In the Enlightenment, mathematicians and astronomers recognized irregular variations in astronomical records and explored means to reduce them. When statistics was developed throughout the nineteenth century, a core line of thought was to treat empirical data as a quantity deflected by independent random errors, to model errors with a probabilistic distribution, and to compute its statistical attributes.[14] Retrieving rational conclusions from data plagued with uncertainty had been a central mandate of statistics. This uncertainty was often described as measurement errors (in astronomy), "personal equations" (in psychology), or variances (in genetics).[15] Yet, statisticians in the first half of the twentieth century did not have a close contact with electrical engineers, statistical physicists, and probability theorists. Connections between statistics of errors and noise studies were slim.

The situation began to change after World War II. By the 1950s, the terms "noise" and "signal" entered the scholarly discussions within the statistical community. These terms provided a new vocabulary for statisticians to talk about the longstanding issues of extracting relevant information from data contaminated by random errors and variances, but with a different methodological toolbox inspired from the prewar sound reproduction, statistical physics, mathematics of stochastic processes, and wartime information

[13] For example, see Michael Nielsen and Isaac Chuang, *Quantum and Quantum Information* (Cambridge, UK: Cambridge University Press, 2010), 353–398.

[14] Stigler, *The History of Statistics* (1986).

[15] For example, see Ronald A. Fisher, "The correlation between relatives on the supposition of Mendelian inheritance," *Philosophical Transactions of the Royal Society of Edinburgh*, 52 (1918), 399–433.

weaponry. In this context, errors and variances were understood more and more commonly in terms of noise.

In 1959, for instance, a review of a book titled *Principles and Applications of Random Noise Theory* appeared in *Journal of the American Statistical Association*. The reviewer Emanuel Parzen at Stanford University informed readers that "this is an exposition of those statistical ideas introduced into engineering during the second world war, whose wartime development was summarized in the classic works of S. O. Rice ('Mathematical Analysis of Random Noise'....) and Norbert Wiener (*The Extrapolation, Interpolation, and Smoothing of Stationary Time Series with Engineering Applications*...)."[16] The same author reviewed another book titled *Topics in the Theory of Random Noise* for the journal *Technometrics* five years later.[17] A casual search on the databases Compendex and Web of Science in major statistical journals in the 1950s–70s yields various articles matching the keyword "noise" in their titles: "Regression with systematic noise" (1964), "Large-sample estimation of an unknown discrete waveform which is randomly repeating in Gaussian noise" (1965), "Prediction of a noise-distorted, multivariate, non-stationary signal" (1967), "Measurement of a wandering signal amid noise" (1967), "On detecting a signal in N stationarily correlated noise series" (1971), "Approximate best linear prediction of a certain class of stationary and non-stationary noise-distorted signals" (1971), "Estimation of models of autoregressive signal plus white noise" (1974).[18]

In these works, "noise" referred not to disturbing sounds, but to abstract random errors that followed certain probability distributions and plagued the "signals" which statisticians aimed to retrieve from empirical time-series data. Some papers went as far as using noise to indicate anomalous electrical currents passing through a cell membrane via a neural transmitter, or irregular rates of chemical reactions. When the MIT mathematician Herman Chernoff

[16] Emanuel Parzen, "Book *Review, Principles and Applications of Random Noise Theory*," *Journal of the American Statistical Association*, 54:286 (June 1959), 525.

[17] Emanuel Parzen, "Book Review, *Topics in the Theory of Random Noise. Vol. 1. General Theory of Random Processes. Nonlinear Transformation of Signals and Noise* by R. L. Stratonovich and R. A. Silverman," *Technometrics*, 6:4 (November 1964), 473–474.

[18] Michael Milder, "Regression with systematic noise," *Journal of the American Statistical Association*, 59:306 (June 1964), 422–428; Melvin Hinich, "Large-sample estimation of an unknown discrete waveform which is randomly repeating in Gaussian noise," *The Annals of Mathematical Statistics*, 36:2 (April 1965), 489–508; Eugene Sobel, "Prediction of a noise-distorted, multivariate, non-stationary signal," *Journal of Applied Probability*, 4:2 (August 1967), 330–342; E.J. Hannan, "Measurement of a wandering signal amid noise," *Journal of Applied Probability*, 4:1 (April 1967), 90–102; Robert Shumway, "On detecting a signal in N stationarily correlated noise series," *Technometrics*, 13:3 (August 1971), 499–519; Marcello Pagano, "Estimation of models of autoregressive signal plus white noise," *The Annals of Statistics*, 2:1 (January 1974), 99–108.

introduced an approach to pattern recognition (i.e. identification of a feature of an individual such as their facial configuration or fingerprint from a large population) in 1980, he claimed that his method worked "in the presence of noise," i.e. random errors offsetting individual patterns.[19]

Today, the notion of informational noise is so embedded in our data-loaded mentality that we barely recall the word's genealogical relationships with sound reproduction or even thermal fluctuations. In the canonical narratives of big data, noise is portrayed as a general challenge for identifying meaningful patterns, making rational forecasting, or retrieving relevant information from an ocean of digital traces gathered through the world-wide web and a heterogeneous variety of networks, systems, and databases administered by the state, corporations, academia, and non-governmental organizations. Like their statistician predecessors in the 1960s–80s, data scientists deliberate on how to "filter noise" and enhance "the signal-to-noise ratio" for big data through a repertoire from the aftermath of wartime cyberneticians' stochastic detection and estimation to the more fashionable machine learning and other techniques in artificial intelligence.[20]

To some observers of our technoscientific culture, this semantic change indicates that noise is not simply an unavoidable buzzing background that haunts data or messages but has definite and graspable mathematical properties. Rather, noise can represent any unwanted, unintended, unexpected, or uncertain components in the process of communicating, knowing, or information gathering. Such a generalization of meaning implies that the handling of noise can be an epistemic issue. On the one hand, selection of signals buried in noise may be more challenging than employing algorithmic procedures or engineering optimization, for the distinction between "wanted" and "unwanted" really depends on the contents of data and the purpose of knowledge- or information-related activities. Philosophical insights are needed for this type of tasks. On the other hand, noise has a potential to

[19] Stuart Bevan, Richard Kullberg, and John Rice, "An analysis of cell membrane noise," *The Annals of Statistics*, 7:2 (March 1979), 237–257; N.R. Goodman, S. Katz, B.H. Kramer, and M.T. Kuo, "Frequency response from stationary noise: Two case histories," *Technometrics*, 3:2 (May 1961), 245–268; Herman Chernoff, "The identification of an element of a large population in the presence of noise," *Annals of Statistics*, 8:6 (November 1980), 1179–1197.

[20] For example, see Hossein Hassani and Emmanuel Sirimal Silva, "Forecasting with Big Data: A review," *Annals of Data Science*, 2:1 (2015), 5–19; Seref Sagiroglu and Duygu Sinanc, "Big Data: A review," *Proceedings of the IEEE International Symposium on Collaborative Technologies and Systems* (2013), 42–47; Annie Waldherr, Daniel Maier, Peter Miltner, and Enrico Günter, "Big Data, big noise: The challenge of finding issue networks on the Web," *Social Science Computer Review*, 35:4 (2017), 427–443; Sheikh Mohammad Idrees, M. Afshar Alam, and Parul Agarwal, "A study of big data and its challenges," *International Journal of Information Technology*, 11:2 (2019), 841–846.

become sources of surprising discoveries, since what is dubbed "unwanted" or "useless" in one context may contain rich or insightful information in another.

Ian Hacking paid attention to this epistemic feature of noise. In his exploration of a philosophy for scientific experimentation, he pointed out the importance of handling noise in experiments, as it was related to the practice of "debugging":[21]

> Debugging is not a matter of theoretically explaining or predicting what is going wrong. It is partly a matter of getting rid of "noise" in the apparatus. Although it also has a precise meaning, "noise" often means all the events that are not understood by any theory. The instrument must be able to isolate, physically, the properties of the entities that we wish to use, and damp down all the other effects that might get in our way.

An even more interesting situation occurred when what was deemed to be noise in the regular operation of an instrument became relevant signals in an unexpected way.

Hacking's example was radio astronomy growing out of anomalies in wireless communications. In the 1930s, Karl Jansky at Bell Labs noted certain hissing sounds that accompanied the reception of the radio telephone between New York and London and declared after experimenting with the direction of reception that the source of such noise came from the center of the Milky Way. Jansky's finding of this "star noise" launched the use of radio telescopes as astronomical instruments. In 1965, Arno Penzias and R.W. Wilson at Bell Labs investigated further into the source of star noise. In so doing, they came across another kind of background noise, which appeared in their radio telescope as a small amount of energy that seemed to be everywhere in space and uniformly distributed. Penzias and Wilson made efforts to eliminate all possible sources of noise they could imagine. Even so, the homogeneous background radiation—corresponding to a temperature of 3°K—remained. Fortunately, around the same time, a theoretical physics group at Princeton University developed the big bang theory to account for the beginning of the universe. A cosmic background radiation of 3°K as a relic of the gigantic energy event at the beginning of time cooling down with the expanding universe was a prediction from the theory. The noise that bothered Penzias and Wilson became

[21] Ian Hacking, *Representing and Intervening: Introductory Topics in the Philosophy of Natural Science* (Cambridge, UK: Cambridge University Press, 1983), 265.

evidence for the cosmological model. Hacking named this situation a "happy meeting" between experiment and theory.[22]

To Hacking, events like the inception of radio astronomy and the identification of the 3°K cosmic background radiation were lucky coincidences in which unexpected noise was turned serendipitously into meaningful signals associated with new discoveries. To some others, however, noise played not an accidental but an *essential* role in epistemic activities at large. Daniel Gethmann re-examined Karl Jansky's and Charles Jenkins's studies of radio astronomical noise in transatlantic wireless telephony in the 1930s. He argued that these cases revealed the centrality of noise in acts of communication in general. Invoking the thoughts of the French philosopher Michel Serres, Gethmann claimed that in a communication process noise functioned effectively as "the third party" in addition to the transmitter and the receiver. Although this "third party" often interfered or interrupted the normal transfer of messages, it could play a positive part by deflecting the existing flow of information and become the conveyer of new messages.[23]

Noise occupied a unique position in Serres's metaphysics. Partly originated from the Leibnitzian intellectual tradition, partly inspired by Shannon's information theory, and partly embedded in the "linguistic turn" of philosophy in the second half of the twentieth century, Serres's program framed all transformations, exchanges, purposeful actions, and deeds of knowing and communicating in the natural and human worlds in terms of networks of transactions. Each link of this gigantic network had a "transmitter" on one side and a "receiver" on the other. This link resembling a Shannonian communication channel was always interjected with an uncertain and disturbing third party similar to noise. Serres used the notion of the "parasite" to describe these interfering factors. The prototype of Serres's parasite was the city rat and the country rat in La Fontaine's fable. In Josué Harari and David Bell's account, what Serres meant by "parasite" was:

> an operator that interrupts a system of exchange. The abusive guest partakes of the host's meal, consumes food, and gives only words, conversations in return. The biological parasite enters an organism's body and absorbs substances meant for the host organism. Noise occurs between two positions in an informational circuit and disrupts messages exchanged between them.

[22] Ibid., 159–160.

[23] Daniel Gethmann, "The aesthetics of the signal: Noise research in long-wave radio communications," *Osiris*, 28 (2013), 64–79.

Superficially, a parasite was a negative entity, a destroyer of order and systems. Profoundly, however, a parasite:[24]

> is an integral part of the system. By experiencing a perturbation and subsequently integrating it, the system passes from a simple to a more complex stage. Thus, by virtue of its power to perturb, the parasite ultimately constitutes, like the *clinamen* and the demon, *the condition of possibility of the system*. In this way the parasite attests from within order the primacy of disorder, it produces by way of disorder a more complex order.

The negative became the positive, the parasite led to a new order for "the system," and noise was turned into information of a more complex type.

It is beyond the scope of this book to delve into Serres's dialectical theory of everything, of which we can see traces leading to Bruno Latour's famous actor–network theory.[25] The point of invoking Serres here is to remind us that by the third quarter of the twentieth century, the abstract conceptualization of informational noise through the Shannonian, Wienerian, thermodynamic, and statistical frameworks had inspired philosophical thinking about the natural, material, and human worlds. Identifying noise as a new source of more sophisticated order, Serres shared the same zeitgeist as Prigogine and his fellow scientists *qua* intellectuals. As Serres asserted:[26]

> We are surrounded by noise. And noise is indistinguishable .... It is our apperception of chaos, our apprehension of disorder, our only link to the scattered distribution of things .... None assure us that we are surrounded by fluctuation and that we are full of fluctuation. And it chases us from chaos; by the horror it inspires in us, it brings us back and calls us to order .... Noise destroys and horrifies. But order and flat repetition are in the vicinity of death. Noise nourishes a new order. Organization, life, and intelligent thought live between order and noise, between disorder and perfect harmony.

[24] Josué Harari and David Bell, "Introduction: Journal à plusieurs voies," in Michel Serres, *Hermes: Literature, Science, Philosophy*, edited by Josué Harari and David Bell (Baltimore, MD: Johns Hopkins University Press, 1982), xvi–xxvii (italics in the original). For Serres's comprehensive development of the notion, see Michel Serres, *The Parasite*, translated by Lawrence Schehr (Baltimore, MD: Johns Hopkins University Press, 1982).

[25] For the connections between Serres's and Latour's thoughts, see Michel Serres, *Conversations on Science, Culture, and Time: Michel Serres with Bruno Latour* (Ann Arbor, MI: University of Michigan Press, 1995).

[26] Serres, *The Parasite* (1982), 126–127.

Serres's discourse about noise bore a stark resemblance to Prigogine's and Feigenbaum's narratives concerning the emerging order from a fluctuating and chaotic world.

Technology

The generalization of noise into an informational concept after World War II not only prompted scientific and philosophical deliberations on the nature of the complicated world or the possibilities and limits of knowledge. It also had deep impacts on the technological landscape related to today's computational form of life. The most immediate postwar engineering legacy from the studies of noise was the development of spread-spectrum communications systems. From Shannon's insight, noise did not have to be a mere obstruction in the process of communication; it could be an important resource as well. If messages were encoded or modulated with a noise-like carrier, then the rate of information transmission or the robustness against disturbances could be enhanced. Parallel to Shannon's theory, technologists in the 1940s–50s experimented with various noise-modulation schemes. Emerging from the American military–industrial complex, these spread-spectrum systems bore a hallmark of the early Cold War: Sylvania's WHYN radio aviation guidance for the US Air Force, Sperry Gyroscope's CYTAC, MIT's Project Hartwell on underwater antisubmarine warning for the US Navy, Lincoln Labs' NOMAC and F9C/Rake for the SAGE air-defense system, JPL's CODOVAC for wireless rocket control.[27]

Spread-spectrum communications did not catch the attention of the general public until Jacobs and Viterbi at Qualcomm advanced CDMA for mobile wireless networks circa 1988. Yet, the grappling with informational noise had already preoccupied engineers decades before the craze for cellular phones. A large part of this development was associated with the emergence of the new area of *signal processing* after World War II. Historian Frederick Nebeker named 1948 the "Annus Mirabilis" of signal processing, the year that witnessed the debut of Shannon's information theory, the Bernard–Pierce–Shannon analysis of PCM, the introduction of Long Playing (LP) phonographic records, and the enactment of the IEEE Signal Processing Society.[28] Shannon's doctrine represented noise as spheres of uncertainty surrounding messages in a

[27] Scholtz, "The origin of spread-spectrum communications" (1982), 822–854.

[28] Frederik Nebeker, *Signal Processing: The Emergence of a Discipline, 1948 to 1998* (New Brunswick, NJ: IEEE History Center, 1998), 13–28; *Fifty Years of Signal Processing: The IEEE Signal Processing*

signal space. This perspective treated the suppression of noise from signals as a problem of error correction via coding designs that placed smartly distinct messages in the signal space. Shannon's approach opened the field of error-correction coding, one of the first applications of information theory.[29]

In signal processing, filter design was under the heaviest influence of the wartime studies of noise. How to separate unwanted noise from wanted signals had been a longstanding issue in telecommunications engineering. In telephone and radio technology in the 1910s–30s, a common approach was to design an electric circuit's frequency response so that it could remove a signal's components outside its bandwidth. The probabilistic method from Wiener's gunfire control and Uhlenbeck's radar detection changed the nature of filter design: noise filtering now involved statistical prediction and estimation that minimized the mean-square error under a given probabilistic distribution of noise. After World War II, the Wiener filter continued to play a significant part in situations when both signals and noise were stochastic. In 1960, Rudolf Kálmán, a Hungarian immigrant with a Ph.D. in electrical engineering from Columbia University, further enriched the technology of noise filtering by developing an adaptive and recursive algorithm for statistical estimation of signals embedded in stochastic noise. Known as the "Kalman filter," this new technique was applied to the Apollo Project and US missile guidance and navigation.[30]

Spread-spectrum communication, error-correction coding, and statistical filtering were computationally intense. The rapid progress of semiconductor microelectronics, automatic computers, and digitization of information technologies after World War II made possible the implementation and employment of these noise-coping techniques. While they started as military projects, by the middle of the Cold War their applications were extended to civilian sectors, and were used in realms from television, satellite communication, computer networks, and mobile phones to compact discs, environmental remote sensing, seismic data processing, and biomedical imaging. From MP3 and JPEG to 5G and streamline, these techniques are built into today's information technologies. In these technologies, noise represents deviations from abstract signals that refer to almost any circumstances.

Society and Its Technologies 1948–1998 (Piscataway, NJ: The IEEE Signal Processing Society, 1998), 2–5; available at https://signalprocessingsociety.org/our-story/society-history (accessed April 8, 2022).

[29] Mathew Valenti, "The evolution of error control coding," available at https://citeseerx.ist.psu.edu/viewdoc/download?doi=10.1.1.104.6249&rep=rep1&type=pdf (accessed April 8, 2022).

[30] Nebeker, *Fifty Years of Signal Processing* (1998), 6–13.

The technological notions of noise since World War II were not limited to tangible hardware and software. Parallel to engineers' noise-coping techniques in microelectronics, digital computing, telecommunications, and signal processing, mathematicians and economists in the second half of the twentieth century developed crucial *theoretical techniques* in the formal representations of noise. As we have seen, Bachelier at the turn of the century attempted to build a mathematical structure of random walks to model fluctuating stock prices. In the 1920s–40s, Kolmogorov et al. in Moscow developed an axiomatic formulation of probability theory and a theory of Markov processes. Based on measure theory, Lévy in Paris and Daniell in Houston expressed discrete or continuous random variables in a functional space. Wiener and his collaborators furthered this idea to construct a representation of Brownian motion in a differential space and developed its generalized harmonic analysis. At the dawn of the war, a rudimentary theory of stochastic processes came to the fore.

In the 1940s, researchers from Japan, the US, and France looked further into the mathematical properties of stochastic processes. These processes were continuous but did not possess common features of continuous functions such as differentiability. A challenge was to analyze not only the curves' probability density functions but the curves as such. In 1944, a Japanese mathematician Kiyosi Itô, a minor official at the National Statistical Bureau and a Ph.D. candidate at the University of Tokyo, published a paper on stochastic integration. He extended the Riemann integral for analytical functions to an integral that took both the integrand and the variable of integration as stochastic processes. Itô's insights enabled the formulation of integral and differential calculus for random processes, instead of relying on equations for their probability density functions (e.g. the Fokker–Planck equation or the Kolmogorov equation) or deterministic equations of motion with a stochastic force (e.g. the Langevin equation). Along this line of thought, Joseph Doob at the University of Illinois Urbana-Champaign in the 1940s–50s developed a theory of the martingale—a Markov process whose expected value equaled its most recent observation. He demonstrated that many kinds of random paths could be decomposed into entities with a martingale-like structure. In the 1940s–60s, stochastic integration and martingales formed a cutting-edge technical repertoire for the mathematical theory of random processes.[31]

These theoretical techniques developed by mid-century mathematicians for stochastic processes found major applications in finance and economics. A

[31] Jarrow and Protter, "A short history of stochastic integration and mathematical finance" (2004), 75–91; Davis and Etheridge, *Louis Bachelier's* Theory of Speculation (2006), chapter 3, "From Bachelier to Kreps, Harrison and Pliska," 80–115.

pioneer of postwar mathematical finance was Paul Samuelson. According to legend, a statistician Jimmy Savage at the University of Chicago "rediscovered" Bachelier's dissertation in the 1950s and wrote postcard notices to recommend it to his economist friends. Samuelson was among the recipients. At the time, the junior MIT professor in economics was working on a theoretical determination of prices for options, a derivative financial product popular in postwar America and Europe. Bachelier's modeling of stock prices with random walks inspired Samuelson. As a veteran of Rad Lab, he was familiar with the properties of Brownian motion and understood the limitation of Bachelier's model. Among other problems, Samuelson found, Bachelier's Brownian motion of price could wander into positive or negative values. But in a real market, the price of a commodity never went below zero. To resolve this problem, Samuelson replaced Bachelier's process with a "geometric" Brownian motion in which a random walk appeared as an exponent of a financial product's price so that it would never become negative.[32]

Samuelson's geometric random walk gained increasing traction and became a core technique among financial economists in the 1950s–70s. To them, modeling commodity prices in a market with stochastic fluctuations independent of their previous values had enormous economic implications. Eugene Fama at the University of Chicago contended in 1970 that a "memoryless" market—the one that Samuelson's geometric Brownian motion characterized—was an "efficient" market, for the prices reflected all relevant information about the market now and gaining more information about its past would not help increase profit. Later, Samuelson and his colleagues adopted more sophisticated mathematics of martingales in lieu of geometric random walks to model option prices. This move was consistent with the so-called "principle of no arbitrage"—it was impossible to make profit from arbitraging the same commodity at two distinct places in a perfect market—proposed by Franco Modigliani and Merton Miller in 1958. The integration of the economic insights of treating prices as Brownian processes in an ideal market with the mathematical apparatuses of stochastic integrodifferential equations and martingales formed the basis of the model for pricing financial derivative commodities that Fisher Black, Myron Scholes, and Robert Merton developed in the 1970s.[33]

[32] Jarrow and Protter, "A short history of stochastic integration and mathematical finance" (2004), 79–81; Davis and Etheridge, *Louis Bachelier's* Theory of Speculation (2006), 1–3; MacKenzie, *An Engine, Not a Camera* (2006), 62–64.

[33] For the details of the development of the Black–Scholes–Merton theory, see MacKenzie, *An Engine, Not a Camera* (2006).

Samuelson, Black, Scholes, Merton, and other researchers in mathematical finance during the 1950s–70s did not use the term "noise" to refer to price volatilities. Yet, as the keyword gained increasing influence in physical and life sciences, information technosciences, statistics, and computing and communication technologies, it entered economists' lexicons. In 1986, Fischer Black, a founder of the Black–Scholes–Merton theory and an analyst at the investment group Goldman Sachs, published a paper titled "Noise" in *Journal of Finance*. Black took noise to be the contrast with information. It was what made observations imperfect, the arbitrary element in expectations, or the information that had not arrived yet. That is, noise epitomized all kinds of uncertainty in economies. He aimed to develop a generic theory of noise that encompassed finance, econometrics, and macroeconomics. Under this framework, noise explained why people traded in a market, why inflation and the rate of exchange did not reflect the actual economic conditions, and why a direct access to these conditions was never possible. To get a feeling for Black's ambitious scope, it is worth quoting the abstract of his paper:

> The effects of noise on the world, and on our views of the world, are profound. Noise in the sense of a large number of small events is often a casual factor much more powerful than a small number of large events can be. Noise makes trading in financial markets possible, and thus allows us to observe prices for financial assets. Noise causes markets to be somewhat inefficient, but often prevents us from taking advantage of inefficiencies. Noise in the form of uncertainty about future tastes and technology by sector causes business cycles, and makes them highly resistant to improvement through government intervention. Noise in the form of expectations that need not follow rational rules cause inflation to be what it is, at least in the absence of a gold standard or fixed exchange rates. Noise in the form of uncertainty about what relative prices would be with other exchange rates makes us think incorrectly that changes in exchange rates or inflation rates cause changes in trade or investment flows or economic activity. Most generally, noise makes it very difficult to test either practical or academic theories about the way that financial or economic markets work. We are forced to act largely in the dark.

The random fluctuations in Bachelier's and Samuelson's mathematical machinery of financial theory were turned into an ontological and epistemological premise of our socioeconomic world.[34]

[34] Fischer Black, "Noise," *The Journal of Finance*, 41:3 (July 1986), 529–543.

Despite the sea changes of the stochastic-process theory from Einstein, Bachelier, Kolmogorov, and Wiener to Itô, Doob, and Samuelson, most "noise technologies" of the twentieth century shared a core underpinning. No matter whether the practical goal was statistical filtering and estimation, channel coding, or modeling the irregular and unknown factors in physical systems, living organisms, or economic markets, the uncertainty was often captured by "purely" random trajectories with a normal distribution. This Brownian fluctuation was the first stochastic process studied by the pioneers of noise research. Throughout the twentieth century, this heritage served as a technical substrate for the fields, but at the same time configured and constrained their possible directions of development.

Individuals did try to conceive something else, but to no avail. A case in point is the French-American mathematician Benoit Mandelbrot. At IBM Watson Labs in the 1960s, Mandelbrot was dissatisfied with the assumption of random processes with a normal distribution, especially with respect to their applicability to describing a financial market. To him, the tails of a normal distribution converged too quickly to zero, implying that "rare events" had a quite low probability. In the real world, preposterous things happened all the time. To model fairly extreme events, he believed that a different type of distribution with much higher tails was called for. Mandelbrot constructed a series of probability distributions with higher values at extrema and other weird properties, which he dubbed "strong noise" or "monsters." Although Mandelbrot made much effort to promote this new type of distribution functions, his ideas were not received well. In the end, the Gaussian random walk continued to prevail.[35]

Writing a History of Noise

Writing a history of noise is challenging, for the subject lacks a "natural" boundary. Although such a history is filled with discoveries and inventions, it is not about unveiling a scientific fact or entity, establishing a theory or model, making an engineering design or technological system—topics that populate the traditional history of science and technology. Instead, the story is more about noise as plights, abnormalities, and "parasites" that cause problems to scientific research, technological development, or common life, and the processes in which these problems were discussed, framed, transformed, tackled, resolved (if lucky), or left aside (if not).

[35] MacKenzie, *An Engine, Not a Camera* (2006), 105–118.

Disciplinary borders did not hold in these processes. The narrative easily traversed across mechanical and electrical engineering, physics, mathematics, computer and information technologies, and later biology, chemistry, earth sciences, economics, finance, and data science. The relevant material cultures and problem situations were diverse: phonographic recording, acoustic metrology, electronic amplifiers, ultramicroscopes, colloids, FM radio, stock markets, functional spaces, radar detection, antiaircraft gunfire control, encrypted communication, and electronic warfare. No unique paradigm, research program, or conceptual scheme dominated the story. While mutual borrowing of ideas, methods, and techniques was frequent, such interactions by no means marked the triumph of one framework over others.

It is also difficult to embed the history of noise within the culture, practice, norms, values, aesthetics, constraints, opportunities, and dynamics of a single organization, institution, social group, intellectual community, geographical region, or political jurisdiction—an approach embraced by many contemporary works in the history of science and technology. The historical actors in this book ranged from the household names of Einstein, Edison, Shannon, and Wiener to individuals like a marginalized French mathematician banished from Paris, a Hollywood star-inventor, a Polish refugee and a Chinese student kept out of their homelands during a war, and anonymous technicians at telephone firms. Their activities occurred in the US, UK, France, Germany, the Netherlands, Switzerland, Austria–Hungary, and Poland. Generally, most historical actors shared the common features of twentieth-century scientists and technologists at large: professional training; working at modern universities, corporate, or governmental laboratories; experience and practice configured by the two world wars. Specifically, their social, cultural, and intellectual backgrounds included the emergence of electrical mass media and the music industry, the indeterminism that prevailed at *fin de siècle* Vienna, shifting goals of statistical physics, endeavors to settle the issue of atomism, the identity of "pure" mathematics with axiomatization and rigor, the American military–industrial complex, and the information revolution in the mid-century. These heterogenous contexts are useful for making sense of the individual episodes in the history of noise. Yet, if we intend to understand that history as a whole, a different perspective is needed.

An approximation to this perspective is what some scholars have called "historical epistemology" and "historical ontology." The former concerns the origin, unfolding, and transformation for the methods of acquiring knowledge and the standards of its validity or reliability. The latter tackles the processes in which categories of beings were formed, developed, altered, and configured.

Treating epistemology and ontology as subjects of historical investigations, these approaches emphasize that the conditions of knowledge and the conditions of existence taken for granted today are by no means inevitable. Rather, they are outcomes of trajectories full of contingencies, ambiguities, choices, and disruptions. Moreover, these approaches often focus not on the in-depth historical contexts of particular times or places in those trajectories, but on their longue-dureé characters. Under the frameworks of historical epistemology and historical ontology, researchers have examined the historicity of, e.g. objectivity, psychological trauma, probability as a style of reasoning, trust in numbers, the notion of facts, and the concept of efficiency.[36]

From this perspective, the history of noise can be seen as the development and evolution of concepts, methods, and techniques to understand, reduce, control, and utilize uncertainty of various kinds. Agitations of the electric current across a resistor or a vacuum-tube amplifier were construed to be identical to the Brownian motions of suspended particles in a liquid. And such electronic noise was extended to model all types of irregularities in telecommunications, control and estimation in engineering systems, signal processing, and statistical data treatments. The Fokker–Planck equation, Langevin equation, stochastic-process theory, generalized harmonic analysis, differential space, threshold detection, and geometric representations of noise as channel capacity were introduced in specific problem situations of statistical mechanics, electrical engineering, mathematical analysis, and computing and information technologies. But they were appropriated, elaborated, and transformed into broader realms of applications that coped with errors, deviations, and fluctuations at large. What is documented in this book is a historical ontology and historical epistemology that witnessed the rise of noise as a new type of being epitomizing uncertainty and the forming of new ways, skills, and assumptions to deal with it.

Yet, if we view the history of noise only in terms of unfolding of concepts, theories, and mathematics, then we miss the centrality of sensory experiences and material practices in this history. People of the late twentieth and early twenty-first centuries have used "noise" to refer to errors, deviations, and fluctuations. Semantically, however, noise is annoying sounds, cacophony, din, disharmonious tones, and intruding aural pollutions. This book aims to

[36] Daston and Galison, *Objectivity* (2007); Lorraine Daston (ed.), *Biographies of Scientific Objects* (Chicago, IL: University of Chicago Press, 1999); Ian Hacking, *Historical Ontology* (Cambridge, MA: Harvard University Press, 2002); Hacking, *Taming of Chance* (1990); Porter, *Trust in Numbers* (1996); Mary Poovey, *The History of the Modern Fact: Problems of Knowledge in the Sciences of Wealth and Society* (Chicago, IL: University of Chicago Press, 1998); Jennifer Karns Alexander, *The Mantra of Efficiency: From Waterwheel to Social Control* (Baltimore, MD: Johns Hopkins University Press, 2008).

demonstrate not only how the sonic sense of noise turned into the informational sense of noise, but also how the former continued to play an important part in the development of the latter. Thus, what mattered in the story were not only the derivations of the Fokker–Planck equation, statistical filtering, or the channel coding theorem, but also the search for a phonographic record material most resilient to scratches, measurements of street din, testing electronic tubes for radio, and designing a noise wheel to scramble wireless signals.

A large part of the conceptual, theoretical, and mathematical results with respect to the informational noise came not from statistics, the longstanding academic discipline to deal with errors in data, but from the research, design, and performance assessment in the sound reproduction technologies of telegraph, telephony, and radio. There was no cognitive or logical relationship between the two senses of noise. The engineering flavor and sonic metaphors in the theoretical treatments of informational noise were contingent upon the peculiar paths of technoscience in the twentieth century: mass electronic media capitalism, the world wars and the military–industrial complex, the information revolution. But they were not simply historical relics. If we learn from Raymond Williams's insight about keywords and take "noise" to be one of them (which I believe it is), then we should take the changing meanings of the term as a reflection of the changing techno-sociocultural circumstances where the past has left marks and continued to influence the present.[37]

The underpinning of sound reproduction is recognizable in our highly computerized and informational technical culture. Many core techniques in contemporary signal processing invoke lexicons and metaphors from electroacoustic engineering dating back to the early twentieth century: filtering, anti-aliasing, sampling rate, echo-cancellation, and equalization. Today, the most popular devices in our digital life are smart *phones*. A more poignant case in point concerns the recent discovery of gravitational waves from the Laser Interferometer Gravitational-Wave Observatory (LIGO). As anthropologist Stefan Helmreich indicated, the scientists working on LIGO data translated interferometric signals into sounds and declared their ability to "listen to" the distorted space-time due to gravitational waves. To him, this odd translation epitomized their "sonic way of knowing and being," as they converted the abstract "signals" generated from experimental and theoretical apparatuses

[37] Williams, *Keywords* (1976), 9–24.

into something they could grasp with embodied experiences.[38] The sonic manifestations of signal and noise did not disappear into history. They reverberate in today's scientific and technological culture. The stories in this book constitute the conceptual, material, methodical, and technical substrates for them to continue to happen.

[38] Stefan Helmreich, "Gravity's reverb: Listening to space-time, or articulating the sounds of gravitational-wave detection," *Cultural Anthropology*, 31:4 (2016), 464–492.

Bibliography

Archival Sources

AT&T Archives, Warren, New Jersey, US.

Claude Shannon Papers, Library of Congress, Washington, DC, US.

DSIR Papers, National Archives, London, UK.

George Uhlenbeck Papers, Manuscript Collection, Library of the University of Michigan, Ann Arbor, Michigan, US.

Historical Records, Rutherford Appleton Laboratory (Space Science Department), Didcot, UK.

Interview with Mortimer Rogoff by David Mindell, November 28, 2003, Cambridge, Massachusetts, US.

Marian Smoluchowski Papers, Manuscript Collections, MS9397, Library of Jagellonian University, Krákow, Poland.

Norbert Wiener Collection (MC22), MIT Archives, Cambridge, Massachusetts, US.

Paul Langevin Papers, Library of ESPCI Paris Tech, Paris, France.

Thomas A. Edison Papers, Rutgers University, New Jersey, US (available in microfilms).

Walter Schottky Papers (NL 100), Deutsches Museum Archives, Munich, Germany.

Published Sources

Aitken, Hugh, *Syntony and Spark: The Origin of Radio*, Princeton, NJ: Princeton University Press, 1976.

—, *The Continuous Wave: Technology and American Radio, 1900-1932*, Princeton, NJ: Princeton University Press, 1985.

Alexander, Jennifer Karns, *The Mantra of Efficiency: From Waterwheel to Social Control*, Baltimore, MD: Johns Hopkins University Press, 2008.

Anonymous, "John B. Johnson, recent Sarnoff Award winner" (obituary), *IEEE Spectrum*, 8:1 (1971), 107.

Appleton, Edward V., Robert A. Watson-Watt, and John F. Herd, "On the nature of atmospherics—II," *Proceedings of the Royal Society of London*, 111 (1926), 613–653.

—, "On the nature of atmospherics—III," *Proceedings of the Royal Society of London*, 111 (1926), 654–677.

Armstrong, Edwin, "Methods of reducing the effect of atmospheric disturbances," *Proceedings of the IRE*, 16:1 (January 1928), 15–26.

—, "A method of reducing disturbances in radio signaling by a system of frequency modulation," *Proceedings of the IRE*, 24:5 (May 1936), 689–740.

Arnold, H.D., and Lloyd Espenschied, "Transatlantic radio telephony," *Journal of the American Institute of Electrical Engineers*, 42 (1923), 815–826.

Aspray, William, "The scientific conceptualization of information: A survey," *IEEE Annals of the History of Computing*, 7:2 (1985), 117–140.

Attali, Jacques, *Noise: The Political Economy of Music*, translated by Brian Massumi, Minneapolis: University of Minnesota Press, 1985.

Austin, Louis W., "The relation between atmospheric disturbances and wave length in radio reception," *Proceedings of the Institute of Radio Engineers*, 9:1 (1921), 28–40.

—, "The reduction of atmospheric disturbances in radio reception," *Proceedings of the Institute of Radio Engineers*, 9:1 (1921), 41–55.

Bachelier, Louis, Théorie de la spéculation," *Annales Scientifiques de l'École Normale Supérieure*, (Ser. 3), 17 (1900), 21–86.

Bailey, Peter, "Breaking the sound barrier: A historian listening to noise," *Body and Society*, 2:2 (1996), 49–66.

Ballantine, Stuart, "Schrot-effect in high-frequency circuits," *Journal of the Franklin Institute*, 206 (1928), 159–167.

Basore, Bennett, *Noise-like Signals and Their Detection by Correlation*, Sc.D. dissertation, Cambridge, MA: MIT, 1952.

Bederson, Benjamin, "Samuel Abraham Goudsmit," *Biographical Memoirs of the National Academy of Sciences*, 90 (2008), 1–29.

Beller, Mara, *Quantum Dialogue: The Making of a Revolution*, Chicago, IL: University of Chicago Press, 1999.

Benjamin, Walter, *The Work of Art in the Age of Mechanical Reproduction*, New York: Penguin, 2008.

Bennett, William R., "Secret telephony as a historical example of spread-spectrum communication," *IEEE Transactions on Communications*, 31:1 (1983), 98–104.

Bensaude-Vincent, Bernadette, "When a physicist turns on philosophy: Paul Langevin (1911-39)," *Journal of the History of Ideas*, 49:2 (1988), 319–338.

Berkovitz, Joseph, "The world according to de Finetti: On de Finetti's probability theory and its application to quantum mechanics," in Y. Ben Menachem and M. Hemmo (eds.), *Probability in Physics* (Dordrecht: Springer, 2012), 249–280.

Berliner, Emile, "The development of the talking machine," *Journal of the Franklin Institute*, 176:2 (August 1913), 189–200.

Bevan, Stuart, Richard Kullberg, and John Rice, "An analysis of cell membrane noise," *The Annals of Statistics*, 7:2 (March 1979), 237–257.

Beyer, Robert, *Sounds of Our Times: Two Hundred Years of Acoustics*, New York: Springer-Verlag, 1999.

Bigg, Charlotte, "Evident atoms: Visuality in Jean Perrin's Brownian motion research," *Studies in the History and Philosophy of Science*, 39 (2008), 312–322.

Bijsterveld, Karin, *Mechanical Sound: Technology, Culture, and Public Problems of Noise in the Twentieth Century*, Cambridge, MA: MIT Press, 2008.

Biquard, Pierre, *Paul Langevin: Scientifique, Éducateur, Citoyen*, Paris: Seghers, 1969.

Black, Fischer, "Noise," *The Journal of Finance*, 41:3 (July 1986), 529–543.

Blackmore, John (ed.), *Ludwig Boltzmann: His Later Life and Philosophy. Book I: A Documentary History*, Dordrecht: Kluwer, 2010.

Boltzmann, Ludwig, "Studien über das Gleichgewicht der lebendige Kraft zwischen bewegten materiellen Punkten," *Sitzungsberichte, Kaiserliche Akademie der Wissenschaften, Wien, Mathematisch-Naturwissenschaftliche Klasse*, 58 (1868), 517–560.

—, "Einige allgemeine Sätze über Wärmegleichgewicht," *Sitzungsberichte, Kaiserliche Akademie der Wissenschaften, Wien, Mathematisch-Naturwissenschaftliche Klasse*, 63 (1871), 679–711.

—, "Weitere Studien über das Wärmegleichgewicht unter Gasmolekülen," *Sitzungsberichte, Kaiserliche Akademie der Wissenschaften, Wien, Mathematisch-Naturwissenschaftliche Klasse*, 66 (1872), 275–370.

—, "Über die Beziehung eines allgemeine mechanischen Satzes zum zweiten Hauptsatze der Wärmetheorie," *Sitzungsberichte, Kaiserliche Akademie der Wissenschaften, Wien, Mathematisch-Naturwissenschaftliche Klasse*, 75 (1877), 67–73.

—, "Über die Beziehung zwischen des zweiten Hauptsatze der mechanischen Wärmetheorie und der Wahrscheinlichkeitsrechnung, respective den Sätzen über das Wärmegleichgewicht," *Sitzungsberichte, Kaiserliche Akademie der Wissenschaften, Wien, Mathematisch-Naturwissenschaftliche Klasse*, 76 (1877), 373–435.

Born, Max, "Max Karl Ernst Ludwig Planck, 1858-1947," *Obituary Notices of Fellows of the Royal Society*, 6:17 (1948), 161–188.

Bouk, Dan, *How Our Days Became Numbered: Risk and the Rise of Statistical Individuals*, Chicago, IL: University of Chicago Press, 2005.

Bower, Calvin, "The transmission of ancient music theory into the Middle Ages," in Thomas Christensen (ed.), *The Cambridge History of Western Music Theory* (Cambridge: Cambridge University Press, 2002), 136–167.

Bown, Ralph, Carl R. Englund, and H.T. Friis, "Radio transmission measurements," *Proceedings of the Institute of Radio Engineers*, 11:2 (1923), 115–153.

Brain, Robert, "Standards and semiotics," in Timothy Lenoir (ed.), *Inscribing Science: Scientific Texts and the Materiality of Communication* (Stanford, CA: Stanford University Press, 1998), 249–284.

Brittain, James, "John R. Carson and conservation of radio spectrum," *Proceedings of the IEEE*, 84:6 (1996), 909–910.

Bromberg, Joan Lisa, "Coherence and noise in the era of the maser," *Perspectives on Science*, 24:1 (2016), 93–111.

Bruce, Robert, *Bell: Alexander Graham Bell and the Conquest of Solitude*, Ithaca, NY: Cornell University Press, 1990.

Brush, Stephen, "A history of random processes: I. Brownian movement from Brown to Perrin," *Archives for History of Exact Sciences*, 5 (1968), 1–36.

—, *Statistical Physics and the Atomic Theory of Matter: From Boyle and Newton to Landau and Onsager*, Princeton, NJ: Princeton University Press, 1983.

Buchwald, Jed, and Andrew Warwick (eds.), *Histories of the Electron: The Birth of Microphysics*, Cambridge, MA: MIT Press, 2004.

Burger, Herman Carel, "On the theory of the Brownian movement and the experiments of Brillouin," *Proceedings of the Huygens Institute—Royal Netherlands Academy of Arts and Sciences (KNAW)*, 20:1 (1918), 642–658.

Cahan, David, *Helmholtz: A Life in Science*, Chicago, IL: University of Chicago Press, 2018.

Callen, Herbert, and Theodore Welton, "Irreversibility and generalized noise," *Physical Review*, 83:1 (1951), 34–40.

Campbell, Norman, "The theory of the 'Schrot-effect,'" *Philosophical Magazine*, 50 (1925), 81–86.

Carson, John, "Notes on the theory of modulation," *Proceedings of the IRE*, 10:2 (February 1922), 57–64.

—, "Method and means for signaling with high-frequency waves," US Patent 1449382 (27 March 1923).

—, "Signal-to-static-interference ratio in radio telephony," *Proceedings of the IRE*, 11 (June 1923), 271–274.

—, "Selective circuits and static interference," *Transactions of the AIEE*, 43 (June 1924), 789–797.

—, "Selective circuits and static interference," *The Bell System Technical Journal*, 4 (April 1925), 265–279.

—, "The reduction of atmospheric disturbances," *Proceedings of the IRE*, 16:7 (July 1928), 966–975.

Carson, John, and Thornton Fry, "Variable frequency electric circuit theory with application to the theory of frequency modulation," *The Bell System Technical Journal*, 16 (October 1937), 513–540.

Carson, John, and Otto Zobel, "Transient oscillations in electric wave-filters," *The Bell System Technical Journal*, 3 (July 1923), 1–52.

Casimir, Hendrik B.G., and Sybren R. de Groot, "Levensbericht van Adriaan Daniël Fokker," *Jaarboek, Huygens Institute—Royal Netherlands Academy of Arts and Sciences (KNAW)* (1972), 114–118.

Chandrasekhar, Subrahmanyan, "Stochastic problems in physics and astronomy," *Reviews of Modern Physics*, 15:1 (1943), 1–89.

—, "Marian Smoluchowski as the founder of the physics of stochastic phenomena," in Subrahmanyan Chandrasekhar, Mark Kac, and Roman Smoluchowski (eds.), *Marian Smoluchowski: His Life and Scientific Work* (Warsaw: Polish Scientific Publishers, 1999), 21–28.

Chernoff, Herman, "The identification of an element of a large population in the presence of noise," *Annals of Statistics*, 8:6 (November 1980), 1179–1197.

Christensen, Thomas (ed.), *The Cambridge History of Western Music Theory*, Cambridge: Cambridge University Press, 2002.

Coen, Deborah, *Vienna in the Age of Uncertainty: Science, Liberalism, and Private Life*, Chicago, IL: University of Chicago Press, 2007.

Coffey, William, Yuri Kalmykov, and John Waldron, *The Langevin Equation: With Applications in Physics, Chemistry, and Electrical Engineering*, Singapore: World Scientific, 1996.

Cohen, Ezechiel Godert David, "George E. Uhlenbeck and statistical mechanics," *American Journal of Physics*, 58 (1990), 619–624.

Cohen, Leon, "The history of noise: On the 100th anniversary of its birth," *IEEE Signal Processing Magazine*, 22 (2005), 20–45.

Conway, Flo, and Jim Siegelman, *Dark Hero of the Information Age: In Search of Norbert Wiener, the Father of Cybernetics*, New York: Basic Books, 2006.

Courtault, Jean-Michel, Yuri Kabanov, Bernard Bru, Pierre Crépel, Isabelle Lebon, and Arnauld Le Marchand, "Louis Bachelier: On the centenary of *Théorie de la Spéculation*," *Mathematical Finance*, 10:3 (2000), 341–353.

Crosby, Murray, "Frequency modulation noise characteristics," *Proceedings of the IRE*, 25:4 (April 1937), 472–514.

Curie, Pierre, *Propriétés Magnétiques des Corps à Diverses Températures*, Ph.D. dissertation, Paris: University of Paris Sorbonne, 1895.

Daston, Lorraine, *Classical Probability in the Enlightenment*, Princeton, NJ: Princeton University Press, 1988.

—, (ed.), *Biographies of Scientific Objects*, Chicago, IL: University of Chicago Press, 1999.

Daston, Lorraine, and Peter Galison, *Objectivity*, Cambridge, MA: Zone Books, 2007.

Davenport, Wilbur, "The MIT Lincoln Laboratory F9C System (for IEEE 1981 Pioneer Award)," *IEEE Transactions on Aerospace and Electronic Systems*, 18:1 (January 1982), 157–158.

Davenport, Wilbur B., Jr., and William L. Root, *An Introduction to the Theory of Random Signals and Noise*, New York: McGraw-Hill, 1958.

Davis, A.H., "An objective noise-meter for the measurement of moderate and loud, steady and impulse noises," *Journal of the Institute of Electrical Engineers*, 83:500 (1938), 249–260.

Davis, Mark, and Alison Etheridge (translators and commentators), *Louis Bachelier's* Theory of Speculation*: The Origin of Modern Finance*, Princeton, NJ: Princeton University Press, 2006.

De Haas-Lorentz, Geertruida, *Over de Theorie van de Brown'sche Beweging en Daarmede Verwante Verschijnselen*, Ph.D. dissertation, Leyden: University of Leyden, 1912.

DeLillo, Don, *White Noise*, New York: Penguin, 1985.

Donal, John S., Jr., "Abnormal shot effect of ions of tungstous and tungstic oxide," *Physical Review*, (Ser. 2) 36 (1930), 1172–1189.

Doob, Joseph, "The Brownian movement and stochastic equations," *Annals of Mathematics*, 43:2 (1942), 351–369.

Dörfel, Günter, "The early history of thermal noise: The long way to paradigm change." *Annalen der Physik*, 524 (2012), A117–A121.

Dörfel, Günter, and Dieter Hoffmann, "Von Albert Einstein bis Norbert Wiener—frühe Ansichten und spate Einsichten zum Phänomen des elektronischen Rauschens." Preprint 301, Berlin: Max Planck Institute for the History of Science, 2005.

Douglas, Susan, *Inventing American Broadcasting, 1899–1922*, Baltimore, MD: Johns Hopkins University Press, 1989.

Dresden, Max, "Obituaries: George E. Uhlenbeck," *Physics Today*, 42:12 (1989), 91–94.

Druon, Maurice, *Dictionnaire de l'Académie française*, 9th edition, Paris: L'Académie française, 2011.

Eargle, John, *The Microphone Book*, Oxford: Focal Press, 2005.

Edison Concert Phonograph ad, *Phonoscope*, 3:2 (February 1899), 5.

Einstein, Albert, "Zur allgemeinen molekularen Theorie der Wärme," *Annalen der Physik*, 14 (1904), 354-362.

—, *Eine neue Bestimmung der Moleküldimensionen*, Ph.D. dissertation, University of Zurich (1905), printed in John Stachel, David Cassidy, Jürgen Renn, and Robert Schulmann (eds.), *The Collected Papers of Albert Einstein*, Volume 2: *The Swiss Years, 1900-1909* (Princeton, NJ: Princeton University Press, 1989), 183–205.

—, "On the movement of small particles suspended in stationary liquids required by the molecular-kinetic theory of heat," in John Stachel, David Cassidy, Jürgen Renn, and Robert Schulmann (eds.), *The Collected Papers of Albert Einstein*, Volume 2 : *The Swiss Years, 1900-1909* (Princeton, NJ: Princeton University Press, 1989), 123–134; article translated by Anna Becker from "Über die von der molekularkinetischen Theorie der Wärme geforderte Bewegung von in ruhenden Flüssigkeiten suspendierten Teilchen," *Annalen der Physik*, 17 (1905): 549–560.

—, "On the theory of Brownian motion," in John Stachel, David Cassidy, Jürgen Renn, and Robert Schulmann (eds.), *The Collected Papers of Albert Einstein*, Volume 2 : *The Swiss Years, 1900-1909* (Princeton, NJ: Princeton University Press, 1989), 180–191; article translated by Anna Becker from "Zur Theorie der Brownschen Bewegung," *Annalen der Physik*, 19 (1906): 371–381.

—, "Über die Gültigkeitsgrenze des Satzes von thermodynamischen Gleichgewicht und über die Möglichkeit einer neuen Bestimmung der Elementarquanta," *Annalen der Physik*, 22 (1907), 569–572.

—, "Theorie der Opaleszenz von homogenen Flüssigkeiten und Flüssigkeitsgemischen in der Nähe des kritischen Zustandes," *Annalen der Physik*, 33 (1910), 1275–1298.

—, "Marian Smoluchowski," *Die Naturwissenschaften*, 50 (1917), 107–108.

Eldar, Avigdor, and Michael Elowitz, "Functional roles for noise in genetic circuits," *Nature*, 467:9 (2010), 167–173.

Exner, Felix, "Notiz zu Brown's Molecularbewegung," *Annalen der Physik*, 2 (1900), 843–847.

Fagen, M.D. (ed.), *A History of Engineering and Science in the Bell System: The Early Years (1875–1925)*, vol. 1, New York: Bell Telephone Laboratories, 1975.

—, *A History of Engineering and Science in the Bell System: National Service in War and Peace (1925–1975)*, vol. 2, New York: Bell Telephone Laboratories, 1978.

Fano, Robert, *The Transmission of Information* (I and II), MIT RLE Technical Report No. 65, March 1949, and No. 149, February 1950.

Fischer, Claude, *America Calling: A Social History of Telephone up to 1940*, Berkeley: University of California Press, 1994.

Fisher, Ronald A., "The correlation between relatives on the supposition of Mendelian inheritance," *Philosophical Transactions of the Royal Society of Edinburgh*, 52 (1918), 399–433.

Fleming, John Ambrose, *The Principles of Electric Wave Telegraphy and Telephony*, New York: Longmans & Green, 1916.

Flock, Warren, "The Radiation Laboratory: Fifty years later," *IEEE Antennas and Propagation Magazine*, 33:5 (1991), 43–48.

Fokker, Adriaan D., *Over Brown'sche Bewegingen in het Stralingsveld, en Waarschijnlijkheids-Beschouwingen in de Stralingstheorie*, Ph.D. dissertation, Leyden: University of Leyden, 1912.

—, "Die mittlere Energie rotierender elektrischer Dipole im Strahlungsfeld," *Annalen der Physik*, 43 (1914), 810–820.

Ford, George, "George Eugene Uhlenbeck, 1900-1988," *Biographical Memoirs of the National Academy of Sciences*, 91 (2009), 1–22.

Foucault, Michel, *Society Must Be Defended*, Paris: Picador, 2003.

Franklin, Allan, A.W.F. Edwards, Daniel J. Fairbanks, Daniel L. Hartl, and Teddy Seinfeld, *Ending the Mendel-Fisher Controversy*, Pittsburgh, PA: University of Pittsburgh Press, 2008.

Free, Edward Elway, "Practical methods of noise measurement," *Journal of the Acoustical Society of America*, 2:1 (1930), 18–29.

Fry, Thornton C., "The theory of the Schroteffekt," *Journal of the Franklin Institute*, 199 (1925), 203–220.

Fürth, Reinhold, "Einige Untersuchungen über Brownsche Bewegung an einem Einzelteilchen," *Annalen der Physik*, 53 (1917), 177–213.

—, "Die Bestimmung der Elektronenladung aus dem Schroteffekt an Glühkathodenröhren," *Physikalische Zeitschrift*, 23 (1922), 354–362.

Gabor, Dennis, "Theory of communication," *Journal of the Institution of Electrical Engineers*, 93:3 (1946), 429–458.

Galilei, Galileo, *Dialogues Concerning Two New Sciences*, translated by Henry Crew and Alfonso de Salvio, Buffalo, NY: Prometheus Books, 1991.

Galison, Peter, "The ontology of the enemy: Norbert Wiener and the cybernetic vision," *Critical Inquiry*, 21:1 (1994), 228–266.

—, *Image and Logic: A Material Culture of Microphysics*, Chicago, IL: University of Chicago Press, 1997.

Galton, Francis, and Henry Watson, "On the probability of extinction of families," *Journal of the Royal Anthropological Institute*, 4 (1875), 138–144.

Garber, Elizabeth, "Maxwell's kinetic theory 1859-70," in Raymond Flood, Mark McCartney, and Andrew Whitaker (eds.), *James Clerk Maxwell: Perspectives on His Life and Work* (Oxford: Oxford University Press, 2014), 139–153.

Gelatt, Roland, *The Fabulous Phonograph: 1877-1977*, New York: Macmillian, 1977.

Gerlach, Walther, and E. Lehrer, "Über die Messung der rotatorischen Brownschen Bewegung mit Hilfe einer Drehwage," *Naturwissenschaften*, 15:1 (1927), 15.

Gethmann, Daniel, "The aesthetics of the signal: Noise research in long-wave radio communications," *Osiris*, 28 (2013), 64–79.

Gibbs, Josiah Willard, *Elementary Principles in Statistical Mechanics: Developed with Especial Reference to the Rational Foundation of Thermodynamics*, New York: Scribner, 1902.

Gigerenzer, Gerd, Zeno Swijyink, Theodore Porter, Lorraine Daston, John Beatty, and Lorenz Krüger, *The Empire of Chance: How Probability Changed Science and Everyday Life*, Cambridge: University of Cambridge Press, 1997.

Gleick, James, *Chaos: Making a New Science*, New York: Penguin Books, 1987.

Goldsmith, Mike, *Discord: The Story of Noise*, Oxford: Oxford University Press, 2012.

Gooday, Graeme, *The Morals of Measurement: Accuracy, Irony, and Trust in Late Victorian Electrical Practice*, Cambridge: Cambridge University Press, 2004.

Goodman, N.R., S. Katz, B.H. Kramer, and M.T. Kuo, "Frequency response from stationary noise: Two case histories," *Technometrics*, 3:2 (May 1961), 245–268.

Gouy, Louis Georges, "Note sur le mouvement brownien," *Journal de Physique*, 7 (1888), 561–564.

Grattan-Guinness, Ivor, *Joseph Fourier, 1768-1830: A Survey of His Life and Work*, Cambridge, MA: MIT Press, 1972.

Green, Burdette, and David Butler, "From acoustics to *Tonpsychologie*," in Thomas Christensen (ed.), *The Cambridge History of Western Music Theory* (Cambridge: Cambridge University Press, 2002), 246–271.

Green, Paul, *The Lincoln F9C Radioteletype System*, Lincoln Laboratories Technical Memorandum No. 61, 14 May 1954.

Grimm, Jacob, and Wilhelm Grimm, *Deutsches Wörterbuch*, Leipzig: Verlag Von S. Hirzel, vol. 4 (1878) and vol. 6 (1885).

Guerlac, Henry, *Radar in World War II*, Washington, DC: American Institute of Physics, 1987.

Hacking, Ian, *Representing and Intervening: Introductory Topics in the Philosophy of Natural Science*, Cambridge, UK: Cambridge University Press, 1983.

—, *The Taming of Chance*, Cambridge, UK: Cambridge University Press, 1990.

—, *Historical Ontology*, Cambridge, MA: Harvard University Press, 2002.

—, *The Emergence of Probability: A Philosophical Study of Early Ideas about Probability Induction and Statistical Inference*, Cambridge, UK: Cambridge University Press, 2006.

Hannan, E.J., "Measurement of a wandering signal amid noise," *Journal of Applied Probability*, 4:1 (April 1967), 90–102.

Harari, Josué, and David Bell, "Introduction: Journal à plusieurs voies," in Michel Serres, *Hermes: Literature, Science, Philosophy*, edited by Josué Harari and David Bell (Baltimore, MD: Johns Hopkins University Press, 1982), ix–xl.

Haraway, Donna, "A cyborg manifesto: Science, technology, and socialist-feminism in the late twentieth century," in *Simians, Cyborgs and Women: The Invention of Nature* (New York: Routledge, 1991), 169–181.

Hart, Liddell, *A History of the Second World War*, London: Weidenfeld Nicolson, 1970.

Hartley, Ralph, "Transmission of information," *The Bell System Technical Journal*, 7 (1928), 535–563.

Hartmann, Carl A., "Über die Bestimmung des elektrischen Elementarquantums aus dem Schroteffekt," *Annalen der Physik*, 65 (1921), 51–78.

Hassani, Hossein, and Emmanuel Sirimal Silva, "Forecasting with Big Data: A review," *Annals of Data Science*, 2:1 (2015), 5–19.

Hawkins, Thomas, *Lesbesgue Theory of Integration: Its Origin and Development*, New York: Chelsea Publishing Company, 1979.

Heilbron, John, *Electricity in the 17th and 18th Centuries: A Study of Early Modern Physics*, Berkeley: University of California Press, 1982.

Heims, Steve, *The Cybernetic Group*, Cambridge, MA: MIT Press, 1991.

Helmholtz, Hermann, *On the Sensation of Tone as a Physiological Basis for the Theory of Music*, translated by Alexander Ellis, New York: Dover, 1954.

Helmreich, Stefan, "Gravity's reverb: Listening to space-time, or articulating the sounds of gravitational-wave detection," *Cultural Anthropology*, 31:4 (2016), 464–492.

Herlinger, Jan, "Medieval canonics," in Thomas Christensen (ed.), *The Cambridge History of Western Music Theory* (Cambridge: Cambridge University Press, 2002), 168–192.

Hertz, Heinrich, "On an effect of ultraviolet light upon the electric discharge," in *Electric Waves* (New York: Dover, 1962), 63–79.

Hinich, Melvin, "Large-sample estimation of an unknown discrete waveform which is randomly repeating in Gaussian noise," *The Annals of Mathematical Statistics*, 36:2 (April 1965), 489–508.

Hobsbawm, Eric, *The Age of Extremes: A History of the World, 1914–1991*, New York: Vintage Books, 1994.

Holmes, Frederic, and Kathryn Olesko, "The images of precision: Helmholtz and the graphic method," in Norton Wise (ed.), *The Values of Precision* (Princeton, NJ: Princeton University Press, 1995), 198–221.

Home Grand Graphophone ad, *Phonoscope*, 3:11 (November 1899), 16.

Hong, Sungook, *Wireless: From Marconi's Black-Box to the Audion*, Cambridge, MA: MIT Press, 2010.

Houdijk, A., and Pieter Zeeman, "The Brownian motion of a thread," *Proceedings of the Huygens Institute—Royal Netherlands Academy of Arts and Sciences (KNAW)*, 28:1 (1925), 52–54.

Houston, Edwin, "The gramophone," *Journal of the Franklin Institute*, 125:1 (January 1888), 44–54.

Howling, D.H., "Surface and groove noise in disc recording media, I and II," *Journal of the Acoustical Society of America*, 31:11 (November 1959), 1463–1472; 31:12 (December 1959), 1626–1637.

Hughes, Thomas, *Rescuing Prometheus: Four Monumental Projects that Changed the World*, New York: Vintage Books, 1998.

Hull, Albert W., "Measurements of high frequency amplification with shielded-grid pliotrons," *Physical Review*, (Ser. 2) 27 (1926), 439–454.

Hull, Albert W., and N.H. Williams, "Determination of elementary charge *e* from measurements of shot-effect," *Physical Review*, (Ser. 2) 25 (1925), 148–150.

Idrees, Sheikh Mohammad, M. Afshar Alam, and Parul Agarwal, "A study of big data and its challenges," *International Journal of Information Technology*, 11:2 (2019), 841–846.

Isaacson, Walter, *Einstein: His Life and Universe*, New York: Simon and Schuster, 2007.

Israel, Paul, *Edison: A Life of Invention*, New York: Wiley, 1998.

Jackson, Henry B., "On the phenomena affecting the transmission of electric waves over the surface of the sea and the earth," *Proceedings of the Royal Society of London*, 70 (1902), 254–272.

Jackson, Myles, *Harmonious Triads: Physicists, Musicians, and Instrument Makers in Nineteenth-Century Germany*, Cambridge, MA: MIT Press, 2006.

James, Ioan, "Claude Elwood Shannon," *Biographical Memoirs of Fellows of the Royal Society*, 55 (2009), 257–265.

Jarrow, Robert, and Philip Protter, "A short history of stochastic integration and mathematical finance: The early years, 1880–1970," *Lecture Notes—Monograph Series, Institute of Mathematical Statistics: A Festschrift for Herman Robin*, 45 (2004), 75–91.

Johnson, John B., "Bemerkung zur Bestimmung des elektrischen Elementarquantums aus dem Schroteffekt," *Annalen der Physik*, (Ser. 4) 67 (1922), 154–156.

—, "The Schottky effect in low frequency circuit," *Physical Review*, 26 (1925), 76–85.

—, "Thermal agitation of electricity in conductors," *Physical Review*, 29 (1927), 50–51, 367–368.

—, "Thermal agitation of electricity in conductors," *Physical Review*, (Ser. 2) 32 (1928), 98–101.

—, "Electronic noise: the first two decades," *IEEE Spectrum* 8:1 (1971), 42–46.

Johnson, John B., and Frederick Llewellyn, "Limits to amplification," *The Bell System Technical Journal*, 14 (1935), 92–94.

Joliot-Curie, Frédérique, "Paul Langevin, 1872-1946," *Obituary Notices of Fellows of the Royal Society*, 7 (1951), 405–419.

Kac, Mark, *Enigmas of Chance: An Autobiography*, New York: Harper & Row, 1985.

—, "Marian Smoluchowski and the evolution of statistical thought in physics," in Subrahmanyan Chandrasekhar, Mark Kac, and Roman Smoluchowski (eds.), *Marian Smoluchowski: His Life and Scientific Work* (Warsaw: Polish Scientific Publishers, 1999), 15–20.

Kaiser, David, *Drawing Theories Apart: The Dispersion of Feynman Diagrams in Postwar Physics*, Chicago, IL: University of Chicago Press, 2005.

Kay, Lily, "Who wrote the book of life: Information and the transformation of molecular biology, 1945-1955," *Science in Context*, 8:4 (1995), 609–634.

—, *Who Wrote the Book of Life: A History of the Genetic Code*, Stanford, CA: Stanford University Press, 2000.

Kaye, George William Clarkson, "The measurement of noise," *Proceedings of the Royal Institution of Great Britain*, 26 (1929–1931), 435–488.

Keegan, John, *The Second World War*, London: Pimlico, 1989.

Keesom, Willem Hendrik, "Spektrophotometrische Untersuchung der Opaleszenz eines einkomponentigen Stoffers in der Nähe des kritischen Zustandes," *Annalen der Physik*, 35 (1911), 591–598.

Keller, Evelyn Fox, *Making Sense of Life: Explaining Biological Development with Models, Metaphors, and Machines*, Cambridge, MA: Harvard University Press, 2002.

—, "Organisms, machines, and thunderstorms: A history of self-organization, part I," *Historical Studies in the Natural Sciences*, 38:1 (2008), 45–75.

—, "Organisms, machines, and thunderstorms: A history of self-organization, part II: complexity, emergence, and stable attractors," *Historical Studies in the Natural Sciences*, 39:1 (2009), 1–31.

Kendall, David George, "Andrei Nikolaevich Kolmogorov," *Bulletin of the London Mathematical Society*, 22 (1990), 31–47.

Kenrick, George, "The analysis of irregular motions with applications to the energy-frequency spectrum of static and telegraphic signal," *Philosophical Magazine* (Series 7), 7:41 (1929), 176–196.

Kern, Stephen, *The Culture of Time and Space, 1880-1918*, Cambridge, MA: Harvard University Press, 2003.

Kevles, Daniel, *The Physicists: The History of a Scientific Community in Modern America*, New York: Knopf, 1977.

Khintchine, Alexander, "Korrelationstheorie der stationären stochastischen Prozesse," *Mathematische Annalen*, 109:1 (1934), 604–615.

Kittler, Friedrich, *Gramophone, Film, Typewriter*, translated by Geoffrey Winthrop-Young and Michael Wutz, Palo Alto, CA: Stanford University Press, 1999.

Klein, Morris, *Mathematical Thought from Ancient to Modern Times*, Volume 3, New York: Oxford University Press, 1972.

Kline, Ronald, *The Cybernetic Moment: Or Why We Call Our Age the Information Age*, Baltimore, MD: Johns Hopkins University Press, 2015.

Knorr Cetina, Karin, *Epistemic Cultures: How the Sciences Make Knowledge*, Cambridge, MA: Harvard University Press, 1999.

Kolmogorov, Andrei, "Über die analytischen Methoden in der Wahrscheinlichkeitsrechnung," *Mathematische Annalen*, 104 (1931), 415–458.

—, *Foundation of the Theory of Probability*, translated by Nathan Morrison, New York: Chelsea, 1956.

Kozanowski, H.N., and N.H. Williams, "Shot effect of the emission from oxide cathodes," *Physical Review*, (Ser. 2) 36 (1930), 1314–1329.

Kragh, Helge, *Quantum Generations: A History of Physics in the Twentieth Century*, Princeton, NJ: Princeton University Press, 1999.

Kramers, Hendrik Anthony, "Levensbericht van L.S. Ornstein," *Jaarboek, Huygens Institute—Royal Netherlands Academy of Arts and Sciences (KNAW)* (1940–1941), 225–231.

Krüger, Lorenz, "Probability as a theoretical concept in physics," *Proceedings of the Biennial Meeting of the Philosophy of Science Association* (1986), 273–287.

Kuhn, Thomas, "Energy conservation as an example of simultaneous discovery," in Marshall Clagett (ed.), *Critical Problems in the History of Science* (Madison: University of Wisconsin Press, 1969), 321–356.

—, "The function of measurement in modern physical science," in *The Essential Tensions: Selected Studies in Scientific Tradition and Change* (Chicago, IL: University of Chicago Press, 1977), 178–224.

—, *Black-Body Theory and the Quantum Discontinuity, 1894-1912*, New York: Oxford University Press, 1978.

Langevin, Paul, "Sur la théorie du magnétisme," *Comptes Rendus des Séances de l'Académie des Sciences*, 139 (1905), 1204–1207.

—, "Sur la théorie du magnétisme," *Journal de Physique Théorique et Appliquée*, 4:1 (1905), 678–693.

—, "Une formule fondamentale de théorie cinétique," *Annales de Chimie et de Physique*, 5 (1905), 245–288.

—, "Sur la théorie du mouvement brownien," *Comptes Rendus des Séances de l'Académie des Sciences*, 146 (1908), 530–533; article translated by Anthony Gythiel and introduced by Don Lemons, "Paul Langevin's 1908 paper 'On the theory of Brownian Motion' ['Sur la théorie du mouvement brownien', C.R. Acad. Sci. (Paris), 146, 530–533 (1908)]," *American Journal of Physics*, 65:11 (1997), 1079–1081.

—, "La théorie électromagnétique et le bleu du ciel," *Bulletin des Séances de la Société Française de Physique, Résumé des Communications* (1911), 80–82.

—, *La Théorie Cinétique du Magnétisme et Les Magnetrons*, Rapport au Conseil Solvay, Paris: Gauthier-Villars, 1912.

—, "La physique du discontinu," in *Conférence à la Société française de Physique*, 1913, published in *Les Progrès de la Physique Moléculaire* (Paris: Gauthier-Villars, 1914), 1–46.

Lawson, James, and George Uhlenbeck (eds.), *Threshold Signals*, New York: McGraw-Hill, 1950.

Leslie, Stuart, *The Cold War and American Science: The Military-Industrial-Academic Complex at MIT and Stanford*, New York: Columbia University Press, 1993.

Lessing, Lawrence, *Man of High Fidelity: Edwin Howard Armstrong*, New York: Bantam, 1969.

Levinson, Norman, Walter Pitts, and W.F. Whitmore, "Recent contributions to the theory of random functions," *Science*, 103:2670 (1946), 283–284.

Littré, Émile, *Dictionnaire de la Langue Française*, Paris: Librairie Hachette, 1883.

Liu, Lydia, *The Freudian Robot: Digital Media and the Future of the Unconscious*, Chicago, IL: University of Chicago Press, 2010.

Llewellyn, Frederick, "A study of noise in vacuum tubes and attached circuits," *Proceedings of the Institute of Radio Engineers*, 18:2 (1930), 243–265.

Loschmidt, Josef, "Über den Zustand des Wärmegleichgewichtes eines Systems von Körpern mit Rucksicht auf die Schwerkraft," *Sitzungsberichte, Kaiserliche Akademie der Wissenschaften, Wien, Mathematisch-Naturwissenschaftliche Klasse*, 73 (1876), 128–142.

MacKenzie, Donald, *An Engine, Not a Camera: How Financial Models Shaped Markets*, Cambridge, MA: MIT Press, 2006.

Madelung, Otfried, "Walter Schottky (1886-1976)," *Festkörperprobleme*, 26 (1986), 1–15.

Maiocchi, Robert, "The case of Brownian motion," *The British Journal for the History of Science*, 23 (1990), 257–283.

Mansell, James, *The Age of Noise in Britain: Hearing Modernity*, Urbana: University of Illinois Press, 2017.

Marconi, Umberto M.B., Andrea Puglisi, Lamberto Rondoni, and Angelo Vulpiani, "Fluctuation-dissipation: Response theory in statistical physics," *Physical Reports*, 461 (2008), 111–273.

Marx, Leo, *The Machine in the Garden: Technology and the Pastoral Idea in America*, Oxford: Oxford University Press, 2000.

Masani, Pepsi (ed.), *Norbert Wiener: Collected Works with Commentaries, Volumes 1 and 2*, Cambridge, MA: MIT Press, 1976 and 1979.

Mauro, Philip, "Recent development of the art of recording and reproducing sounds," *Phonoscope*, 3:9 (September 1899), 7–8.

Maxwell, James Clerk, "Illustrations of the dynamical theory of gases," *Philosophical Magazine*, 19 (1860), 19–32, 20 (1860), 21–37.

—, "On the viscosity or internal friction of air and other gases," *Philosophical Transactions of the Royal Society*, 156 (1866), 249–268.

McKean, Henry, "Mark Kac," *Biographical Memoirs of the National Academy of Sciences*, 59 (1990), 215–235.

Michel, Franz, and Oskar Pfetscher, "Siemens-Röhren haben Tradition: Vorgeschichte und Werden der Siemens-Röhrenfabrik," in Siemens & Halske Aktiengesellschaft, *50 Jahre Entwicklung und Fertigung von Elektronenröhren im Hause Siemens*, reprinted from *Siemens-Zeitschrift*, 36:2 (1962), 6–21.

Middleton, David, "S.O. Rice and the theory of random noise: Some personal recollections," *IEEE Transactions on Information Theory*, 34:6 (1988), 1367–1373.

Milder, Michael, "Regression with systematic noise," *Journal of the American Statistical Association*, 59:306 (June 1964), 422–428.

Miller, Dayton C., *The Science of Musical Sounds*, New York: Macmillian, 1922.

Millman, S. (ed.), *A History of Engineering and Science in the Bell System*, Vol. 5, *Communications Science (1925-1980)*, New York: Bell Telephone Laboratories, 1984.

Mills, Mara, "Deafening: Noise and the engineering of communication in the telephone system," *Grey Room*, 43 (2011), 119–142.

Mindell, David, "Opening Black's box: Rethinking feedback's myth of origin," *Technology and Culture*, 41:3 (2000), 405–434.

—, *Between Human and Machine: Feedback, Control, and Computer before Cybernetics*, Baltimore, MD: Johns Hopkins University Press, 2002.

Monod, Jacques, *Chance and Necessity: An Essay on the Natural Philosophy of Modern Biology*, translated by Austryn Wainhouse, New York: Vintage Books, 1971.

Montroll, Elliott, "On the Vienna School of statistical thought," *American Institute of Physics Conference Proceedings*, 109:1 (1984), 1–10.

Nebeker, Frederik, *Signal Processing: The Emergence of a Discipline, 1948 to 1998*, New Brunswick, NJ: IEEE History Center, 1998.

Nielsen, Michael, and Isaac Chuang, *Quantum and Quantum Information*, Cambridge, UK: Cambridge University Press, 2010.

Niss, Martin, "Brownian motion as a limit to physical measuring processes: A chapter in the history of noise from the physicists' point of view," *Perspectives on Science*, 24:1 (2016), 29–44.

Nolan, Catherine, "Music theory and mathematics," in Thomas Christensen (ed.), *The Cambridge History of Western Music Theory* (Cambridge: Cambridge University Press, 2002), 272–304.

Norton, John, and John Earman, "Exorcist XIV: The wrath of Maxwell's demon. Part I: From Maxwell to Szilard," *Studies in the History and Philosophy of Modern Physics*, 29 (1998), 435–471.

—, "Exorcist XIV: The wrath of Maxwell's demon. Part II: From Szilard to Landauer and beyond," *Studies in the History and Philosophy of Modern Physics*, 30 (1999), 1–40.

Nyquist, Harry, "Certain factors affecting telegraphic speed," *The Bell System Technical Journal*, 3 (1924), 324–346.

—, "Thermal agitation in conductors," *Physical Review*, (Ser. 2) 29 (1927), 614.

—, "Thermal agitation of electric charge in conductors," *Physical Review*, (Ser. 2) 32 (1928), 110–113.

—, "Certain topics in telegraph transmission theory," *Transactions of the American Institute of Electrical Engineers*, 43 (1928), 412–422.

Oliver, Bernard, John Pierce, and Claude Shannon, "The philosophy of PCM," *Proceedings of the Institute of Radio Engineers*, 36 (1948), 1324–1331.

Ornstein, Leonard, "On the Brownian motion," *Proceedings of the Huygens Institute—Royal Netherlands Academy of Arts and Sciences (KNAW)*, 21:1 (1919), 96–108.

Ornstein, Leonard, and Herman Carel Burger, "Zur Theorie des Schroteffektes," *Annalen der Physik*, (Ser. 4) 70 (1923), 622–624.

Owens, Larry, "The counterproductive management of science in the Second World War: Vannevar Bush and the Office of Scientific Research and Development," *The Business History Review*, 68:4 (1994), 515–576.

Pagano, Marcello, "Estimation of models of autoregressive signal plus white noise," *The Annals of Statistics*, 2:1 (January 1974), 99–108.

Pais, Abraham, *Subtle is the Lord: The Science and the Life of Albert Einstein*, New York: Oxford University Press, 2005.

Pantalony, David, "Seeing a voice: Rudolph Koenig's instruments for studying vowel sounds," *The American Journal of Psychology*, 117:3 (2004), 425–442.

—, *Altered Sensations: Rudolph Koenig's Acoustical Workshop in Nineteenth Century Paris*, Dordrecht: Springer, 2009.

Papoulis, Anthanasios, *Probability, Random Variables and Stochastic Processes*, New York: McGraw-Hill, 2002.

Parzen, Emanuel, "Book Review, *Principles and Applications of Random Noise Theory*," *Journal of the American Statistical Association*, 54:286 (June 1959), 525.

—, "Book Review, *Topics in the Theory of Random Noise. Vol. 1. General Theory of Random Processes. Nonlinear Transformation of Signals and Noise* by R. L. Stratonovich and R. A. Silverman," *Technometrics*, 6:4 (November 1964), 473–474.

Pearson, Karl, "The problem of the random walk," *Nature*, 77 (1905), 294.

Perrin, Jean Baptiste, "Mouvement brownien et réalité moléculaire," *Annales de Chimie et de Physique*, series 8, 18 (1909), 1–114.

—, "Mouvement brownien et molécules," *Journal de Physique: Théorique et Appliquée et de Physique*, 9:1 (1910), 5–39.

Phillips, V.J., *Waveforms: A History of Early Oscillography*, Bristol: Adam Hilger, 1987.

Pickering, Andrew, *The Cybernetic Brain: Sketches of Another Future*, Chicago, IL: University of Chicago Press, 2011.

Planck, Max, "Über einen Satz der statistischen Dynamik und seine Erweiterung in der Quantentheorie," *Sitzungsberichte der Königlich Preussischen Akademie der Wissenschaften*, 24 (1917), 324–341.

Poincaré, Henri, "Sur le problème des trios corps et les équations de dynamique," *Acta Mathematica*, 13 (1889), 1–270.

Pólya, George, "Über eine Aufgabe der Wahrscheinlichkeitrechnung betreffend die Irrfahrt im Strassennetz," *Mathematische Annalen*, 84 (1921), 149–160.

Polyphone ad, *Phonoscope*, 3:8 (August 1899), 2.

Poovey, Mary, *The History of the Modern Fact: Problems of Knowledge in the Sciences of Wealth and Society*, Chicago, IL: University of Chicago Press, 1998

Porter, Theodore, *Trust in Numbers: The Pursuit of Objectivity in Science and Public Life*, Princeton, NJ: Princeton University Press, 1996.

Potter, Ralph K., "High-frequency atmospheric noise," *Proceedings of the Institute of Radio Engineers*, 19:10 (1931), 1731–1765.

Price, Robert, "Further notes and anecdotes on spread-spectrum origins," *IEEE Transactions on Communications*, 31:1 (1983), 85–97.

—, "A conversation with Claude Shannon: One man's approach to problem solving," *IEEE Communication Magazine*, 22:5 (1984), 123–125.

Price, Robert, and Paul Green, *An Anti-Multipath Communication System*, Lincoln Laboratories Technical Memorandum No. 65, November 9, 1956.

Purrington, Robert, *Physics in the Nineteenth Century*, New Brunswick, NJ: Rutgers University Press, 1997.

Rabinbach, Anson, *The Human Motor: Energy, Fatigue, and the Origins of Modernity*, Berkeley: University of California Press, 1990.

Reich, Leonard, *The Making of American Industrial Research: Science and Business at GE and Bell, 1876-1926*, Cambridge, UK: Cambridge University Press, 1985.

Renn, Jürgen, "Einstein's controversy with Drude and the origin of statistical mechanics: A new glimpse from the 'love letters,'" *Archives for History of Exact Sciences*, 51 (1997), 315–354.

Rhodes, Richard, *Hedy's Folly: The Life and Breakthrough Invention of Hedy Lamarr, the Most Beautiful Woman in the World*, New York: Vintage Books, 2011.

Rice, Stephen O., "Mathematical analysis of random noise," *The Bell System Technical Journal*, 23 (July 1944), 282–332.

—, "Mathematical analysis of random noise," *The Bell System Technical Journal*, 24 (January 1945), 46–156.

Rogoff, Mortimer, "Spread-spectrum communications (for IEEE 1981 Pioneer Award)," *IEEE Transactions on Aerospace and Electronic Systems*, 18:1 (January 1982), 154–155.

Rosenblueth, Arturo, Julian Bigelow, and Norbert Wiener, "Behaviour, purpose, and teleology," *Philosophy of Science*, 10:1 (1943), 18–24.

Saad, Theodore A., "The story of the MIT Radiation Laboratory," *IEEE Aerospace and Electronic Systems Magazine*, 5:10 (1990), 46–51.

Sagiroglu, Seref, and Duygu Sinanc, "Big Data: A review," *Proceedings of the IEEE International Symposium on Collaborative Technologies and Systems* (2013), 42–47.

Schafer, Raymond Murray, *The Soundscape: Our Sonic Environment and the Tuning of the World*, Rochester, VN: Destiny Books, 1977.

Schatzberg, Eric, *Technology: Critical History of a Concept*, Chicago, IL: University of Chicago Press, 2018.

Scholtz, Robert A., "The origin of spread-spectrum communications," *IEEE Transactions on Communications*, 30:5 (1982), 822–854.

Schottky, Walter, "Über spontane Stromschwankungen in verschiedenen Elektrizitätsleitern," *Annalen der Physik*, (Ser. 4) 57 (1918), 541–567.

—, "Zur Berechnung und Beurteilung des Schroteffektes," *Annalen der Physik*, (Ser. 4) 68 (1922), 157–176.

—, "Jena 1912 und das U3/2-Gesetz: Eine Reminiszenz aus der Vorzeit der Elektronenröhren," in Siemens & Halske Aktiengesellschaft, *50 Jahre Entwicklung und Fertigung von Elektronenröhren im Hause Siemens*, reprinted from *Siemens-Zeitschrift*, 36:2 (1962), 22–24.

Schwartz, Mischa, "Armstrong's invention of noise-suppressing FM," *IEEE Communications Magazine*, 47:4 (April 2009), 20–23.

—, "Improving the noise performance of communication systems: 1920s to early 1930s," *IEEE Communications Magazine*, 47:12 (December 2009), 16–20.

Segal, Irving Ezra, "Norbert Wiener," *Biographical Memoirs of the National Academy of Sciences of the United States of America*, 61 (1992), 389–436.

Serchinger, Reinhard, "Walter Schottky und die Forschung bei Siemens," in Ivo Schneider, Helmuth Trischler, and Ulrich Wengenroth (eds.), *Oszillationen: Naturwissenschaftler und Ingenieure zwischen Forschung und Markt* (Munich: R. Oldenbourg Verlag and Deutsches Museum, 2000), 167–209.

Serres, Michel, *The Parasite*, translated by Lawrence Schehr, Baltimore, MD: Johns Hopkins University Press, 1982.

—, *Conversations on Science, Culture, and Time: Michel Serres with Bruno Latour*, Ann Arbor: University of Michigan Press, 1995.

Seth, Suman, *Crafting the Quantum: Arnold Sommerfeld and the Practice of Theory: 1890–1926*, Cambridge, MA: MIT Press, 2010.

Shannon, Claude, "A mathematical theory of cryptography," Technical Memoranda, AT&T Bell Telephone Laboratories, September 1, 1945.

—, "A mathematical theory of communication," *The Bell System Technical Journal*, 27 (1948), 379–423, 623–656.

—, "Communication in the presence of noise," *Proceedings of the Institute of Radio Engineers*, 37 (1949), 10–21.

—, "Communication theory of secrecy systems," *The Bell System Technical Journal*, 28 (1949), 656–715.

Shumway, Robert, "On detecting a signal in N stationarily correlated noise series," *Technometrics*, 13:3 (August 1971), 499–519.

Simon, Mark K., Jim K. Omura, Robert A. Scholtz, and Barry K Levitt, *Spread Spectrum Communications*, vol. 1, Rockville, MD: Computer Science Press, 1985, chapter 2, "The historical origins of spread-spectrum communications," 39–134.

Simpson, John, and Edmund Weiner (eds.), *The Oxford English Dictionary*, Oxford: Clarendon Press, 1989.

Smilor, Raymond, "Toward an environmental perspective: The anti-noise campaign, 1893-1932," in Martin Melosi (ed.), *Pollution and Reform in American Cities* (Austin: University of Texas Press, 1980), 135–151.

Smoluchowski, Marian, "Über den Temperaturesprung bei Wärmeleitung in Gasen," *Sitzungsberichte, Kaiserliche Akademie der Wissenschaften, Wien, Mathematisch-Naturwissenschaftliche Klasse*, 107: Abt. I a (1898), 304–329.

—, "Über Unregelmäβigkeiten in der Verteilung von Gasmolekülen und deren Einfluβ auf Entropie und Zustandgleichung," in *Festschrift Ludwig Boltzmann* (Leipzig: Johann Ambrosius Barth, 1904), 626–641.

—, "Sur le chemin moyen parcouru par les molécules d'un gaz et sur son rapport avec la théorie de la diffusion," *Bulletin de l'Académie des Sciences de Crocovie: Classe des Sciences Mathématiques et Naturelles* (1906), 202–213; reprinted in Wladysiaw Natanson and Jan Stock (eds.), *Pisma Mariana Smoluchowskiego (The Works of Marian Smoluchowski)* (Kraków: Academy of Sciences and Letters, 1924), 479–489.

—, "Zur kinetischen Theorie der Brownschen Molekularbewegung und der Suspensionen," *Annalen der Physik*, 21 (1906), 756–780.

—, "Molekular-kinetische Theorie der Opaleszenz von Gasen im kritischen Zustande sowie einiger verwandter Erscheinungen," *Annalen der Physik*, 25 (1908), 205–226.

—, "On opalescence of gases in the critical state," *Philosophical Magazine*, 23 (1911), 165–173.

—, "Über Brownsche Molekularbewegung unter Einwirkung äußerer Kräfte und deren Zusammenhang mit der verallgemeinerten Diffusionsgleichung," *Annalen der Physik*, 48 (1915), 1103–1112.

—, "Drei Vorträge über Diffusion, Brownsche Molekularbewegung und Koagulation von Kolloidteilchen," *Physikalische Zeitschrift*, 17 (1916), 557–571, 587–599, reprinted in Wladysiaw Natanson and Jan Stock (eds.), *Pisma Mariana Smoluchowskiego (The Works of Marian Smoluchowski)* (Kraków: Academy of Sciences and Letters, 1924), 530–594.

Smoluchowski, Roman, "Life of Marian Smoluchowski," and "Chronological table of Marian Smoluchowski's life," both in Subrahmanyan Chandrasekhar, Mark Kac, and Roman Smoluchowski (eds.), *Marian Smoluchowski: His Life and Scientific Work* (Warsaw: Polish Scientific Publishers, 1999), 9–14; 129–130.

Sobel, Eugene, "Prediction of a noise-distorted, multivariate, non-stationary signal," *Journal of Applied Probability*, 4:2 (August 1967), 330–342.

Sommerfeld, Arnold, "Zum Andenken an Marian Smoluchowski," *Physikalische Zeitschrift*, 18 (1917), 533–539.

Spalding, Keith, *An Historical Dictionary of German Figurative Language*, Oxford: Basil Blackwell, 1974.

Stachel, John, David Cassidy, Jürgen Renn, and Robert Schulmann, "Einstein on the nature of molecular forces," in John Stachel, David Cassidy, Jürgen Renn, and Robert Schulmann (eds.), *The Collected Papers of Albert Einstein*, Volume 2 : *The Swiss Years, 1900-1909* (Princeton, NJ: Princeton University Press, 1989), 3–8.

—, "Einstein on the foundations of statistical physics," in John Stachel, David Cassidy, Jürgen Renn, and Robert Schulmann (eds.), *The Collected Papers of Albert Einstein*, Volume 2 : *The Swiss Years, 1900-1909* (Princeton, NJ: Princeton University Press, 1989), 41–55.

—, "Einstein's dissertation on the determination of molecular dimensions," in John Stachel, David Cassidy, Jürgen Renn, and Robert Schulmann (eds.), *The Collected Papers of Albert Einstein*, Volume 2 : *The Swiss Years, 1900-1909* (Princeton, NJ: Princeton University Press, 1989), 170–182.

Staley, Richard, *Einstein's Generation: The Origin of the Relativity Revolution*, Chicago, IL: University of Chicago Press, 2009.

Sterne, Jonathan, *The Audible Past: Cultural Origins of Sound Reproduction*, Durham, NC: Duke University Press, 2003.

Stigler, Stephen, *The History of Statistics: The Measurement of Uncertainty before 1900*, Cambridge, MA: Belknap Press of Harvard University Press, 1986.

Stoletov, Aleksandr, "Sur une sorte de courants electriques provoques par les rayons ultraviolets," *Comptes Rendus de l'Académie des Sciences*, 106 (1888), 1149.

Stöltzner, Michael, "Vienna indeterminism: Mach, Boltzmann, Exner," *Synthese*, 119:1 (1999), 85–111.

Strutt, John (Lord Rayleigh), *The Theory of Sound*, vol. 1 and vol. 2, London: Macmillan, 1877 and 1888.

Sullivan, Edward, and Timothy Weithers, "Louis Bachelier: The father of modern option pricing theory," *The Journal of Economic Education*, 22:2 (1991), 165–171.

Therrien, Charles, "The Lee-Wiener legacy: A history of the statistical theory of communication," *IEEE Signal Processing Magazine*, 19:6 (2002), 33–44.

Thompson, Emily, "Machines, music and the quest for fidelity: Marketing the Edison Phonograph in America, 1877-1925," *Musical Quarterly*, 79 (Spring 1995), 131–171.

—, *The Soundscape of Modernity: Architectural Acoustics and the Culture of Listening in America, 1900–1933*, Cambridge, MA: MIT Press, 2002.

Uhlenbeck, George, "Reminiscences of Professor Paul Ehrenfest," *American Journal of Physics*, 24 (1956), 431–433.

—, "Fifty years of spin: personal reminiscences," *Physics Today*, 29:6 (1976), 43–48.

Uhlenbeck, George, and Samuel Goudsmit, "A problem in Brownian motion," *Physical Review*, 34 (1929), 145–151.

Uhlenbeck, George, and Leonard Ornstein, "On the theory of the Brownian motion," *Physical Review*, 36 (1930), 823–841.

Van Cittert-Eymers, J.G., "Burger, Herman Carel," in Charles Coulston Gillispie and Frederic Lawrence Holmes (eds.), *Dictionary of Scientific Biography*, 2 (New York: Scribner, 1970), 600–601.

Van Lear, George, "The Brownian motion of the resonance radiometer," *Review of Scientific Instruments*, 4 (1933), 21–27.

Van Lear, George, and George Uhlenbeck, "The Brownian motion of strings and elastic rods," *Physical Review*, 38 (1931), 1583–1598.

Von Plato, Jan, *Creating Modern Probability: Its Mathematics, Physics, and Philosophy in Historical Perspective*, Cambridge: Cambridge University Press, 1998.

Waldherr, Anne, Daniel Maier, Peter Miltner, and Enrico Günter, "Big Data, big noise: The challenge of finding issue networks on the Web," *Social Science Computer Review*, 35:4 (2017), 427–443.

Wang, Ming Chen, *A Study of Various Solutions of the Boltzmann Equation*, Ph.D. dissertation, University of Michigan, 1942.

—, "Zhuanshun jiushi zai" (「轉瞬九十載」), *Wuli* (《物理》), 35:3 (2006), 174–182.

Wang, Ming Chen, and George Uhlenbeck, "On the theory of the Brownian motion II," *Review of Modern Physics*, 17:2&3 (April–July 1945), 323–342.

Watson-Watt, Robert A., "The origin of atmospherics," *Nature*, 110 (1922), 680–681.

—, "Directional observations of atmospherics—1916-1920," *Philosophical Magazine*, 45:269 (1923), 1010–1026.

—, "Directional observations of atmospheric disturbances, 1920-21," *Proceedings of the Royal Society of London, A*, 102:717 (1923), 460–478.

—, "Atmospherics," *Proceedings of the Physical Society of London*, 37 (1924–1925), 23D–31D.

—, "The directional recording of atmospherics," *Journal of the Institution of Electrical Engineers*, 64 (1926), 596–610.

—, *Three Steps to Victory: A Personal Account by Radar's Greatest Pioneer*, London: Odhams Press, 1957.

Watson-Watt, Robert A., and Edward V. Appleton, "On the nature of atmospherics—I," *Proceedings of the Royal Society of London*, 103 (1923), 84–102.

Watson-Watt, Robert A., John F. Herd, and L.H. Bainbridge-Bell, *Applications of the Cathode Ray Oscillograph in Radio Research*, London: H.M. Stationery Office, 1933.

Watson-Watt, Robert A., John F. Herd, and F.E. Lutkin, "On the nature of atmospherics—V," *Proceedings of the Royal Society of London*, 162 (1937), 267–291.

Weber, Max, *The Protestant Ethics and the Spirit of Capitalism*, New York: Schribner, 1958.

Weekes, Karen, "Consuming and dying: Meaning and the marketplace in Don DeLillo's *White Noise*," *Literature Interpretation Theory*, 18 (2007), 285–302.

Weinstein, P. (ed.), *J. C. Poggendorffs Biographisch-Literarisches Handwörterbuch*, Band 5 (Berlin: Verlag Chemie, 1922), 403.

Welsh, Walter, and Leah Brodbeck Stenzel Burt, *From Tinfoil to Stereo: The Acoustic Years of the Recording Industry, 1877–1929*, Gaineseville: University of Florida Press, 1994.

Whitaker, Andrew, "Maxwell's famous (or infamous) demon," in Raymond Flood, Mark McCartney, and Andrew Whitaker (eds.), *James Clerk Maxwell: Perspectives on His Life and Work* (Oxford: Oxford University Press, 2014), 163–186.

Whittaker, John M., *Interpolatory Function Theory*, Cambridge: Cambridge University Press, 1935.

Wiener, Norbert, "The mean of a functional of arbitrary elements," *Annals of Mathematics*, 2:22 (1920), 66–72.

—, "The average of an analytical functional," *Proceedings of the National Academy of Science*, 7:9 (1921), 253–260.

—, "The average of an analytic function and the Brownian movement," *Proceedings of the National Academy of Science*, 7:10 (1921), 294–298.

—, "Differential-space," *Journal of Mathematics and Physics*, 2 (1923), 130–174.

—, "The average value of a functional," *Proceedings of the London Mathematical Society*, 22 (1924), 454–467.

—, "On the representation of functions by trigonometrical integrals," *Mathematical Zeitschrift*, 24 (1925), 575–617.

—, "The harmonic analysis of irregular motion," *Journal of Mathematics and Physics*, 5 (1926), 99–121.

—, "The harmonic analysis of irregular motion (second paper)," *Journal of Mathematics and Physics*, 5 (1926), 158–189.

—, "The spectrum of an arbitrary function," *Proceedings of the London Mathematical Society*, 2:27 (1928), 483–496.

—, "Generalized harmonic analysis," *Acta Mathematica*, 55 (1930), 117–258.

—, "The homogeneous chaos," *American Journal of Mathematics*, 60:4 (1938), 897–936.

—, *Cybernetics, Or Control and Communication in the Animal and the Machine*, Cambridge, MA: MIT Press, 1948.

—, *Extrapolation, Interpolation, and Smoothing of Stationary Time Series*, Cambridge, MA: The Technology Press of MIT, and New York: John Wiley & Sons Inc., 1949.

—, *Ex-Prodigy: My Childhood and Youth*, Cambridge, MA: MIT Press, 1964.

—, *A Life in Cybernetics—Ex-Prodigy: My Childhood and Youth, and I am a Mathematician: The Later Life of a Prodigy*, Cambridge, MA: MIT Press, 2018.

Wiener, Norbert, Armand Siegel, Bayard Rankin, and William Ted Martin, *Differential Space, Quantum Systems, and Prediction*, Cambridge, MA: MIT Press, 1966.

Wilcox, Leonard, "Baudrillard, DeLillo's White Noise, and the end of heroic narrative," in Hugh Ruppensburg and Tim Engles (eds.), *Critical Essays on Don DeLillo* (New York: G.K. Hall &6 Co., 2000), 196–212.

Wildes, Karl, and Nilo Lindgren, *A Century of Electrical Engineering and Computer Science at MIT, 1882–1982*, Cambridge, MA: MIT Press, 1986.

Williams, N.H., and S. Huxford, "Determination of the charge of positive thermions from measurements of shot effect," *Physical Review*, (Ser. 2) 33 (1929), 773–788.

Williams, N.H., and H.B. Vincent, "Determination of electronic charge from measurements of shot-effect in aperiodic circuits," *Physical Review*, (Ser. 2), 28 (1926), 1250–1264.

Williams, Raymond, *Keywords: A Vocabulary of Culture and Society*, New York: Oxford University Press, 1985.

Willink, Bastiaan, "Origin of the Second Golden Age of Dutch Science after 1860: Intended and unintended consequences of educational reform," *Social Studies of Science*, 21:3 (August 1991), 503–526.

Wise, Norton (ed.), *The Values of Precision*, Princeton, NJ: Princeton University Press, 1995.

—, *Aesthetics, Industry, and Science: Hermann von Helmholtz and the Berlin Physical Society*, Chicago, IL: University of Chicago Press, 2018.

Wittje, Roland, "Concepts and significance of noise in acoustics: Before and after the Great War," *Perspectives on Science*, 24:1 (2016), 7–28.

—, *The Age of Electroacoustics*, Cambridge, MA: MIT Press, 2016.

Wright, Aaron Sidney, *More than Nothing: A History of the Vacuum in Theoretical Physics, 1925-1980*, Oxford: Oxford University Press, forthcoming.

Yeang, Chen-Pang, "Tubes, randomness, and Brownian motions: Or, how engineers learned to start worrying about electronic noise," *Archive for the History of Exact Sciences*, 65:4 (2011), 437–470.

—, "The sound and shapes of noise: Measuring disturbances in early twentieth-century telephone and radio engineering," *ICON: Journal of the International Committee for the History of Technology*, 18 (2012), 63–85.

—, *Probing the Sky with Radio Waves: From Wireless Technology to the Development of Atmospheric Science*, Chicago, IL: University of Chicago Press, 2013.

—, "Two mathematical approaches to random fluctuations," *Perspectives on Science*, 24:1 (2016), 45–72.

—, "From modernizing the Chinese language to information science: Chao Yuen Ren's route to cybernetics," *Isis*, 108:3 (2017), 553–580.

Zermelo, Ernst, "Ueber einen Satz der Dynamik und die mechanische Wärmetheorie," *Annalen der Physik*, (Ser. 3), 57 (1896), 485–494.

Sources from the World-Wide Web

"Adriaan Daniël Fokker," in Huygens-Fokker Foundation, Centre for Microtonal Music website, http://www.huygens-fokker.org/whoswho/fokker.html (accessed April 8, 2022).

Google Ngrams:

https://books.google.com/ngrams/graph?content=noise+information&year_start=1800&year_end=2019&corpus=26&smoothing=3&direct_url=t1%3B%2Cnoise%20information%3B%2Cc0#t1%3B%2Cnoise%20information%3B%2Cc0 (accessed April 8, 2022).

https://books.google.com/ngrams/graph?content=noise+signal&year_start=1800&year_end=2019&corpus=26&smoothing=3&direct_url=t1%3B%2Cnoise%20signal%3B%2Cc0 (accessed April 8, 2022).

https://books.google.com/ngrams/graph?content=noise&year_start=1800&year_end=2019&corpus=26&smoothing=3&direct_url=t1%3B%2Cnoise%3B%2Cc0 (accessed April 8, 2022).

Hochfelder, David, "Oral history: Paul Green," by IEEE History Center, Interview #373, October 15, 1999, https://ethw.org/Oral-History:Paul_Green (accessed April 8, 2022).

Kuhn, Thomas, "Oral history transcript of Dr. George Uhlenbeck, Session I," interview on 30 March 1962, Niels Bohr Library and Archives, American Institute of Physics, Washington, D.C.; available at https://www.aip.org/history-programs/niels-bohr-library/oral-histories/4922-1 (accessed April 8, 2022).

—, "Oral history transcript of Dr. George Uhlenbeck, Session V," interview on December 9, 1963, Niels Bohr Library and Archives, American Institute of Physics, Washington, D.C.; available at https://www.aip.org/history-programs/niels-bohr-library/oral-histories/4922-5 (accessed April 8, 2022).

Nebeker, Frederick, *Fifty Years of Signal Processing: The IEEE Signal Processing Society and Its Technologies 1948–1998* (Piscataway, NJ: The IEEE Signal Processing Society, 1998), 2–5; available at https://signalprocessingsociety.org/our-story/society-history (accessed April 8, 2022).

Perrin, Jean Baptiste, "The discontinuous structure of matters," Nobel Lecture in Physics, 1926, available at the Nobel Prize website: https://www.nobelprize.org/prizes/physics/1926/perrin/lecture/ (accessed April 8, 2022).

Price, Robert, "Oral history: Claude E. Shannon," IEEE History Center, Interview #423, July 28, 1982, available at https://ethw.org/Oral-History:Claude_E._Shannon (accessed April 8, 2022).

"Research for Progress: The Development of the Central Research Laboratory," available at the Siemens website: https://new.siemens.com/global/en/company/about/history/stories/research-laboratory.html (accessed April 8, 2022).

Slepian, David, “Stephen O. Rice 1907–1986,” in *Memorial Tributes: National Academy of Engineering*, Volume 4, available at: https://www.nae.edu/189134/STEPHEN-O-RICE-19071986?layoutChange=Print (accessed April 8, 2022).

Tonn, Jenna, “Gender,” *Encyclopedia of the History of Science* (March 2019), online journal from Carnegie Mellon University, available at: https://lps.library.cmu.edu/ETHOS/article/id/20/ (accessed October 12, 2022).

Valenti, Mathew, “The evolution of error control coding,” available at https://citeseerx.ist.psu.edu/viewdoc/download?doi=10.1.1.104.6249&rep=rep1&type=pdf (accessed April 8, 2022).

Van Berkel, K., “Uhlenbeck, George Eugene (1900–1988),” *Biografisch Woordenboek van Netherland*, 6 (2013), available at http://resources.huygens.knaw.nl/bwn1880-2000/lemmata/bwn6/uhlenbeck (accessed April 8, 2022).

Index

Figures and footnotes are indicated by an f and n following the page number.

C

X